Textbook of Botany

Textbook of Botany

Edited by **Isabelle Nickel**

SYRAWOOD
PUBLISHING HOUSE
New York

Published by Syrawood Publishing House,
750 Third Avenue, 9th Floor,
New York, NY 10017, USA
www.syrawoodpublishinghouse.com

Textbook of Botany
Edited by Isabelle Nickel

International Standard Book Number: 978-1-68286-354-1 (Hardback)

Printed in China.

Contents

Preface

Botany is a vast field of study that focuses on understanding structural, chemical and biological processes in plants. The book focuses upon some of the significant aspects of botany such as plant taxonomy, plant metabolism and physiology, plant morphology, biochemistry, molecular genetics, etc. It strives to provide a comprehensive overview about this discipline and to help develop a better understanding of the latest advances within this field. This book is a vital tool for all researching and studying botany.

Various studies have approached the subject by analyzing it with a single perspective, but the present book provides diverse methodologies and techniques to address this field. This book contains theories and applications needed for understanding the subject from different perspectives. The aim is to keep the readers informed about the progress in the field; therefore, the contributions were carefully examined to compile novel researches by specialists from across the globe.

Indeed, the job of the editor is the most crucial and challenging in compiling all chapters into a single book. In the end, I would extend my sincere thanks to the chapter authors for their profound work. I am also thankful for the support provided by my family and colleagues during the compilation of this book.

Editor

1

A zoom into the nanoscale texture of secondary cell walls

Tobias Keplinger[1,2], Johannes Konnerth[3], Véronique Aguié-Béghin[4,5], Markus Rüggeberg[1,2], Notburga Gierlinger[1,2] and Ingo Burgert[1,2*]

Abstract

Background: Besides classical utilization of wood and paper, lignocellulosic biomass has become increasingly important with regard to biorefinery, biofuel production and novel biomaterials. For these new applications the macromolecular assembly of cell walls is of utmost importance and therefore further insights into the arrangement of the molecules on the nanolevel have to be gained. Cell wall recalcitrance against enzymatic degradation is one of the key issues, since an efficient degradation of lignocellulosic plant material is probably the most crucial step in plant conversion to energy. A limiting factor for in-depth analysis is that high resolution characterization techniques provide structural but hardly chemical information (e.g. Transmission Electron Microscopy (TEM), Atomic Force Microscopy (AFM)), while chemical characterization leads to a disassembly of the cell wall components or does not reach the required nanoscale resolution (Fourier Tranform Infrared Spectroscopy (FT-IR), Raman Spectroscopy).

Results: Here we use for the first time Scanning Near-Field Optical Microscopy (SNOM in reflection mode) on secondary plant cell walls and reveal a segmented circumferential nanostructure. This pattern in the 100 nm range was found in the secondary cell walls of a softwood (spruce), a hardwood (beech) and a grass (bamboo) and is thus concluded to be consistent among various plant species. As the nanostructural pattern is not visible in classical AFM height and phase images it is proven that the contrast is not due to changes in surfaces topography, but due to differences in the molecular structure.

Conclusions: Comparative analysis of model substances of casted cellulose nanocrystals and spin coated lignin indicate, that the SNOM signal is clearly influenced by changes in lignin distribution or composition. Therefore and based on the known interaction of lignin and visible light (e.g. fluorescence and resonance effects), we assume the elucidated nanoscale structure to reflect variations in lignification within the secondary cell wall.

Keywords: Secondary cell walls, Scanning near field optical microscopy, Atomic force microscopy, Diffraction limit, Lignification

Background

Structure, chemistry and mechanics of the polymer assembly of secondary cell walls have been studied for decades, because of the high social, environmental and economic relevance of the plant material in classical applications (e.g. wood, paper). More recent research activities are driven by the utilization of lignocellulosic biomass for bioenergy [1-3] and novel biomaterials [4]. The organization of secondary cell walls at the micro-level, with its three-layered and multilamellar structure for wood species and bamboo respectively, is well understood. However we are still lacking knowledge on the organization at the nanoscale in particular about the spatial arrangement and interaction of the different polymers within the cell wall. Most secondary cell walls of xylem cells are made up of the three dominating cell wall polymers cellulose, lignin and hemicelluloses. Cellulose fibrils with a diameter of 3-4 nm are arranged in bigger agglomerates (fibril bundles) with a size of 20-25 nm and are embedded in a matrix consisting of lignin and hemicelluloses [5,6]. During the last decades different cell wall models on the spatial arrangement of the

* Correspondence: iburgert@ethz.ch
[1]Institute for Building Materials, ETH Zurich, Zurich, Switzerland
[2]Applied Wood Research Laboratory, Empa - Swiss Federal Laboratories for Material Testing and Research, Duebendorf, Switzerland
Full list of author information is available at the end of the article

macromolecules within the secondary cell walls have been developed. The micellar theory of Nägeli [7] (developed in the 19th century) refers to a spatial distribution of the macromolecules in a concentric lamellae structure due to alternating circumferential layers of higher cellulose and higher lignin content. This theory has been supported in various studies based on visible/ultraviolet microscopy, the delamination behavior of wood fibers, the distribution and form of pores after selective removal of lignin as well as electron microscopy and atomic force microscopy studies [8-11]. A slightly different model was supposed by Kerr and Goring [12] in terms of concentric arrangement of batches of higher cellulose and lignin content resulting in a segmented lamella structure. Based on high resolution field emission scanning electron microscopy studies on fractured wood samples and cell wall degradation by fungi, an alternative radial agglomeration of cellulose structures was proposed [13,14]. More recently, electron microscopy and atomic force microscopy (AFM) studies supported a random texture without any structured arrangement of the wood components [15,16].

All introduced models are mainly based on electron microscopy studies [11,12,14,17] or on AFM experiments [5,9,10,16]. Both methods are excellent in providing structural information with high resolution, but their potential to provide chemical information is quite limited. Alternative techniques that are strong in chemical analysis of cell walls such as Raman spectroscopy [18] are limited in the required spatial resolution due to the diffraction limit (Rayleigh criterion). To probe structure and chemistry on the nano-level at the same time, scanning probe microscopy has to be combined with spectroscopic techniques [19,20]. One possible combination is Scanning Near-Field Optical Microscopy (SNOM), which has already been developed in the late nineteen-twenties based on the first considerations on breaking the diffraction limit in microscopy [21]. A sample is scanned by a gold or silver coated optical tip, into which a laser is coupled in, with a subwavelength aperture at the end of the optical fiber. The generated signal is not affected by the constraints of diffraction and provides additional photo-optical and thus chemical information. It is based on evanescent waves, which are characterized by rapidly decaying amplitudes very close to the surface (nm range) [22-24]. Although SNOM has been used for the characterization of various biological samples such as proteins, lipids, polysaccharides, DNA and cells [25,26], intact secondary plant cell walls have not been analyzed yet. Therefore the aim was to reveal new insights into the chemical structure of secondary plant cell walls with the help of Scanning Near Field Optical Microscopy by zooming into the nanoscale texture of cell walls of a hardwood, a softwood and a grass.

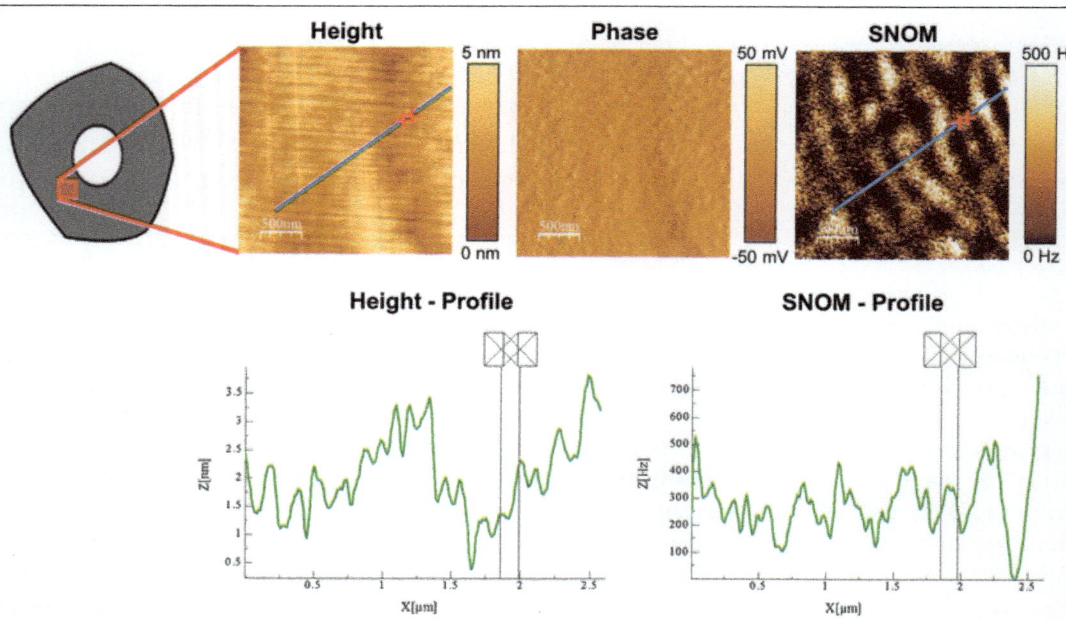

Figure 1 Height/Phase/SNOM Image of a cross section of beech and the corresponding SNOM and height profiles. The height and the phase images of the cross section of beech wood (measurement position within the secondary wall (grey) is illustrated) show a very homogeneous surface in contrast to the SNOM image which reveals a clearly segmented circumferential lamellar structure and elongated nodule like features of different sizes. The height and SNOM profile indicate that the SNOM signal does not go along with the height signal (exclusion of height artifacts) and additionally the achieved resolution with SNOM of 120 nm is determined by using the FWHM (Full Width at Half Maximum) of the SNOM signal peak marked in the images and in the profiles.

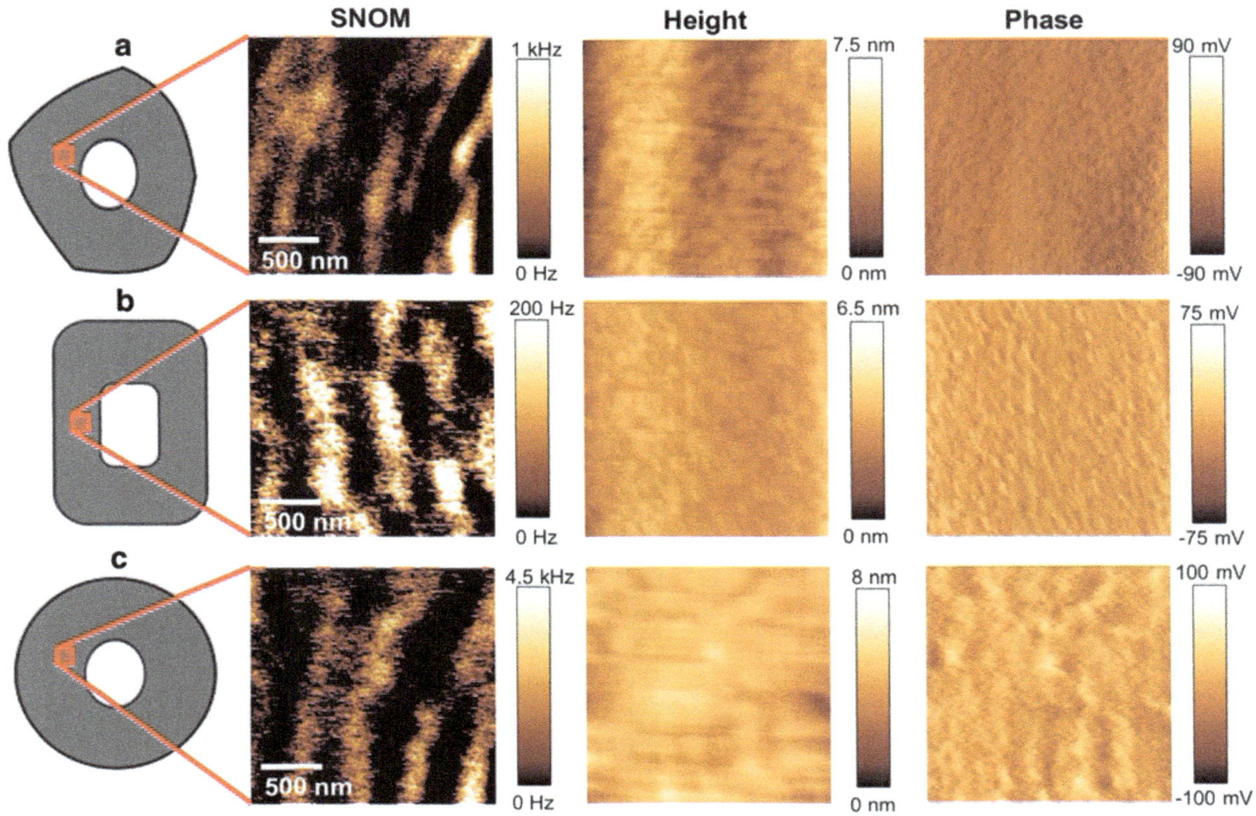

Figure 2 SNOM/Height/Phase images of beech (a), spruce (b) and bamboo (c). The SNOM images of the cross sections of the three plant species (positions of the measurements are illustrated) consistently show a circumferential segmentation, in contrast to the homogeneous height and phase images, which points to compositional changes at the nano-level of the secondary cell wall.

Results and discussion

Figure 1 shows AFM height and phase as well as a SNOM image and the corresponding SNOM/height profile of the marked line of the secondary cell wall layer of a cross section of beech. The height image and the phase image indicate a homogeneous surface (roughness of the height image: 1.1 nm) with only some marginal scratches in the height image which are a result of the microtome cutting. In contrast the SNOM image reveals clearly a segmented circumferential lamellar structure with stripe like features of varying sizes (about 100-200 nm in thickness and 200-700 nm in width).

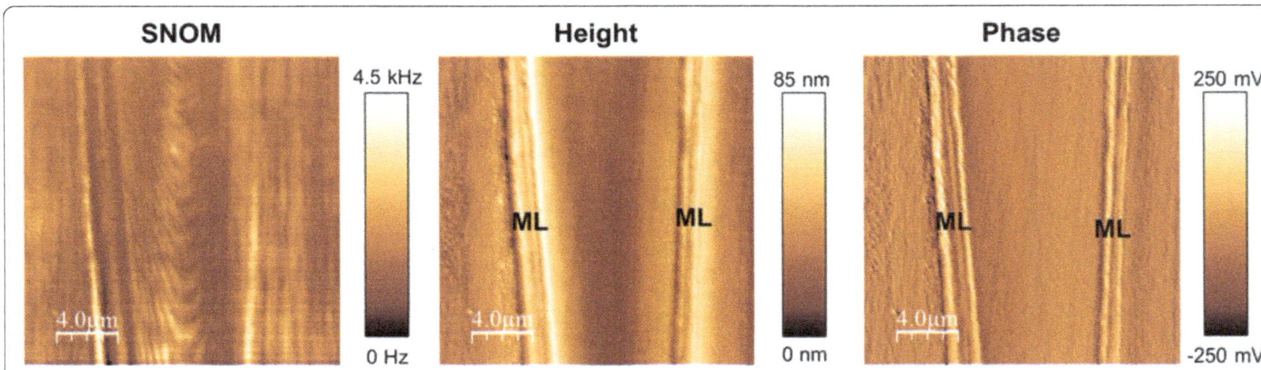

Figure 3 SNOM/Height/Phase images of a longitudinal section of spruce. In the height and phase image of the longitudinal section a homogeneous structure apart from the middle lamella (ML) is revealed. The SNOM image of the slightly tilted longitudinal cut shows curl like structures which are in good agreement with the circumferential segmentation in the cross sections.

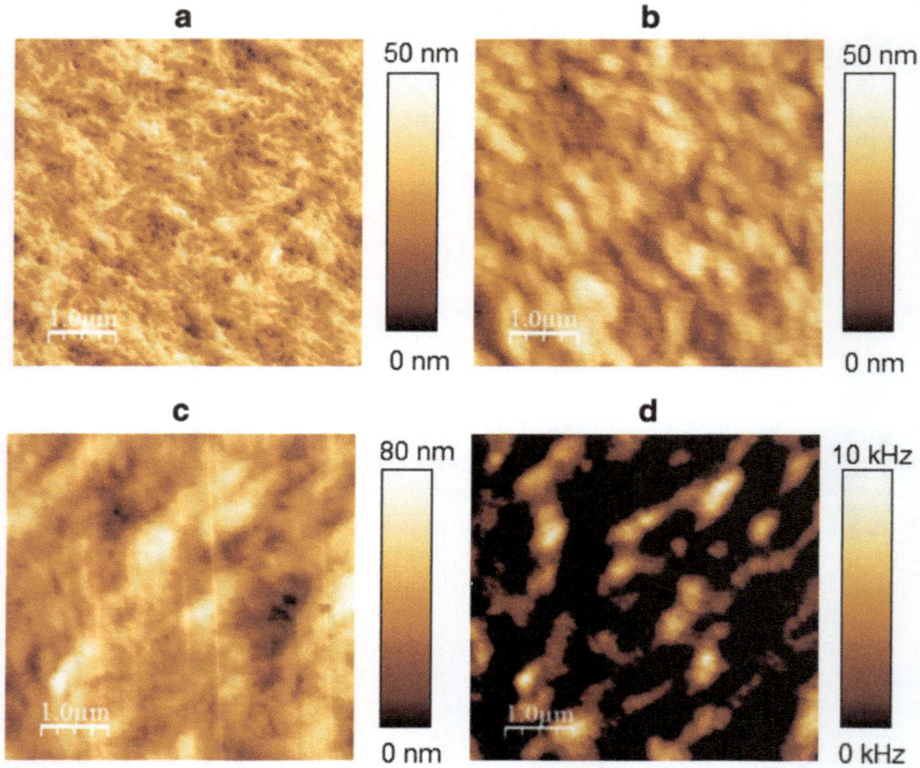

Figure 4 Height images of the network of casted cellulose nanocrystals measured with an AFM (a)/SNOM (b) tip and Height (c)/SNOM (d) image of the bilayer system (cellulose nanocrystals and lignin). In the height image of the casted cellulose, measured with the AFM tip, the nanocrystal network is clearly visualized **(a)**; in contrast to the topography of the same sample revealed by the SNOM tip **(b)**. The height image of the bilayer **(c)** reveals the typically formed globular structures of lignin together with the underlying network of the nanocrystals. In contrast in the SNOM image of the bilayer **(d)** only the globules of the lignin macromolecules and their agglomerations can be seen.

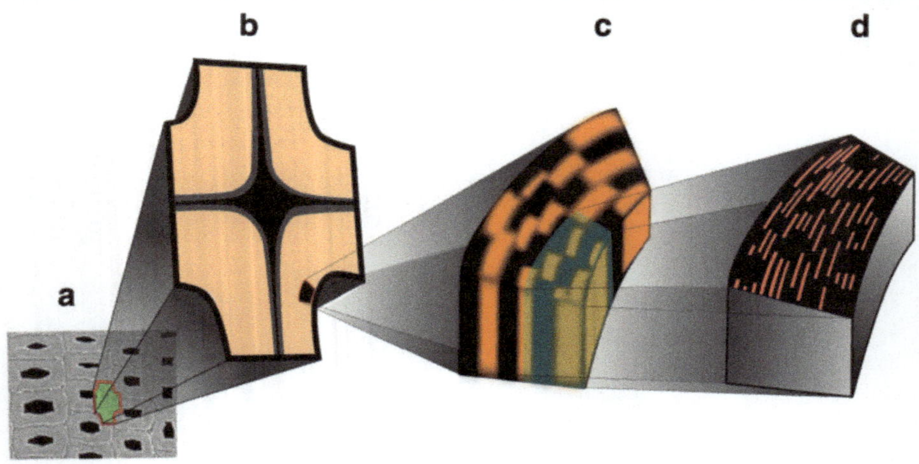

Figure 5 Model of the macromolecular assembly based on the SNOM measurements. In Figure 5**a** latewood cells of spruce are shown and in 5**b** an illustrated zoom in the layered structure of the secondary cell wall. 5**c** reflects the measured segmented lamellar structure in the S2 layer and 5**d** illustrates the presumably underlying structure (red: cellulose fibrils; black: matrix consisting of lignin and hemicelluloses) according to the cell wall model introduced by Kerr and Goring [12].

SNOM images of cross sections of beech (Figure 2a), spruce (2b) and bamboo (2c) cell walls indicate that the circumferential segmentation in light and dark batches is a common feature for different dicot and monocot species, despite differences in cell organization and chemical composition. The cell wall of both wood species possesses a segmented concentric lamellar pattern with a size of around 125 nm in thickness and 200 to 750 nm in width. In bamboo the same pattern is visible but with a slightly larger width (500-800 nm) and thickness of the segments (ca. 150 nm). Another remarkable feature is that the elucidated structures appear differently curved in dependence of cell form and positioning of the SNOM scanning (compare beech at different positions in Figure 1 and Figure 2a). This is a strong indication for the superimposed circumferential pattern of the cellulose/lignin architecture in the cell walls, which is also supported by the fact that the structures can be seen in other anatomical directions. In Figure 3 the SNOM, height and phase images of a longitudinal cut of spruce wood are shown. The SNOM image reveals a curl like structure of the secondary cell wall due to the slightly tilted cutting direction. This structure is not visible in the height and phase image proofing that the contrast is based on differences in the chemical composition/structure. To gain a deeper understanding of the SNOM signal origin in secondary cell wall analysis, we conducted a comparative investigation on a bilayer of cellulose nanocrystals and spin coated DHP (DeHydrogenation-Polymer)-lignin on top. Figure 4a shows the AFM-height image of the network of cellulose nano-crystals after casting on a quartz substrate–measured with a standard tuning fork AFM tip and Figure 4b the same sample measured with a SNOM tip. It is obvious that the spatial resolution with the SNOM tip is smaller compared to the resolution with the AFM tip–nevertheless the overall roughness remains identical (same z-range of the scan). The reduced resolution can be explained as the tip diameter of the SNOM is a couple of times larger (≈100 nm) and in addition the Q-factor is lower.

Figure 4c shows the topography of the self-assembled lignin model compounds on top of the cellulose nanocrystal film. The images indicate that the lignin forms globular structures (bigger z-range compared to Figure 4b) in the size of 250-500 nm on top of the cellulose nanocrystals which is in good agreement with the results of previous publications [27-29]. Remarkably, the underlying structure of the cellulose nanocrystals is still visible. In contrast, the SNOM image of the bilayer (Figure 4d) reveals only the individual globules of lignin macromolecules and their agglomerations.

In general SNOM offers the possibility to use all the contrast mechanisms of conventional confocal microscopy e.g.: birefringence, elastic scattering, Raman effect and fluorescence [30]. In view of the specific chemical composition of the cell wall components it is most likely that lignin contributes more strongly to the SNOM signal than cellulose and hemicelluloses, due to its interaction with light (autofluorescence and resonance effects) [31-33]. Hence, based on the observations on the model substances it seems reasonable to assume that the SNOM signal predominantly visualizes changes in the lignin amount or composition with high sensitivity. Therefore the nanostructural pattern observed within the cell walls of beech, spruce and bamboo is supposed to be due to variations in lignification, either in amount or chemical structure. The achieved resolution of around 100 nm (limited to the size of the aperture), most probably does not reflect the exact size of the regions of lignin variations. A smaller segmented lamellar structure with structures in the 10 nm range as introduced in the model by Kerr and Goring [12] would presumably result in a similar pattern, since the sizes of the batches cannot be precisely determined (Figure 5).

Conclusions

The SNOM images provide for the first time sub-diffraction limited chemical information on the spatial distribution of the cell wall components in secondary cell walls. The circumferential pattern of cellulose and lignin rich regions is likely to be the result of spatial constraints and progression of the cell wall formation process. The superimposed circumferential structure is due to the cellulose fibril spinning process by cellulose synthase complexes [34], which lay down the cellulose scaffold in a circumferential manner. The subsequent lignification in a radical polymerization process is likely to locally produce segments of higher and lower degrees of lignin or alterations in the chemical structure [35-37], which become evident in the SNOM images. A striking observation is the incidence of the distribution pattern among different plant species, including a representative of hardwoods, softwoods and grasses. This points to a rather universal principle of spatial cellulose and lignin assembly in secondary cell walls, and is therefore highly relevant for the understanding of cell wall structure and its enzymatic degradation for energy conversion from lignocellulosic raw materials in general.

Methods

Air dried specimens of spruce, beech (provided by Oliver Vay, Kompetenzzentrum Holz GmbH, Linz, Austria) and bamboo (from an experimental forest in Miaoshanwu Nature Reserve, Zhejiang Province, China) with dimensions of ~10×10×10 mm^3 (radial/tangential/longitudinal) were glued onto AFM specimen discs (orientation of the fibrils was taken into account) and afterwards on cylinders due to fixation reasons. A highly smooth surface (1.5 ×

1.5 mm^2) was cut using a ultramicrotome (Reichert Jung–Ultracut) with a diamond knife (DiATOME) and the following cutting parameters: speed: 1 mm/s, clearance angle: 8° and cutting angle: 45°. Sections of initially 2 μm reduced gradually to approximately 50 nm were removed in order to minimize mechanical impact of cutting forces on the surface. Both, wood (beech and spruce) and bamboo samples were not embedded, to avoid any potential influence by the embedding material.

Cellulose nano-crystals (CNCs) were prepared from ramie (*Boehmeria nivea*) fibers and synthetic lignins (Dehydrogenation Polymers - DHP) were synthetisized by peroxidase/hydrogen peroxide mediated polymerization of coniferyl alcohol (guaiacyl G unit), as previously described [27]. 200 μg of CNC was applied to a clean quartz slide (3.14 cm^2) from CNC aqueous suspension (0.1%, w/v) and dried under vacuum for 12 h. DHP solution (2 gL^{-1}) in dioxan/water (9/1, v/v) was then deposited by spin-coating on CNC film with a spin-coater (Speedline Technologies, USA). A volume of 50 μL was deposited on a stationary solid slide, which was then accelerated at 1260 rpm/s and spun at 3000 rpm for 40 s. Thickness and refractive index of CNC film before DHP deposit were respectively 270 nm ± 5 nm and 1.550, measured by spectroscopic ellipsometry (Uvisel, Horiba Jobin Yvon, Palaiseau, France).

SNOM measurements were performed with a MultiView 2000™ instrument (Nanonics Imaging, Jerusalem, Israel) in reflection configuration in dry state (scheme of the setup in Figure 6). Images shown are a selection out of ≈ 100 SNOM measurements. The tuning fork AFM was operated in phase-feedback and for coupling a gold coated glass tip with an aperture between 50-150 nm (typical resonant frequency of the tip around 35 kHz and a Q factor of around 500) to the 532 nm laser an optical fiber was used. For the AFM measurement in Figure 4a a tuning fork AFM probe with a tip diameter of 20 nm, a resonant frequency of 33.18 kHz and a Q-factor of ≈ 1520 was used. As a detector an Avalanche Photo diode is needed/used (see Figure 6), as the SNOM signal is extremely small, which makes the signal detection highly challenging.

Size of the generated AFM/SNOM images varies between 20 × 20 μm^2- 2.5 × 2.5 μm^2 at a resolution of 256 × 256 pixels and 12 sub-steps between each pixel. Further information on the spatial resolution of the SNOM and AFM mode of the instrument can be found on www.nanonics.co.il. For processing the acquired data were exported from the measurement software (Quartz, Faraday Instruments) to WSxM 4.0 develop 11.6 software. All images were parabola flattened, equalized and the contrast was adjusted for visibility reasons [38].

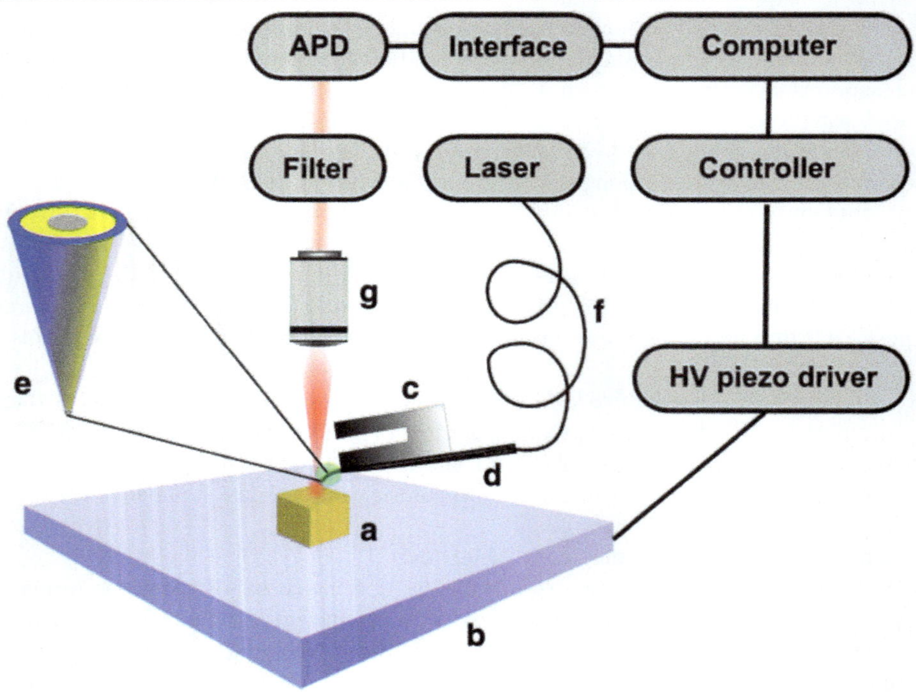

Figure 6 Scheme of the SNOM-setup showing the arrangement of the main components. The sample **(a)** is placed on a piezo-scanning stage **(b)** and scanned with a tuning fork SNOM probe **(c)**. Through the SNOM tip **(d)** (zoom **(e)** showing the inner Au/Cr coating) the laser light coupled out of an optical fibre **(f)** is coming in contact with the sample and reflected back through an objective **(g)** and filter before the signal is detected with an APD.

Abbreviations

AFM: Atomic Force Microscopy; CNC: Cellulose Nano-Crystals; DHP: Dehydrogenation Polymers; FT-IR: Fourier Transform Infra-Red Spectroscopy; SNOM: Scanning Near-Field Optical Microscopy; TEM: Transmission Electron Microscopy.

Competing interests

The authors declare that they have no competing interests.

Authors' contributions

TK, JK, NG and MR performed experiments and analyzed data. VAB provided samples. TK, NG and IB co-wrote the paper. All authors discussed results and commented on the manuscript. All authors read and approved the final manuscript.

Acknowledgements

We thank the SNF for funding in the framework of R'Equip and NRP66, as well BAFU (Bundesamt für Umwelt) and Lignum, Switzerland for the support of the Wood Materials Science professorship. Johannes Konnerth acknowledges the financial support for the STSM at ETH by COST Action FP0802. Correspondence and requests for materials should be addressed to I.B. (iburgert@ethz.ch).

Author details

[1]Institute for Building Materials, ETH Zurich, Zurich, Switzerland. [2]Applied Wood Research Laboratory, Empa - Swiss Federal Laboratories for Material Testing and Research, Duebendorf, Switzerland. [3]Department of Materials Science and Process Engineering, BOKU-University of Natural Resources and Life Science, Konrad Lorenz Strasse 24, A-3430, Tulln, Vienna, Austria. [4]INRA-UMR-614 Fractionnement des Agro-Ressources et Environnement (FARE), F-51686, Reims, France. [5]Université de Reims Champagne-Ardenne, UMR614 Fractionnement des Agro-Ressources et Environnement (FARE), F-51686, Reims, France.

References

1. Ding S-Y, Liu Y-S, Zeng Y, Himmel ME, Baker JO, Bayer EA: **How does plant cell wall nanoscale architecture correlate with enzymatic digestibility?** *Science* 2012, **338**:1055–1060.
2. Dixon RA: **Break down the walls.** *Nature* 2013, **493**:36–37.
3. Himmel ME, Ding SY, Johnson DK, Adney WS, Nimlos MR, Brady JW, Foust TD: **Biomass recalcitrance: engineering plants and enzymes for biofuels production.** *Science* 2007, **315**:804–807.
4. Paris O, Burgert I, Fratzl P: **Biomimetics and biotemplating of natural materials.** *MRS Bulletin* 2010, **35**:219–225.
5. Fahlen J, Salmen L: **Pore and matrix distribution in the fiber wall revealed by atomic force microscopy and image analysis.** *Biomacromolecules* 2005, **6**:433–438.
6. Frey-Wyssling A: **The ultrastructure of wood.** *Wood Sci Technol* 1968, **2**:73–83.
7. Nägeli C: *Die Micellartheorie.* Leipzig: Akademische Verlagsgesellschaft M.B.H; 1928.
8. Clarke SH: **Fine structure of the plant cell wall.** *Nature* 1938, **142**:899–904.
9. Fahlen J, Salmen L: **On the lamellar structure of the tracheid cell wall.** *Plant Biol* 2002, **4**:339–345.
10. Fahlen J, Salmen L: **Cross-sectional structure of the secondary wall of wood fibers as affected by processing.** *J Mater Sci* 2003, **38**:119–126.
11. Fromm J, Rockel B, Lautner S, Windeisen E, Wanner G: **Lignin distribution in wood cell walls determined by TEM and backscattered SEM techniques.** *J Struct Biol* 2003, **143**:77–84.
12. Kerr AJ, Goring DAI: **The Ultrastructural Arrangement of the wood cell wall.** *Cellul Chem Technol* 1975, **9**:563–573.
13. Schwarze F, Engels J: **Cavity formation and the exposure of peculiar structures in the secondary wall (S-2) of tracheids and fibres by wood degrading basidiomycetes.** *Holzforschung* 1998, **52**:117–123.
14. Zimmermann T, Sell J, Eckstein D: **SEM studies on the tension-fracture surfaces of spruce samples.** *Holz Als Roh-Und Werkstoff* 1994, **52**:223–229.
15. Donaldson L, Frankland A: **Ultrastructure of iodine treated wood.** *Holzforschung* 2004, **58**:219–225.
16. Zimmermann T, Thommen V, Reimann P, Hug HJ: **Ultrastructural appearance of embedded and polished wood cell walls as revealed by atomic force microscopy.** *J Struct Biol* 2006, **156**:363–369.
17. Singh AP, Daniel G: **The S2 layer in the tracheid walls of Picea abies wood: inhomogeneity in lignin distribution and cell wall microstructure.** *Holzforschung* 2001, **55**:373–378.
18. Gierlinger N, Keplinger T, Harrington M: **Imaging of plant cell walls by confocal Raman microscopy.** *Nat Protoc* 2012, **7**:1694–1708.
19. Flores SM, Toca-Herrera JL: **The new future of scanning probe microscopy: combining atomic force microscopy with other surface-sensitive techniques, optical and microscopy fluorescence techniques.** *Nanoscale* 2009, **1**:40–49.
20. Yarbrough JM, Himmel ME, Ding SY: **Plant cell wall characterization using scanning probe microscopy techniques.** *Biotechnol Biofuels* 2009, **2**.
21. Synge EH: **A suggested method for extending microscopic resolution into the ultra-microscopic region.** *Philos Mag* 1928, **6**:356–362.
22. Ash EA, Nicholls G: **Super-resolution aperture scanning microscope.** *Nature* 1972, **237**:510–512.
23. Betzig E, Trautman JK, Harris TD, Weiner JS, Kostelak RL: **Breaking the diffraction barrier–optical microscopy on a nanometric scale.** *Science* 1991, **251**:1468–1470.
24. Lewis A, Lieberman K: **Near-field optical imaging with a non-evanescently excited high brightness light-source of subwavelength dimensions.** *Nature* 1991, **354**:214–216.
25. Lewis A, Taha H, Strinkovski A, Manevitch A, Khatchatouriants A, Dekhter R, Ammann E: **Near-field optics: from subwavelength illumination to nanometric shadowing.** *Nat Biotechnol* 2003, **21**:1377–1386.
26. Dunn RC: **Near-field scanning optical microscopy.** *Chem Rev* 1999, **99**:2891–2927.
27. Hambardzumyan A, Molinari M, Dumelie N, Foulon L, Habrant A, Chabbert B, Aguié-Béghin V: **Structure and optical properties of plant cell wall bio-inspired materials: cellulose-lignin multilayer nanocomposites.** *C R Biol* 2011, **334**:839–850.
28. Micic M, Radotic K, Jeremic M, Leblanc RM: **Study of self-assembly of the lignin model compound on cellulose model substrate.** *Macromol Biosci* 2003, **3**:100–106.
29. Micic M, Jeremic M, Radotic K, Leblanc RM: **A comparative study of enzymatically and photochemically polymerized artificial lignin supramolecular structures using environmental scanning electron microscopy.** *J Colloid Interface Sci* 2000, **231**:190–194.
30. Hecht B, Sick B, Wild UP, Deckert V, Zenobi R, Martin OJF, Pohl DW: **Scanning near-field optical microscopy with aperture probes: fundamentals and applications.** *J Chem Phys* 2000, **112**:7761–7774.
31. Barsberg S, Matousek P, Towrie M: **Structural analysis of lignin by resonance Raman spectroscopy.** *Macromolecular Biosci* 2005, **5**:743–752.
32. Albinsson B, Li SM, Lundquist K, Stomberg R: **The origin of lignin fluorescence.** *J Mol Struct* 1999, **508**:19–27.
33. Donaldson LA, Radotic K: **Fluorescence lifetime imaging of lignin autofluorescence in normal and compression wood.** *J Microsc* 2013, **251**:178–187.
34. Cosgrove DJ: **Growth of the plant cell wall.** *Nat Rev Mol Cell Biol* 2005, **6**:850–861.
35. Terashima N, Fukushima K: **Heterogeneity in formation of lignin 11: an autoradiographic study of the heterogeneous formation and structure of pine lignin.** *Wood Sci Technol* 1988, **22**:259–270.
36. Terashima N, Fukushima K, Sano Y, Takabe K: **Heterogeneity in formation of lignin 10: visualization of lignification process in differentiating xylem of pine by microautoradiography.** *Holzforschung* 1988, **42**:347–350.
37. Joseleau JP, Ruel K: **Study of lignification by noninvasive techniques in growing maize internodes - An investigation by Fourier transform infrared cross-polarization magic angle spinning C-13-nuclear magnetic resonance spectroscopy and immunocytochemical transmission electron microscopy.** *Plant Physiol* 1997, **114**:1123–1133.
38. Horcas I, Fernandez R, Gomez-Rodriguez JM, Colchero J, Gomez-Herrero J, Baro AM: **WSXM: a software for scanning probe microscopy and a tool for nanotechnology.** *Rev Sci Instrum* 2007, **78**.

Keel petal incision: a simple and efficient method for genetic crossing in *Medicago truncatula*

Vijaykumar Veerappan[1], Khem Kadel[1], Naudin Alexis[1], Ashley Scott[1], Igor Kryvoruchko[2], Senjuti Sinharoy[2], Mark Taylor[2], Michael Udvardi[2] and Rebecca Dickstein[1*]

Abstract

Background: Genetic crossing is an essential tool in both forward and reverse genetic approaches to understand the biological functions of genes. For *Medicago truncatula* (barrel medic) various crossing techniques have been used which differ in the methods used to dissect the female parent's unopened flower bud to remove immature anthers for prevention of self-pollination. Previously described methods including front, side or back incision methods may damage the flower bud, impeding successful fertilization and/or seed development because they may allow pollen to dislodge and floral organs to desiccate after crossing, all of which diminish the success rates of crossing.

Results: We report the keel petal incision method for genetic crossing in *M. truncatula* ecotype R108 and demonstrate successful crosses with two other *M. truncatula* ecotypes, A17 and A20. In the method presented here, an incision is made along the central line of the keel petal from the bottom 1/3rd of the female parent's flower bud to its distal end. This allows easy removal of anthers from the flower bud and access for cross-pollination. After pollination, the stigma and the deposited pollen from the male donor are covered by the keel petal, wing petals and standard petal, forming a natural pouch. The pouch prevents dislodging of deposited pollen from the stigma and protects the internal floral organs from drying out, without using cling-film or water-containing chambers to maintain a humid environment. The keel petal incision method showed an approximate 80% success rate in the *M. truncatula* R108 ecotype and also in other ecotypes including Jemalong A17 and A20.

Conclusions: Our keel petal incision protocol shows marked improvement over existing methods with respect to the ease of crossing and the percentage of successful crosses. Developed for the *M. truncatula* R108 ecotype, the protocol has been demonstrated with A17 and A20 ecotypes and is expected to work with other ecotypes. Investigators of varying experience have achieved genetic crosses in *M. truncatula* using this method.

Keywords: Legume, Genetic crossing, Barrel medic, *Medicago truncatula*, Artificial hybridization, Keel petal

Background

Medicago truncatula (barrel medic) is an important model legume species extensively used to study symbiotic interactions with soil rhizobia and arbuscular mycorrhizal fungi leading to symbiotic nitrogen fixation and mycorrhization respectively [1-3]. It has also found use as a model for studies on secondary metabolism [4-6], plant pathogens [7,8], leaf development [9-11] and other processes. *M. truncatula*, in the Galegoid clade of the Papilionoideae legume subfamily, is closely related to economically important crops also in the Galegoid clade including alfalfa (*M. sativa*), pea (*Pisum sativum*) and lentil (*Lens culinaris*) as well as crop legumes in the Phaseoloid clade including soybean (*Glycine max*) and common bean (*Phaseolus vulgaris*) [12]. In the past two decades, extensive genomic and genetic resources have been developed for *M. truncatula*, making it an elite legume model amenable for functional genomics as well as for genetic studies to identify key regulators in important processes. These resources include a global gene expression atlas [13] and a nearly complete genome sequence [14]. *M. truncatula*'s genome may contain up to 50,000 protein coding genes, most of unknown biological function [14]. To facilitate genetic studies in *M. truncatula*, several populations of mutants have been developed [15], including those treated with ethyl methane sulfonate (EMS) [16], fast neutron bombardment

* Correspondence: beccad@unt.edu
[1]Department of Biological Sciences, University of North Texas, 1155 Union Circle #305220, Denton, Texas 76203, USA
Full list of author information is available at the end of the article

(FNB) [17-19] and a near saturation insertion mutant population carrying the tobacco *Tnt1* retrotransposon [20-23].

An efficient technique for genetic crossing in *M. truncatula* is essential in our efforts to characterize mutants and to help identify their defective genes. Genetic crossing is commonly used to cross-fertilize mutants with their wild-type parents and to outcross mutants to different ecotypes to map the mutation of interest. Mutant populations generated using EMS, FNB or *Tnt1* transposition harbor multiple mutations in the same plant which may require backcrossing to the wild-type parent to remove extraneous mutations not linked to the mutant phenotypes under study and to investigate the mode of inheritance of phenotypes/traits. Back- and out-crossing provide valuable tools to identify causal mutations underlying mutant phenotypes by co-segregation/linkage analysis. The *M. truncatula* genome contains many gene families with redundant functions due to gene duplications [12]. To study the functions of redundant genes, double mutants can be created by genetic crossing [24]. Genetic crossing is also essential for linkage mapping of natural variations and for genome-wide association studies [25,26].

M. truncatula flowers are hermaphroditic, with both male and female organs in the same flowers. A *M. truncatula* flower is comprised of a calyx, a large standard petal, two small wing petals on both sides of the flower and a fused keel petal which cover both male and female floral organs (Figure 1A, B). The stigma is surrounded by eight fused anthers (Figure 1B). *M. truncatula* is a self-pollinated species and fertilization occurs when the flower is still closed (cleistogamy). When the flower reaches maturity, its anthers grow close to the stigma surface and rupture to release the pollen grains (anthesis) on the stigma surface which allows the flower to self-fertilize. When performing a genetic cross, it is important to artificially open the closed flower bud to visualize and access the anthers and stigma before anthesis. In a cross, for the flower bud serving as a female, all the anthers are removed before they release pollen. To artificially fertilize the female flower, pollen grains from mature anthers of a male donor are deposited onto the stigma surface of the female flower to allow cross-pollination to occur.

Several detailed genetic crossing techniques have been described for *M. truncatula*, which differ by how the immature flower bud is opened to allow removal of anthers and to allow access to the stigma for cross-pollination (Figure 2) [27-29]. These methods also differ in how the flower is treated post cross-pollination to assure a successful cross. We call these: (1) the front cut method [27-29]; (2) the side cut method, also called the pouch method [28,29]; and (3) back cut method (a variation of the side cut method; Mark Taylor, unpublished data). In the front cut method, the flower buds are cut vertically along the central line of the curved side of the standard petal towards its tip from the proximal one third of the flower bud (Figure 2A). In the side cut (Figure 2B) and back cut (Figure 2C) methods, similar cuts are made on the side and back of the flower buds respectively. The side cut method is time consuming and in all three procedures, the incision may damage the flower bud, rendering the female flower unusable. After cross-pollination using either the front or back cut methods, it is difficult to close the flower bud completely, which may lead to mechanical loss of deposited pollen and/or desiccation of internal floral organs.

Another crossing method employs male-sterile mutants as females. The *mtapetala* (*tap*) mutant, found in the *M. truncatula* Jemalong A17 ecotype [16], has been extensively used for this purpose [16,30-33]. However the *tap* mutation is not found in other ecotypes.

Here we report an improved genetic crossing technique called keel petal incision (Figure 2D; Additional file 1: Video S1; youtu.be/wDtRHWg1LBM), initially developed for the *M. truncatula* R108 ecotype. The keel petal incision method allows easy access to an immature flower for anther removal and artificial pollination. It also preserves floral morphological features permitting the standard and wing petals to close to their natural positions after cross pollination. This allows fertilization and embryo development to occur under protection of the flower petals and obviates the need for maintaining the pollinated flower in cling-film or humidified containers. Keel petal incision is a simple and efficient crossing technique with high success rates (>80%) in the R108 ecotype and in other ecotypes, A17 and A20.

Results and discussion

Plant growth and selection of optimal female and male flowers

For the female in the crosses, we grew *M. truncatula* plants until they started flowering and began to develop one or two pods. Plants that have one or two developing pods will contain flowers in various stages of development as illustrated (Figure 3). For the female parent in the cross, very young flower buds (Figure 3A, B) are not optimal. While they have less chance of self-pollination than older flowers because they have immature anthers, they also typically have less mature stigmas as well. Older flowers (Figure 3D) have more mature stigmas, but also an increased likelihood of self-pollination caused by anther rupture during anther removal. Intermediate flower buds have more mature stigmas and not-yet ripe anthers (Figure 3C) and offer the best choice for a successful cross with fewer chances of self-pollination. For the male parent in the cross, flowers just about to open (Figure 3E, H) are the best choice. The stigma column and anthers can be triggered to open by applying slight pressure to the flower. Alternatively, fresh open (Figure 3F, K) flowers can serve as male parent/pollen donor. Rupture of anthers is verified

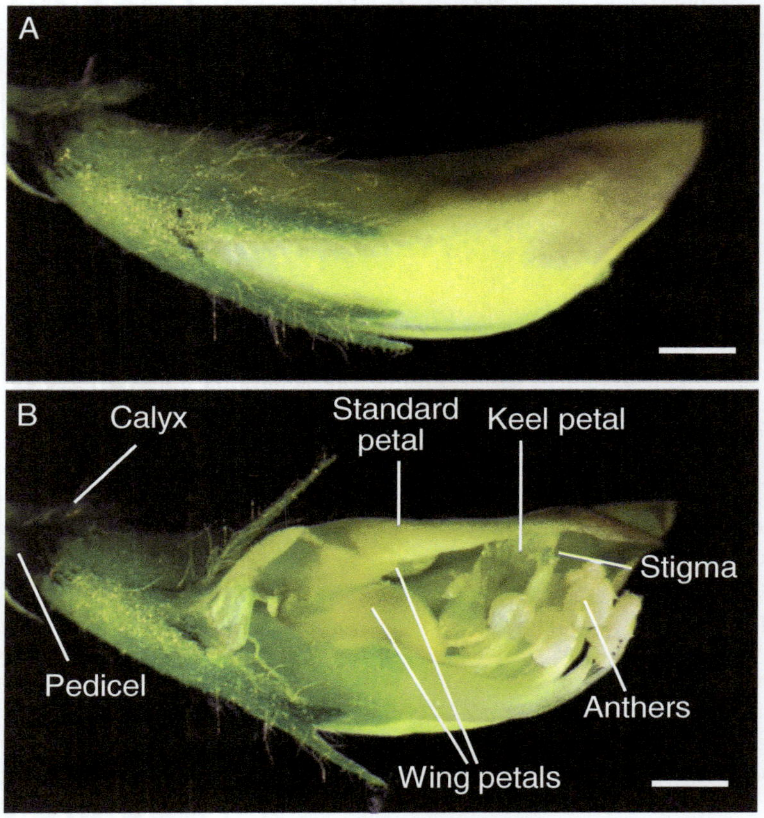

Figure 1 ***M. truncatula*** **flower structure. A**, Unopened flower bud. **B**, Cut flower bud showing internal reproductive organs. Bars = 0.5 mm.

under the dissecting microscope to make sure pollen grains are visible on the anther surface.

Anther removal and artificial pollination

After selecting an unopened flower bud as a female parent (Figure 3C), we removed the leaves, flowers, pods and/or shoots within three to four inches of the bud, leaving at most one trifoliate leaf adjacent to the chosen flower bud for ease of manipulation. Because *M. truncatula* flowers and their internal floral organs are small, it was not feasible to perform crosses using the naked eye. Hence, we performed crossing under a dissecting microscope; a magnifying glass or a magnifying binocular headset would also work for this purpose. The flower serving as female was mounted horizontally on its side on a dissecting microscope stage oriented so that the tip of the standard petal faced towards the base of the microscope and the opening of the standard petal faced the dominant hand of the person performing the cross (Figure 4A; Additional file 1: Video S1). The flower bud was secured to the microscope stage using cellophane tape on the pedicel. Subsequently, the flower bud was held in place using a set of forceps on the calyx, while another set of forceps was used to gently lift the standard petal to access the underlying keel petal. Using the sharp tip of a scalpel, the keel petal was gently cut at the bottom third of the flower bud and incision was made along the central ridge of the keel petal all the way to the distal end of the flower (Figure 4B). During this process, care was taken to assure that the tip of the scalpel blade cut only the keel petal and did not extend far inside the flower bud so as

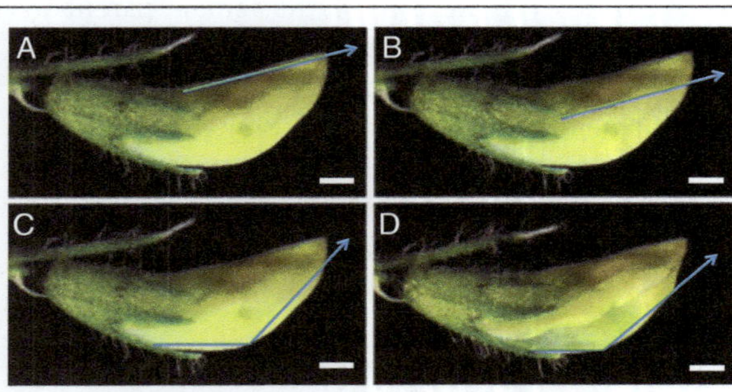

Figure 2 Position of incisions used to artificially open the female flower bud to perform genetic crossing. A, Front cut method. **B**, Side cut method. **C**, Back cut method. **D**, Keel petal incision method. Arrows indicate the sites of the cut/incision made on the unopened flower buds to perform anther removal and artificial pollination. Bars = 0.5 mm.

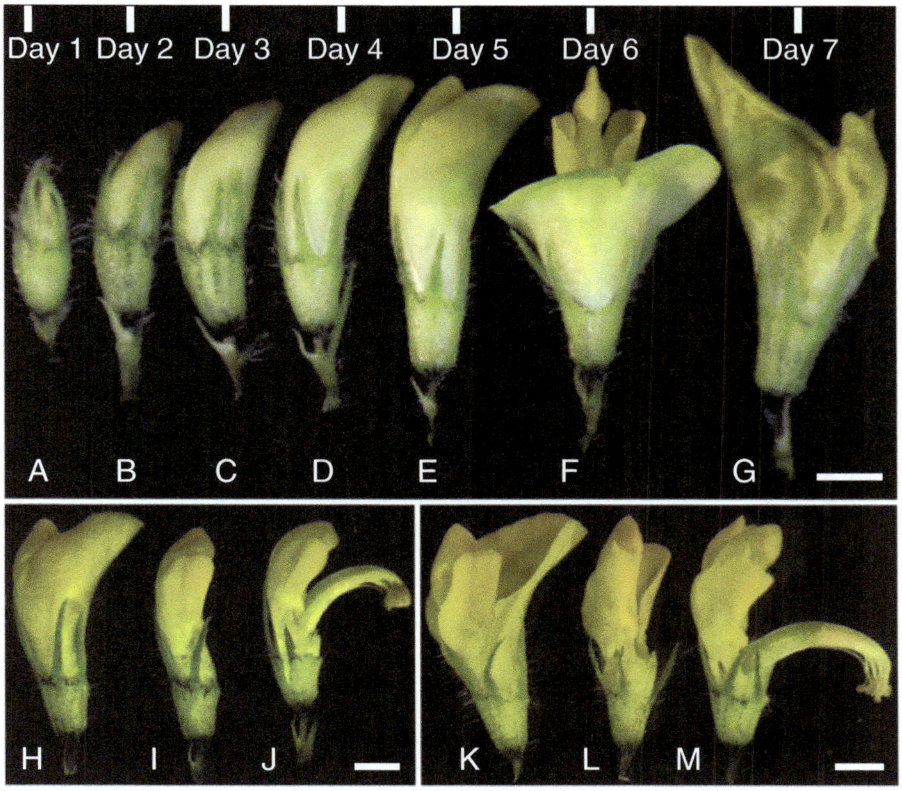

Figure 3 ***M. truncatula*** **flower development and manipulation of pollen donor flowers. A-G**, Flowers from different developmental stages were collected, starting from when flower buds are first visible (Day 0). Flowers at day 3 are optimal to use as the female parent, while those at day 5 are best as the male parent. **H**, Day 5 flower. **I**, Day 5 flower with standard petal removed. **J**, Flower as in **I**, with stigma column exposed, showing anthers. **K**, Day 6 flower, rotated approximately 140° from flower in **F. L**, Day 6 flower with standard petal removed. **M**, flower as in **L**, rotated approximately 110°. Bar = 0.5 mm in top panel; bars = 1 mm in lower panels.

not to damage the stigma column underneath the keel petal and not to rupture the unopened anthers.

Subsequently, the half of the cut keel petal, facing the investigator, along with the wing and standard petals were pushed upwards with the forceps, allowing visualization and access to the anthers and stigma inside the flower bud (Figure 4C). At this point in preparing the flower for pollination, we checked whether the anthers were dehisced. If the anthers were already ruptured, we discarded that flower bud and chose another as the female crossing parent. If the anthers were not yet dehisced, we removed all eight of them by cutting the anther filaments away from the stigma column using sharp forceps tips. Alternatively, anther removal may be accomplished by suction using vacuum applied through a micropipette tip [16]. During anther removal, it is important not to rupture them and to avoid having the anthers contact the stigma. After anther removal, we ascertained that all the anthers were removed (Figure 4D) and visualized the tip of the stigma to verify that there was no accidental deposition of pollen grains on the stigma surface or inside the flower bud during anther removal. In the instances where we saw that this had occurred, we discarded that flower and started over with a fresh one.

To expose the anthers from a male pollen donor in the cross, we triggered a flower that was about to open (Figure 3E, H) with gentle mechanical pressure until the style and anthers popped open. Alternatively, we have used forceps to remove the petals around the anthers in the pollen donor flower (Figure 3I). We found that using forceps to remove the standard petal first and then the wing and keel petals does not damage the anthers. Once the flower is triggered (Figure 3J) or the petals were removed, we observed the anthers using the dissecting microscope to ascertain that the anther was mature and dehisced with silvery white pollen grains released from ruptured anther sacs visible. This can also be seen using other magnification devices. Already-open flowers also served as pollen donors (Figure 3F, K). In these instances, the standard petal was removed with forceps (Figure 3L) exposing the style and anthers (Figure 3M; flower turned approximately 110° from Figure 3L).

The mature anthers with attached pollen were then gently placed on the tip of the stigma of the female

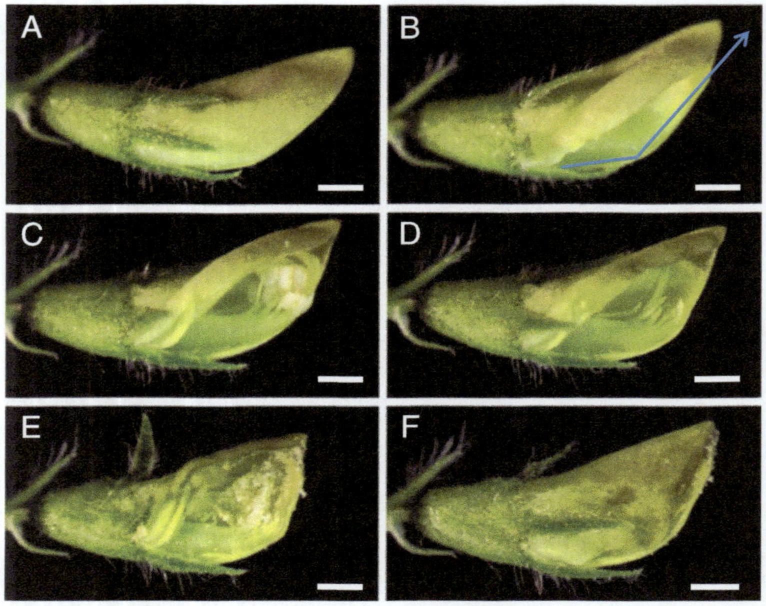

Figure 4 Steps in genetic crossing using the keel petal incision method. **A**, Unopened optimal female flower bud. **B**, Flower bud showing the keel petal under the standard petal indicating the site of incision on the central line of the keel petal. **C**, Flower bud showing the internal floral organs including anthers and stigma underneath the keel petal after making an incision. **D**, Flower bud showing stigma and style after the removal of anthers. **E**, Stigma covered with mature pollen after artificial cross-pollination. **F**, Closed flower bud after deposition of pollen on stigma. Bars = 0.5 mm.

flower multiple times to deposit the pollen grains. At the end of this process, the entire tip of the stigma was surrounded by pollen (Figure 4E). There is variation in pollen quality and quantity from individual flowers; pollen quality may be negatively affected by insect infestation as well. In the case of insufficient pollen from one flower, we used more than one flower from the male pollen donor plant.

After pollination, the flower was closed by pushing the top of the standard petal back into its original position to cover the wing petal, keel petal and stigma, forming a pouch around the stigma that looks similar to the

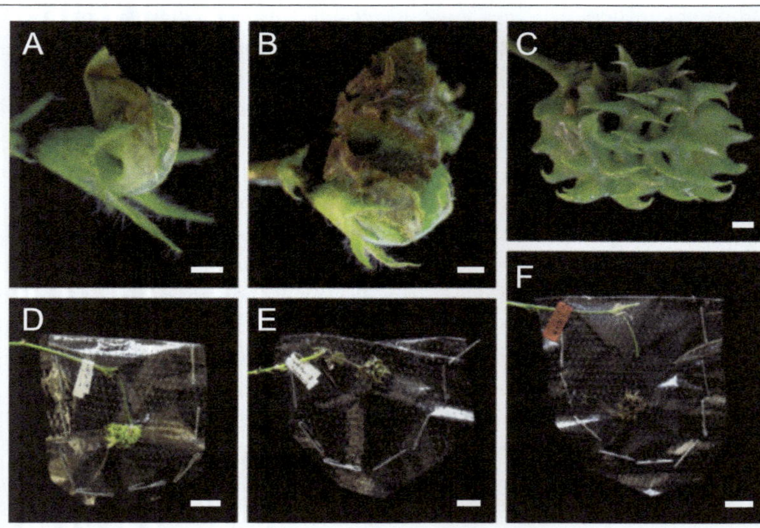

Figure 5 Development of pods from successful crosses. **A**, Curling of artificially cross-pollinated flower bud 3 days post hybridization. **B**, Developing pod 6 days post hybridization. **C**, Developing pod 20 days post-hybridization. **D**, Developing pod, 35 days post-hybridization wrapped with microperforated polythene sheet. **E**, Ripening yellow pod, 42 days post-hybridization. **F**, Completely matured darkened pod fallen off the pedicel, 50 days post-hybridization. Bars = 0.5 mm **(A-C)**. Bars = 5 mm **(D-F)**.

unopened flower bud before crossing (Figure 4F). This prevented pollen from dislodging from the stigma during subsequent movement of the plant and protected the stigma and pollen from desiccation. The floral petal pouch obviates the need for protecting the pollinated flower with an artificial covering. These coverings promote fungal and bacterial growth on the flowers that may cause the flower to decay and drop. The pouch also does away with the need to place the cross-pollinated flower bud inside humid containers, which may also cause flower damage. After the cross-pollination was completed, the cellophane tape holding the pedicel was carefully removed and the pollinated flower was labeled.

Seed pod development

To distinguish between artificially pollinated and self-pollinated pods, we monitored the artificially pollinated flower buds daily and trimmed away new shoots growing adjacent to the pollinated flowers. Successful crosses were visualized by slight curling of pollinated flowers from the 3rd day onwards (Figure 5A). Most of the time, the cross-pollinated flowers continued to curl (Figure 5B) and develop into mature pods (Figure 5C). Once the pods from successful crosses developed into medium size (Figure 5C), we wrapped them using micro-perforated polythene sheets stapled in place (Figure 5D) to prevent the pods from falling off after maturity.

To increase the size of the pods from cross-pollination and number of seeds per cross, we removed all pods on the plant resulting from self-pollination. Removing pods promotes more flower formation and prolongs the duration of flowering. After all the crosses were completed, shoots that do not carry pods from cross-pollination were trimmed. We observed developing pods turning yellow (Figure 5E) and darkening and dropping from the pedicel at maturity (Figure 5F).

We air-dried the pods from successful crosses at room temperature for 2–3 weeks. After this, we collected seeds from the pods and proceeded to confirm the crosses in the F_1 generation. We found that larger pods typically contain 5–6 seeds/pod whereas the smaller pods only carried 1–3 seeds/pod. In general, crosses performed on younger plants resulted in larger pods whereas older plants produced smaller pods.

Table 1 Estimation of crossing success via seed pod formation

Crossing parents, wild-type as female	Attempted crosses	Seed pods formed	Success rate,%
R108 X A17	9	6	67
R108 X A20	14	13	93
R108 X NF10796	17	12	71
R108 X NF11014	22	19	86
R108 X NF11044	27	22	81
R108 X NF11166	7	7	100
R108 X NF11217	11	8	73
R108 X NF1320	38	29	76
R108 X NF1320-29-3	17	17	100
R108 X NF1320-BC1-F3	21	16	76
R108 X NF4619	5	4	80
R108 X NF8324	31	23	74
A17 X R108	43	41	95
A17 X NF11044	91	82	90
A17 X NF11217	30	26	87
A17 X NF8324	18	12	67
A20 X R108	70	63	90
A20 X NF11044	58	49	84
A20 X NF11217	11	10	91
A20 X NF1320	15	9	60
A20 X NF8324	29	21	72
Sub-total	584	489	84
Crossing parents, mutant as female	**Attempted crosses**	**Seed pods formed**	**Success rate, %**
NF10796 X R108	14	9	64
NF11014 X R108	16	11	69
NF11044 X R108	28	22	79
NF11166 X R108	4	4	100
NF11217 X R108	16	11	69
NF1320 X R108	30	23	77
NF8324 X A20	7	5	71
NF8324 X R108	34	27	79
Sub-total	149	112	75
Total crosses	733	601	82

For each cross, the first parent listed in the cross served as the female. All crosses were performed by two graduate students, one with one year and the other with two years of research experience. Success rates were calculated based on whether a pod formed in the cross-pollinated flower buds ten days post-pollination.

Cross-pollination confirmation

As a first test of the success of cross-pollination by the keel petal incision method, we counted the number of seed pods that formed after cross pollination. We estimated an overall success rate of 82%, with slightly higher rates of pod formation when a wild-type parent was the female in the cross (Table 1).

Confirmation of successful crossing relied on phenotyping and genotyping the progeny. Our labs employ forward genetics to identify new legume genes that control the development of symbiotic nitrogen-fixing root nodules using the *M. truncatula Tnt1* insertion mutant collection in the R108 ecotype background. *Tnt1* mutants contain multiple *Tnt1* insertions and there are no naturally occurring *Tnt1* inserts in any

M. truncatula ecotype [21,34]. To study the mode of inheritance of mutant phenotypes and to test the co-segregation of candidate *Tnt1* insertions with nodule phenotypes, we backcrossed two particular mutants with ineffective, white or brown non-fixing nodules (Nod + Fix-), NF11217 and NF10796, to their wild-type R108 using the keel petal incision method.

For NF11217 backcrosses, the mutant plant was used as the female parent and the wild-type R108 served as the male pollen donor. For NF10796 backcrosses, wild-type served as the female and the mutant plant was used as a male parent. Typically, it is advisable to perform reciprocal crosses by switching male and female parents in separate crosses, because the mutant phenotypes might be governed by maternal or paternal inheritance.

When recessive *Tnt1* mutants were used as female, successful crosses were confirmed by wild-type phenotype progeny in the F_1 population. Nodules of NF11217 mutant (female parent), wild-type (male parent) and NF11217 × wild-type-BC_1F_1 progeny were assessed for visual phenotypes fifteen days after inoculation of roots with *Sinorhizobium meliloti*. The nodules from wild-type parents are pink (Figure 6A) because of the presence of leghemoglobin associated with efficient symbiotic nitrogen fixation, whereas the NF11217 mutant plants show white nodules, defective in symbiotic nitrogen fixation (Figure 6B). All the BC_1F_1 plants tested showed pink nodules similar to the wild-type (Figure 6C-H), indicating that cross-pollination was successful and suggesting the mutation in NF11217 is recessive.

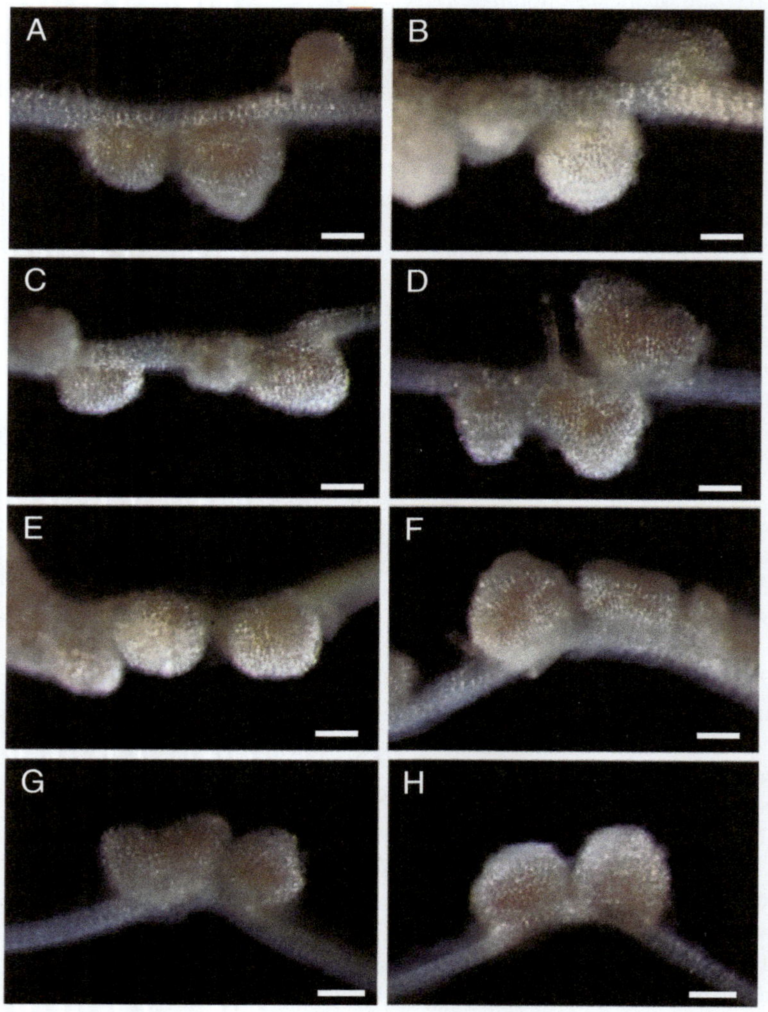

Figure 6 Confirmation of successful crosses by nodule phenotypes of the progeny. *M. truncatula* plants were grown on an aeroponic chamber, as described in Materials and methods. Five days after germination, plants were starved for nitrogen for five days and then inoculated with *Sinorhizobium meliloti*. Root nodules were examined 15 days post-inoculation. **A**, Wild-type R108. **B**, NF11217 mutant. **C-H**, NF11217 X wild-type R108-BC_1F_1. The R108 plant **(A)** has pink wild-type nodules while those of NF11217 are white and ineffective **(B)**. Each plant from the BC_1F_1 progeny **(C-H)** has pink nodules indicative of a successful cross. Bars = 250 μm.

When wild-type was used as female, success of the crosses was verified by genotyping the F_1 plants for the presence of *Tnt1* transgene. For the case of NF10796 serving as the male in a cross with a wild-type R108 female, PCR (polymerase chain reaction) genotyping of BC_1F_1 progeny using *Tnt1* specific primers was performed. Our results showed the presence of *Tnt1* insertions in the R108 x NF10796-BC_1F_1 plants (Figure 7A) indicating crossing success. As expected, there was no amplification of *Tnt1* sequences in the wild-type R108 (Figure 7A).

As a further test of the utility of the keel petal incision method, we attempted out-crossing of *Tnt1* Nod+ Fix- mutant NF8324 in the R108 background, to both the A17 and A20 wild-type ecotypes. In these cases, we used the A17 or A20 wild-types as females and NF8324 as a male in the crosses. F_1 plants arising from the crosses were genotyped by PCR for the presence of *Tnt1* sequences in the progeny, indicating a successful cross. As can be seen by the examples (Figure 7B, C), the keel petal incision method also worked for these other *M. truncatula* ecotypes. However, F_1 progeny from inter-ecotype crosses involving R108 were observed to have developmental phenotypes (not shown); thus, special care must be taken with these progeny.

We examined ninety BC_1F_1 plants for nodulation phenotypes resulting from backcrosses involving Nod+/Fix- mutants female parents with wild-type R108 as male and found almost all of them had wild-type nodules, indicating successful cross-pollination and absence of self-pollination (Table 2). With the reciprocal situation using the wild-type as female and mutant as male, we tested forty plants from the progeny and obtained success rates of approximately 87% overall using the R108, A17 and A20 ecotypes, as verified by presence of *Tnt1* sequences in the BC_1F_1 and F_1 progeny (Table 2).

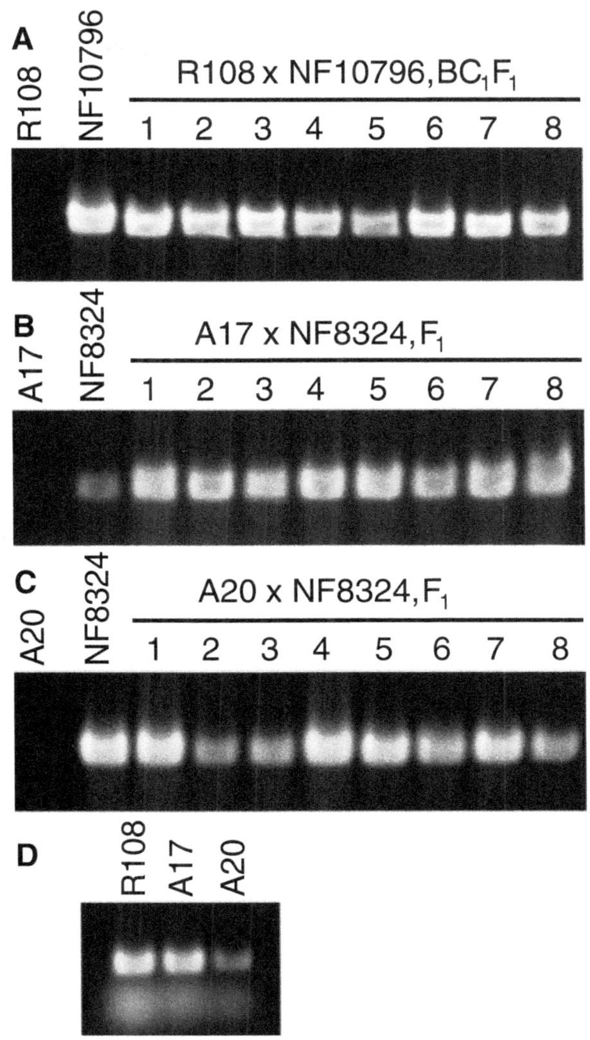

Figure 7 Confirmation of successful crosses by genotyping. **A-C**, Wild-type R108, NF10796 and NF8324, R108 x NF10796-BC_1F_1, A17 x NF8324-F_1 and A20 x NF8324-F_1 plants were PCR genotyped using *Tnt1* transposon specific primers. **D**, Primers for the *MtIRE-like* gene were used in the PCR amplifications as positive control for the wild-type genomic DNA. *Tnt1* DNA is not found in wild-type R108, A17 or A20 plants but can be detected in all F_1 plants indicating successful crosses **(A-C)**. Stained gel images have been cropped to show the PCR products of 612 bp from *Tnt1* and 422 bp from *MtIRE-like* amplifications respectively.

Conclusions

Crossing is essential for genetics. Here we described a new crossing technique for *M. truncatula* that involves incising the keel petal of the female flower to expose the flower's stigma and anthers, remove its anthers and permit an artificial cross. After the cross is complete, the standard, wing and keel petals can be arranged similarly to their original positions in the unopened flower bud, protecting the floral organs from desiccation and removing the need for extraneous materials to maintain humidity for the fertility of the flower and deposited pollen. We demonstrated that this method can be used for the *M. truncatula* R108, A17 and A20 ecotypes and expect that it will find utility in other ecotypes. This method has been successfully used by investigators of varying experience, including junior graduate students and undergraduates in our labs.

Materials and methods

Plant growth

M. truncatula seeds were scarified with sulfuric acid, treated with bleach, imbibed and vernalized as described [16,28]. Shorter vernalization (1–4 days at 4°C) results in delayed flowering and a larger plant with higher numbers of flowers. Longer vernalization (4–14 days at 4°C) results in earlier flowering accompanied by smaller plant stature and reduced flower numbers. If plants are grown exclusively for crossing, longer vernalization is preferred to

Table 2 Confirmed successful crosses

Crossing parents, mutant as female	Total number of plants tested	Number of confirmed plants by nodule phenotyping	Success rate,%
NF11217 X R108	28	28	100
NF11044 X R108	31	31	100
NF8324 X R108	17	17	100
NF1320 X R108	14	11	78
Crossing parents, wild-type as female	**Total number of plants tested**	**Number of confirmed plants by PCR genotyping**	**Success rate,%**
R108 X NF10796	8	8	100
R108 X NF8324	13	11	84
A17 X NF8324	8	8	100
A20 X NF8324	11	8	72

For each cross, the first parent listed in the cross was used as the female. For cases where the mutant served as female in the cross, successful crossing was confirmed by the appearance of pink nodules in the BC_1F_1 progeny. When wild-type served as female, including inter-ecotype crosses, crosses were confirmed in the BC_1F_1, or F_1 progeny by PCR genotyping using *Tnt1* specific primers, as in Figure 7.

shorten the time to flowering. Details of the effect of vernalization on flowering time and growth stature can be found in [28]. Different vernalization times were used and did not appear to affect crossing success. After vernalization, seeds were germinated at 25°C in the dark; in our hands, seeds germinated within 12–24 hours. Subsequently, the plants were screened for nodulation phenotypes 15 days post-inoculation with *S. meliloti* using an aeroponic chamber [30,35], or by growing the plants on a Turface/vermiculite mixture [23]. After phenotyping, the seedlings were grown in medium size (four inch; ten cm) pots so that they are easy to handle while performing crosses. To avoid waterlogged roots, we grew *M. truncatula* in a peat-based potting medium with Turface (Profile, Buffalo Grove, Ill) mixed in (3:1), at 22°C under16:8 hr light:dark cycles. Plants were irrigated as needed and fertilized with a commercial general purpose fertilizer once a week.

Crosses and PCR confirmation

The crosses were performed under a Nikon model C-PS stereo microscope (Nikon, Tokyo, Japan). Images and video clips of *M. truncatula* flowers and crossing techniques were obtained using Leica MZ10F (Leica, Buffalo Grove, IL) microscope. Two pairs of fine tip forceps (HL-14 #5, www.buyincoins.com) and a straight-edge scalpel (scalpel blade handle 9303 #3, and scalpel blade 9311 #11, both from www.microscopesamerica.com) were used for keel petal incision, the removal of anthers from the unopened female flower bud and artificial cross-pollination (Additional file 2: Figure S1). Mature pods from the successful cross-pollinations were wrapped using micro-perforated polythene sheets (MP1120160T, www.prismpak.com).

The following primers were used for PCR genotyping to confirm crosses: *Tnt1*-F, GCATTCAAACTAGAAGA CAGTGCTACC and *Tnt1*-R, TGTAGCACCGAGATA CGGTAATTAACAAGA [34] (*Tnt1*, Genbank:X13777). *MtIRE-like* specific primers [36] (*MtIRE-like*, Genbank: AY770392, Genbank:AC122727) were used as a control to confirm that extracted DNA from wild-type plants was of sufficient quality for PCR using *MtIRE*-F, CCAAATC GTTGAAAGCTCGTTCACAACTCC and *MtIRE*-R, CG TCTTGACCAGCAAACACGACACG.

Additional files

Additional file 1: Video S1. The keel petal incision crossing method for *M. truncatula*. The method is narrated step-by-step: youtu.be/wDtRHWg1LBM.

Additional file 2: Figure S1. Forceps and scalpel used in crossing. Two pairs of fine tip forceps, e.g., HL-14 #5, www.buyincoins.com, and a straight-edge scalpel, e.g., scalpel blade handle 9303 #3, and scalpel blade 9311 #11, both from www.microscopesamerica.com, were used for keel petal incision, the removal of anthers from the unopened female flower bud and cross-pollination. Similar forceps and scalpels are available from other vendors. Bar = 5 cm.

Abbreviations

BC: Back cross; DNA: Deoxyribonucleic acid; EMS: Ethyl methane sulfonate; FNB: Fast neutron bombardment; *M. truncatula*: *Medicago truncatula*; PCR: Polymerase chain reaction.

Competing interests

The authors declare that they have no competing interests.

Authors' contributions

VV designed the experiments and made the supplemental video. VV, KK, NA, AS, IK, SS and MT carried out the experiments and edited the manuscript. MU edited the manuscript. VV and RD coordinated the experiments, and wrote and edited the manuscript. All authors read and approved the final manuscript.

Acknowledgements

This work was supported by National Science Foundation grant IOS 1127155 to MU and RD and by USDA-NIFA grant 2010-65115-20384 to MU. AS was partially supported by Department of Education grant P217A120023 to Dr. D Elrod and UNT's McNair Scholars Program. Video and audio equipment used in the production of the supplemental video for this article were provided by the University of North Texas Media Library.

Author details

[1]Department of Biological Sciences, University of North Texas, 1155 Union Circle #305220, Denton, Texas 76203, USA. [2]Plant Biology Division, The Samuel Roberts Noble Foundation, Ardmore, OK 73401, USA.

References

1. Cook DR: **Medicago truncatula - a model in the making!** *Curr Opin Plant Biol* 1999, **2**:301–304.
2. Kouchi H, Imaizumi-Anraku H, Hayashi M, Hakoyama T, Nakagawa T, Umehara Y, Suganuma N, Kawaguchi M: **How many peas in a pod? Legume genes responsible for mutualistic symbioses underground.** *Plant Cell Physiol* 2010, **51**:1381–1397.
3. Barker DG, Bianchi S, London F, Dattee Y, Duc G, Essad S, Flament P, Gallusci P, Genier G, Muel X, Tourneur J, Denarie J, Huguet T: ***Medicago truncatula*, a**

model plant for studying the molecular genetics of the *Rhizobium*-legume symbiosis. *Plant Mol Biol Report* 1990, **8**:40–49.
4. Dixon RA, Liu C, Jun JH: **Metabolic engineering of anthocyanins and condensed tannins in plants.** *Curr Opin Biotechnol* 2013, **24**:329–335.
5. Verdier J, Zhao J, Torres-Jerez I, Ge S, Liu C, He X, Mysore KS, Dixon RA, Udvardi MK: **MtPAR MYB transcription factor acts as an on switch for proanthocyanidin biosynthesis in *Medicago truncatula*.** *Proc Natl Acad Sci USA* 2012, **109**:109.
6. Zhao J, Dixon RA: **The 'ins' and 'outs' of flavonoid transport.** *Trends Plant Sci* 2010, **15**:72–80.
7. Uppalapati SR, Ishiga Y, Doraiswamy V, Bedair M, Mittal S, Chen J, Nakashima J, Tang Y, Tadege M, Ratet P, Chen R, Schultheiss H, Mysore KS: **Loss of abaxial leaf epicuticular wax in *Medicago truncatula irg1/palm1* mutants results in reduced spore differentiation of anthracnose and nonhost rust pathogens.** *Plant Cell* 2012, **24**:353–370.
8. Uppalapati SR, Marek SM, Lee H-K, Nakashima J, Tang Y, Sledge MK, Dixon RA, Mysore KS: **Global gene expression profiling during *Medicago truncatula–Phymatotrichopsis omnivora* interaction reveals a role for jasmonic acid, ethylene, and the flavonoid pathway in disease development.** *Mol Plant Microbe Interactions* 2009, **22**:7–17.
9. Tadege M, Lin H, Bedair M, Berbel A, Wen J, Rojas CM, Niu L, Tang Y, Sumner L, Ratet P, McHale NA, Madueño F, Mysore KS: ***STENOFOLIA* regulates blade outgrowth and leaf vascular patterning in *Medicago truncatula* and *Nicotiana sylvestris*.** *Plant Cell* 2011, **23**:2125–2142.
10. Lin H, Niu L, McHale NA, Ohme-Takagi M, Mysore KS, Tadege M: **Evolutionarily conserved repressive activity of WOX proteins mediates leaf blade outgrowth and floral organ development in plants.** *Proc Natl Acad Sci USA* 2013, **110**:366–371.
11. Ge L, Peng J, Berbel A, Madueño F, Chen R: **Regulation of compound leaf development by *PHANTASTICA* in *Medicago truncatula*.** *Plant Physiol* 2014, **164**:216–228.
12. Zhu H, Choi H-K, Cook DR, Shoemaker RC: **Bridging model and crop legumes through comparative genomics.** *Plant Physiol* 2005, **137**:1189–1196.
13. Benedito VA, Torres-Jerez I, Murray JD, Andriankaja A, Allen S, Kakar K, Wandrey M, Verdier J, Zuber H, Ott T, Moreau S, Niebel A, Frickey T, Weiller G, He J, Dai X, Zhao PX, Tang Y, Udvardi MK: **A gene expression atlas of the model legume *Medicago truncatula*.** *Plant J* 2008, **55**:504–513.
14. Young ND, Debellé F, Oldroyd GED, Geurts R, Cannon SB, Udvardi MK, Benedito VA, Mayer KFX, Gouzy J, Schoof H, Van de Peer Y, Proost S, Cook DR, Meyers BC, Spannagl M, Cheung F, De Mita S, Krishnakumar V, Gundlach H, Zhou S, Mudge J, Bharti AK, Murray JD, Naoumkina MA, Rosen B, Silverstein KAT, Tang H, Rombauts S, Zhao PX, Zhou P, *et al*: **The *Medicago* genome provides insight into the evolution of rhizobial symbioses.** *Nature* 2011, **480**:520–524.
15. Tadege M, Wang TL, Wen J, Ratet P, Mysore KS: **Mutagenesis and beyond! Tools for understanding legume biology.** *Plant Physiol* 2009, **151**:978–984.
16. Penmetsa RV, Cook DR: **Production and characterization of diverse developmental mutants of *Medicago truncatula*.** *Plant Physiol* 2000, **123**:1387–1398.
17. Wang HL, Li G, Chen R: **Fast Neutron Bombardment (FNB) Mutagenesis for Forward and Reverse Genetic Studies in Plants.** In *Floriculture, Ornamental and Plant Biotechnology: Advances and Topical Issues (1st Edition). Volume 1.* Edited by Silva JT. Isleworth, UK: Global Science Books; 2006:629–639.
18. Rogers C, Wen J, Chen R, Oldroyd G: **Deletion-based reverse genetics in *Medicago truncatula*.** *Plant Physiol* 2009, **151**:1077–1086.
19. Rogers C, Oldroyd GED: **Fast Neutron Mutagenesis for Functional Genomics.** In *The Handbook of Plant Functional Genomics.* Edited by Kahl G, Meksem K. Weinheim, Germany: Wiley-VCH; 2008:291–306.
20. Tadege M, Ratet P, Mysore KS: **Insertional mutagenesis: a Swiss Army knife for functional genomics of *Medicago truncatula*.** *Trends Plant Sci* 2005, **10**:229–235.
21. Tadege M, Wen J, He J, Tu H, Kwak Y, Eschstruth A, Cayrel A, Endre G, Zhao PX, Chabaud M, Ratet P, Mysore KS: **Large-scale insertional mutagenesis using the *Tnt1* retrotransposon in the model legume *Medicago truncatula*.** *Plant J* 2008, **54**:335–347.
22. Cheng X, Wen J, Tadege M, Ratet P, Mysore KS: **Reverse genetics in *Medicago truncatula* using *Tnt1* insertion mutants.** *Methods Mol Biol* 2011, **678**:179–190.
23. Pislariu CI, Murray JD, Wen J, Cosson V, Muni RRD, Wang M, Benedito VA, Andriankaja A, Cheng X, Jerez IT, Mondy S, Zhang S, Taylor ME, Tadege M, Ratet P, Mysore KS, Chen R, Udvardi MK: **A *Medicago truncatula* tobacco retrotransposon insertion mutant collection with defects in nodule development and symbiotic nitrogen fixation.** *Plant Physiol* 2012, **159**:1686–1699.
24. Floss DS, Levy JG, Lévesque-Tremblay V, Pumplin N, Harrison MJ: **DELLA proteins regulate arbuscule formation in arbuscular mycorrhizal symbiosis.** *Proc Natl Acad Sci USA* 2013, doi:10.1073/pnas.1308973110.
25. Yeoh CC, Balcerowicz M, Zhang L, Jaudal M, Brocard L, Ratet P, Putterill J: **Fine mapping links the *FTa1* flowering time regulator to the dominant *Spring1* locus in Medicago.** *PLoS One* 2013, **8**:e53467. doi:53410.51371/journal.pone.0053467.
26. Jaudal M, Yeoh CC, Zhang L, Stockum C, Mysore KS, Ratet P, Putterill J: **Retroelement insertions at the Medicago *FTa1* locus in *spring* mutants eliminate vernalisation but not long-day requirements for early flowering.** *Plant J* 2013, **76**:580–591.
27. Pathipanawat W, Jones RAC, Sivasithamparam K: **An improved method for artificial hybridization in annual *Medicago* species.** *Aust J Agric Res* 1994, **45**:1329–1335.
28. Chabaud M, Lichtenzveig J, Ellwood S, Pfaff T, Journet E-P: **Vernalization, Crossings and Testing for Pollen Viability.** In *The Medicago Truncatula Handbook.* Edited by Mathesius U, Journet E-P, Sumner LW. 2006. http://www.noble.org/MedicagoHandbook/. ISBN 0-9754303-1-9.
29. Taylor M, Blaylock L, Nakashima J, McAbee D, Ford J, Harrison M, Udvardi M: ***Medicago Truncatula* Hybridization – Supplemental Videos.** In *The Medicago Truncatula Handbook, Version 2011.* Edited by Mathesius U, Journet E-P, Sumner LW. Ardmore, Oklahoma, USA: The Samuel Roberts Noble Foundation; 2011:1–5. vol. http://www.noble.org/MedicagoHandbook/. pp. 1–5.
30. Veereshlingam H, Haynes JG, Sherrier DJ, Penmetsa RV, Cook DR, Dickstein R: ***nip*, a symbiotic *Medicago truncatula* mutant that forms root nodules with aberrant infection threads and plant defense-like response.** *Plant Physiol* 2004, **136**:3692–3702.
31. Schnabel E, Mukherjee A, Smith L, Kassaw T, Long S, Frugoli J: **The *lss* supernodulation mutant of *Medicago truncatula* reduces expression of the *SUNN* gene.** *Plant Physiol* 2010, **154**:1390–1402.
32. Schnabel EL, Kassaw TK, Smith LS, Marsh JF, Oldroyd GED, Long SR, Frugoli JA: **The *ROOT DETERMINED NODULATION1* gene regulates nodule number in roots of *Medicago truncatula* and defines a highly conserved, uncharacterized plant gene family.** *Plant Physiol* 2011, **157**:328–340.
33. Penmetsa RV, Frugoli JA, Smith LS, Long SR, Cook DR: **Dual genetic pathways controlling nodule number in *Medicago truncatula*.** *Plant Physiol* 2003, **131**:998–1008.
34. D'Erfurth I, Cosson V, Eschstruth A, Lucas H, Kondorosi A, Ratet P: **Efficient transposition of the *Tnt1* tobacco retrotransposon in the model legume *Medicago truncatula*.** *Plant J* 2003, **34**:95–106.
35. Journet E-P, de Carvalho-Niebel F, Andriankaja A, Huguet T, Barker DG: **Rhizobial Inoculation and Nodulation of *Medicago Truncatula*.** In *The Medicago Truncatula Handbook.* Edited by Mathesius U, Journet E-P, Sumner LW. 2006. http://www.noble.org/MedicagoHandbook/. ISBN 0-9754303-1-9.
36. Pislariu CI, Dickstein R: **An *IRE*-like AGC Kinase Gene, *MtIRE*, has unique expression in the invasion zone of developing root nodules in *Medicago truncatula*.** *Plant Physiol* 2007, **144**:682–694.

3

Suitable transfection methods for single particle tracing in plant suspension cells

Janett Göhring[1], Nick Fulcher[2], Kurt Schilcher[3], Andrea Barta[1] and Jaroslaw Jacak[1,3*]

Abstract

Background: A multitude of different imaging systems are already available to image genetically altered RNA species; however, only a few of these techniques are actually suitable to visualize endogenous RNA. One possibility is to use fluorescently-labelled and hybridization-sensitive probes. In order to yield more information about the exact localization and movement of a single RNA molecule, it is necessary to image such probes with highly sensitive microscope setups. More challenges arise if such experiments are conducted in plant cells due to their high autofluorescence and demanding transfection procedures.

Results: Here, we report *in planta* imaging of single RNA molecules using fluorescently labeled molecular beacons. We tested three different transfection protocols in order to identify optimal conditions for transfection of fluorescent DNA probes and their subsequent detection at the single molecule level.

Conclusions: We found that an optimized heat shock protocol provided a vastly improved transfection method for small DNA molecules which were used for subsequent single RNA molecule detection in living plant suspension cells.

Background

Information about the distribution and spatio-temporal dynamics of distinct RNA molecules helps to gain a deeper understanding of a multitude of complex biological processes (e.g., post-transcriptional regulation of gene expression via splicing [1,2], effects of RNA interference [3], or RNA's complex interaction with chromatin [4]). *In vivo* imaging of RNA has traditionally required usage of overexpressed tagged proteins, saturated probes, or other reporter systems [5-9]. Often, it is challenging to show that these artificially introduced reporters do not have a negative effect on cellular functions, or a biasing effect on raw data. Labelling using hybridization-sensitive probes would, therefore, provide a nearly unperturbed examination of single endogenous mRNA molecules. In many standard biochemical experiments, bulk populations are analyzed instead of populations of single cells. Subtle and often very interesting effects will only be observable in small subpopulations [10]. A prerequisite for such population statistics is, of course, to refrain from following the dogma of the 'representative cell' as the outcome of analysis.

There are several advantages that come with studying biological processes at a single molecule level. Single molecule tracing experiments give accurate information about forces that determine the characteristic motion of an individual molecule. Single molecule diffusion parameters calculated within the fragmented subcellular space (e.g. organelles/compartments [11]) can be a powerful tool to decipher cellular processes (e.g. protein interactions in cell signaling [12,13]) or to characterize cell subpopulations [14]. Typically, determination of a single particle trace of an RNA, i.e. one molecule's diffusion in real time, provides information about direction of movement, step size, and whether there are confined volumes allowing movement within the observed cell volume [15,16]. By analyzing enough molecules, it can also lead to insights about the percentage of mobile fractions and which kind of movement pattern they are following.

Recently, we successfully used fluorescent hybridization-sensitive probes (Molecular Beacons, MBs) to monitor the distribution of mRNA molecules in living plant cells [17].

* Correspondence: jaroslaw.jacak@jku.at
[1]Max F. Perutz Laboratories, Medical University of Vienna, Dr. Bohrgasse 9/3, Vienna, Austria
[3]Upper Austria University of Applied Sciences, Campus Linz, Linz 4020, Austria
Full list of author information is available at the end of the article

A simple workflow followed by a robust statistical pipeline led to the establishment of a tool which provides quantitative information on *in vivo* RNA localization and abundance. Due to their special conformation, MBs only fluoresce upon binding to their target sequence. This feature makes MBs the ideal tool for imaging of endogenous RNA in living plant cells as it provides a better signal-to-noise ratio; this is beneficial since autofluorescence in plants is, compared to other organisms, drastically increased [18]. However, a severe disadvantage of MBs is the fact that they need to be delivered into the cell. Hitherto used techniques include microinjection and biolistic bombardment [19,20] which are both laborious, time consuming, and expensive. Moreover, cells become considerably stressed via these invasive transfections. More advanced transfection protocols for mammalian cells make use of nanopore technologies and microfluidics [21,22]. But, nonetheless, microinjection remains the only convenient method for transiently transfecting intact plant tissue.

In order to use MBs for RNA imaging in living plant protoplasts, we needed to establish a gentle transfection protocol. We chose to test polyethylenglycole (PEG), electroporation, and heat shock mediated transfection methods to investigate the localization of a specific MB against the transcripts of RS2Z33, a plant-specific SR protein, which has been described elsewhere [23,24].

Methods

MB design

Molecular Beacons are DNA oligonucleotides with a stem loop structure, whose 5'end is conjugated to a fluorophore (Atto550), and its 3'end to a quencher (BHQ2). The loop sequence targets the RNA of interest (RS2Z33 transcripts), whereas the self-complementary stem keeps the quencher and the fluorophore in close proximity, suppressing any signal. If the MB binds to the target, the stem opens and the fluorophore emits a signal. The MB was designed according to Gohring et al. [17]. 33mRNA_EJ6 (5'-Atto550-GATCGCGCGTGATCGGCTGTAGCTTCG GCCGCGATC-BHQ2-3').

Cell Culture, Isolation of Protoplasts, and PEG transfection

Arabidopsis cell suspension culture protoplasts (derived from 10d old Col-0 seedlings) were prepared and immediately used for PEG transformation as described in Lorkovic et al. [25].

Electroporation

The optimization procedure included:

- Variation of the electroporation buffer in respect to osmolarity, electrical conductivity, and resistance.
- Variation of the cell density, washing buffer (PIB or GM buffer, for final buffer composition see below), length of incubation on ice after electroporation
- Variation of electrical parameters: voltage, capacity, transferred energy, pulse lengths and intervals.

Further details can be found in the Additional file 1: Data S1. The criteria for the experimental outcome were subjective and included the evaluation of the cellular morphology, presence of cell debris, cell viability, and the state of transfection.

Two hours after protoplast isolation, cells were transfected by electroporation. Approximately 1×10^5 protoplast cells were washed with 0.275 M Calcium-nitrate and pelleted in a microcentrifuge (150 g for 2 minutes). For electroporation, 0.7 M mannitol (~700 mOsm) was mixed to a final concentration of 500 nM MB. A final volume of 800 µL in a 4 mm cuvette (Biozym Scientific, Oldendorf) was pulsed with following settings of the device Easyject Optima, EquiBio: 240 V, 75 µF, 2 pulses with the length of approximately 7.5 ms and an interval of 20 s. Subsequently, cells were fed with 0.34 M GM-buffer [3163 mg/L Gamborg B5 powder including vitamins (Duchefa), 170 mM D-glucose, 170 mM D-mannitol, 1 mg/L 2.4D, pH 5.5 adjusted with KOH], incubated at room temperature, and kept in the dark for 24 h. Shortly before measurement, cells were washed twice with GM-buffer.

Heat shock

The optimization procedure included:

- Variation of the osmolarity of the PIB buffer (see below), in order to test the mechanical stress limits of the cellular membrane.
- Variation of the cell density, heat shock temperature, length, and subsequent length of incubation on ice.
- Variation of the washing buffers (PIB or GM buffer, for final buffer composition see below).

Further details can be found in the Supplementary Data S1. The criteria for experimental outcome were subjective and included the evaluation of the cellular morphology, presence of cell debris, cell viability and the state of transfection.

Two hours after protoplast isolation, cells were transfected by heat shock. The protocol was adjusted from Hicks et al. [26]. Approximately 1×10^5 protoplast cells were resuspended in 200 µL 2×PIB buffer [2 mM MgAc, 50 mM KAc, 5 mM NaAc, 2 mM PMSF, 20 mM HEPES pH7.2, 1 mM DTT, 225 mM mannitol, 125 mM Spermin, 125 mM Spermidin]. Cells were incubated on ice in the dark for 10 minutes and subsequently pelleted in a microcentrifuge (150 g for 2 minutes). The pellet was resuspended in a solution containing 500 nM MBs in 2xPIB buffer and heat shocked for 30 minutes at 28°C in the dark. Afterwards, cells were immediately put on ice for 4 minutes and 100 µL of 1xPIB were added as fast as possible. Eventually, cells were fed with 0.34 M GM-buffer [3163 mg/L Gamborg B5 powder including vitamins (Duchefa), 170 mM D-glucose, 170 mM D-mannitol, 1 mg/L 2.4D, pH 5.5 adjusted with KOH], incubated at room temperature, and kept in the dark for 24 h. Shortly before measurement, cells were washed twice with GM-buffer.

Imaging of single molecules and analysis

The images were taken on a modified Olympus IX81 inverted microscope. The samples were illuminated through an Olympus UApo N 100×/1.49 NA oil objective with diode laser at 532 nm (Cobolt Calypso 100 TM). The signal acquisition was carried out on an Andor iXonEM + 897 (back illuminated) EMCCD (160 nm pixel size). The experiments were performed using excitation powers of 0.025 kW/cm2 at 532 nm. The samples were illuminated for 5 ms with 35 ms delay time. The illumination protocols were timed with a custom made LabView® based control software. Filter: Overlay (642/532), Dichroic filter: Cy3/Cy5, Emission-filter: Cy3/Cy5 + Bandpass 595/50 (Chroma). The cells were imaged in two illumination configurations, the widefield and highly inclined and laminated optical light sheet (HILO) illumination. The light sheet illumination reduces background fluorescence within the cell, originating from scattered light or other fluorescent molecules [27].

Analysis

Given the heterogeneous background, we have chosen a detection method based on the isotropic undecimated wavelet transform (IUWT) [28,29]. Wavelet thresholding offers a robust solution for the detection of small bright features, e.g. detection of sub-cellular structures labeled by fluorescent dyes. Since a fluorescent dye can be considered a point light source, its image is the point spread function (PSF) of the optical system that can be approximated by a two-dimensional Gaussian shape.

The typical noise model includes the photon count noise, following a Poisson distribution, and an additive Gaussian read-out noise:

$$n \sim Poi(N) * Gauss(0, \sigma),$$

With n the number of detected photo-electrons and N the number of emitted photons.

For an EMCCD camera, the model including the effect of the multiplier can be approximated (as described in [30]) by the formula:

$$p(n, N) \sim Gauss\left(GN, FG\sqrt{(N)}\right) = \frac{1}{\sqrt{2\pi F^2 G^2 N}} e^{\left(\frac{-(n-GN)^2}{2F^2G^2N}\right)}$$

Where F represents the excess noise and G the average multiplier gain.

In order to deal with the heterogeneity of the noise, a variance stabilizing transform is applied prior to the IUWT [31]. This step significantly reduces the background and improves the quality of single molecule localization. The method relies on a successive fitting of the noise with a constant kernel wavelet matrix [32]. Subsequently, the fitted background image is subtracted from the original image. After several iterative repeats position of candidates for Gaussian fitting are chosen [28,33,34]. Subsequently, the detected molecule's position was determined with

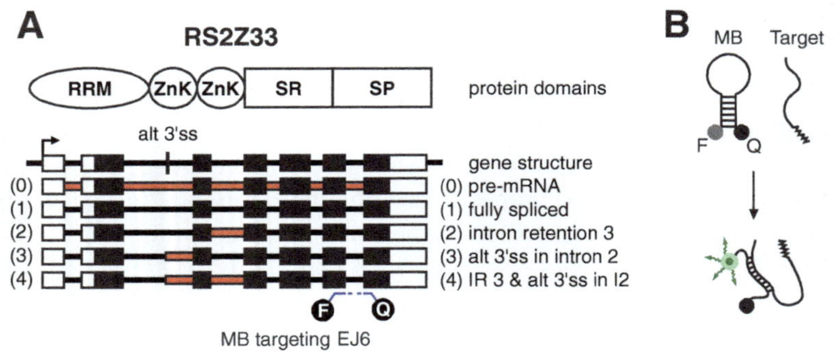

Figure 1 The scheme depicts the protein and gene structure of RS2Z33 and its transcripts. A) The Molecular Beacon was designed to bind to exon junction 6 of RS2Z33; i.e. all splice variants of RS2Z33 are targeted. Red – intronic sequences; Black – exonic sequences; white – untranslated regions; RRM – RNA recognition motif; ZnK – Zink knuckle; SR – Serine/Arginine rich domain; SP – Serine/Proline-rich domain; alt 3'-ss – alternative 3' splice site; IR – intron retention; I2 – intron 2; MB – Molecular Beacon; EJ6 – exon junction 6; F – fluorophore; Q – quencher. **B)** Illustration of the work principle of MB.

subpixel accuracy by least square fitting of a 2D Gaussian function in the neighborhood of the detected significant pixels [33-36].

Results

The aim of this study is to determine the most suitable transfection method for introducing MBs into living plant suspension cells, preferably reducing the applied stress to the cells. It needs to allow subsequent imaging of MB-labeled RNA at the single molecule level; an optimization for probe concentration and light scattering was, therefore, necessary (strongly depends on debris that attaches to membranes of healthy cells). We decided to test three different transfection approaches: PEG transfection, electroporation and heat shock. In each experiment, wild type *Arabidopsis thaliana* cell culture protoplasts were transfected with fluorescently labeled MBs (Atto550-dye and BHQ2-quencher), which target the exon junction 6 of RS2Z33 (see Figure 1 for a detailed gene structure and the splice variants targeted by the described MB). Exon junction 6 of RS2Z33 is present in all known transcripts of the gene and, thus, can be used as ideal MB target for demonstrating single molecule imaging.

For plant suspension cells, standard transient PEG transfection protocols [37] yield high transfection efficiency and sufficient viability, and compared to the mentioned alternatives have low material demands (e.g., microinjection, bombardment). Figure 2 displays four representative plant protoplasts after transfection of the MB using PEG 6000. Specific signal of the MB was detected on a subcellular level and demonstrates that small oligonucleotides can be introduced into living plant cells using PEG as DNA carrier. The MBs, however, remained unspecifically attached to the PEG and stayed in an open state which introduced strong fluorescence artefacts inside the cell (Figure 2, right panel) as well as fluorescent labelling of the vacuolar membrane in strongly transfected cells (Figure 2, first row). To ensure, that the fluorescent signal originates from the MBs, we measured the autofluorescence of the plant cells which were either transfected only with PEG 6000 (Additional file 2: Figure S1) or with PEG and the MB targeting EJ6 of RS2Z33 (Figure 2 and Additional file 2: Figure S1). None of the cells displayed any specific fluorescence signal arising from PEG itself, but multiple fluorescent artefacts were observed in cells transfected with MBs via the PEG procedure. We were surprised to find that the standard transformation protocol using PEG (polyethylene glycol) is not suited for introducing MBs into living suspension cells. PEG is an unspecifically binding DNA carrier and, unfortunately, triggers the MB to assume open conformation which results in strong fluorescent emission. Another outcome of this experiment is that PEG seems to unevenly distribute throughout the living plant cell, forming fluorescent artefacts and attach to the vacuolar membrane. Since

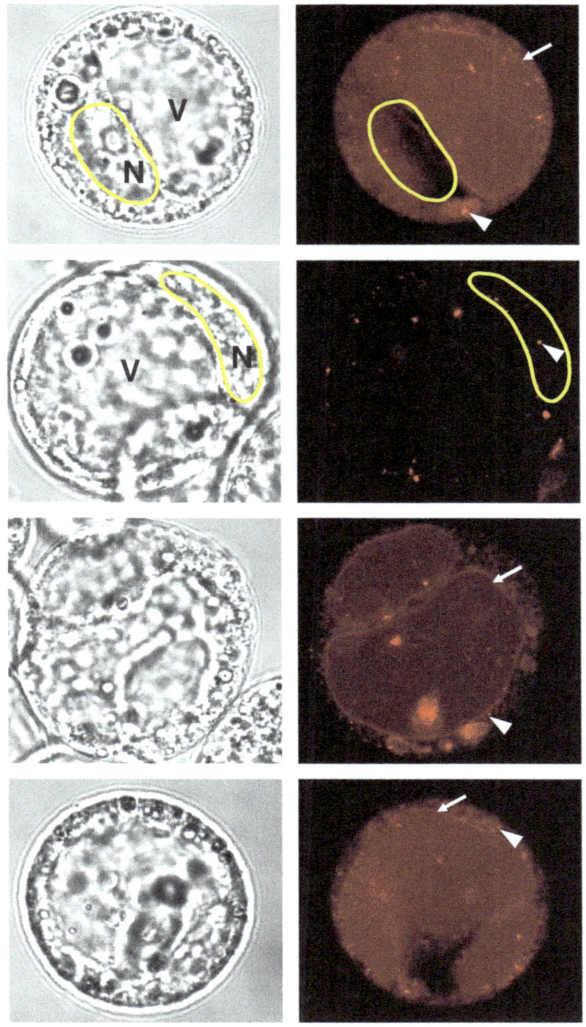

Figure 2 Transfection of MBs via the standard PEG protocol leads to unspecific conglomerates within the cell. Representative images of protoplasts transfected with a MB labeled with an Atto550 fluorophore targeting exon junction 6 (Ej6) mRNA of RS2Z33. Arrowheads label foci of PEG-MB conglomerates. Arrows depict vacuolar membranes which are also fluorescently labeled by PEG-MBs. Note the difference between the strongly (1st, 3rd and 4th row) and poorly transfected cells (2nd row). On the left side are the transmission images with visible nucleus (marked yellow). V - vacuole; N - nucleus.

PEG is used as standard transfection substrate, care should be taken depending on the experimental design and question.

Another way to transform plant protoplasts is by electroporation [38,39]. Interestingly, none of the commercially available electroporation systems are adapted for *A. thaliana* protoplastic cells. Therefore, we needed to optimize the protocol in respect to electroporation buffers, used DNA mass, pulse length, and recovery (for more details and the final protocol see methods). Since

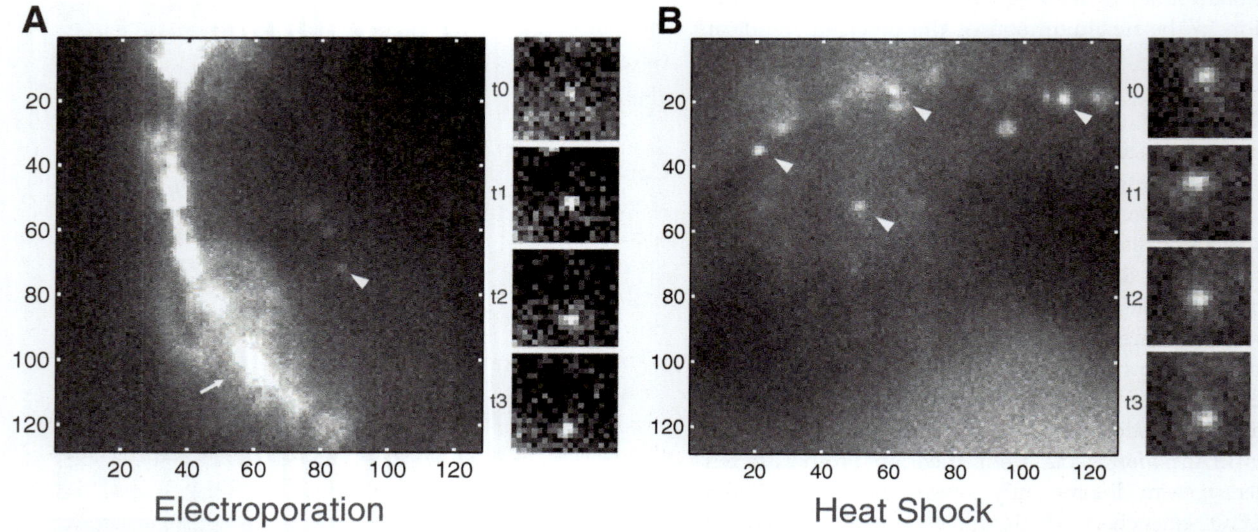

Figure 3 Electroporation of MBs allows the detection of single molecules and single molecule tracing. Cells were illuminated via HILO to reduce the background fluorescence. **A)** The overview image shows an electroporated cell. Notice the high fluorescent signal arising from cell debris attached to the cellular membrane. The right panel contains details of a detected single molecule signal in consecutive time frames t0 to t3 (SNR 6,7 ± 2) (5 ms illumination and 30 ms delay between each time frame). **B)** Heat shock transfection of MBs allows the detection of single molecules. The overview image shows a heat shock treated cell. Since the signal-to-noise-ratio of the detected single molecules is 10.1 ± 0.23, the heat shock performed better than electroporation. The right panel contains details of a detected single molecule signal in consecutive time frames t0 to t3 (5 ms illumination and 30 ms delay between each time frame). The arrowheads points to signals of single molecules and the arrow demarks the signal arising from extracellular debris.

MBs are by far smaller than plasmids, which are usually used for transfecting cells, we aimed to increase transfection efficiency for small DNA molecules and additionally to minimize cellular stress. With the optimized parameters, the cells showed good transfection efficiency which allowed confocal imaging [17]. High heterogeneity of the transfection efficiencies was observed; however, we could show that cells with low transfection efficiency are suitable for imaging of single molecules (Figure 3A). The density of molecules is critical for single molecule

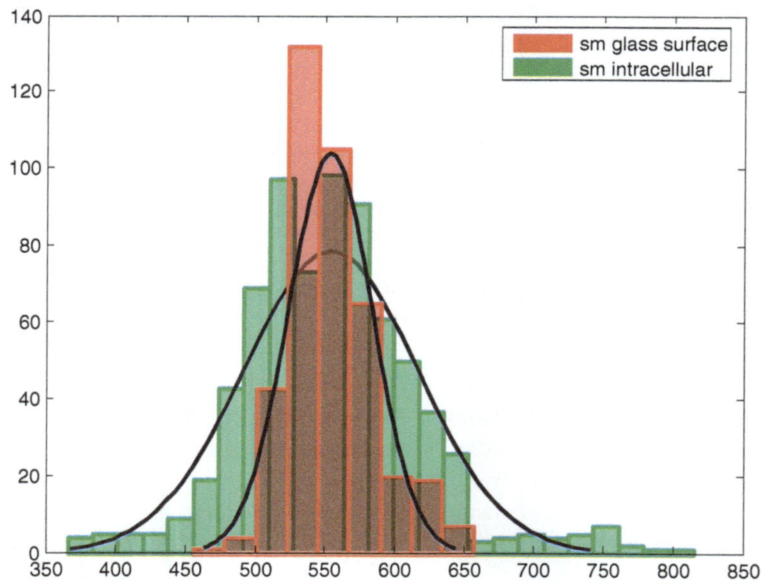

Figure 4 Distribution of single MB signals. Statistical distribution of fluorescence intensity counts per fluorescent spot during 5 ms illumination time, obtained from: sparsely and randomly distributed MBs on glass and (red) from single fluorescent spots inside plant cells (green). The single molecule signal distributions have a high similarity, consequently the signal inside the cells originated from fluorescent MBs. Number of cells: 21, Number of signals: 1000. sm - single molecule.

tracking since higher concentrations impede the identification of the path of one distinct molecule. In order to identify the origin of the signal peaks inside a cell, their average intensities were compared to signals of individual MBs on glass. For the comparison, MBs were immobilized on a glass-substrate. Due to the DNA-glass interaction, a certain percentage of the MB opened and became fluorescent. The intensities of the opened MBs have been compared to the signals obtained inside a cell. The comparison showed a high similarity of the single molecule intensity distributions (Pearson correlation coefficient is 0.3) (Figure 4). Due to the 3D single molecule MB signal distribution inside the cell, the histogram is broadened relative to the signals analyzed on a glass slide. Such intensity histogram broadening can be attributed to the deviation of the single molecule position in the focal plane. Eventually, the MB-transfected cells were measured using a single-molecule sensitive setup with HILO illumination. However, cell debris of dead cells attached to the cell surface, and increased fluorescence and light scattering along the cell membrane led to a strong extracellular background (Figure 3A). This led to a reduced single molecule signal. The high single-molecule brightness and the low background noise within a diffraction limited region obtained within an area next to the single molecule allow for reliable (signal-to-background noise ratio 7 ± 2, SNR averaged over all signals) detection of fluorescent MB molecules. In summary, electroporation enables transfection of MBs allowing subsequent single molecule imaging. However, the heterogeneity in transfection efficiency as well as background and light scattering led to a strong impact on single molecule SNR. Therefore, we decided to search for a gentler transfection method, to reduce extracellular cell debris and to yield just a small amount of labeled mRNAs within the cell (i.e. low transfection efficiency).

The transfection of MBs via heat shock turned out to be a compromise. We optimized the heat shock protocol (oriented along the permeability protocol from Hicks et al. [26]) in respect of used buffers, heat shock, and recovery time, as well as amount of DNA initially used per cell (see Methods for more details). The investigated cells exhibited good viability with relatively low transfection efficiency (with clearly separated single molecule signals). The detected fluorescent signals were at or below the detection limit of a standard laser scanning microscopy setup. We observed relatively high transfection homogeneity within the cell population, good viability, and a low fluorescent background (comparable to non-transfected cells) (Figure 3B). The average SNR of detected MB single molecule signals in cells was 9.7 ± 2.4 (SNR averaged over all signals), which is sufficient for any single molecule fluorescence experiment e.g. single molecule tracking (Figure 3 and Additional file 3: Figure S2). The transfection efficiency shows a clear dependence on concentration and incubation time, which can be used for a relatively precise adjustment of MB concentration. Consequently, this method seems to be a good approach for transfection of cells with MBs permitting single molecule analysis.

Conclusion

Herein, we report the establishment of two transfection methods for *A. thaliana* suspension protoplasts optimized for measuring Molecular Beacons with single molecule sensitivity. The standard PEG-based transfection method is not applicable for fluorescent microscopy since PEG-bound MBs remain in an open state and lead to a considerable increase in background fluorescence. Electroporated cells also tend to exhibit high background due to the increased amount of extracellular debris. However, transfection via heat shock turned out to be the optimal transfection method for MBs. The transfection has low material requirements and a smaller work load compared to other described methods e.g., microinjection or bombardment. Heat shocked cells are vital and not influenced in their morphology. The extracellular background is reduced and cells exhibit low transfection efficiencies which are best suited for single molecule measurement. Moreover, we report for the first time the successful measurement of single molecules within living plant suspension cells.

Additional files

Additional file 1: Data S1. Optimization protocols for the electroporation and heat shock.

Additional file 2: Figure S1. PEG does not lead to increased fluorescence levels within the measured channel. **A)** The cells transfected only with PEG serve as negative control. **B)** Cells transfected with Atto550-conjugated MBs and PEG. Although a very low sensitivity widefield setup (far below single molecule sensitivity) has been used (Axiovert 200M with AxioCam MRm, standard filter set), fluorescent PEG-MB conglomerates (arrowheads) can easily be observed inside the cell. **C)** A transmission image (TM) of the cells transfected with PEG and MBs. MB – Molecular Beacon against exon junction 6 of RS2Z33; TM – transmission image; PEG – polyethylenglykole.

Additional file 3: Figure S2. Movie of a heat shock transfected protoplast. The movie from the fluorescent microscope shows a zoom on the nucleus and surrounding area. Single molecules, appearing as bright dots, corresponding to Atto550-labeled MB in open State bound to the target RNA. The illumination time is 5 ms; time between each frame is 30 ms.

Competing interests

The authors declare no competing interests.

Authors' contributions

JG and JJ designed the study, carried out the optimization of the transfection methods and performed single molecule detection, and drafted the manuscript. NF, AB and KS conceived of the study, participated in its coordination, and helped to draft the manuscript. Since KS passed away during the preparation of the manuscript, only the remaining authors read and approved the final manuscript.

Funding
This work was supported by grants to AB of the Austrian Science Fund (FWF) [DK W1207; I-254; SFB RNA-REG F43-P10].

Author details
[1]Max F. Perutz Laboratories, Medical University of Vienna, Dr. Bohrgasse 9/3, Vienna, Austria. [2]Gregor Mendel Institute, Dr. Bohrgasse 3, Vienna, Austria. [3]Upper Austria University of Applied Sciences, Campus Linz, Linz 4020, Austria.

References

1. So MK, Gowrishankar G, Hasegawa S, Chung JK, Rao J: **Imaging target mRNA and siRNA-mediated gene silencing in vivo with ribozyme-based reporters.** *Chembiochem* 2008, **9**:2682–2691.
2. Bhaumik S, Walls Z, Puttaraju M, Mitchell LG, Gambhir SS: **Molecular imaging of gene expression in living subjects by spliceosome-mediated RNA trans-splicing.** *Proc Natl Acad Sci U S A* 2004, **101**:8693–8698.
3. Chang E, Zhu MQ, Drezek R: **Novel siRNA-based molecular beacons for dual imaging and therapy.** *Biotechnol J* 2007, **2**:422–425.
4. van den Bogaard PT, Tyagi S: **Using molecular beacons to study dispersal of mRNPs from the gene locus.** *Methods Mol Biol* 2009, **464**:91–103.
5. Stockley PG, Stonehouse NJ, Murray JB, Goodman ST, Talbot SJ, Adams CJ, Liljas L, Valegard K: **Probing sequence-specific RNA recognition by the bacteriophage MS2 coat protein.** *Nucleic Acids Res* 1995, **23**:2512–2518.
6. Daigle N, Ellenberg J: **LambdaN-GFP: an RNA reporter system for live-cell imaging.** *Nat Methods* 2007, **4**:633–636.
7. Ozawa T, Natori Y, Sato M, Umezawa Y: **Imaging dynamics of endogenous mitochondrial RNA in single living cells.** *Nat Methods* 2007, **4**:413–419.
8. Tyagi S, Kramer FR: **Molecular beacons: probes that fluoresce upon hybridization.** *Nat Biotechnol* 1996, **14**:303–308.
9. Hasegawa S, Gowrishankar G, Rao J: **Detection of mRNA in mammalian cells with a split ribozyme reporter.** *Chembiochem* 2006, **7**:925–928.
10. Muramoto T, Cannon D, Gierlinski M, Corrigan A, Barton GJ, Chubb JR: **Live imaging of nascent RNA dynamics reveals distinct types of transcriptional pulse regulation.** *Proc Natl Acad Sci U S A* 2012, **109**:7350–7355.
11. Gombos I, Crul T, Piotto S, Gungor B, Torok Z, Balogh G, Peter M, Slotte JP, Campana F, Pilbat AM, Hunya A, Toth N, Literati-Nagy Z, Vigh L Jr, Glatz A, Brameshuber M, Schütz GJ, Hevener A, Febbraio MA, Horvath I, Vigh L: **Membrane-lipid therapy in operation: the HSP co-inducer BGP-15 activates stress signal transduction pathways by remodeling plasma membrane rafts.** *Plos One* 2011, **6**:e28818.
12. Axmann M, Huppa JB, Davis MM, Schutz GJ: **Determination of interaction kinetics between the T cell receptor and peptide-loaded MHC class II via single-molecule diffusion measurements.** *Biophys J* 2012, **103**:L17–L19.
13. Serge A, de Keijzer S, Van Hemert F, Hickman MR, Hereld D, Spaink HP, Schmidt T, Snaar-Jagalska BE: **Quantification of GPCR internalization by single-molecule microscopy in living cells.** *Integr Biol Quant Biosci nano macro* 2011, **3**:675–683.
14. Brameshuber M, Schutz GJ: **Detection and quantification of biomolecular association in living cells using single-molecule microscopy.** *Methods Enzymol* 2012, **505**:159–186.
15. Shav-Tal Y, Darzacq X, Shenoy SM, Fusco D, Janicki SM, Spector DL, Singer RH: **Dynamics of single mRNPs in nuclei of living cells.** *Science* 2004, **304**:1797–1800.
16. Vargas DY, Raj A, Marras SA, Kramer FR, Tyagi S: **Mechanism of mRNA transport in the nucleus.** *Proc Natl Acad Sci U S A* 2005, **102**:17008–17013.
17. Gohring J, Jacak J, Barta A: **Imaging of endogenous messenger RNA splice variants in living cells reveals nuclear retention of transcripts inaccessible to nonsense-mediated decay in Arabidopsis.** *The Plant cell* 2014, **26**:754–764.
18. Chapman S, Oparka KJ, Roberts AG: **New tools for in vivo fluorescence tagging.** *Curr Opin Plant Biol* 2005, **8**:565–573.
19. Seki M, Iida A, Morikawa H: **Transient expression of the beta-glucuronidase gene in tissues of Arabidopsis thaliana by bombardment-mediated transformation.** *Mol Biotechnol* 1999, **11**:251–255.
20. Ueki S, Lacroix B, Krichevsky A, Lazarowitz SG, Citovsky V: **Functional transient genetic transformation of Arabidopsis leaves by biolistic bombardment.** *Nat Protoc* 2009, **4**:71–77.
21. Li N, Wong PK: **Transfection of molecular beacons in microchannels for single-cell gene-expression analysis.** *Bioanalysis* 2010, **2**:1689–1699.
22. Nelson EM, Kurz V, Shim J, Timp W, Timp G: **Using a nanopore for single molecule detection and single cell transfection.** *Analyst* 2012, **137**:3020–3027.
23. Lopato S, Forstner C, Kalyna M, Hilscher J, Langhammer U, Indrapichate K, Lorkovic ZJ, Barta A: **Network of interactions of a novel plant-specific Arg/Ser-rich protein, atRSZ33, with atSC35-like splicing factors.** *J Biol Chem* 2002, **277**:39989–39998.
24. Kalyna M, Lopato S, Barta A: **Ectopic expression of atRSZ33 reveals its function in splicing and causes pleiotropic changes in development.** *Mol Biol Cell* 2003, **14**:3565–3577.
25. Lorkovic ZJ, Hilscher J, Barta A: **Use of fluorescent protein tags to study nuclear organization of the spliceosomal machinery in transiently transformed living plant cells.** *Mol Biol Cell* 2004, **15**:3233–3243.
26. Hicks GR, Smith HM, Lobreaux S, Raikhel NV: **Nuclear import in permeabilized protoplasts from higher plants has unique features.** *Plant cell* 1996, **8**:1337–1352.
27. Tokunaga M, Imamoto N, Sakata-Sogawa K: **Highly inclined thin illumination enables clear single-molecule imaging in cells.** *Nat Methods* 2008, **5**:159–161.
28. Olivo-Marin JC: **Extraction of spots in biological images using multiscale products.** *Pattern Recogn* 2002, **35**:1989–1996.
29. Starck JL, Fadili J, Murtagh F: **The undecimated wavelet decomposition and its reconstruction.** *Ieee T Image Process* 2007, **16**:297–309.
30. Hynecek J, Nishiwaki T: **Excess noise and other important characteristics of low light level imaging using charge multiplying CCDs.** *IEEE Trans On Electron Devices* 2003, **50**:239–245.
31. Muresan L, Jacak J, Klement EP, Hesse J, Schutz GJ: **Microarray analysis at single-molecule resolution.** *IEEE Trans Nanobioscience*, **9**:51–58.
32. Benjamini Y, Hochberg Y: **Controlling the False Discovery Rate: A Practical and Powerful Approach to Multiple Testing.** *J R Stat Soc* 1995, **57**:289–300.
33. Muresan L, Jacak J, Klement EP, Hesse J, Schutz GJ: **Microarray Analysis at Single-Molecule Resolution.** *Ieee T Nanobiosci* 2010, **9**:51–58.
34. Sams M, Silye R, Gohring J, Muresan L, Schilcher K, Jacak J: **Spatial cluster analysis of nanoscopically mapped serotonin receptors for classification of fixed brain tissue.** *J Biomed Opt* 2014, **19**:011021.
35. Jacak J, Schnidar H, Muresan L, Regl G, Frischauf A, Aberger F, Schutz GJ, Hesse J: **Expression analysis of multiple myeloma CD138 negative progenitor cells using single molecule microarray readout.** *J Biotechnol* 2013, **164**:525–530.
36. Hesse J, Jacak J, Kasper M, Regl G, Eichberger T, Winklmayr M, Aberger F, Sonnleitner M, Schlapak R, Howorka S, Muresan L, Frischauf AM, Schütz GJ: **RNA expression profiling at the single molecule level.** *Genome Res* 2006, **16**:1041–1045.
37. Mathur J, Koncz C: **PEG-mediated protoplast transformation with naked DNA.** *Methods Mol Biol* 1998, **82**:267–276.
38. Miao Y, Jiang L: **Transient expression of fluorescent fusion proteins in protoplasts of suspension cultured cells.** *Nat Protoc* 2007, **2**:2348–2353.
39. Bates GW: **Genetic transformation of plants by protoplast electroporation.** *Mol Biotechnol* 1994, **2**:135–145.

4

Rapid determination of leaf area and plant height by using light curtain arrays in four species with contrasting shoot architecture

Dimitrios Fanourakis[1], Christoph Briese[1], Johannes FJ Max[1,2], Silke Kleinen[1], Alexander Putz[1], Fabio Fiorani[1*], Andreas Ulbrich[1,3] and Ulrich Schurr[1]

Abstract

Background: Light curtain arrays (LC), a recently introduced phenotyping method, yield a binary data matrix from which a shoot silhouette is reconstructed. We addressed the accuracy and applicability of LC in assessing leaf area and maximum height (base to the highest leaf tip) in a phenotyping platform. LC were integrated to an automated routine for positioning, allowing *in situ* measurements. Two dicotyledonous (rapeseed, tomato) and two monocotyledonous (maize, barley) species with contrasting shoot architecture were investigated. To evaluate if averaging multiple view angles helps in resolving self-overlaps, we acquired a data set by rotating plants every 10° for 170°. To test how rapid these measurements can be without loss of information, we evaluated nine scanning speeds. Leaf area of overlapping plants was also estimated to assess the possibility to scale this method for plant stands.

Results: The relation between measured and calculated maximum height was linear and nearly the same for all species. Linear relations were also found between plant leaf area and calculated pixel area. However, the regression slope was different between monocotyledonous and dicotyledonous species. Increasing the scanning speed stepwise from 0.9 to 23.4 m s^{-1} did not affect the estimation of maximum height. Instead, the calculated pixel area was inversely proportional to scanning speed. The estimation of plant leaf area by means of calculated pixel area became more accurate by averaging consecutive silhouettes and/or increasing the angle between them. Simulations showed that decreasing plant distance gradually from 20 to 0 cm, led to underestimation of plant leaf area owing to overlaps. This underestimation was more important for large plants of dicotyledonous species and for small plants of monocotyledonous ones.

Conclusions: LC offer an accurate estimation of plant leaf area and maximum height, while the number of consecutive silhouettes that needs to be averaged is species-dependent. A constant scanning speed is important for leaf area estimations by using LC. Simulations of the effect of varying plant spacing gave promising results for method application in sets of partly overlapping plants, which applies also to field conditions during and after canopy closure for crops sown in rows.

Keywords: Leaf area estimation, Maximum height estimation, Overlapping plants, Phenotyping, Sensor speed, Sensor-to-plant system, Shoot profile, Shoot silhouette

* Correspondence: f.fiorani@fz-juelich.de
[1]IBG-2: Plant Sciences, Institute for Bio- and Geosciences, Forschungszentrum Jülich, 52425 Jülich, Germany
Full list of author information is available at the end of the article

Background

The capacity to quantitatively explore plant phenotypes (from single cells to the whole plant) and their dynamic responses to a changing environment is a necessary requirement for genetic and physiological research by crop breeders, agricultural industry, and academia. Although molecular profiling technologies now enable the generation of a large amount of data with decreasing costs, the understanding of the link between genotype and phenotype still remains fragmented [1]. Insufficient technical and conceptual capacity of the plant scientific community to probe existing genetic resources and unravel environmental effects limits faster progress in this field [2]. To address this challenge, several automated phenotyping platforms have been developed from academia or commercial sources during the past decade (reviewed by [3-5]). The development of phenotyping applications for non-invasive assessment of the dynamics of plant biomass development is a cornerstone for this effort.

In phenotyping platforms the most commonly used method of assessing shoot biomass is by acquiring digital images of the plants, following their positioning at a specified orientation towards a camera under defined illumination conditions [1,6,7]. Following image acquisition, digital image processing enables the extraction of plant features from the image background based on colour and brightness analysis [8]. The main limitations of biomass assessment by using imaging methods such as colour imaging in 2D spatial dimensions are: a) overlapping leaves and stems lead to underestimation of shoot area and often restrict this application to a given plant size or developmental stage, and b) segmenting the images requires rather sophisticated processing pipelines [6,9,10]. Light curtain arrays (LC) are a recently introduced phenotyping technology, which has been used successfully to assess canopy height in the field [11,12]. The setup consists of a pair of parallel bars, one radiating and the other receiving the emitted light (Figure 1A). In this way, the sensor records whether or not the light beams are interrupted by an object. By scanning the crop of interest, LC produce a binary data set of the plants' profile (Figure 1B). Unlike imaging methods, the plant distance to the sensor and illumination conditions during measurement do not affect the data. For a given geometry of each emitter and receiver arrays, no calibration of the sensors is needed and considerably less data processing steps as compared to imaging methods, are required (Figure 1B–E; Figure 2). However, a systematic approach to address the potential of LC in assessing leaf area of individual plants or sets of plants in greenhouse cultivation was, to our knowledge, not previously conducted.

Phenotyping platforms may be operated using plant-to-sensor or sensor-to-plant measurement routines depending if plants are moved to the sensor or vice-versa, respectively. With the exception of small rosette plants [13-15], there are certain limitations in implementing imaging methods as a sensor-to-plant approach. Not only several cameras would need to be moved at a defined orientation above the plant, but also light conditions need to be strictly controlled during imaging. This need partly explains that in existing platforms, dedicated to phenotypic evaluation of plants of different sizes, a plant-to-sensor approach is implemented, where plants are moved (manually or automatically) to dedicated imaging stations (e.g. [6,16,17]). Contrary to this, LC can be more easily implemented using either plant-to-sensor or sensor-to-plant methodologies. For example, LC may offer an alternative in assessing large size plants (e.g., taller than 1 m) by employing different plant cultivation systems.

The aim of this study was to investigate whether or not LC can accurately estimate plant biomass in addition to maximum height (base to the highest leaf tip) in species with contrasting shoot architecture. To test the applicability of this method to field scale, we also estimated the biomass of plants spaced at different distances where increasing overlaps took place.

Results and discussion

A constant scanning speed is critical for estimating leaf area

The sensor positioning system was a prototype developed for shoot phenotyping. As the scanning speed determines the system throughput and the scalability of the approach to larger number of plants, we tested by recording plant silhouettes (also referred as profiles [11], see Figure 1B) whether or not it affects the acquired data in the speed range of 0.9 to 23.4 m min^{-1}. The assessment was carried out in rapeseed plants covering a range of both leaf area (64–350 cm^2) and maximum height (10–20 cm). Maximum height refers to the length from the base to the highest leaf tip (Figure 1C–E). We observed that calculated pixel area exponentially decreased as scanning speed increased (Additional file 1: Figure S1A). Instead, the calculated maximum height remained constant (Additional file 1: Figure S1B).

Plant silhouettes were recorded at each scanning speed for a given position (referred as starting position), as well as for 17 view angles, each differing by 10° from the previously measured one. These values were further averaged. Figure 3 shows the correlation between calculated pixel area at different scanning speeds, as described above, and leaf area measured destructively. This correlation was always highly significant ($R^2 \geq 0.95$). However, the relation between these parameters strongly depended on the scanning speed.

LC have been previously used to determine canopy height in the field [11,12]. Here we show that for maximum height estimations by using LC, a wide range of

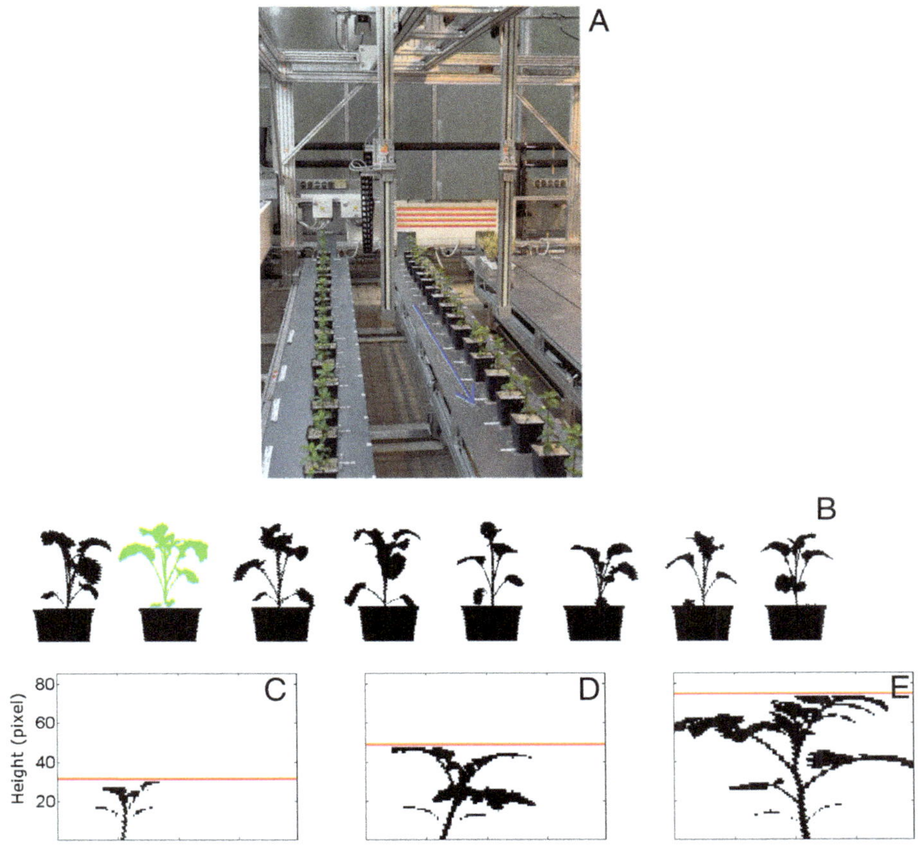

Figure 1 Working principle of light curtain arrays (LC) and shoot silhouette extraction. (A) LC are scanning a row of plants. The light barriers, arranged on the two vertical poles, emit and receive light beams (examples of which are shown by the red horizontal lines). The blue arrow depicts the movement of LC during the scan. Pot height is 13 cm. **(B)** Plant profile of a row of rapeseed plants. The green colour indicates the segmentation step during which the plant silhouette is separated from the pot silhouette. **(C–E)** Estimation of maximum height (base to the highest leaf tip; depicted by the vertical line) from the silhouette of tomato plants differing in size.

constant (0.9–23.4 m min^{-1}) or even varying scanning speeds can be used without any noticeable effects on the obtained data. Instead, a constant scanning speed is essential for estimating plant leaf area. Moreover, in this case, a speed-specific relation between calculated pixel area and measured leaf area needs to be obtained in each case.

Leaf area estimation by means of pixel area is improved as more silhouettes are averaged and/or the larger the view angle between them

Experiments were conducted in four species with contrasting shoot architecture, including two monocotyledonous (barley, maize) and two dicotyledonous (rapeseed, tomato) ones. Plants were grown from two days following transplanting up to four weeks. Because the information that can be extracted by a single silhouette may be limited due to overlapping (i.e. not visible at a given angle) plant parts, several silhouettes were recorded. At first, plant silhouette was determined at the starting position. The pot was then rotated by 10° with respect to the horizontal and the next plant silhouette was measured. Subsequently, the same procedure was repeated for 16 times, covering in total the 0 to 170° angle range.

Figure 4 shows the correlation between all acquired silhouettes versus the calculated pixel area, where each vertical cloud of points represents a single plant. In panel A, a single plant is represented by 18 points. For panel B, three means of 16 consecutive positions (corresponding to 0–150°, 10–160° and 20–170° plant rotation) were calculated for each plant. These three means nearly overlapped (Figure 4B).

The effect of averaging an increasing number of consecutive profiles (1–16) as well as of using larger angles (10–90°) between consecutive profiles on the correlation coefficient between calculated plant pixel area and measured leaf area is given in Figure 5. This correlation coefficient was very high ($R^2 \geq 0.938$) when a single silhouette was employed in both rapeseed and maize, and further increased ($R^2 \geq 0.97$) when more consecutive silhouettes

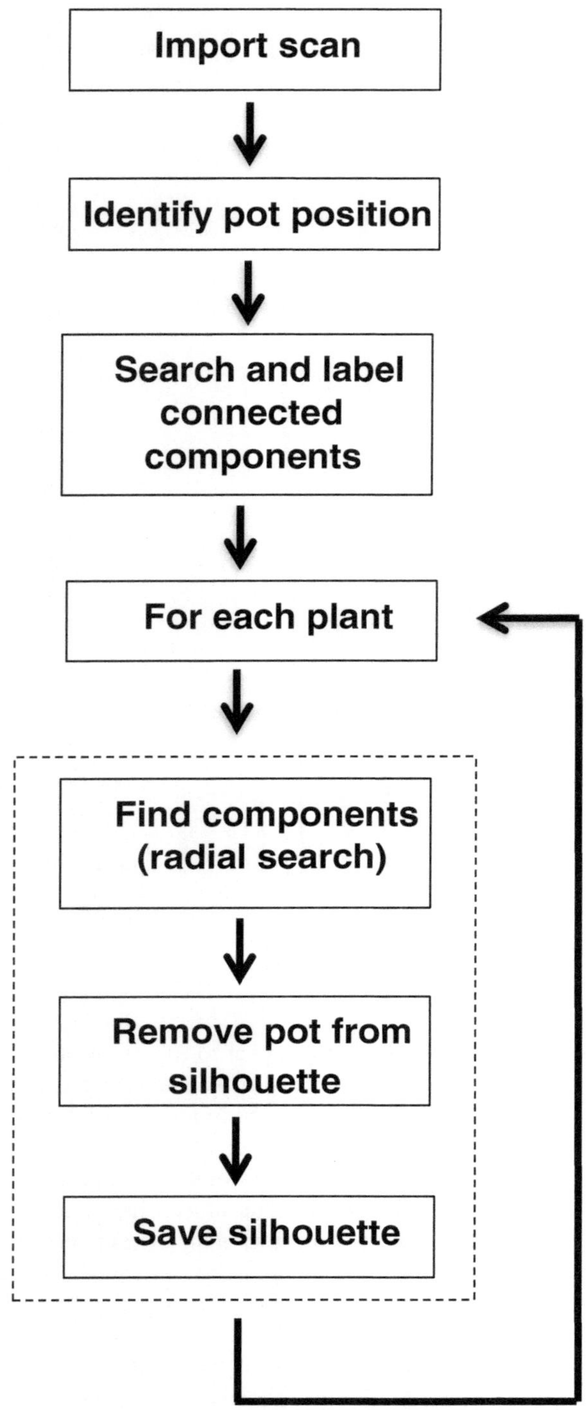

Figure 2 Flow diagram of the data processing pipeline employed to extract a single shoot silhouette from a scan of several plants (see Figure 1B).

were averaged and/or larger angles were taken between the consecutive silhouettes (Figures 5A, B). Similar results were obtained for tomato and barley (Additional file 2: Figure S2A, B).

The maximum deviation to the mean quantifies the extent to which the calculated value of a given silhouette deviates from the average value of 16 consecutive silhouettes and expresses the maximum error that can occur. In other words, this parameter expresses the percentage of the difference of the most distant value in each vertical cloud of points (see Figure 4A), representing a single plant, with respect to the average. Although the correlation coefficient between calculated plant pixel area and measured leaf area is very high ($R^2 \geq 0.938$) when taking into account only one silhouette, the maximum deviation to the mean in calculated pixel area of a single silhouette is 17 and 23% for rapeseed and maize, respectively (Figure 5C, D). The maximum deviation to the mean rapidly decreased as more consecutive silhouettes are taken into account and/or using larger angles between them (Figure 5C, D). For instance, the maximum deviation to the mean was about 3% when four profiles differing by 50° were averaged in both studied species. The same trend was observed when analysing tomato or barley (Additional file 2: Figure S2C, D).

The correlation coefficient between calculated and measured maximum plant height was close to 0.99 (Figure 6A, B). This correlation coefficient was hardly affected by averaging more plant silhouettes or increasing the angles between them. Unlike leaf area estimations, for maximum plant height the maximum deviation to the mean was low (≤6%) for a single profile (Figure 6C, D). Averaging more profiles or increasing the angle between them strongly decreased the maximum deviation to the mean between calculated and measured maximum height. Analysing maximum plant height in tomato and barley gave similar results (Additional file 3: Figure S3).

Despite the high correlation coefficient (≈0.94, and 0.85 for barley) between calculated plant pixel area and measured leaf area in all four species when considering a single silhouette, we found that the maximum deviation to the mean (expressed as percentage) ranged between 17 and 29% (corresponding to rapeseed and barley, respectively). In other words, a single silhouette may result in up to 29% mis-estimation of plant pixel area, which will lead to a respective error in the calculation of leaf area. We found that this issue is readily solved by averaging an increasing number of consecutive profiles and/or using larger angles between them. We observed that the number of such consecutive profiles that needs to be averaged was not dependent on plant size within the measured range (an example is given in Additional file 4: Figure S4). For any given maximum deviation to the mean, the number of needed profiles at several angle options between them can be calculated. Plant rotation may be performed manually or automatically (by using a rotating stage). Another possibility is to use multiple pairs of LC that capture the plant profiles at different angles. By adopting this approach, several silhouettes at

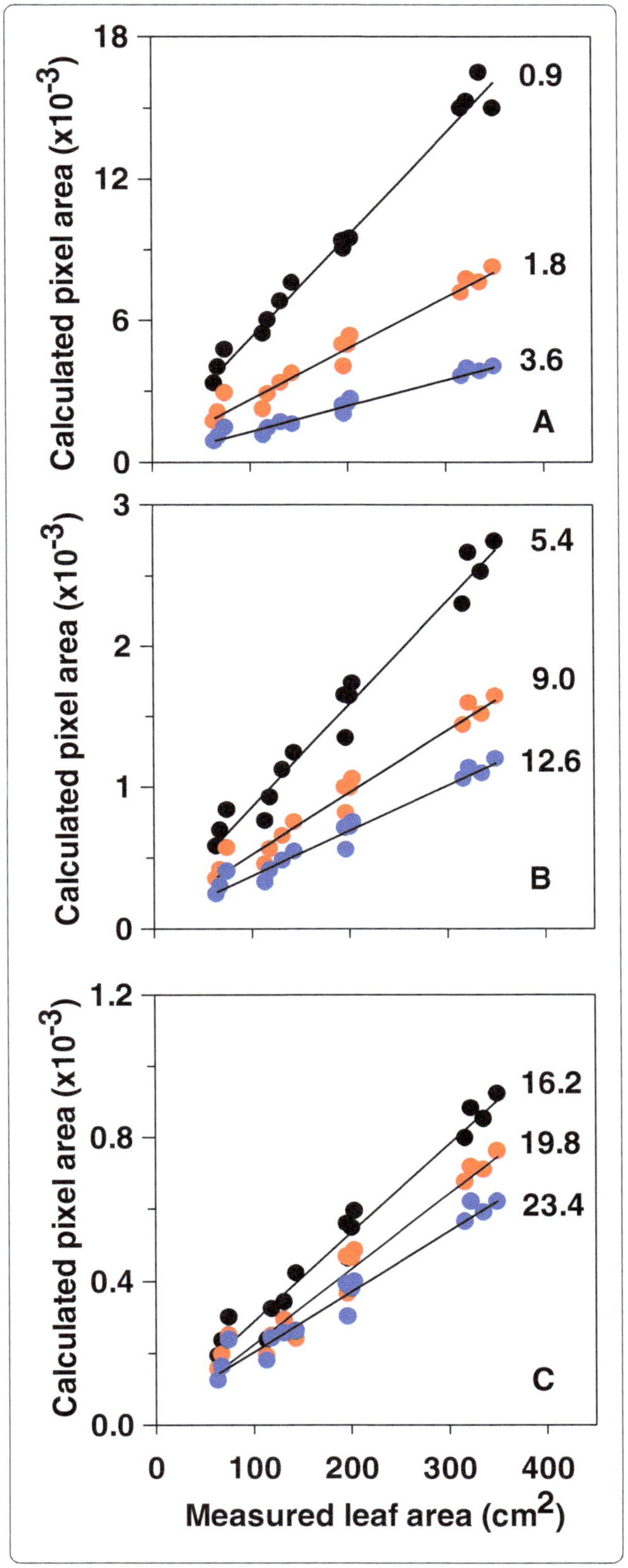

Figure 3 Calculated plant pixel area (average of 18 consecutive silhouettes differing by 10°) as a function of measured plant leaf area at different scanning speeds in rapeseed. Values in each panel indicate an increasing scanning speed during measurement and are plotted with a different symbol colour. **A** Scanning speed 0.9, 1.8, 3.6 m min^{-1}. **B** Scanning speed 5.4, 9.0, 12.6 m min^{-1}. **C** Scanning speed 16.2, 19.8, 23.4 m min^{-1}. Correlation coefficients ranged between 0.9589 and 0.9865 (slopes all significant at $P < 0.0001$). Leaf area ranged between 64 and 350 cm^2 (n = 15).

different angles can be recorded within a single routine that is amenable to automation.

When considering a single silhouette the correlation coefficient (approximately 0.97, and 0.83 for barley) between calculated and measured maximum height was higher in all four species compared with the one estimating leaf area. Consequently, the maximum error in the estimation of maximum height was comparatively lower (<10%, and 15% for barley) and it was reduced using the same approach, as mentioned above.

When scanning a row of (non-rotating) plants on a cultivation table in the greenhouse or in the field, one most probably will encounter a difference from the real total pixel area of these plants. Since this difference for each individual plant is sometimes positive and other times negative, most probably averaging a large number of plants will be closer to the real total pixel area than estimated by adding individual errors.

Calculated and measured values of both maximum height and leaf area were linearly related, the latter being sensitive to shoot architecture

By plotting the calculated plant pixel area against the measured leaf area, it became apparent that the assessed dicotyledonous species (rapeseed, tomato) largely overlapped (Figure 7A). Data of the monocotyledonous species (maize, barley) showed the same trend. However, we observed that the relation between calculated plant pixel area and measured leaf area was different between monocotyledonous and dicotyledonous species (Figure 7A). Calculated pixel area was also significantly correlated with measured shoot dry weight (Additional file 5: Figure S5).

Unlike leaf area, dicotyledonous and monocotyledonous species showed similar relation between calculated and measured maximum height (Figure 7B). In all cases, calculated maximum height was the same as the one measured manually with the exception of barley for which a slight underestimation was observed for plants around 30 cm tall.

The relation between calculated pixel area and measured leaf area apparently depends on shoot architecture. Monocotyledonous and dicotyledonous species showed distinct differences in this relation. Instead, maximum height estimations were not species-dependent. In barley,

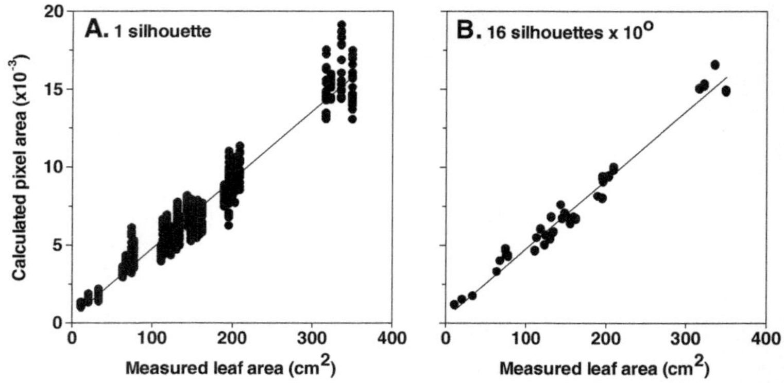

Figure 4 Calculated plant pixel area by using a single silhouette (A) or averaging 16 consecutive silhouettes differing by 10° (B) as a function of measured plant leaf area in rapeseed. Correlation coefficients were 0.9382 and 0.9717 for **A** and **B**, respectively (both slopes significant at $P < 0.0001$). Thirty five plants were assessed. Measurements were conducted at a constant scanning speed of 0.9 m min^{-1}.

an underestimation of maximum height was observed. The pixels connecting plant parts are not always continuous (Figure 1C–E). In this species missing pixels for the very thin profile of leaf tips at the position where the maximum height is estimated may account for this underestimation.

The pixel area of several overlapping plants is indicative of total leaf area

A simulation study was conducted to evaluate the effect of plant spacing distance on the calculated pixel area (an example is shown in Figure 8). Simulations were conducted using pairs of plants, spaced between 0 and 20 cm, with a

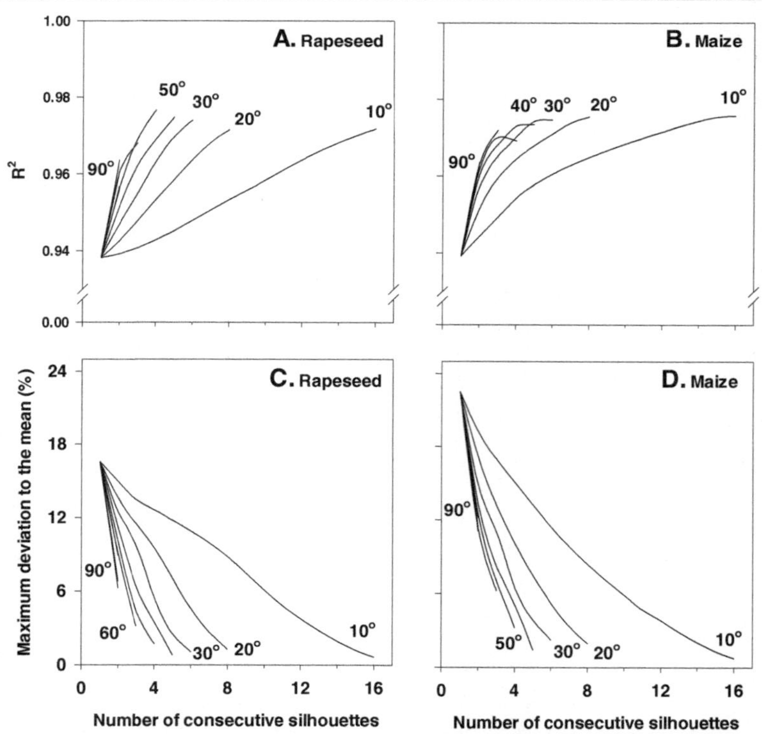

Figure 5 Correlation coefficient between calculated plant pixel area versus measured plant leaf area (A, B), and the maximum deviation to the mean (expressed as percentage; C, D) as a function of number of consecutive silhouettes that are taken into account as well as the angle between them in two species. Leaf area ranged between 12 and 350 cm^2 for rapeseed (n = 35), and between 7 and 317 cm^2 for maize (n = 31). Measurements were conducted at a constant scanning speed of 0.9 m min^{-1}.

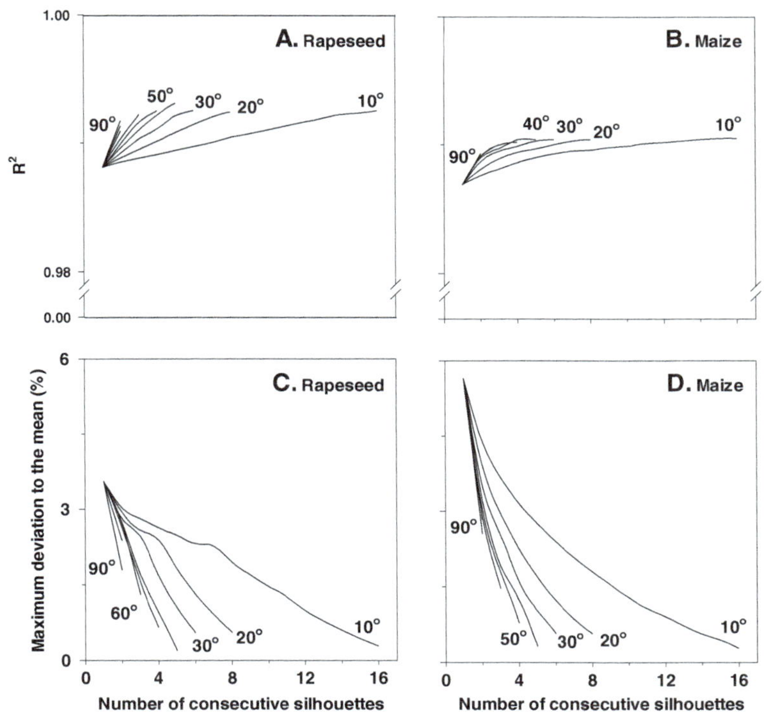

Figure 6 Correlation coefficient between calculated and measured maximum plant height (base to the highest leaf tip; A, B), and the maximum deviation to the mean (expressed as percentage; C, D) as a function of number of consecutive silhouettes that are taken into account as well as the angle between them in two species. Maximum plant height ranged between 3 and 20 cm for rapeseed (n = 35), and between 7 and 68 cm for maize (n = 31). Measurements were conducted at a constant scanning speed of 0.9 m min^{-1}.

4 cm step (Additional file 6: Figure S6 and Additional file 7: Figure S7). In each spacing distance, the total pixel area (i.e. of both plants) was calculated for every view angle of both plants yielding 18 × 18 combinations. The total pixel area was then compared to the value at which the two plants were spaced at a distance where no overlapping parts occurred. Moreover, we calculated the width of the overlapping area, namely the maximum distance of any overlapping pixel in the horizontal direction. Simulations included three pairs of plants, each pair having small, medium or large leaf area (Additional file 6: Figure S6 and Additional file 7: Figure S7).

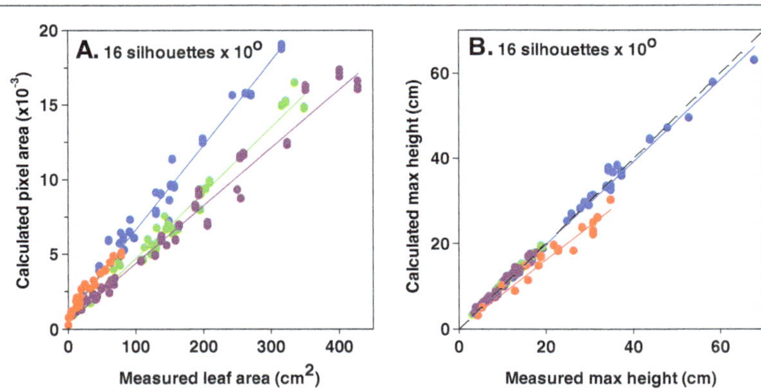

Figure 7 Calculated plant pixel area (A) and calculated max height (base to the highest leaf tip; B) versus measured values in four species (maize, blue; barley, red; tomato, purple; rapeseed, green). Correlation coefficients for all species ranged between 0.9542 and 0.9925 ($P < 0.0001$). For maximum height a single regression model was highly significant ($P < 0.0001$). For leaf area the slope of a single regression model is significant for monocyledonous species ($P < 0.0001$) but not for dicotyledonous ones ($P = 0.34$). The dashed line depicts the 1:1 relationship. Measurements were conducted at a constant scanning speed of 0.9 m min^{-1}.

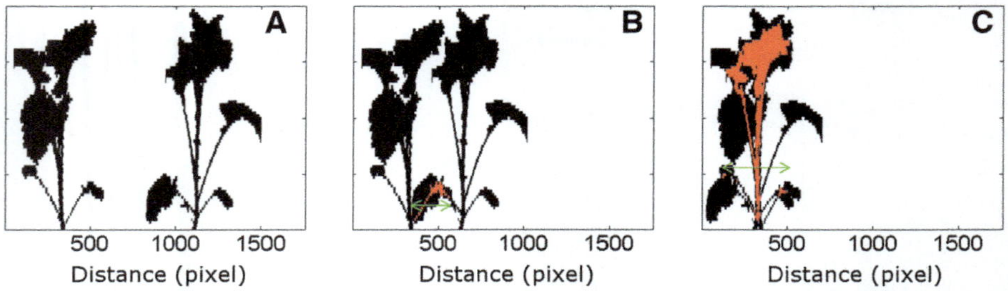

Figure 8 Overlapping width (the maximum distance of any overlapping pixel in the horizontal direction), depicted by the arrow, as spacing distance decreases. Red areas represent the overlap. **A** No overlap between silhouettes. **B** Moderate overlap between silhouettes. **C** Maximum overlap between silhouettes.

At 12 cm spacing distance, the width of the overlapping area was generally limited (≤8 cm), resulting in a very small underestimation of total pixel area (≤5%; Additional file 6: Figure S6 and Additional file 7: Figure S7). Decreasing the plant distance from 8 cm to lower values resulted in a higher width of the overlapping area, leading to increased pixel area underestimation. At the same spacing distance, both the width of the overlapping area and the pixel area underestimation increased with plant size in rapeseed and tomato (Additional file 6: Figure S6 and Additional file 7: Figure S7). Instead, the pixel area underestimation was higher in smaller maize and barley plants.

Attempting to scale this method to a typical field plant array, one may (and will) encounter overlapping parts of neighbouring plants, reducing the accuracy of estimation. This depression of pixel area, as a result of overlaps, depends on spacing distance, plant size, and shoot geometry. Spacing distance 0 cm represents a situation in which plants are placed in two different rows behind each other, and the LC scan both rows simultaneously. Scanning two rows at the same time (i.e. spacing distance 0 cm) produced the largest underestimation in all four species. Against expectations, two plants behind each other resulted in underestimation which was far less than 100%, since their structures were never identical; meaning that in none of the different orientations of a plant covered the other plant. A very small spacing distance of 4 cm resulted in less underestimation of plant pixel area, as compared to 0 cm, but the magnitude was still considerable. At 0 or 4 cm spacing distances, the larger the plant, the higher was the underestimation of pixel area in rapeseed and tomato. Contrary to these findings, the highest underestimation was observed in small maize and barley plants. At more realistic spacing distances (≥8 cm) like those used in the field in row sowing, the underestimation was rather small in all cases for the range of plant size that we examined.

Conclusions

Light curtain arrays (LC) are a phenotyping tool yielding a binary data set of the shoot profile. It can be employed as either a sensor-to-plant (as in this study) or a plant-to-sensor system. Here we found that LC give a rapid estimate of leaf area and maximum height on individual plants of four species with contrasting shoot architecture. The accuracy of this estimation is improved as more plant profiles, following rotation, are averaged and/or the larger the angle between the acquired profiles. Leaf area estimation is strongly influenced by the scanning speed, whereas maximum height estimation is insensitive. Based on these experiments, we generally expect that up to early tillering stage for monocotyledonous species and until the 6th leaf stage in dicotyledonous ones, the estimation of leaf area can be sufficiently accurate. At later developmental stages, the current method may not be necessarily suitable for precise estimation of leaf area. However, regarding measurement of maximum height, LC are only limited by the vertical (and customizable) dimension of the sensor. By using LC the biomass of several plants placed in a row can be calculated, even when those partly overlap opening the window to additional field application.

Methods

Plant material and growth conditions

Experiments included two dicotyledonous (*Brassica rapa* cv. Campino, *Lycopersicon esculentum* cv. Harzfeuer) and two monocotyledonous (*Zea mays* cv. Badischer Gelber, *Hordeum vulgare* cv. Barke) species. Seeds were sown, germinated, and plants were grown up to the cotyledons stage (i.e. fully open cotyledons and before the appearance of the first leaf) or the second leaf stage for the dicotyledonous and monocotyledonous species, respectively. This period ranged from twelve to fourteen days, depending on the species. Subsequently, seedlings were transplanted in 2 L pots containing commercially available soil ('Nullerde' Archut Erzeugnisse GmbH, Vechta, Germany), which were randomly distributed over a glasshouse compartment, located in central part of Germany (Jülich, 50.9°N). The day and night set points were 19 and 17°C for air temperature, while for relative air humidity these were 60 and 50%, respectively. Supplementary light was provided by high-pressure sodium lamps (SON-T Agro, 250 W, Philips, Eindhoven, The Netherlands) at 50 μmol m^{-2} s^{-1} photosynthetic photon flux density for 16 h per day (from 0600 to 2200 hours). This light intensity was recorded during the nocturnal period at 50 cm from the root-to-shoot interface.

The plants were watered weekly with 100 ml of a full strength Hoagland solution [18]. Two to three days following watering with Hoagland solution, plants were flushed with rainwater (approximately 10% drainage) to prevent salt accumulation. Measurements were conducted between two days and four weeks after transplanting.

Sensor positioning system

The sensor positioning system, employed for the experiment, was a prototype developed for phenotyping purposes. It consists of five laterally movable table elements and an overhead loadbearing system. The layout of the tables enables their movement into different arrangements in which they are closer or further apart, allowing an optimal use of the available space, as explained below. During operation, the tables are automatically arranged in a way that gaps on both sides of a given table are created. In this way, it is possible for the overhead loadbearing system to scan that table, carrying the plants designated for the measuring cycle. The remaining four tables are positioned without gaps. The same principle is used to enable access for the personnel. In this way the space, which would be required for passageways between conventional greenhouse tables, is made available for experimentation. The overhead load bearing system consists of an aluminum framework, designed to minimize shadowing. It has a payload capacity of 50 kg. It can move in both x and y directions (accuracy of ± 1 mm), as well as in the z direction (accuracy of ± 2 mm). The x and y movement is due to a laterally movable crossbeam, equipped with a longitudinal carriage rail. The movement in the z direction is enabled through two cantilever arms, equipped with carriage rails. The technical conception and realization was performed by a company (MK-Maschinenbau Kitz GmbH, Troisdorf, Germany).

LC measurements

The LC system is a parallel set of light barriers at a distance of 2.5 mm, which are arranged on two vertical poles (INFRASCAN 5000, Sitronic GmbH, Austria). The vertical support poles are 1.59 m long (1.43 m of which contain light barriers), spaced at a (horizontal) distance of 55 cm (Figure 1A). One set of light barriers is emitting infrared light (950 nm), while the other one is receiving the emitted light. In this way, the sensor detects if any of these beams are interrupted by an object placed between the emitting and receiving sides.

Before measurement, sixteen plants were placed in a row on a bench (Figure 1A). To minimize overlaps, pots were spaced at 30 cm. The light barriers were guided along the row of plants, in a way that the emitter was placed from one side of the measurement bench (perpendicular to the horizontal), while the receiver was placed on the other side (Figure 1A). The bottom edge of the light barriers was located at the bench level, allowing monitoring of both pots and plants. Care was taken that no plant parts were below the upper pot edge. All measurements took place at a constant sensor speed.

At first, the plant silhouette was recorded (Figure 1B). Subsequently, the plants were rotated by 10° to the horizontal, and another scan was conducted. In total 18 scans were acquired, starting from the initial position (0°) up to 170° with a 10° rotation for each run. Plants were not rotated by 180°, since this would result in the same silhouette as the initial position (0°). View angles larger than 180° were also not acquired, since the yielded data have been already recorded (e.g. 190° the same silhouette as 10°, 200° the same silhouette as 20°, etc.).

After all plant silhouettes were recorded, the number of leaves (≥1 cm), leaf area, leaf and plant (aboveground) dry mass were recorded. Leaf area was determined with a leaf area meter (LI-COR Model Li-3100, LI-COR, Lincoln, NE), and dry weight was assessed after drying the tissue for 24 h at 80°C.

Plant pixel area and maximum height estimation

The LC yield a binary data set, where the values 0 and 1 correspond to continuous or interrupting light path of a given photodiode. The methodology employed to extract a single shoot silhouette from a scan of several plants is described in Figure 2. At first, individual plant position was identified from the scans including several plants (Figure 1B). This was done by first defining the pot center. Secondly, small size plant parts (e.g. stems and petioles) that were not connected to the silhouette were included (see Figure 1C–E). To identify which data belong to a given plant, a radial search within a radius of 25 pixels was conducted, beginning at the bottom center of each pot position. In this way, all pixels belonging to that area are added to the silhouette. Finally, the pixels belonging to the pot were removed from the silhouette (Figure 1B). Starting at the bottom of the silhouette, we defined that the upper edge of the pot is at the position where the number of pixels in two consecutive rows differed more than 80%. With this approach, any plant parts below the upper edge of the pot are also removed. Plant pixel area was taken as the number of pixels belonging to a given silhouette, while maximum plant height was taken as the maximum pixel value in the y-axis (Figure 1C–E). The analysis was conducted by using the MATLAB program (R2012b, MathWorks, Natick, MA).

Simulation of spacing distance

To evaluate the potential of LC in assessing plant leaf area of overlapping plants, simulations were conducted at variable spacing distances. Two representative small, medium and large plants per species were selected. In each pair of plants, the spacing distance was decreased from 20 cm (where no overlaps occurred) down to 0 cm (maximum overlapping areas), with a step of 4 cm. In each spacing distance, simulations included all different angle combinations (18 × 18 view angles). In each spacing distance and angle combination, the pixel area of the two plants was assessed and this was compared to the sum of pixel areas of the two individual plants.

Additional files

Additional file 1: FigureS1. The effect of scanning speed on calculated plant pixel area (average of 18 consecutive silhouettes differing by 10°) and calculated maximum plant height (base to the highest leaf tip; average of 18 consecutive silhouettes differing by 10°) in rapeseed. Both are expressed as a percentage of the value at lowest scanning speed (0.9 m min^{-1}). Leaf area ranged between 64 and 350 cm^2, while maximum plant height varied between 10 and 20 cm. The SEM bars are not visible, because the SEM is smaller than the symbol (n = 15).

Additional file 2: FigureS2. Correlation coefficient between calculated plant pixel area versus measured plant leaf area (A, B), and the maximum deviation to the mean (expressed as percentage; C, D) as a function of number of consecutive silhouettes that are taken into account as well as the angle between them in two species. Leaf area ranged between 13 and 429 cm^2 for tomato (n = 36), and between 2 and 80 cm^2 for barley (n = 29). Measurements were conducted at a constant scanning speed of 0.9 m min^{-1}.

Additional file 3: FigureS3. Correlation coefficient between calculated and measured maximum plant height (base to the highest leaf tip; A, B), and the maximum deviation to the mean (expressed as percentage; C, D) as a function of number of consecutive silhouettes that are taken into account as well as the angle between them in two species. Maximum plant height ranged between 4 and 18 cm for tomato (n = 36), and between 4 and 35 cm for barley (n = 29). Measurements were conducted at a constant scanning speed of 0.9 m min^{-1}.

Additional file 4: FigureS4. Maximum deviation to the mean (expressed as percentage) as a function of calculated pixel area (indicative of plant size) in two species. Six consecutive silhouettes, following plant rotation at different angles, were averaged. Leaf area ranged between 12 and 350 cm^2 for rapeseed (n = 35), and between 7 and 317 cm^2 for maize (n = 31). Measurements were conducted at a constant scanning speed of 0.9 m min^{-1}.

Additional file 5: FigureS5. Calculated plant pixel area versus measured dry weight in four species (maize, blue; barley, red; tomato, purple; rapeseed, green). Correlation coefficients ranged between 0.9517 and 0.9764 (slopes all significant at $P < 0.0001$). Measurements were conducted at a constant scanning speed of 0.9 m min^{-1}.

Additional file 6: FigureS6. Underestimation of calculated plant pixel area as a function of width of the overlapping area at different plant distances (referred by different colours) in two species. The width of the overlapping area refers to the maximum distance of any overlapping pixel in the horizontal direction (see Figure 8). Simulations were conducted by using a pair of small (A, B; 68 and 78/ 77 and 79 cm^2 leaf area), medium (C, D; 201 and 210/ 153 and 158 cm^2 leaf area) and large (E, F; 317 and 323/ 244 and 271 cm^2 leaf area) rapeseed and maize plants, respectively. Measurements were conducted at a constant scanning speed of 0.9 m min^{-1}.

Additional file 7: FigureS7. Underestimation of calculated plant pixel area as a function of width of the overlapping area at different plant distances (referred by different colours) in two species. The width of the overlapping area refers to the maximum distance of any overlapping pixel in the horizontal direction (see Figure 8). Simulations were conducted by using a pair of small (A, B; 69 and 70/ 17 and 18 cm^2 leaf area), medium (C, D; 189 and 194/ 34 and 39 cm^2 leaf area) and large (E, F; 255 and 260/ 63 and 64 cm^2 leaf area) tomato and barley plants, respectively. Measurements were conducted at a constant scanning speed of 0.9 m min^{-1}.

Abbreviation

LC: Light curtain arrays.

Competing interests

The authors declare that they have no competing interests.

Authors' contributions

DF performed the experimental work, acquired the silhouettes, carried out the data analysis and interpretation, and wrote the manuscript. CB developed the software tools for handling the silhouettes, and performed the simulations. JFJM and AU were responsible for the development and description of the sensor positioning system, as well as were involved in the development of the LC. SK and AP wrote the routine for sensor positioning, and were responsible for the hardware setup. FF and US supervised the study. All authors contributed in reading, editing and approving the final manuscript.

Acknowledgements

This research received the financial support of the Forschungszentrum Jülich in the Helmholtz Association. This work was partly funded by the European Plant Phenotyping Network (EPPN -European Union FP7 Capacities Programme) through a post-doctoral grant to DF (grant agreement 284443), by the German Plant Phenotyping Network (DPPN - BMBF Fz. 031A053), and by the AgroHort med project (EU fund for regional development, EU-EFRE-NRW AZ: B0833.01.09).

Author details

[1]IBG-2: Plant Sciences, Institute for Bio- and Geosciences, Forschungszentrum Jülich, 52425 Jülich, Germany. [2]Present address: Center of Applied Biology, Department of Soil Science and Plant Nutrition, Hochschule Geisenheim University, 65366 Geisenheim, Germany. [3]University of Applied Sciences Osnabrück, 49076 Osnabrück, Germany.

References

1. Fiorani F, Schurr U: **Future scenarios for plant phenotyping.** *Annu Rev Plant Biol* 2013, **64:**267–291.
2. Pieruschka R, Poorter H: **Phenotyping plants: genes, phenes and machines.** *Funct Plant Biol* 2012, **39:**813–820.
3. Furbank RT, Tester M: **Phenomics – technologies to relieve the phenotyping bottleneck.** *Trends Plant Sci* 2011, **16:**635–644.
4. Araus JL, Cairns JE: **Field high-throughput phenotyping: the new crop breeding frontier.** *Trends Plant Sci* 2014, **19:**52–61.
5. Dhondt S, Wuyts N, Inzé D: **Cell to whole-plant phenotyping: the best is yet to come.** *Trends Plant Sci* 2013, **18:**428–439.
6. Golzarian MR, Frick RA, Rajendran K, Berger B, Roy S, Tester M, Lun DS: **Accurate inference of shoot biomass from high-throughput images of cereal plants.** *Plant Methods* 2011, **7:**2.
7. Walter A, Studer B, Kolliker R: **Advanced phenotyping offers opportunities for improved breeding of forage and turf species.** *Ann Bot* 2012, **110:**1271–1279.
8. Fiorani F, Rascher U, Jahnke S, Schurr U: **Imaging plants dynamics in heterogenic environments.** *Curr Opin Biotechnol* 2012, **23:**227–235.
9. Green JM, Appel H, MacNeal Rehrig E, Harnsomburana J, Chang JF, Balint-Kurti P, Shyu CR: **PhenoPhyte: a flexible affordable method to quantify 2D phenotypes from imagery.** *Plant Methods* 2012, **8:**1–12.

10. Paproki A, Sirault X, Berry S, Furbank R, Fripp J: **A novel mesh processing based technique for 3D plant analysis.** *BMC Plant Biol* 2012, **12**:63.
11. Montes J, Technow F, Dhillon B, Mauch F, Melchinger A: **High-throughput non-destructive biomass determination during early plant development in maize under field conditions.** *Field Crops Res* 2011, **121**:268–273.
12. Busemeyer L, Mentrup D, Möller K, Wunder E, Alheit K, Hahn V, Maurer HP, Reif JC, Würschum T, Müller J, Rahe F, Ruckelshausen A: **BreedVision – A Multi-Sensor Platform for Non-Destructive Field-Based Phenotyping in Plant Breeding.** *Sensors* 2013, **13**:2830–2847.
13. Granier C, Aguirrezabal L, Chenu K, Cookson SJ, Dauzat M, Hamard P, Thioux JJ, Rolland G, Bouchier-Combaud S, Lebaudy A, Muller B, Simonneau T, Tardieu F: **PHENOPSIS, an automated platform for reproducible phenotyping of plant responses to soil water deficit in *Arabidopsis thaliana* permitted the identification of an accession with low sensitivity to soil water deficit.** *New Phytol* 2006, **169**:623–635.
14. Jansen M, Gilmer F, Biskup B, Nagel KA, Rascher U, Fischbach A, Briem S, Dreissen G, Tittmann S, Braun S, De Jaeger I, Metzlaff M, Schurr U, Scharr H, Walter A: **Simultaneous phenotyping of leaf growth and chlorophyll fluorescence via GROWSCREENFLUORO allows detection of stress tolerance in *Arabidopsis thaliana* and other rosette plants.** *Funct Plant Biol* 2009, **36**:902–914.
15. Arvidsson S, Perez-Rodriguez P, Mueller-Roeber B: **A growth phenotyping pipeline for *Arabidopsis thaliana* integrating image analysis and rosette area modeling for robust quantification of genotype effects.** *New Phytol* 2011, **191**:895–907.
16. Reuzeau C, Pen J, Frankard V, Wolf J, Peerbolte R, Broekaert W, van Camp W: **TraitMill: a discovery engine for identifying yield-enhancement genes in cereals.** *Mol Plant Breed* 2005, **3**:753–759.
17. Hartmann A, Czauderna T, Hoffmann R, Stein N, Schreiber F: **HTPheno: an image analysis pipeline for high-throughput plant phenotyping.** *BMC Bioinforma* 2011, **12**:148.
18. Hoagland DR, Arnon DI: **The water-culture method for growing plants without soil.** *Calif Agric Exp Sta Cir* 1938, **3**:346–347.

5

LEAF-E: a tool to analyze grass leaf growth using function fitting

Wannes Voorend[1,2,3], Peter Lootens[3], Hilde Nelissen[1,2], Isabel Roldán-Ruiz[3], Dirk Inzé[1,2] and Hilde Muylle[3*]

Abstract

In grasses, leaf growth is often monitored to gain insights in growth processes, biomass accumulation, regrowth after cutting, etc. To study the growth dynamics of the grass leaf, its length is measured at regular time intervals to derive the leaf elongation rate (LER) profile over time. From the LER profile, parameters such as maximal LER and leaf elongation duration (LED), which are essential for detecting inter-genotype growth differences and/or quantifying plant growth responses to changing environmental conditions, can be determined. As growth is influenced by the circadian clock and, especially in grasses, changes in environmental conditions such as temperature and evaporative demand, the LER profiles show considerable experimental variation and thus often do not follow a smooth curve. Hence it is difficult to quantify the duration and timing of growth. For these reasons, the measured data points should be fitted using a suitable mathematical function, such as the beta sigmoid function for leaf elongation.

In the context of high-throughput phenotyping, we implemented the fitting of leaf growth measurements into a user-friendly Microsoft Excel-based macro, a tool called LEAF-E. LEAF-E allows to perform non-linear regression modeling of leaf length measurements suitable for robust and automated extraction of leaf growth parameters such as LER and LED from large datasets. LEAF-E is particularly useful to quantify the timing of leaf growth, which forms an important added value for detecting differences in leaf growth development. We illustrate the broad application range of LEAF-E using published and unpublished data sets of maize, *Miscanthus* spp. and *Brachypodium distachyon*, generated in independent experiments and for different purposes. In addition, we show that LEAF-E could also be used to fit datasets of other growth-related processes that follow the sigmoidal profile, such as cell length measurements along the leaf axis.

Given its user-friendliness, ability to quantify duration and timing of leaf growth and broad application range, LEAF-E is a tool that could be routinely used to study growth processes following the sigmoidal profile.

Keywords: Leaf elongation rate, Non-linear regression, Leaf length, Cell length, Growth zone

Background

Leaf growth has been monitored in a wide variety of grass species such as maize, rice, wheat, barley, *Lolium*, *Miscanthus*, *Sorghum* and *Brachypodium* making use of leaf length measurements taken at regular time intervals during development [1-8]. Parameters derived from these measurements such as leaf elongation rate (LER) and leaf elongation duration (LED) have been shown to be major determinants of individual and whole plant leaf area [9-14] and can be used to explain differences in final leaf length in response to environmental conditions and/or between genotypes [3,4,15].

In plant growth modeling, there is a growing consensus that approaches applying linear and exponential models are inadequate [16]. A linear fit assumes a constant LER over a longer period during leaf development [1,3,9,10] and an exponential or a log-linear relation assumes a constant relative elongation rate (RER). These assumptions limit the utility of the models, as both LER and RER may vary with environmental conditions and developmental stage [16]. The polynomial model does cope with variations in LER and RER during leaf development. However, polynomial functions tend to make spurious upward or downward predictions, especially at the extremes of the data [16,17]. Nonlinear regression is a more suitable

* Correspondence: hilde.muylle@ilvo.vlaanderen.be
[3]Plant Sciences Unit – Growth and Development, Institute for Agricultural and Fisheries Research (ILVO), Caritasstraat 21, 9090 Melle, Belgium
Full list of author information is available at the end of the article

strategy to describe leaf growth and to accommodate temporal variation in growth rates [16].

The beta sigmoid function, first used to describe whole plant growth [18], has been successfully applied to model the growth pattern of a single grass leaf [7,19]. Yin and coworkers [18] compared the performance of the beta sigmoid function with that of some other widely used sigmoid functions, such as Gompertz, Weibull and Richards to analyze datasets from maize, pea and wheat and concluded that the beta sigmoid function is unique in dealing with determinate growth [18]. This is due to the prediction of a zero growth rate at both the start and end of the determinate growth period which is characterized by three sub-phases: an early exponential growth phase, an approximately linear growth phase, followed by a steadily decelerating growth phase [20]. Furthermore, in contrast to other functions, the beta sigmoid function incorporates biologically relevant parameters and is highly flexible for describing various asymmetrical sigmoidal patterns [18].

In the context of high-throughput leaf phenotyping, there is a need for user-friendly tools that provide rapid and robust analysis of growth parameters from large datasets. Non-linear regression using function fitting is currently imbedded in statistical work packages such as SAS and R rendering the calculation, extraction and visualization of specific leaf growth parameters, such as LED, from large datasets difficult and time-consuming.

Here, we describe LEAF-E, a nonlinear regression-based tool for analyzing grass leaf growth data. The tool can be used to derive biologically relevant parameters such as final leaf length, maximal LER, LED but also parameters for the quantification of the timing of leaf growth, an important asset of this tool. To allow for the analysis of large datasets, the fitting procedure was automated in a user-friendly Microsoft Excel macro, which is innovative. We show how the application of this tool can assist data analysis and interpretation of experiments in which different genotypes or the response of single genotypes to different growth conditions are compared. For this purpose, we quantified and compared leaf growth parameters in published and unpublished datasets of three grass species: *Zea mays* (maize), *Brachypodium distachyon* and *Miscanthus spp.*

Results and discussion

Fitting of kinematic individual leaf length measurements using the beta sigmoid function

Goodness of fit of the beta sigmoid function for maize, Brachypodium and Miscanthus leaf growth

First, we investigated to what extent the beta sigmoid function can be used to accurately fit leaf length measurements in function of thermal time or growing degree days (°Cd) in the three species considered here. Thermal time was used since plant growth and development are more closely related to accumulated mean daily temperature above a base value in the absence of other limiting conditions [21,22]. Equation 1 was used to fit length measurements of the 4th leaf over thermal time of nine B104 maize plants (dataset 1a, Figure 1). This resulted in R^2 values ranging from 0.9970 to 0.9989 with a mean value of 0.9981. Function fitting of leaf length measurements in *Miscanthus* and *Brachypodium* (datasets 2 and 3, respectively) rendered similar results: an

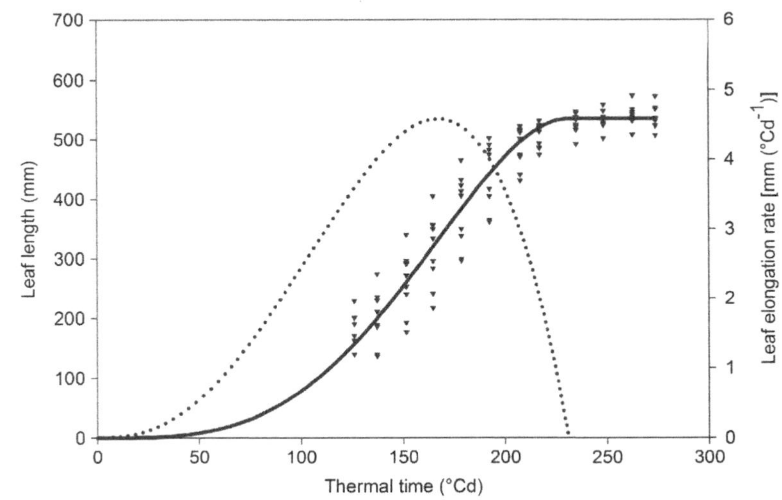

Figure 1 Leaf length and LER of maize B104 non-transgenic plants. Triangles represent length measurements of the 4th leaf of nine non-transgenic maize plants from dataset1a. The measurements of each individual leaf were fitted and biologically relevant growth parameters were extracted with LEAF-E. The S-shaped leaf length curve and bell-shaped LER curve are function plots using the mean values of the function parameters for the nine non-transgenic maize plants.

overall mean R^2-value of 0.9931, ranging from 0.9669 to 0.9989 (n = 18) for the two *Miscanthus* species, and an overall mean R^2-value of 0.9932, ranging from 0.9871 to 0.9993 (n = 36) for the four *Brachypodium* inbred lines. Plots of the fittings and R^2-values of individual plants of all datasets can be found in Additional file 1. A linear regression analysis of the measured leaf lengths versus the estimated value for those respective points in thermal time resulted in an R^2 value of 0.9986 for maize (dataset 1a), 0.9951 for *Miscanthus* (dataset 2) and 0.9940 for *Brachypodium*, indicating a good fit of the data for the three species investigated (Additional file 2 A, D and G). Nevertheless, analysis of the residuals showed a remaining sinusoidal pattern which indicates that part of the data could not be explained by the model (Additional file 3). The higher R^2 value for the fittings of the maize dataset over the *Miscanthus* and *Brachypodium* datasets might be due to the more controlled environment of the growth chamber for maize as compared to the greenhouse for *Miscanthus* and *Brachypodium*. This indicates the importance of conducting growth experiments in a controlled environment and/or close monitoring of important growth regulating factors such as temperature, preferably at the level of the plant apex.

Based on these results, we can conclude that the beta sigmoid function is able to accommodate leaf growth measurements of three grass species with very distinct phenotypic characteristics. Maize and *Miscanthus spp.* both possess a 'C4' metabolism, however, maize is an annual crop characterized by one stem, whereas *Miscanthus spp.* are rhizomatous perennials that form numerous tillers. *Brachypodium* is a small, annual 'C3' plant used as a model for several temperate grain crops such as wheat and barley [23]. Based upon these findings and the results obtained previously in *L. perenne* [7], we can conclude that the beta sigmoid function is probably of broad application for describing leaf growth in both C3 and C4 grass species. For that reason, the beta sigmoid function was further used to develop an automated fitting procedure in an excel macro that we called LEAF-E and to derive biologically relevant parameters from the different datasets.

Biologically relevant function parameters

The Excel macro that we designed, automatically generates the fitted function parameters (the final leaf length *Lm*, the moment of maximal leaf elongation rate *tm*, and the moment at which leaf growth ceases *te*) and all additional parameters in the form of a table. For visual inspection, it also generates a graph showing the original data points, the fitted growth curve and the function parameters for each biological replicate (see Additional files 4 and 5). Although the tool can be used to fit the measurements of several replicates jointly, fitting data of individual leaves allows statistical analysis, such as estimation of averages and standard deviation values, for all growth parameters, and comparison of genotypes or treatments, a strategy that is both straightforward and statistically correct [17]. The advantage of using the beta sigmoid function in the form of Auzanneau and coworkers [7] is that the function parameters (*Lm, tm* and *te*) themselves are biologically relevant when assessing grass leaf growth. This represents a clear advantage over functions that are based on parameters with no biologically relevant meaning or parameters that are difficult to interpret visually such as the Weibull equation [18]. In addition, parameters that allow for the quantification of the timing of leaf growth, such as *tm* and *te*, can be estimated, which is not trivial using other methodologies such as a log-linear, polynomial or linear fit.

We used the nine B104 maize plants from dataset 1a to illustrate how leaf growth can be analyzed using LEAF-E (Figure 1, Table 1). The final length (*Lm*) of the 4th leaf in the nine B104 maize plants was 535 ± 6 mm on average. This value was attained after 231 ± 5°Cd (*te*), which, in this experiment, is equivalent to 16.5 days after sowing. The moment at which the LER was maximal, *tm*, was 167 ± 5°Cd, or 12.0 days after sowing.

Flexibility to extract additional biologically relevant information from the dataset

The major advantage of fitting a continuous function to the data is that for any given thermal time *t* (°Cd), the leaf length *L* (mm) can be estimated and vice versa (Figure 2).

For example, the leaves of the nine B104 maize plants of dataset 1a (Table 1) reached 100 mm (*t100*) at 108 ± 4°Cd. In this experiment this was just before the 4th leaf emerged from the pseudo-stem or whorl (the spiral arrangement of leaves forming a cylindrical structure from where newly formed leaves emerge), approximately 8 days after sowing. Knowing that the average final length (*Lm*) is 535 ± 6 mm, we can state that a considerable share of at least 19% of the final maize leaf length is hidden in the pseudo-stem of this genotype. For *t20%*, the moment at which the 4th leaf reaches 20% of its final length, we obtained an estimate of 111 ± 4°Cd or 7.9 days after sowing. Exactly 50% of the final leaf length was attained at on average 153 ± 5°Cd (*t50%*), which is not significantly different from the moment of the maximal LER, *tm* at 167 ± 5°Cd. The leaf reached 90% of its final leaf length (*t90%*) at 203 ± 5°Cd or 14.5 days after sowing. The use of parameters that are independent of final leaf length, such as *t20%*, *t50%* and *t90%*, can however be meaningful when comparing genotypes that differ inherently in final leaf lengths. Likewise, we calculated a time window between reaching *20%* and *90%* of the final leaf length. The parameter was named [*LED(20%-90%)*] and corresponded to 92 ± 2°Cd or 6.7 days in this experiment (dataset 1a).

Table 1 Effect of *GA20ox1* overexpression on maize leaf elongation

Growth parameter			*AtGA20ox1* OE (mean ± SE)	Control (mean ± SE)	Difference in mean(+)	
	Lm	(mm)	743 ± 13	535 ± 6	38.9%	***
	LERmax	($mm°C^{-1}d^{-1}$)	6.2 ± 0.2	4.6 ± 0.1	34.2%	***
Thermal time points	t100	(°Cd)	107 ± 2	108 ± 4	−1.7%	NS
	t20%	(°Cd)	121 ± 3	111 ± 4	8.7%	NS
	t50%	(°Cd)	165 ± 3	153 ± 5	7.6%	*
	tm	(°Cd)	180 ± 4	167 ± 5	7.7%	*
	t90%	(°Cd)	217 ± 4	203 ± 5	6.7%	*
	te	(°Cd)	246 ± 5	231 ± 5	6.3%	*
Leaf elongation durations	LED(100-e)	(°Cd)	139 ± 4	123 ± 2	13.3%	**
	LED(20%-90%)	(°Cd)	96 ± 2	92 ± 2	4.2%	NS
	LED(20%-e)	(°Cd)	125 ± 30	120 ± 2	4.0%	NS

A segregating population produced by backcrossing (BC) a transgenic plant overexpressing the *Arabidopsis thaliana GIBBERELLIC ACID 20 OXIDASE1 (GA20ox1) gene* to the wild-type line B104 was analysed for leaf growth. The results are based on the analysis of eleven transgenic and nine non-transgenic BC1 plants. Lm: final leaf length; LERmax: maximal leaf elongation rate; t20%, t50%, t90%, te: time points at which the leaf reaches 20%, 50%, 90% and 100% of the final leaf length, respectively; t100: time point at which the leaf reaches 100 mm; tm: time point at which the leaf reaches LERmax; LEDs: leaf elongation durations between above stated thermal time points.

+Statistical significance based on student t-test of non-transgenic plants (n = 9) vs *GA20ox1* overexpression (n = 11), *$p < 0.05$, ** $p < 0.01$, ***$p < 0.001$, NS non-significant. Applied base temperature for thermal time calculation = 10°C, Mean of overall R^2 values = 0.9983 (0.9970-0.9991).

Depending on the objective of a particular experiment, several other useful parameters can be extracted from the growth curve. Parameters can be deduced from the model, even at time points before empirical evidence can be taken (for maize e.g. t100 and in some cases also t20%) using non-destructive measurements of leaf length over time, as the leaf is then still hidden in the pseudo-stem. However, these are approximations and will need validation using destructive measurements of leaf elongation when one wants to investigate early leaf development, i.e. when the leaf is still hidden in the pseudo-stem [24,25].

Leaf elongation rate and steady-state growth

Similarly, for any given thermal time t (°Cd), the LER (mm/°Cd) can be estimated using the first derivative of

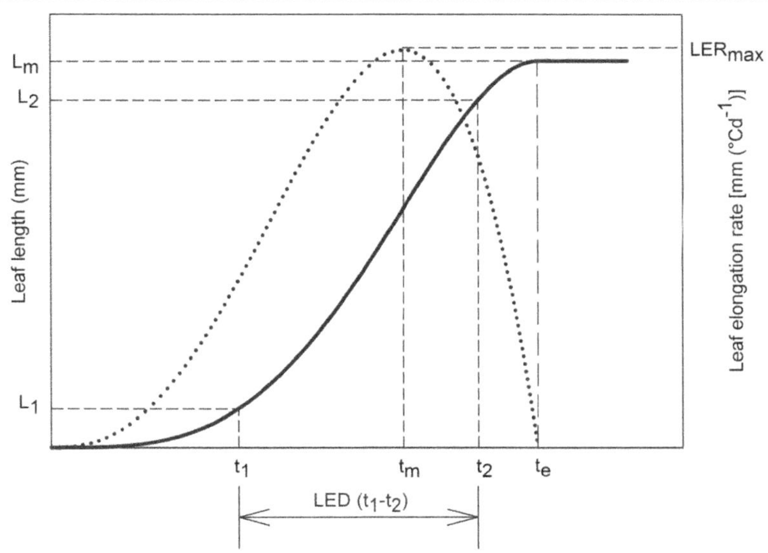

Figure 2 Deriving leaf growth parameters from the fitted leaf length and LER curve using LEAF-E. The leaf length curve (S-shaped) is generated by fitting Equation 1 to measurements of a single leaf. Based upon the leaf length curve the final leaf length L_m can be calculated as well as the thermal time t needed to reach any given leaf length L, e.g. *te* is the thermal time needed to reach L_m. As a result, LEDs can be calculated between every desirable pair of thermal time points, e.g. LED(t1-t2). The first derivative of Equation 1 renders a bell-shaped LER curve from which maximal leaf elongation rate *LERmax*, occurring at the thermal time point *tm*, can be extracted.

the beta sigmoid function (Equation 2). The maximum of this bell-shaped curve, is denoted as the maximal LER or *LERmax* (Figure 2). For the nine B104 maize plants (dataset 1a), we estimated a *LERmax* of 4.6 ± 0.1 mm/°Cd, equivalent to an impressive 2.7 mm/h or 64 mm/day.

Often, the LER profile is derived from calculations of leaf length increases between consecutive measurements divided by the respective thermal time interval. These calculations were performed on a plant-by-plant basis and can be viewed in Additional file 1. The LERmax is then defined as the maximal value in that LER profile. A linear regression analysis of the estimated LER and LERmax to the calculated LER and LERmax, respectively showed high R^2 values (Additional file 2), indicating a linear relationship between values obtained by the two approaches. The R^2 values of the linear regression were highest for the maize dataset (dataset 1a), 0.9067 and 0.9228 for LER and LERmax respectively (see Additional file 2 for R^2 values of all linear regression analyses).

Several studies have assumed that during the growth of a grass leaf a period of constant LER can be defined [5,15,25,26]. This assumption of steady-state growth in grass species such as maize has often been used as an acceptable simplification of the actual growth process [5,15,25,26], although it has been argued that this steady-state growth period is relatively short compared to the total leaf growth period [8,25,27]. LEAF-E allows to quantify a steady state growth period. This can be defined as a period in leaf development for which the LER is relatively stable, for example the thermal time window between the points at which LER has a value of 90% or 95% of LERmax. For the maize B104 plants of dataset 1a, the thermal time window for which the LER was higher than 90% and 95% of LERmax was 34.5 ± 0.6°Cd or 2.5 days (from 10.7 until 13.1 days after sowing) and 49.1 ± 0.9°Cd or 3.5 days (from 10.1 until 13.6 days after sowing), respectively. We therefore conclude that, for the investigated dataset, a relatively stable LER is found for a time-span of 2.5 to 3.5 days during leaf growth.

Effect of *GA20ox1* overexpression on maize leaf elongation

Comparison of transgenic maize plants overexpressing the GA biosynthesis gene *GA20ox1* with non-transgenic plants in a previous study, demonstrated that altering GA levels specifically affects the size of the division zone, resulting in proportional changes in leaf and whole plant growth rates [28]. This published dataset was used to validate the LEAF-E tool.

Overexpression of *GA20ox1* in maize resulted into significantly longer leaves (*Lm*) and higher maximum leaf elongation rates (*LERmax*) (39% and 34% respectively, $p < 0.001$). This corresponded very well with the 38% increase in LER reported earlier by Nelissen and coworkers [28]. However, in addition to the calculation of *Lm* and *LERmax*, LEAF-E also facilitates analysis of parameters that describe the timing of leaf growth. For example, leaves of transgenic plants took slightly, but significantly more time to reach their full length (*te*) than those of non-transgenic plants (Figure 3A and Table 1). As a consequence, also *t50%*, *t90%* and *te*, as well as *tm*, occurred significantly later in the transgenics. This means that leaves of transgenic plants took slightly more time to reach 50%, 90% and 100% of their final leaf length.

Using LEAF-E to reanalyze dataset 1a, previous findings were confirmed but this analysis allowed a more detailed study of the timing of leaf growth, thereby facilitating the detection of dissimilarities such as a shift in attaining *LERmax* in *GA20ox1* overexpressing plants, that could not be quantified previously.

Variation in leaf growth behavior in two *Miscanthus* species

We investigated the leaf growth characteristics of two genotypes of *Miscanthus* belonging to different species with high potential as bio-energy crops, but with contrasting phenotypic characteristics. *M. sinensis 'Goliath'* is characterized by high shoot densities, whereas *M. x giganteus* produces less, but thicker and taller shoots [29,30].

When comparing leaf growth characteristics of *M. sinensis 'Goliath'* and *M. x giganteus*, we found that *M. sinensis 'Goliath'* had significantly longer leaves than *M. x giganteus* (Table 2, Figure 3B). The leaves of *M. sinensis 'Goliath'* on average attained a length of 1140 ± 46 mm, whereas leaves of *M. x giganteus* on average became 923 ± 26 mm in length (Table 2). The longer leaves of *M. sinensis 'Goliath'* plants cannot be explained by significant changes in *LERmax* nor by an extended elongation period (no significant differences for LED values). However, our analysis revealed that leaves of *M. sinensis 'Goliath'* plants display a very strong initial growth compared to *M. x giganteus*. Parameters *t20%*, *t50%* and *tm* are all attained sooner (Table 2). Leaves of *M. sinensis 'Goliath'* reached 50% of their final leaf length at 230 ± 8°Cd, or 64°Cd sooner than *M. x giganteus* plants, at which point the leaves of the last were just emerging from the whorl.

The analysis with LEAF-E shows that the leaf growth pattern in these two *Miscanthus spp.* is significantly different and that the differences in total leaf length can be attributed to contrasting early leaf growth.

Variation in leaf growth behavior in different *Brachypodium distachyon* inbred lines

Fifty *Brachypodium distachyon* inbred lines are currently being used for a study of natural diversity, which is led by 'The International Brachypodium Initiative' [23]. We analyzed the leaf growth behavior of four diploid

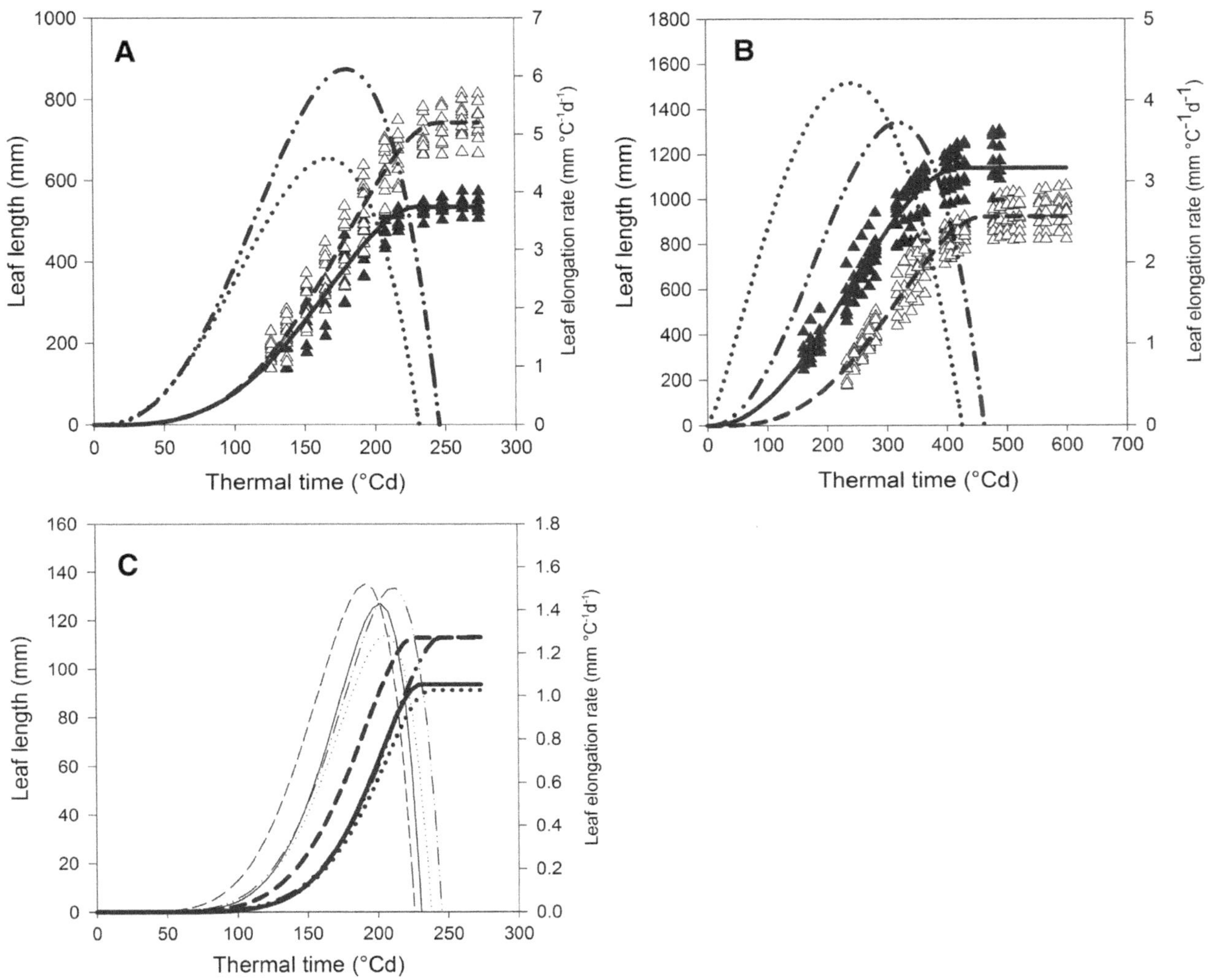

Figure 3 Analysis of the leaf elongation datasets of maize, *Miscanthus* and *Brachypodium* using LEAF-E. (A) Leaf length measurements of transgenic (white triangle) and non-transgenic (black triangle) plants of a segregating population produced by backcrossing a transgenic plant overexpressing the *GA20ox1 gene* to the wild-type line B104 maize, including leaf length (s-shaped) and LER (bell-shaped) function plots for both groups using the mean values of the function parameters. **(B)** Leaf length measurements of *Miscanthus sinensis* 'Goliath' (black triangle) and *M. x giganteus* (white triangle) plants, including leaf length and LER function plots for both groups using the mean values of the function parameters. **(C)** Leaf length and LER function plots for *Brachypodium distachyon* inbred lines Bd3-1 (full line), Bd21 (dotted), Bd21-3 (dashed) and Bd2-3 (dash dotted) using the mean values of the function parameters. For the sake of clarity, the individual leaf length measurements are not shown in this case.

Brachypodium inbred lines that are part of that study: Bd21, Bd21-3, Bd2-3 and Bd3-1 (Figure 3C).

Leaf growth analysis of these four genotypes with LEAF-E revealed distinct leaf growth characteristics. Based upon final leaf length, two groups can be distinguished. Bd21 and Bd3-1 have short leaves and Bd21-3 and Bd2-3 have long leaves (Figure 3C, Table 3). The length of the leaves is determined by both LER and LED. For Bd21, a low LER is probably the underlying factor of the shorter leaves (Table 3). This is in contrast to Bd3-1 which, like Bd21, has short leaves but a *LERmax* that is similar to those of the genotypes with longer leaves (Table 3). Parameters for LED are the smallest for Bd3-1. Thus, the leaf of Bd3-1 is short, most likely due to a short growing period. Bd21-3 and Bd2-3 both have long leaves, a high *LERmax* and similar LED. Despite these similarities, based upon our analysis, we can conclude that the 3rd leaf of Bd21-3 plants starts and finishes its growth significantly earlier in thermal time than that of Bd2-3 (Table 3). These results suggest that different mechanisms might drive leaf growth in different accessions. In *Arabidopsis* it was found that at least five different mechanisms contribute to final leaf size [31].

Table 2 Comparison of leaf elongation in two *Miscanthus* genotypes from different species

Growth parameter			*M. Sinensis* 'Goliath' (mean ± SE)	*M. x giganteus* (mean ± SE)	Difference in mean$^{(+)}$	
	Lm	(mm)	1140 ± 46	923 ± 26	−217	**
	LERmax	($mm°C^{-1}d^{-1}$)	4.2 ± 0.2	3.8 ± 0.2	−0.4	NS
Thermal time points	t100	(°Cd)	95 ± 4	167 ± 4	71	***
	t20%	(°Cd)	141 ± 7	207 ± 5	65	***
	t50%	(°Cd)	230 ± 8	294 ± 8	64	***
	tm	(°Cd)	240 ± 10	320 ± 8	80	***
	t90%	(°Cd)	350 ± 10	400 ± 13.8	50	NS
	te	(°Cd)	425 ± 10	461 ± 18	37	NS
Leaf elongation duration	LED(100-e)	(°Cd)	329 ± 8	295 ± 18	−35	NS
	LED(20%-90%)	(°Cd)	209 ± 5	194 ± 11	−15	NS
	LED(20%-e)	(°Cd)	283 ± 7	255 ± 16	−29	NS

The results are based on the analysis of nine *M. sinensis* 'Goliath' and eight *M. x giganteus* plants. Lm: final leaf length; LERmax: maximal leaf elongation rate; t20%, t50%, t90%, te: time points at which the leaf reaches 20%, 50%, 90% and 100% of the final leaf length, respectively; tm: time point at which the leaf reaches LERmax; LEDs: leaf elongation durations between above stated thermal time points.

+Statistical significance based on student t-test of *M. sinensis* 'Goliath' (n = 9) vs *M. x giganteus* (n = 8), **p < 0.01, ***p < 0.001, NS non-significant. Applied base temperature for thermal time calculation = 8°C, Mean of overall R^2 values = 0.9932 (0.9695-0.9989).

Fitting of cell length measurements along the leaf axis of maize overexpressing *GA20ox1* using the beta sigmoid function

The cell length profile along the longitudinal axis of an actively growing grass leaf also displays a sigmoid pattern [8]. This sigmoidal profile is determined by the spatial distribution of cells in different stages of differentiation along the leaf axis: a number of dividing cells of small size at the leaf base, a stretch of cells that undergo elongation and thus increase in length when being pushed towards the leaf tip, and finally the tip of the leaf that is made up of cells that have reached their final length. We applied LEAF-E, adapted to use the extended version of the beta sigmoid function (Equation 3), to fit cell length measurements of dataset 1b. We found that fitting was successful and resulted in overall R^2-values ranging from 0.8420 up to 0.8749 for cell length measurements of the *GA20ox1* overexpressing and control plants. Knowing that cell lengths can vary considerably, even for adjacent cells of the same cell file in one leaf (Figure 4), these R^2-values are noticeably high. However, a goodness of fit assessment on cell length measurements is, to our knowledge, not reported in literature, making any comparison difficult. We explored the suitability of three other commonly used sigmoidal functions, namely the Weibull, the Logistic and the Gompertz functions for fitting cell length data of the control plants. The fits of cell length data resulted in R^2-values of 0.8504 ± 0.0054 for

Table 3 Comparison of leaf elongation in four *Brachypodium* inbred lines

Growth parameter			Bd3-1 (mean ± SE)$^{(+)}$	Bd21 (mean ± SE)$^{(+)}$	Bd21-3 (mean ± SE)$^{(+)}$	Bd2-3 (mean ± SE)$^{(+)}$
	Lm	(mm)	$94^{a} \pm 3$	$91^{a} \pm 1$	$113^{b} \pm 3$	$113^{b} \pm 2$
	LERmax	($mm°C^{-1}d^{-1}$)	$1.44^{ab} \pm 0.05$	$1.29^{a} \pm 0.03$	$1.52^{b} \pm 0.05$	$1.51^{b} \pm 0.03$
Thermal time points	t20%	(°Cd)	$162^{a} \pm 3$	$163^{a} \pm 2$	$148^{b} \pm 3$	$166^{a} \pm 2$
	t20	(°Cd)	$164^{a} \pm 3$	$166^{a} \pm 2$	$145^{b} \pm 3$	$163^{a} \pm 2$
	t50%	(°Cd)	$189^{a} \pm 2$	$193^{a} \pm 2$	$178^{b} \pm 3$	$197^{a} \pm 2$
	tm	(°Cd)	$202^{a} \pm 2$	$206^{a} \pm 2$	$192^{b} \pm 3$	$211^{a} \pm 2$
	t90%	(°Cd)	$217^{ab} \pm 1$	$222^{bc} \pm 2$	$210^{a} \pm 4$	$229^{c} \pm 2$
	te	(°Cd)	$230.8^{ab} \pm 0.8$	$238^{bc} \pm 2$	$226^{a} \pm 4$	$245^{c} \pm 2$
Leaf elongation durations	LED(20-e)	(°Cd)	$67^{a} \pm 3$	$72^{a} \pm 1$	$81^{b} \pm 2$	$82^{b} \pm 2$
	LED(20%-90%)	(°Cd)	$69^{a} \pm 2$	$75^{ab} \pm 1$	$78^{b} \pm 1$	$79^{b} \pm 1$
	LED(20%-e)	(°Cd)	$55^{a} \pm 2$	$59^{ab} \pm 1$	$62^{b} \pm 1$	$63^{b} \pm 1$

Bd3-1 (n = 7), Bd21 (n = 10) plants, Bd21-3 (n = 7) plants, Bd2-3 (n = 10). Lm: final leaf length; LERmax: maximal leaf elongation rate; t20%, t50%, t90%, te: time points at which the leaf reaches 20%, 50%, 90% and 100% of the final leaf length, respectively; tm: time point at which the leaf reaches LERmax; LEDs: leaf elongation durations between above stated thermal time points.

+Statistical significance indicated with distinct letters based on ANOVA and Scheffé Post hoc test (p < 0.05) between lines Bd3-1 (n = 7), Bd21 (n = 10) plants, Bd21-3 (n = 7) plants, Bd2-3 (n = 10), applied base temperature for thermal time calculation = 11°C, Mean of overall R^2 values = 0.9993 (0.9871-0.9993).

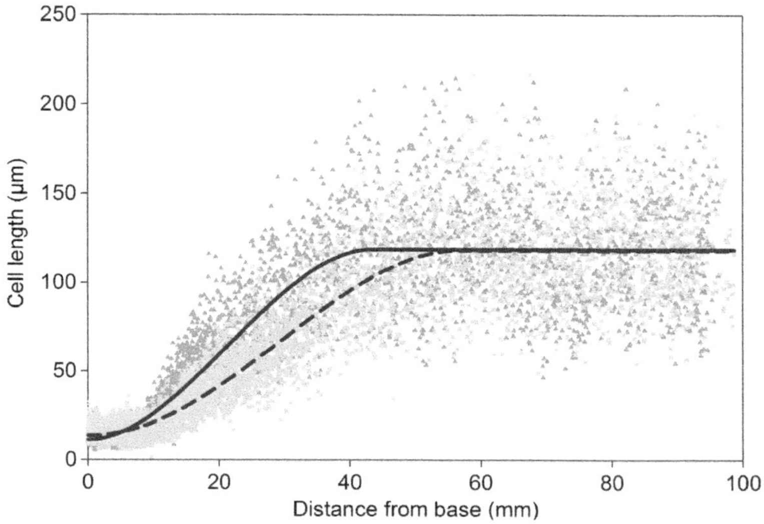

Figure 4 Effect of *GA20ox1* overexpression on the cell length profile of the 4th leaf in maize. The cell length profile along the axis of the 4th leaf is shown for three non-transgenic (triangle, dark gray) and three transgenic (x, light gray) plants. The S-shaped curves are function plots using the mean values of the fitted function parameters for non-transgenic (full) and transgenic (dashed) profiles.

Weibull, 0.8485 ± 0.0055 for logistics, 0.8499 ± 0.0053 for beta sigmoid and 0.8495 ± 0.0054 for Gompertz. The goodness of fit was also evaluated by linear regression of the estimated to the measured values (Additional file 6) and again found highly similar R^2-values. For the beta sigmoid function, we investigated the deviation of the data points from the fitted curve and could detect a slight underestimation of cell lengths in the very first 3 mm from the leaf base (Additional file 3D). This is because the cells in the first few millimeter of the division zone are slightly longer than those in the middle of the division zone; this hyperbolic profile at the leaf basis cannot be fitted with the functions that we used here, since they assume an increment in cell lengths. Additionally, the value of the residuals increased with the distance from the leaf base (Additional files 3D and 6). For this dataset, the use of the beta sigmoid function in the form of Equation 3 is preferred over the other investigated functions because the function parameters themselves provide relevant biological information. However, the exploration of mathematical functions that are more appropriate for fitting cell lengths along the longitudinal axis of grass leaves should be the topic of future research.

The cell length profile along the leaf axis in maize plants overexpressing *GA20ox1* (dataset 1b) was analyzed by Nelissen and coworkers [28] using a polynomial fit approach similar to Rymen and coworkers [3]. Using LEAF-E, we determined that the size of the division zone, defined as the stretch of cells near the leaf base by which function fitted cell lengths do not exceed 40 µm (obtained by DAPI-staining of cells along the proximal-distal axis), is on average 33% longer ($p < 0.01$) in *GA20ox1* overexpressing leaves as compared to non-transgenic leaves (Figure 4, Table 4). We estimated that the elongation zone, defined as the distance between the end of the division zone and the function parameter *Pe* (position in the leaf where cells reach their maximal length, equivalent to *te* for leaf elongation), is 29% longer ($p < 0.01$) in *GA20ox1* overexpressing leaves. Moreover, we can state that the position where maximal cell elongation occurs (*Pm*, equivalent to *tm* for leaf elongation) is situated 8 mm (or 38%) further away from the leaf base in *GA20ox1* overexpressing leaves as compared to non-transgenic plants and that the mature cell length *Lm* (mm) did not differ ($p = 0.92$). These findings are in accordance with results obtained earlier by Nelissen and coworkers [28]. The analysis using LEAF-E thus allowed a more detailed study of the cell length profile, which facilitated the detection of a longer elongation zone in *GA20ox1* overexpressing plants and a shift in the position of maximal cell elongation, that could not be quantified previously. Nevertheless, it might still be necessary to validate the estimation of the size of the division zone using DAPI staining or a similar method, as was performed here. We believe that function fitting, using LEAF-E, can be very useful and might even be a necessary step for the analysis of cell length profiles in the grass leaf in future experiments.

Conclusions

Here, we provide a tool that we named LEAF-E for fast and straightforward analysis of grass leaf growth data using nonlinear regression in a simple Microsoft Excel

Table 4 Effect of GA20ox1 overexpression on the maize cell length profile

Parameter			*AtGA20ox1* OE (mean ± SE)	Control (mean ± SE)	Difference in mean(+)	
Cell length	Lb	(μm)	13.2 ± 1.7	10.9 ± 0.3	22%	NS
	Lm	(μm)	118.1 ± 0.6	118.5 ± 3.8	0%	NS
Position along the leaf axis	Pm	(mm)	29.5 ± 2.1	21.3 ± 0.8	38%	**
	Pe	(mm)	56.6 ± 2.5	43.4 ± 0.3	−30%	**
Zones along the leaf axis	division zone (cell < =40 μm)	(mm)	19.5 ± 0.5	14.6 ± 0.3	33%	**
	elongation zone (cell > 40 μm - Pe)	(mm)	37.1 ± 1.0	28.7 ± 0.9	−29%	**

A segregating population produced by backcrossing (BC) a transgenic plant overexpressing the *Arabidopsis thaliana GIBBERELLIC ACID 20 OXIDASE1 (GA20ox1)* gene to the wild-type line B104 was analyzed for the cell lengths profile along the leaf axis. The results are based on the analysis of three transgenic and three non-transgenic BC1 plants. Lb: initial cell length Lm: final cell length; Pm: position along the leaf axis with maximal cell elongation rate; Pe: position along the leaf axis where cells reach their final length.
+Statistical significance based on student t-test of non-transgenic plants (n = 3) vs *GA20ox1* overexpression (n = 3), $^{**}p < 0.01$, NS non-significant.

format. We automated the fitting procedure using Excel 2010 and the Solver function (32 bit). The results of fitting the beta sigmoid function to the leaf length measurements are stored in tabular form and can easily be analyzed in standard statistical software programs in search of differences due to the applied treatments or to explore inter-genotypic or inter-population differences.

We applied LEAF-E to three datasets containing leaf length measurements of maize, *Miscanthus* and *Brachypodium*. In *Miscanthus* and *Brachypodium*, we have shown that LEAF-E is an appropriate tool for data analysis and that the analyzed species and genotypes display distinct leaf growth characteristics. In maize, the changes in both leaf elongation and cell length profile along the leaf axis as a result of enhanced GA levels, previously demonstrated by Nelissen and coworkers [28], were confirmed. In addition, we demonstrated that using LEAF-E, the timing of leaf growth can be studied, thereby facilitating the detection of dissimilarities in the timing of leaf growth that could not be quantified using other approaches. Furthermore, analysis with LEAF-E allows for a robust calculation of *LERmax* and LED, which leads to reliable detection of significant changes. Therefore, we propose LEAF-E as an excellent tool for comparing leaf growth behavior in different genotypes or to analyze the response of specific genotypes to a treatment.

Methods

Datasets used for validation

Dataset 1a

The dataset on leaf elongation of the maize B104 inbred line overexpressing the *Arabidopsis thaliana GIBBERELLIC ACID 20 OXIDASE1* (*GA20ox1*) gene, previously described by Nelissen and coworkers [28] was reanalyzed here. A segregating population produced by backcrossing the overexpression line (hemizygous for the transgenic event) to the wild-type B104 inbred line and consisting of 9 non-transgenic and 11 transgenic plants was used.

To determine leaf elongation rates, the length of the 4th leaf of transgenic and non-transgenic plants was measured daily until complete development, as previously described [28]. The plants were grown in a growth chamber at 24°C. Here we used a base temperature of 10°C for thermal time (Growing degree days, GDD) calculations. For further details about growth conditions see [28].

Dataset 1b

The same segregating population, used to generate dataset 1a, was previously used by Nelissen and coworkers [28] for the analysis of cell lengths along the leaf axis, based upon methods previously described [32]. In short, the 4th leaf was harvested two days after appearance from the pseudo-stem (stem-like structure composed of concentric rolled or folded blades and sheaths that surround the growing point). At this time point, the ligule is only a few mm away from the base of the plant. The length of cell files adjacent to stomatal rows along the proximal-distal axis was measured using a DIC microscope (AxioImager, Zeiss, USA), and image analysis software (AxioVision, Zeiss, USA). The size of the division zone was determined as the distance between the base and the most distally observed mitotic figure in DAPI-stained leaves along the proximal-distal axis, with a fluorescence microscope (AxioImager, Zeiss, USA). Here we reanalyzed the cell length measurements.

Dataset 2

Eight *Miscanthus* x *giganteus* and nine *M. sinensis* 'Goliath' plants were grown at 20°C (average temperature over the measuring period was 19.1°C) in a greenhouse, in Melle, Belgium, in September 2012 with no supplementary light. Plants were grown from rhizome cuttings in 2-l pots and were hand-watered and not fertilized during the experiment. The rhizomes were excavated during the winter of 2011 and stored in a cold room at 3°C until the start of the experiment. The length of the 4th leaf was measured five times a week (from leaf tip to soil level).

Based on our observations at this developmental stage, the contribution of internode elongation to the height measurements as leaf lengths can be neglected and transition to flowering has not yet occurred. The measurements were spread over a time period of approximately four weeks. The calculation of thermal time was based on the average air temperature in the greenhouse taking into account a base temperature of 8°C, based on Farrell and coworkers [33].

Dataset 3

The *Brachypodium distachyon* inbred lines Bd21, Bd2-3 and Bd3-1 were provided by David F. Garvin from the USDA-ARS (Minnesota, US), and line Bd21-3 was provided by Richard Sibout from INRA-IJPB (Versailles, France). Plants were grown in rootrainers (Haxnicks®, UK) in biological replicates (n = 10, 10, 7, 9 respectively) in a greenhouse at an average temperature of 21°C in Melle, Belgium, August 2012 with no supplementary light. To calculate the thermal time, a base temperature of 10°C was used. Fertilizer was added with the water supply: conductivity Ec = 1mS/cm; water soluble fertilizer Poly-feed (Haifa, Belgium) (N, P_2O_5, K_20; 20:5:20 + 3 MgO). Measurements were taken from the tip of the 3rd leaf to its basal level on a daily basis, for a period of 10 days. Based on our observations at this developmental stage, the contribution of internode elongation to the height measurements as leaf lengths can be neglected and transition to flowering has not yet occurred.

A mathematical function for fitting leaf length measurements

For the estimation of leaf growth parameters we used the beta sigmoid function for determinate growth, inspired by the Euler integral, in the form of Equation 1. This function was used previously by Auzanneau and coworkers [7] and Verdenal and coworkers [19] to model leaf growth after cutting in *Lolium perenne*. In short, the leaf length *L* (mm) at a given moment in development *t* (°Cd) is determined by final leaf length *Lm* (mm) and three particular points in leaf development, expressed as units of thermal time or growing degree days (°Cd). Thermal time is a summation of cumulative differences between daily mean temperature and a specified base temperature, below which the plant does not grow or grows very slowly [22]. These thermal time points are the moment at which leaf growth starts *t0* (°Cd), the moment of maximal leaf growth rate *tm* (°Cd) and the moment at which leaf growth ceases *te* (°Cd). Estimations of *t0* often result in negative values that are biologically not relevant [7]. Therefore, in the experiments in which seedlings were involved (maize and *Brachypodium*), we assumed that *t0* = 0 was at the moment of sowing. In the case of *Miscanthus*, *t0* = 0 was assumed to be at the moment of potting the rhizomes (no visible leaves at this stage).

$$L = Lm.\left(1+\frac{(te-t)}{(te-tm)}\right).\left(\frac{(t-t0)}{(te-t0)}\right)^{\frac{(te-t0)}{(te-tm)}} \quad (1)$$

Equation 1. Beta sigmoid function for fitting leaf length, modified from [7]. Function is applicable for $t0 \leq t \leq te$ and $t0 \leq tm < te$. For $t > te$, Equation 1 is reduced to $L = Lm$.

The leaf elongation rate (LER) at any given moment in leaf development *t* (°Cd) can be calculated from the LER function (Equation 2), which is the first derivative of Equation 1. From this equation we determined the maximum leaf elongation rate or *LERmax* (mm/°Cd), as the LER at *tm*.

$$\frac{dL}{dt} = Lm.\left(\frac{\left(1+\frac{(te-t)}{(te-tm)}\right).\left(\frac{(t-t0)}{(te-t0)}\right)^{\left(\frac{(te-t0)}{(te-tm)}-1\right)}}{(te-tm)} - \frac{\left(\frac{(t-t0)}{(te-tc)}\right)^{\frac{(te-t0)}{(te-tm)}}}{(te-tm)}\right) \quad (2)$$

Equation 2. Leaf elongation rate function, modified from [7].

As Equation 1 is a continuous function, it allows calculating the leaf length *L* (mm) at any given moment in the leaf elongation period *t*, and vice versa (Figure 2). Therefore, in addition to the parameters *Lm*, *t0*, *tm* and *te*, we estimated a set of parameters that can be biologically relevant. For the maize dataset 1a we estimated the time point *t100*, the moment at which the leaf length is 100 mm. The *t100* time point, which is early in development, was chosen since it is close to the moment at which the leaf emerges from the pseudo-stem in maize non-transgenic B104 plants. Furthermore, we estimated *t20%*, *t50%* and *t90%* (°Cd), which are the moments at which the leaf reaches 20%, 50% and 90% of its final length, respectively. For *Brachypodium* (dataset 2), we replaced the *t100* parameter by *t20* (°Cd), the moment at which the leaf reaches 20 mm in length, to accommodate the smaller size of the *Brachypodium* leaf. These extra parameters allow the comparison of responses to different treatments or detect inter-genotypic differences in leaf growth development.

Expressing leaf growth as durations of thermal time, the leaf elongation duration or LED, allows describing leaf growth in a fluctuating environment [15].

On that account, various LEDs were explored. For example, *LED(20%-90%)* defines the leaf growth duration between reaching 20% and 90% of its final length. Parameters *LED(100-e)* and *LED(20%-e)* were defined similarly.

A mathematical function for fitting cell length measurements

Often, analysis of leaf growth involves cell length measurements along the leaf axis, providing insight in the sizes of dividing, elongating and mature cells, and of the leaf zones encompassing these three cell types [3,8,9,15,24,25]. Therefore, we assessed the fitting of cell length measurements along the leaf axis in maize with the beta sigmoid function. However, since in this case cells in the division zone have an initial length before proceeding to the elongation phase and eventually toward mature cells, a different equation had to be used as Equations 1 and 2 assume zero values at the beginning of growth. Yin and coworkers [18] describe an extended version of the beta sigmoid function that allows taking into account the initial length of the cell, *Lb* (mm) (Equation 3). In this case the data points do not represent a time series, but a positional-series of cell lengths along the leaf axis, taking the leaf base as position zero. For ease of interpretation, the symbols of the variable *t* and parameters *tm* and *te* from the original equation have been converted into the positions *p* (mm) along the leaf axis, starting from the base towards the tip. The position at which maximal cell elongation occurs is denoted as *Pm* (mm), and the position at which the cells cease to elongate is denoted as *Pe* (mm) (see Equation 3).

$$L = Lb + (Lm - Lb).\left(1 + \frac{(Pe - p)}{(Pe - Pm)}\right).\left(\frac{p}{Pe}\right)^{\frac{Pe}{(Pe - Pm)}} \qquad (3)$$

Equation 3. Extended version of the beta sigmoid function modified from [18] to fit cell length measurements along the leaf axis.

LEAF-E: function fitting using a Microsoft excel spreadsheet and the SOLVER function

The fitting of leaf growth data using the beta sigmoid function was performed using Excel 2010 and the Solver function (32 bit) according to Brown [34]. The automation of the procedure, as described below, in the form of a macro is innovative. Each row in the datasheet contains the data of one individual leaf (ordered in a time series), the starting values of the parameters of the model, and the formulae to extract the necessary statistical components for the calculation of the least square estimates following Neter and coworkers [35]. First, the macro checks for non-empty rows. When a non-empty row is found, the model is fitted to the data by minimizing the sum of squares of the errors iteratively, and changing the starting values of the parameters at each step. Per row (=leaf), all values described in the previous sections are calculated and stored in a tabular form for further statistical analysis. In addition, a graph showing the original data points, the fitted growth curve and the function parameters is generated automatically. This enables the evaluation of the correlation coefficient and visual interpretation of the goodness of fit. It also provides an easy way to check for miss fits or errors in the data. Miss fitting can occur when the Solver function fails to minimize the sum of squares of the errors using a particular set of starting values. Accordingly, the procedure can be repeated using more appropriate starting values. This way of working guarantees a fast analysis, robust estimation of growth parameters and provides a table in standardized format containing the resulting parameters and derived parameters per leaf. The output data can easily be analyzed in search of differential responses using standard statistical software, depending on the experimental setup. We gave this Excel tool the name LEAF-E. LEAF-E is included as additional file to this article (Additional files 4 and 5, with a user manual on the first sheet). Additional file 4 includes test data of *M. sinensis* 'Goliath' leaf growth and default starting values prior running the macro and Additional file 5 includes the results of the fitting procedure and graphs after running the automated fitting procedure.

Statistical analysis

The statistical analysis of the derived growth parameters comprised a student t-test on datasets containing only two genotypes (datasets 1a, 1b and 2) and an ANOVA followed by post hoc Scheffé tests for datasets containing more than two genotypes (dataset 3). All these analyses were carried out in the software package *STATISTICA* version 11 (Statsoft Inc., USA). For the analysis of the goodness of fit of cell length data of dataset 1b, the fitting of the beta sigmoid and three other commonly used growth functions (Weibull, Gompertz and Logistic) and linear regression of the predicted to the observed values was carried out in *SIGMAPLOT* version 12_5 (Systat Software Inc., USA).

Additional files

Additional file 1: Function fitting of leaf length measurements of maize, *Miscanthus* and *Brachypodium* using LEAF-E on a plant-by-plant basis. The PowerPoint presentation shows a plot for every single plant of every dataset, showing the individual leaf length measurements, the fit of the beta sigmoid function and its R^2-value, the estimated LER curve and the calculated LER (calculated as leaf length increase between two consecutive measurements divided by the respective thermal time interval). Maize GA: transgenic plant overexpressing the *GA20ox1 gene*, Maize control: wild-type B104 line, Bd: *Brachypodium distachyon*, F(t): fitted curve plotted in thermal time, R^2 F(t): R^2 value of the fit, Fler(t): LER curve, plotted in thermal time, LER calc: calculated LER (see above).

Additional file 2: Scatter plots and linear regression for datasets of leaf length measurements. Plots show estimated to measured leaf lengths (A, D, G), estimated LER to calculated LER (calculated as leaf length increase between two consecutive measurements divided by the respective thermal time interval) (B, E, H) and estimated *LERmax* to *LERmax* determined as maximal value of the profile of calculated LER (C, F, I), all on a plant-by-plant basis, for datasets of maize (dataset 1a: A, B, C), *Miscanthus* (dataset 2: D, E, F) and *Brachypodium* (dataset 3: G, H, I).

Additional file 3: Distribution of the residuals for the fitting of the beta sigmoid function to leaf length data of maize (A), *Miscanthus* spp. (B) and *Brachypodium* spp. (C) and cell length data of maize (D) using LEAF-E.

Additional file 4: The Microsoft excel spreadsheet named LEAF-E with test data (*M. sinensis* 'Goliath leaf length measurements) and default starting values prior running the automated fitting procedure.

Additional file 5: The Microsoft excel spreadsheet named LEAF-E with test data (*M. sinensis* 'Goliath leaf length measurements), graphs and results of the fitted growth parameters after running the automated fitting procedure. The first sheet contains a user manual.

Additional file 6: Comparison of goodness of fit of the Beta sigmoid (A), Weibull (B), Gompertz (C) and Logistic (D) functions for the cell length profile of the 4th leaf in maize. The plots show the linear regression of estimated versus measured cell lengths of all data points of dataset 1b.

Competing interests
The authors declare that they have no competing interests.

Authors' contributions
WV wrote this manuscript and was involved in conception and design of LEAF-E and acquisition, analysis and interpretation of the data. PL was involved in conception and design of LEAF-E and is responsible for the implementation and optimization of the tool as a macro in Microsoft Excel and critical revision of the manuscript. HN was involved in providing data, interpretation of data and critical revision of the manuscript. IR-R has made substantial contributions to conception and design of LEAF-E, interpretation of data and critical revision of the manuscript. DI was involved in critical revision of the manuscript and has given final approval of the version to be published. HM has made substantial contributions to conception and design of LEAF-E, was involved in interpretation of data and critical revision of the manuscript and has given final approval of the version to be published. All authors read and approved the final manuscript.

Acknowledgements
We thank the Agency for Innovation by Science and Technology in Flanders (IWT) for funding and Simon Fonteyne (ILVO, Belgium) for providing leaf measurements of *Miscanthus* spp. We also like to thank David Garvin (USDA-ARS, US) and Richard Sibout (INRA-IJPB, France) for providing Brachypodium seed stocks.

Author details
[1]Department of Plant Systems Biology, VIB, Technologiepark 927, 9052 Gent, Belgium. [2]Department of Plant Biotechnology and Bioinformatics, Ghent University, Technologiepark 927, 9052 Gent, Belgium. [3]Plant Sciences Unit – Growth and Development, Institute for Agricultural and Fisheries Research (ILVO), Caritasstraat 21, 9090 Melle, Belgium.

References
1. Fricke W, McDonald AJS, Mattson-Djos L: **Why do leaves and leaf cells of N-limited barley elongate at reduced rates?** *Planta* 1997, **202:**522–530.
2. Munns R, James RA: **Screening methods for salinity tolerance: a case study with tetraploid wheat.** *Plant Soil* 2003, **253:**201–218.
3. Rymen B, Fiorani F, Kartal F, Vandepoele K, Inzé D, Beemster GTS: **Cold nights impair leaf growth and cell cycle progression in maize through transcriptional changes of cell cycle genes.** *Plant Physiol* 2007, **143:**1429–1438.
4. Clifton-Brown JC, Jones MB: **The thermal response of leaf extension rate in genotypes of the C_4-grass *Miscanthus*: an important factor in determining the potential productivity of different genotypes.** *J Exp Bot* 1997, **48:**1573–1581.
5. Lafarge T, Tardieu F: **A model co-ordinating the elongation of all leaves of a sorghum cultivar was applied to both Mediterranean and Sahelian conditions.** *J Exp Bot* 2002, **53:**715–725.
6. Verelst W, Bertolini E, De Bodt S, Vandepoele K, Demeulenaere M, Pè ME, Inzé D: **Molecular and physiological analysis of growth-limiting drought stress in *Brachypodium distachyon* leaves.** *Mol Plant* 2013, **6:**311–322.
7. Auzanneau J, Huyghe C, Escobar-Gutiérrez AJ, Julier B, Gastal F, Barre P: **Association study between the gibberellic acid insensitive gene and leaf length in a Lolium perenne L. synthetic variety.** *BMC Plant Biol* 2011, **11:**183.
8. Parent B, Conejero G, Tardieu F: **Spatial and temporal analysis of non-steady elongation of rice leaves.** *Plant Cell Environ* 2009, **32:**1561–1572.
9. Fiorani F, Beemster GTS, Bultynck L, Lambers H: **Can meristematic activity determine variation in leaf size and elongation rate among four *Poa* species? A kinematic study.** *Plant Physiol* 2000, **124:**845–855.
10. Arredondo JT, Schnyder H: **Components of leaf elongation rate and their relationship to specific leaf area in contrasting grasses.** *New Phytol* 2003, **158:**305–314.
11. Bultynck L, Ter Steege MW, Schortemeyer M, Poot P, Lambers H: **From individual leaf elongation to whole shoot leaf area expansion: a comparison of three *Aegilops* and two *Triticum* species.** *Ann Bot* 2004, **94:**99–108.
12. Reymond M, Muller B, Tardieu F: **Dealing with the genotype x environment interaction via a modelling approach: a comparison of QTLs of maize leaf length or width with QTLs of model parameters.** *J Exp Bot* 2004, **55:**2461–2472.
13. Chenu K, Chapman SC, Hammer GL, McLean G, Ben Haj Salah H, Tardieu F: **Short-term responses of leaf growth rate to water deficit scale up to whole-plant and crop levels: an integrated modelling approach in maize.** *Plant Cell Environ* 2008, **31:**378–391.
14. Dignat G, Welcker C, Sawkins M, Ribaut JM, Tardieu F: **The growths of leaves, shoots, roots and reproductive organs partly share their genetic control in maize plants.** *Plant Cell Environ* 2013, **36:**1105–1119.
15. Tardieu F, Reymond M, Hamard P, Granier C, Muller B: **Spatial distributions of expansion rate, cell division rate and cell size in maize leaves: a synthesis of the effects of soil water status, evaporative demand and temperature.** *J Exp Bot* 2000, **51:**1505–1514.
16. Paine CET, Marthews TR, Vogt DR, Purves D, Rees M, Hector A, Turnbull LA: **How to fit nonlinear plant growth models and calculate growth rates: an update for ecologists.** *Methods Ecol Evol* 2012, **3:**245–256.
17. Motulsky HJ, Ransnas LA: **Fitting curves to data using nonlinear regression: a practical and nonmathematical review.** *FASEB J* 1987, **1:**365–374.
18. Yin X, Goudriaan J, Lantinga EA, Vos J, Spiertz HJ: **A flexible sigmoid function of determinate growth.** *Ann Bot* 2003, **91:**361–371.
19. Verdenal A, Combes D, Escobar-Gutiérrez AJ: **A study of ryegrass architecture as a self-regulated system, using functional-structural plant modelling.** *Funct Plant Biol* 2008, **35:**911–924.
20. Goudriaan J, Van Laar HH: *Modelling Potential Crop Growth Processes: Textbook with Exercises.* Dordrecht; Boston: Kluwer Academic Publishers; 1994.
21. Neild R, Newman J: **Growing season Characteristics and Requirements in the Corn Belt.** In *Natl Corn Handb 40.* West Lafayette, IN 47907: Purdue University Cooperative Extension Service; 1990.
22. Atwell B, Kriedemann P, Turnbull C: *Plants in Action: Adaptation in Nature, Performance in Cultivation.* MacMillan Education Australia Pty Ltd: Melbourne, Australia; 1999.
23. International Brachypodium Initiative: **Genome sequencing and analysis of the model grass *Brachypodium distachyon*.** *Nature* 2010, **463:**763–768.
24. Ben-Haj-Salah H, Tardieu F: **Temperature affects expansion rate of maize leaves without change in spatial distribution of cell length (analysis of the coordination between cell division and cell expansion).** *Plant Physiol* 1995, **109:**861–870.
25. Muller B, Reymond M, Tardieu F: **The elongation rate at the base of a maize leaf shows an invariant pattern during both the steady-state elongation and the establishment of the elongation zone.** *J Exp Bot* 2001, **52:**1259–1268.
26. Bouchabké O, Tardieu F, Simonneau T: **Leaf growth and turgor in growing cells of maize (*Zea mays* L.) respond to evaporative demand under moderate irrigation but not in water-saturated soil.** *Plant Cell Environ* 2006, **29:**1138–1148.
27. Durand J-L, Schäufele R, Gastal F: **Grass leaf elongation rate as a function of developmental stage and temperature: morphological analysis and modelling.** *Ann Bot* 1999, **83:**577–588.
28. Nelissen H, Rymen B, Jikumaru Y, Demuynck K, Van Lijsebettens M, Kamiya Y, Inze D, Beemster GT: **A local maximum in gibberellin levels regulates maize leaf growth by spatial control of cell division.** *Curr Biol* 2012, **22:**1183–1187.
29. Greef JM, Deuter M: **Syntaxonomy of *Miscanthus x giganteus* GREEF et DEU.** *Angew Bot* 1993, **67:**87–90.

30. Hodkinson TR, Renvoize SA, Chase MW: **Systematics of *Miscanthus*.** *Asp Appl Biol* 1997, **49:**189–198.
31. Gonzalez N, Vanhaeren H, Inze D: **Leaf size control: complex coordination of cell division and expansion.** *Trends Plant Sci* 2012, **17:**332–340.
32. Rymen B, Coppens F, Dhondt S, Fiorani F, Beemster GTS: **Kinematic Analysis of Cell Division and Expansion.** In *Plant Developmental Biology (Methods in Molecular Biology, Vol 655).* Edited by Hennig L, Köhler C. New York: Humana Press; 2010:203–227. [Walker JM (Series Editor).
33. Farrell AD, Clifton-Brown JC, Lewandowski I, Jones MB: **Genotypic variation in cold tolerance influences the yield of *Miscanthus*.** *Ann Appl Biol* 2006, **149:**337–345.
34. Brown AM: **A step-by-step guide to non-linear regression analysis of experimental data using a Microsoft Excel spreadsheet.** *Comput Methods Programs Biomed* 2001, **65:**191–200.
35. Neter J, Kutner M, Nachtsheim C, Wasserman W: *Applied Linear Statistical Models.* 4th edition. McGraw-Hill Education: Columbus; 1996.

6

AGROBEST: an efficient *Agrobacterium*-mediated transient expression method for versatile gene function analyses in *Arabidopsis* seedlings

Hung-Yi Wu[1,2], Kun-Hsiang Liu[3,4], Yi-Chieh Wang[1], Jing-Fen Wu[1], Wan-Ling Chiu[3,4,5], Chao-Ying Chen[2], Shu-Hsing Wu[1], Jen Sheen[3,4] and Erh-Min Lai[1,2,3,4*]

Abstract

Background: Transient gene expression via *Agrobacterium*-mediated DNA transfer offers a simple and fast method to analyze transgene functions. Although *Arabidopsis* is the most-studied model plant with powerful genetic and genomic resources, achieving highly efficient and consistent transient expression for gene function analysis in *Arabidopsis* remains challenging.

Results: We developed a highly efficient and robust *Agrobacterium*-mediated transient expression system, named AGROBEST (*Agrobacterium*-mediated enhanced seedling transformation), which achieves versatile analysis of diverse gene functions in intact *Arabidopsis* seedlings. Using β-glucuronidase (GUS) as a reporter for *Agrobacterium*-mediated transformation assay, we show that the use of a specific disarmed *Agrobacterium* strain with *vir* gene pre-induction resulted in homogenous GUS staining in cotyledons of young *Arabidopsis* seedlings. Optimization with AB salts in plant culture medium buffered with acidic pH 5.5 during *Agrobacterium* infection greatly enhanced the transient expression levels, which were significantly higher than with two existing methods. Importantly, the optimized method conferred 100% infected seedlings with highly increased transient expression in shoots and also transformation events in roots of ~70% infected seedlings in both the immune receptor mutant *efr-1* and wild-type Col-0 seedlings. Finally, we demonstrated the versatile applicability of the method for examining transcription factor action and circadian reporter-gene regulation as well as protein subcellular localization and protein–protein interactions in physiological contexts.

Conclusions: AGROBEST is a simple, fast, reliable, and robust transient expression system enabling high transient expression and transformation efficiency in *Arabidopsis* seedlings. Demonstration of the proof-of-concept experiments elevates the transient expression technology to the level of functional studies in *Arabidopsis* seedlings in addition to previous applications in fluorescent protein localization and protein–protein interaction studies. In addition, AGROBEST offers a new way to dissect the molecular mechanisms involved in *Agrobacterium*-mediated DNA transfer.

Keywords: *Agrobacterium*, *Arabidopsis*, Transient transformation, Gene expression, Innate immunity, Gain-of-function

Background

Agrobacterium-mediated DNA transfer is currently the most facile and versatile method to deliver gene constructs into the nucleus for gene function analysis in diverse plant species [1-3]. Although stable integration of physiologically active and regulated transgenes is the ultimate goal, transient gene expression via *Agrobacterium*-mediated DNA transfer in different plant tissues offers a simple and fast method to analyze transgene functions, which is amenable for high-throughput screens. The transient expression assay is also ideal for systematic dissection of the exquisite and complex processes of *Agrobacterium*–plant interactions and DNA transfer events [4-7].

Agrobacterium tumefaciens is a soil phytopathogen that naturally infects plant wound sites and causes crown gall disease via delivery of transferred (T)-DNA from bacterial cells into host plant cells through a bacterial type IV

* Correspondence: emlai@gate.sinica.edu.tw
[1]Institute of Plant and Microbial Biology, Academia Sinica, Taipei 11529, Taiwan
[2]Department of Plant Pathology and Microbiology, National Taiwan University, Taipei 10617, Taiwan
Full list of author information is available at the end of the article

secretion system (T4SS) [8]. Although *Agrobacterium* is considered a wound-associated pathogen, it can transfer DNA into diverse host cells or tissues under unwounded conditions [9-13]. Interestingly, most of the *Arabidopsis* mutants that are resistant to *Agrobacterium* transformation identified by root explant assays remain highly transformable by floral dip transformation [14]. The mechanisms and plant factors involved in *Agrobacterium*-mediated transformation may differ between wounded and unwounded cells or different tissues. However, the mechanisms underlying *Agrobacterium* infection in unwounded cells/tissues have not been explored.

In plant biology research, *Arabidopsis* mesophyll-protoplast transfection [15,16] and *Agrobacterium*-mediated leaf infiltration in *Nicotiana benthamiana* [17] are the well-established and commonly used platforms for transient gene expression analysis. The *Arabidopsis* mesophyll-protoplast transient expression system allows for versatile and high-throughput analyses of diverse gene functions and signal transduction pathways; advanced skills with training and practice are essential for successful use of this powerful tool for gene function studies [16,18,19]. *Agrobacterium*-mediated transient expression methods by leaf infiltration have been developed for a wide range of plants including *Nicotiana*, lettuce, tomato, and *Arabidopsis* [20-23]. However, the use of 4- to 5-week-old adult plants with manual infiltration has limited application in high-throughput analyses. Furthermore, although *Arabidopsis* is the most-studied model plant with superbly annotated genome sequences and powerful genetic and genomic resources mostly available for the Columbia (Col) accession, achieving highly efficient and consistent transient expression in Col by adult leaf infiltration is challenging [22,24].

The use of young seedlings for *Agrobacterium*-mediated transient expression assays will greatly simplify and amplify the power of the method. Indeed, *Agrobacterium*-mediated transient expression in *Arabidopsis* seedlings has been recently developed for fast and robust analysis of protein subcellular localization and protein–protein interactions [25-27]. The system's requirement for high-density *Agrobacterium* cells and vacuum infiltration [27] or chemical treatment (e.g., the addition of surfactant Silwet L-77) [26] to achieve high cellular transformation efficiency could induce innate immunity and stress responses in plants, which globally alters cellular, physiological, and signaling processes and severely retards growth [28,29]. Thus, developing a system that circumvents a plant defense barrier may be a key to enhance transient expression efficiency in *Arabidopsis* seedlings. Furthermore, such a fast, robust, and highly efficient transient expression system could support gain-of-function studies of diverse genes and signaling pathways *in planta*.

Pattern-triggered immunity (PTI) induced by a microbe- or pathogen-associated molecular pattern (MAMP or PAMP) is the first line of active defense in both plants and animals against pathogens [28-30]. Previous studies have suggested that *Agrobacterium*-mediated transformation efficiency may be compromised when plants recognize *Agrobacterium* MAMPs by corresponding pattern-recognition receptors (PRRs) to trigger PTI and block *Agrobacterium* infection [22,24]. The elongation factor Tu (EF-Tu) receptor mutant *efr-1*, which cannot sense EF-Tu MAMP, showed increased *Agrobacterium*-mediated transient expression efficiency, as did transgenic *Arabidopsis* expressing a potent bacterial effector AvrPto to suppress PTI signaling with agroinfiltration of 4- to 5-week-old leaves [22,24]. However, whether these immune-compromised *Arabidopsis* plants are amenable to increase *Agrobacterium*-mediated transient expression efficiency in young seedlings has not been tested. Defining the condition for reliable and highly efficient transformation in healthy Col-0 seedlings will be extremely valuable but has never been achieved.

In this study, we systematically investigated various biological factors and growth variances to define a combination of key factors that contribute to the unprecedentedly high transient transformation and reporter gene expression efficiency in *Arabidopsis* seedlings. As a result of these investigations, we developed an optimized AGROBEST (*Agrobacterium*-mediated enhanced seedling transformation) method that enabled high transient transformation and expression efficiency in both *efr-1* mutant and Col-0 *Arabidopsis* seedlings. Importantly, we demonstrated the versatile applicability of AGROBEST in gain-of-function studies for the MYB75 transcription factor in specific target-gene activation and for *GIGANTEA* (*GI*) reporter gene expression regulated by the *Arabidopsis* circadian clock. The AGROBEST method is a fast, simple, reliable, and versatile tool for systematic gene function analysis and a new tool for dissecting the *Agrobacterium*-mediated DNA transfer processes.

Results

Cotyledons of young *Arabidopsis* EF-TU receptor mutant is highly susceptible to *Agrobacterium*-mediated transient transformation

Environmental and biological factors such as growth conditions, host plants, and *Agrobacterium* strains can affect the transformation efficiency. We first evaluated the transient expression efficiency of selected *Arabidopsis* ecotypes and mutants defective in pattern-recognition receptors (PRRs) with a disarmed *A. tumefaciens* strain C58C1(pTiB6S3ΔT-DNA) [31] containing a pCH32 helper plasmid [32] and abbreviated as C58C1(pTiB6S3ΔT)^H. The T-DNA vector pBISN1 harboring the *gusA-intron* [12] was transformed into C58C1(pTiB6S3ΔT)^H to infect 4-d-old seedlings, and β-glucuronidase (GUS) activity was determined to monitor transient expression efficiency at 3 days post-infection (dpi). We consistently observed 100% of analyzed EF-Tu receptor mutant *efr-1* seedlings were successfully transformed, with strong and homogenous GUS staining in cotyledons, with 4-fold higher GUS activity in *efr-1* than wild-type Col-0 seedlings (Figure 1A and B, Additional file 1: Table S1). The flagellin receptor mutant, *fls2*, and the Ws ecotype that is highly susceptible to *Agrobacterium* transformation in the root explant [14] and a natural *fls2* variant [33] showed similar transient GUS expression efficiency as the Col-0, so the *fls2* mutant contributes little to enhancing *Agrobacterium*-mediated transient transformation (Figure 1A and B). Our seedling transient expression results confirm and further support that EFR but not FLS2 is an important factor limiting *Agrobacterium*-mediated transient expression efficiency previously observed by agroinfiltration of *Arabidopsis* adult leaves [24,34,35].

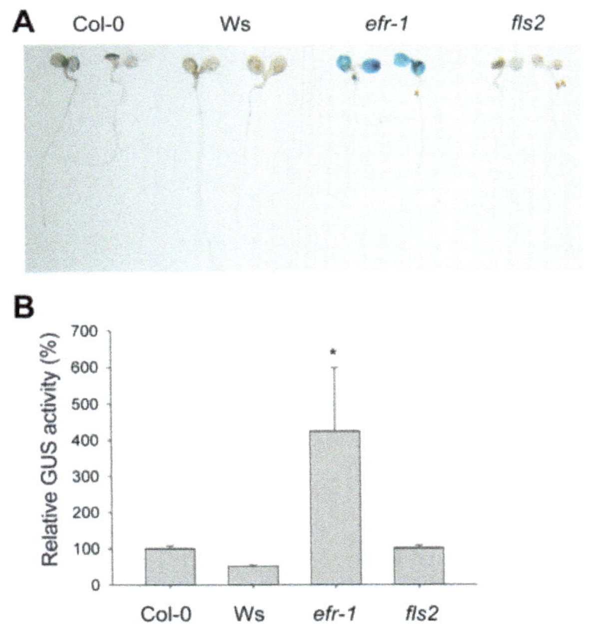

Figure 1 Transient transformation assays in different *Arabidopsis* ecotype/genotypes. Four-day-old *Arabidopsis* seedlings were infected with *Agrobacterium* strain C58C1(pTiB6S3ΔT)^H carrying pBISN1, which was pre-incubated in AB-MES (pH5.5) supplemented with 200 μM acetosyringone (AS) to induce *vir* gene expression. Seedlings were co-cultivated with pre-induced *A. tumefaciens* cells with final OD_{600} = 0.02 in the MS medium (1/2 MS, 0.5% sucrose (w/v), pH 5.5) containing 50 μM AS and determined for transient GUS expression levels by overnight GUS staining **(A)** and quantitative GUS activity **(B)** at 3 dpi. The GUS activity obtained from Col-0 seedlings was set to 100% and that of Ws, *efr-1*, and *fls2* is relative to that of Col-0. Data are mean ± SD GUS activity from two biological replicates. Similar results were obtained from at least two independent experiments. Values significantly different from that obtained with Col-0 are denoted (*P = 0.058 by Student's t test).

Buffered medium at pH 5.5 with AB salts is critical for high transient expression efficiency

To exploit this transient expression system for higher efficiency, we tested several factors including pre-induction and co-cultivation conditions. Pre-induction with acetosyringone (AS) in AB-MES medium (ABM50 and ABM200 methods) and continuous addition of AS to stimulate *vir* gene expression during the infection process are required for efficient transient GUS expression. Because AB-MES medium (pH 5.5) is the optimized medium for *vir* gene induction [36,37], we tested whether mixing AB-MES (pH 5.5) with an equal volume of commonly used plant culture MS medium (1/2 MS, 0.5% sucrose (w/v), pH 5.5), named ABM-MS (1/2 AB-MES, 1/4 MS, 0.25% sucrose (w/v), pH 5.5) in the presence of AS could produce high transient expression efficiency. Strikingly, GUS activity was strongly expressed in all seedlings and was 20-fold higher with co-cultivation in ABM-MS than in MS medium alone (Figure 2A, Additional file 1: Table S1). To avoid over-staining, the reaction time for histological GUS staining shown in Figure 2 was limited to 6 hr instead of overnight for the result in Figure 1A.

Key components in AB-MES are AB salt (17.2 mM K_2HPO_4, 8.3 mM NaH_2PO_4, 18.7 mM NH_4Cl, 2 mM KCl), minerals (1.25 mM $MgSO_4$, 100 μM $CaCl_2$, 10 μM $FeSO_4$), glucose (2% w/v), and buffering with MES (50 mM) to pH 5.5. We thus tested whether one of these components is responsible for the increased transient expression efficiency. The addition of AB salts with MES buffered at pH 5.5 in MS medium was sufficient to result in comparable levels of GUS expression as with ABM-MS (Figure 2B). Therefore, AB salts alone, pH 5.5 buffered by MES, or both, are critical for the increased transient expression efficiency. Strikingly, all MS media with the addition of AB salts buffered with MES or sodium phosphate at pH 5.5 showed comparable and strong GUS activity as that with ABM-MS (Figure 2C). However, omitting AB salts resulted in ~50% reduction in GUS activity, and no GUS activity was detected with MS medium buffered with sodium phosphate at pH 7.0 in the presence or absence of AB salts. Thus, buffered pH at 5.5 and the presence of AB salts in MS co-cultivation medium are the two key factors for this high transient expression efficiency. We named this optimized infection method AGROBEST (*Agrobacterium*-mediated enhanced seedling transformation).

Disarmed *Agrobacterium* strain C58C1(pTiB6S3ΔT)^H enables highly efficient AGROBEST-mediated transient expression in Col-0 seedlings

Next, we tested whether the AGROBEST method optimized with *efr-1* seedlings could also improve *Agrobacterium*-mediated transient transformation in wild-type Col-0 seedlings. Because the use of C58C1(pTiB6S3ΔT)^H as

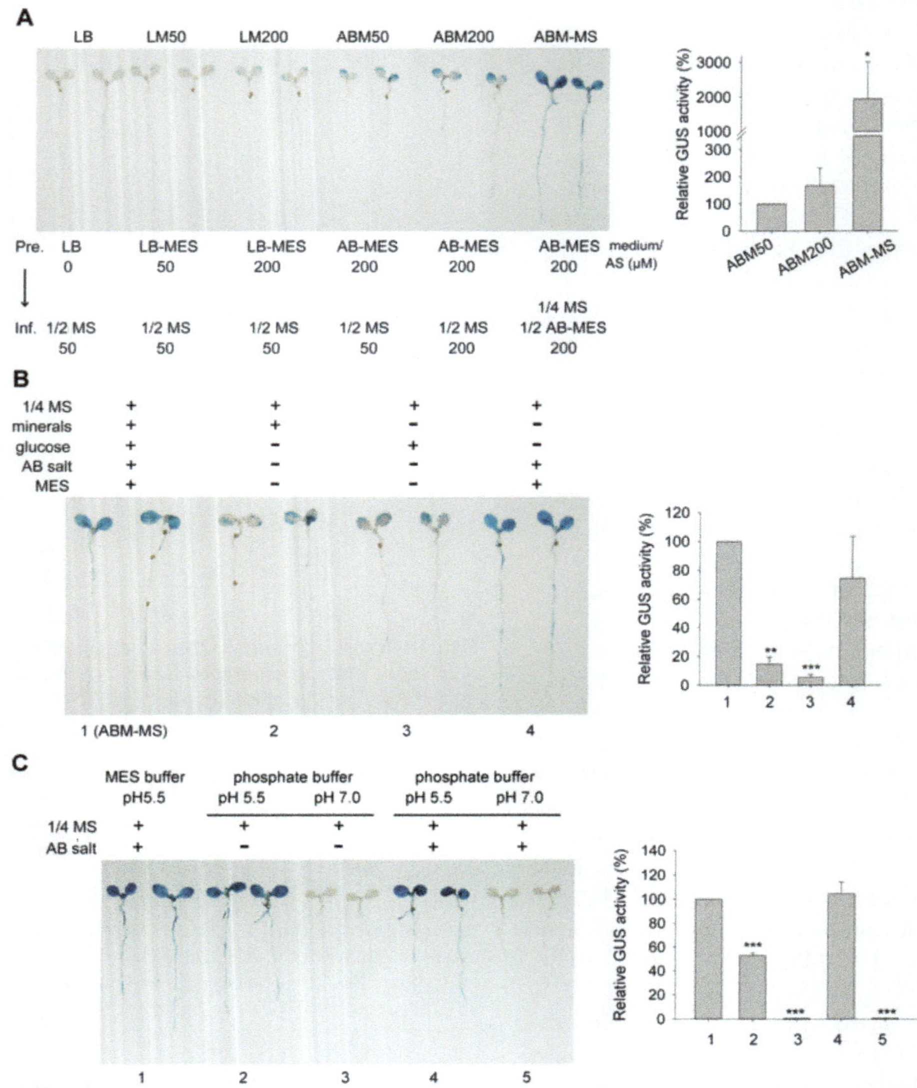

Figure 2 Optimization of *Agrobacterium* pre-culture and infection media for efficient transient expression efficiency. Four-day-old *Arabidopsis efr-1* seedlings infected with *Agrobacterium* C58C1(pTiB6S3ΔT)^{H} carrying pBISN1 were grown in various pre-culture and co-cultivation media to test their effects on transient GUS expression efficiency measured by GUS staining and quantitative GUS activity. **(A)** Various pre-culture and infection media in the absence or presence of *vir* gene inducer AS at the indicated concentration. **(B)** Effect of factors in AB-MES medium on increased transient expression efficiency. **(C)** Effect of AB salts, pH and buffering systems on transient GUS expression efficiency. Data for relative quantitative GUS activity are mean ± SD of 3 independent experiments. Values significantly different from that infected by ABM50 **(A)** or condition 1 **(B and C)** are denoted (*$P < 0.05$, **$P < 0.01$, ***$P < 0.005$ by Student's *t* test).

compared with other disarmed or virulent *A. tumefaciens* strains produced higher transient expression levels with leaf agroinfiltration of various plants [23], we also tested whether C58C1(pTiB6S3ΔT)^{H} is a more superior strain in our system. We compared C58C1(pTiB6S3ΔT)^{H} with the wild-type virulent strain C58 or C58-derived disarmed strain GV3101(pMP90) [38] for their transient expression efficiency in *efr-1* and Col-0 seedlings using both sub-optimal ABM50 and optimized AGROBEST methods. Remarkably, Col-0 seedlings infected by all transfer-competent strains achieved significantly higher transient expression efficiency by AGROBEST than ABM50 (Figure 3A and B). Moreover, Col-0 seedlings infected by AGROBEST showed higher transient expression than *efr-1* seedling infected by ABM50 (Figure 3A and B). No GUS stains could be detected in control seedlings without infection (MOCK) or infected with Δ*virB2*, a strain lacking the key component of the type IV secretion system (T4SS) essential for T-DNA/effector translocation [8,39]. Therefore, the GUS activity detected was indeed from T-DNA gene expression inside the plant cells. Strikingly, 5- to 15-fold higher GUS activity was observed in *efr-1* or Col-0

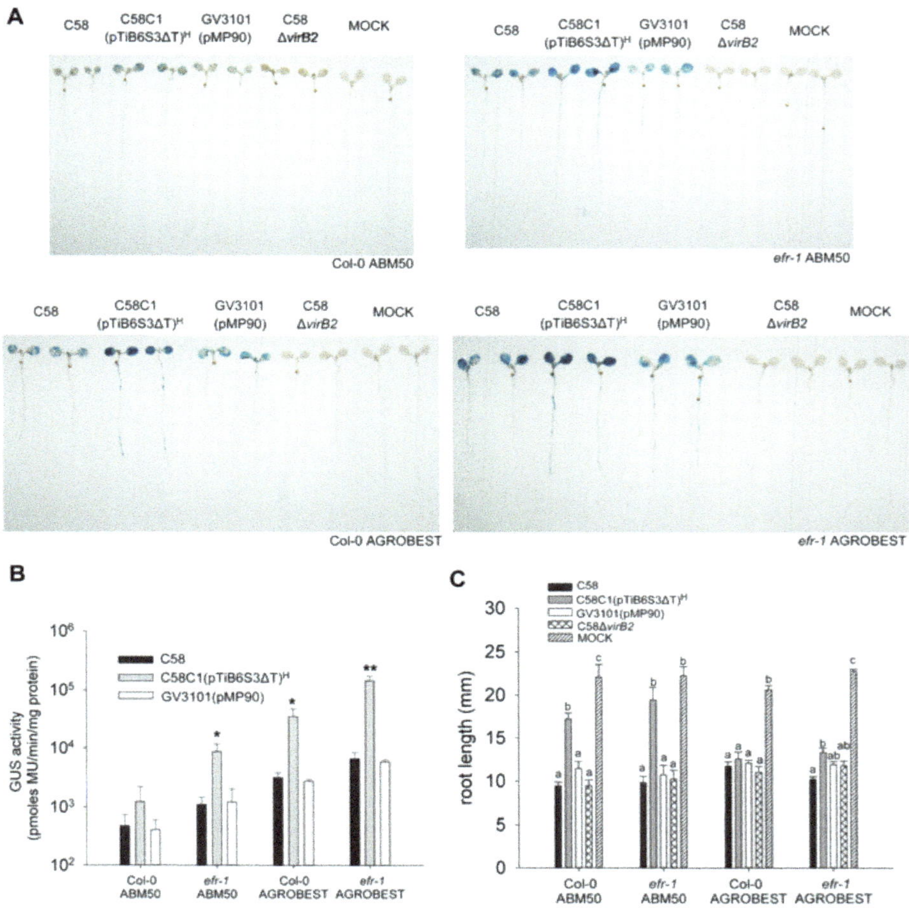

Figure 3 Transient transformation of the *Arabidopsis* seedlings by various *Agrobacterium* strains. Four-day-old *Arabidopsis* Col-0 and *efr-1* seedlings infected with different *Agrobacterium* stains carrying pBISN1 by ABM50 or ABM-MS (named as AGROBEST) were compared by GUS staining **(A)**, quantitative GUS activity **(B)**, and root length **(C)** at 3 days post-inoculation (dpi). Data for quantitative GUS activity are mean ± SD of at least 4 biological replicates from 2 independent experiments. Values significantly different from that infected with wild-type C58 are denoted (*$P < 0.05$, **$P < 0.01$ by Student's *t* test). Data for root length measurement are mean ± SEM of 4-6 biological replicates from 2–4 independent experiments. Statistics was analyzed by ANOVA and means annotated with the same letter (a-c) are not significantly different; those with different letters are significantly different ($P < 0.05$). Seedlings grown in the same co-cultivation medium without *Agrobacterium* infection are indicated (MOCK).

seedlings infected with C58C1(pTiB6S3ΔT)^H than with C58 or GV3101(pMP90) (Figure 3A and B). The root length was significantly shorter for *Arabidopsis* seedlings infected with C58, Δ*virB2*, or GV3101(pMP90) at 3 dpi in all combinations or with C58C1(pTiB6S3ΔT)^H by the AGROBEST method as compared with uninfected seedlings (MOCK) (Figure 3C). Notably, seedlings infected with C58C1(pTiB6S3ΔT)^H showed no or little inhibition of root elongation in *efr-1* seedlings with the AMB50 method, which achieves fair although not the highest transient expression efficiency.

AGROBEST achieves higher transient expression efficiency than existing methods in both *efr-1* and Col-0 seedlings

We also compared AGROBEST with previously developed methods [26,27] for their transient expression efficiency in both Col-0 and *efr-1* seedlings. Remarkably, all Col-0 seedlings infected by the AGROBEST showed ~10-fold increased transient expression efficiency than with two existing methods, the FAST method [26] and the method by Marion et al. [27] with either GUS (Figure 4A) or luciferase (*LUC2*) (Figure 4B) used as reporters. Interestingly, both AGROBEST and the Marion et al. method achieved significantly higher transient expression activity in *efr-1* than in Col-0, *efr-1* seedlings remained poorly transformed by the FAST method (Figure 4A and B). As a result, AGROBEST conferred at least 40-fold and 3-fold higher transient expression efficiency in *efr-1* seedlings than with FAST and the Marion et al. methods, respectively (Figure 4A and B). However, we detected no increased transient expression activity in infected seedlings of the dexamethasone (DEX)-induced

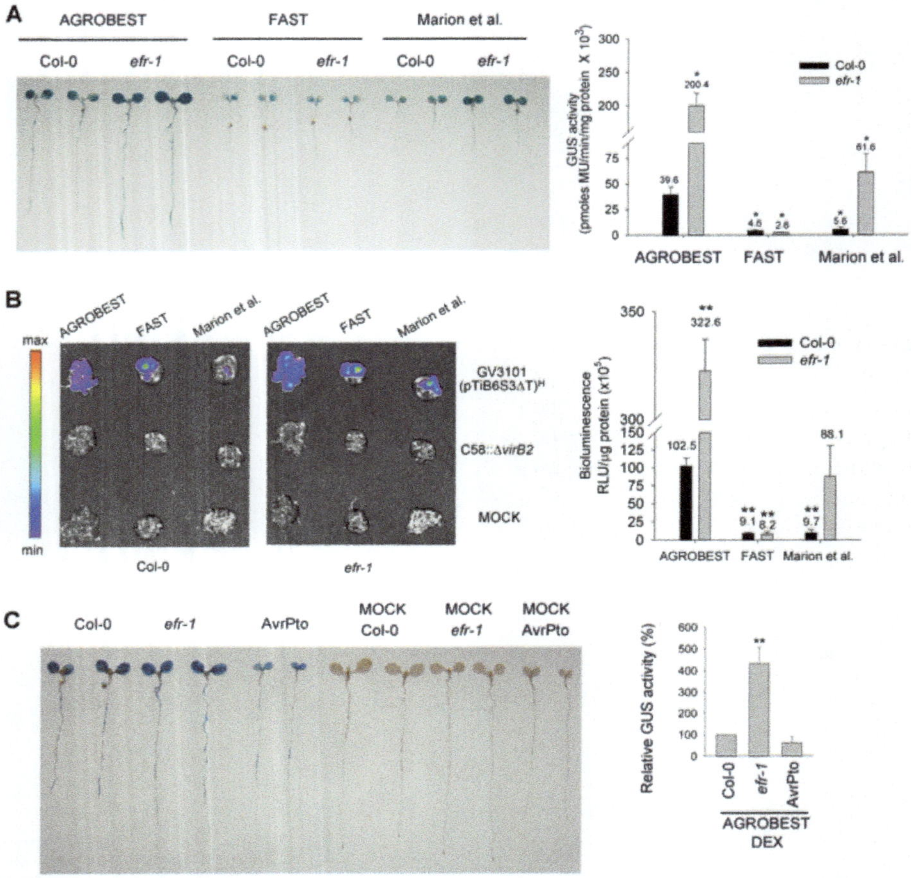

Figure 4 AGROBEST enables high transient expression levels in Col-0. Four-day-old *Arabidopsis* seedlings were infected with *Agrobacterium* strain C58C1(pTiB6S3ΔT)^H carrying pBISN1 **(A and C)** or 35S::*LUC2* **(B)**, and transient expression activity was determined at 3 dpi. **(A)** Transient GUS expression efficiency of Col-0 and *efr-1* seedlings by AGROBEST, FAST and Marion et al. methods. Data for quantitative GUS activity are mean ± SD of 3 biological replicates. Values significantly different from those obtained with Col-0 by AGROBEST are denoted (*$P < 0.05$ by Student's *t* test). **(B)** Transient luciferase expression efficiency of Col-0 and *efr-1* seedlings by AGROBEST, FAST and Marion et al. methods. Seedlings infected by C58Δ*virB2* carrying 35S::*LUC2* were used as a background control and those without *Agrobacterium* infection are indicated as MOCK. Luciferase activity of Col-0 obtained by AGROBEST was set to 100% and that of others is relative to activity of Col-0 by AGROBEST. Data are mean ± SD of 3 biological replicates. Values significantly different from those obtained with Col-0 by AGROBEST are denoted (**$P < 0.01$, by Student's *t* test). **(C)** Transient GUS expression efficiency of Col-0, AvrPto transgenic line, and *efr-1* by AGROBEST. For dexamethasone (DEX) treatment, 3-d-old seedlings were treated with 10 μM DEX for 1 day and the following 3 days infected by the AGROBEST method. Quantitative GUS activity from DEX-induced Col-0 seedlings by AGROBEST was set to 100% and that of others is relative to activity of DEX-induced Col-0 seedlings with AGROBEST. Data are mean ± SD GUS activity from 4 repeats (2 biological repeats from each of 2 independent experiments). Values significantly different from that obtained with Col-0 are denoted (**$P < 0.01$ by Student's *t* test). Seedlings grown in the same co-cultivation medium without *Agrobacterium* infection are indicated (MOCK).

AvrPto transgenic line than in Col-0 seedlings with the AGROBEST method (Figure 4C), despite a significantly higher transient expression efficiency than Col-0 detected in adult leaves by agroinfiltration [22]. The AvrPto transgenic line germinated at the same rate and grew to a similar size as Col-0 and *efr-1*, but growth was arrested with the addition of DEX at 3 days old. This finding is consistent with previous studies showing that overexpression of AvrPto can also interfere with growth hormone signals and trigger cell death by interrupting the diverse functions of BAK1 and BKK1 in multiple receptor complexes, not restricted to PRRs [40].

Impact of seedling age and infection time on transient expression efficiency of AGROBEST in *efr-1* seedlings

Because the highest transient expression efficiency in *efr-1* seedlings can be achieved by infection with C58C1 (pTiB6S3ΔT)^H by AGROBEST, we chose this combination to test the versatility and applicability of AGROBEST. For example, dissecting the minimal infection time (from 1–5 days) and range of seedling age (from 3- to 6-d-old) applicable for efficient transient expression is of interest. We tested different ages of *Arabidopsis efr-1* seedlings infected at different dpi and noted that GUS signals were barely detectable at 1 dpi but gradually reached a plateau at 3 or 4 dpi (Figure 5A). When 5- or 6-d-old seedlings were infected, we observed transient GUS expression in true leaves at 3 or 4 dpi. Although strong GUS staining could still be detected in seedlings at 4 or 5 dpi, these seedlings often showed bleached lesions in cotyledons (Figure 5B), which explained the lack of GUS expression in part of the cotyledons at 4 or 5 dpi (Figure 5A). At 3 dpi, the bleached lesions were more visible when the transformation was performed with 5- or 6-d-old seedlings than with 3- or 4-d-old seedlings. Thus, the use of younger seedlings for AGROBEST may be more desirable for maintaining plants in healthy and physiological conditions.

To test the minimal infection time for GUS detection and to avoid plant damage due to prolonged *Agrobacterium* infection, the co-cultivation medium was replaced with fresh medium containing antibiotics (100 μM Timentin) at 1 or 2 dpi to inhibit bacterial growth. In 4-d-old seedlings, we detected low levels of GUS signals with an additional 2 or 3 days of cultivation after Timentin treatment at 1 dpi (Figure 6). Importantly, with Timentin treatment at 2 dpi, seedlings with 1 to 3 days of additional cultivation remained healthy (without bleached lesions) and showed strong GUS signals in cotyledons. Because *Agrobacterium* cells were mostly killed when true leaves emerged from infected seedlings, the newly grown true leaves were not efficiently transformed. Therefore, the use of 4-d-old seedlings infected for 2 days followed by an additional 1 to 3 days of cultivation with Timentin is the optimal condition to transiently express genes for functional studies.

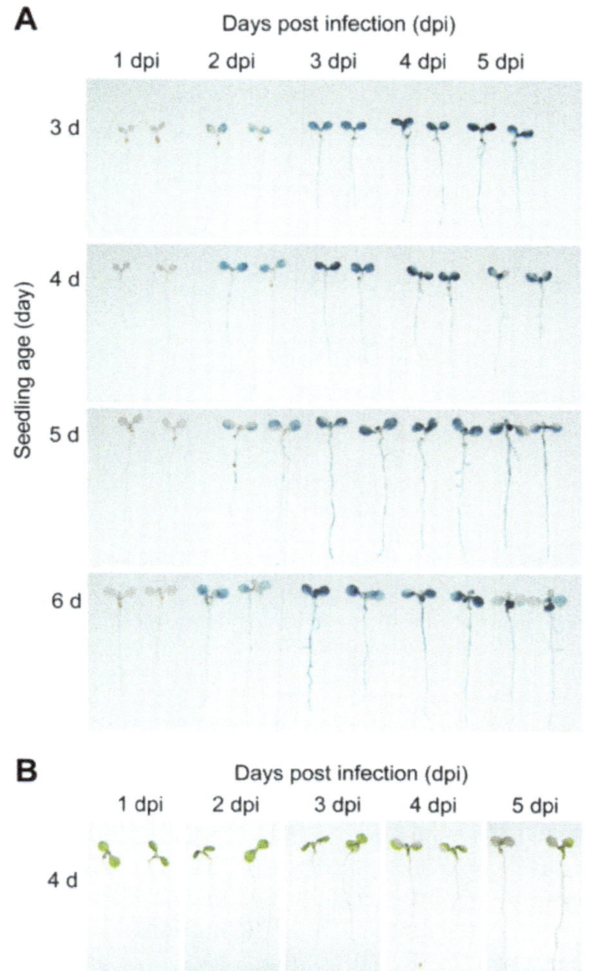

Figure 5 Impact of seedling age and infection time on transient expression. (A and B) Different ages of *Arabidopsis efr-1* seedlings were infected with C58C1(pTiB6S3ΔT)[H] carrying pBISN1 by the AGROBEST method and analyzed for GUS activity **(A)** and morphologic features **(B)** at different dpi.

Widespread transient transformation events in different organs and cell types

The high transient expression efficiency with AGROBEST was mostly evident with strong and homogeneous GUS signals detected in cotyledons of 100% infected Col-0 or *efr-1* seedlings (Figures 3A, 4A and 7A, Additional file 1: Table S1). When 7-d-old seedlings were used for infection, strong GUS signals were also detected in true leaves, as shown in *efr-1* seedlings (Figure 7B). However, in roots, GUS signals could be detected in ~70% of Col-0 or *efr-1* seedlings infected by AGROBEST (Additional file 1: Table S1) and mostly appeared in lateral root initiation sites or in the elongation zone (Figure 7C and D). In addition to analyzing the GUS reporter, we determined the expression of fluorescent proteins as reporters at cellular and subcellular levels using *efr-1* seedlings. With expression of the Venus-intron or NLS-RFP driven by the CaMV 35S promoter, fluorescent protein signals were widely detected in cotyledon cells (Figure 7E and F), mainly in epidermal pavement cells but also in guard cells and mesophyll cells (Figure 7G-I). For roots, epidermal cells consistently showed fluorescent protein signals (Figure 7J). Therefore, the AGROBEST seedling transformation system allows for high transient gene expression and, potentially, functional analysis in diverse tissues and cell types in *Arabidopsis* seedlings.

Studies of protein subcellular localization and protein–protein interactions

Because *Arabidopsis* plants are less amenable for transient expression analysis, both fluorescent protein localization and bimolecular fluorescence complementation (BiFC) studies are often conducted in protoplasts via transfection or in *N. benthamiana* leaves via agroinfiltration because of the high transient expression efficiency [16,17]. Here, we co-infected two *A. tumefaciens* strains carrying a binary vector for 35S::Venus-intron or 35S::NLS-RFP in *efr-1* seedlings and detected both cytoplasmic and nuclear fluorescence signals for Venus and nuclear localization of

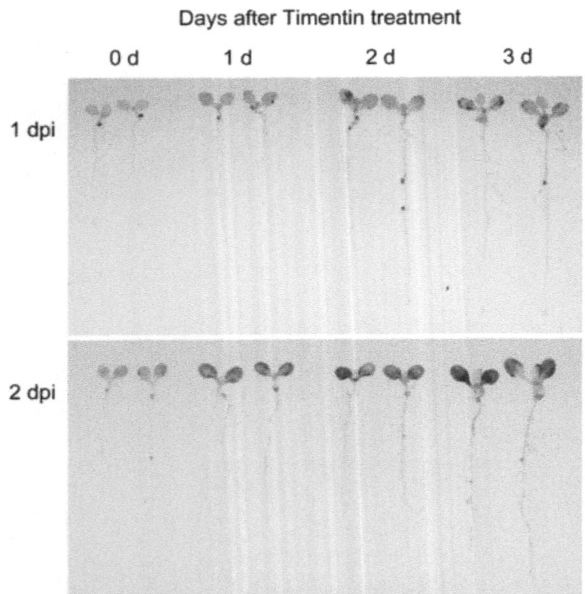

Figure 6 Impact of Timentin treatment on transient GUS expression efficiency. Four-day-old *Arabidopsis efr-1* seedlings were infected with *Agrobacterium* C58C1(pTiB6S3ΔT)^{H} carrying pBISN1 by the AGROBEST method at 1 or 2 dpi before Timentin treatment. GUS staining was performed at 0 to 3 days after Timentin treatment.

NLS-RFP in separate or the same cells (Figure 7K). Our assay is also feasible for BiFC studies, which is supported by the interaction of two known interacting proteins, F-box protein TIR1 (transport inhibitor response 1) and ASK1 (*Arabidopsis* Skp1-like protein) [41], in the nucleus (Figure 7L). Thus, AGROBEST is an ideal system for subcellular localization and protein-protein interaction studies.

AGROBEST for the expression analysis of a circadian clock reporter gene

Encouraged by the high transient expression efficiency with AGROBEST, we next tested its applicability in gene function/regulation study in physiological contexts. Most *Arabidopsis* genes express rhythmically under various thermocycles, photocycles, or circadian clock conditions [42]. Reporter genes driven by promoters of the circadian genes are commonly used to monitor the regulation of circadian genes. To test whether circadian rhythm could be monitored in transiently transformed seedlings, a circadian reporter (*GI::LUC2*) constructed by fusing the promoter of the circadian gene *GIGANTEA* (*GI*) with the luciferase gene (*LUC2*) [43] was used. Four-day-old *Arabidopsis efr-1* seedlings were infected with *Agrobacterium* delivering *GI::LUC2* for 3 days under 16-h/8-h light/dark cycles and then transferred to MS medium in the presence of 100 μM Timentin and 0.5 mM luciferin under continuous light to monitor real-time bioluminescence for 5 days. In contrast to constant low levels of bioluminescence from seedlings infected with a vector control, *Arabidopsis* seedlings infected with *Agrobacterium* delivering *GI::LUC2* showed clear circadian oscillation at slightly lengthened period for at least 5 days (Figure 8). The observed transiently expressed *GI* circadian cycle is indistinguishable from the stable *GI* expression in *GI::LUC2* transgenic *Arabidopsis* plant (TP) [43], although with lower amplitude. The comparable circadian oscillation between the stable and transient expression of the *GI::LUC2* indicated that the slightly longer period we observed was unlikely a result of the *Agrobacterium* infection. This result indicated the applicability of AGROBEST for transient expression of circadian rhythm reporter in *Arabidopsis* seedlings without detectable interference by *Agrobacterium* infection.

AGROBEST for functional assays of transcription factor MYB75

Next, we tested AGROBEST for gain-of-function studies. For a proof of concept, we transiently expressed a transcription factor MYB75 because of its well-established function in anthocyanin accumulation by upregulating a key gene encoding chalcone synthase (CHS) in the anthocyanin synthesis pathway [44]. Four-day-old *efr-1 Arabidopsis* seedlings were infected for 3 days after Timentin treatment for an additional 3 days to determine the effect on *MYB75* transient expression. *MYB75* mRNA level in infected seedlings was 60-, 400-, and 200-fold higher when the expression was driven by single (1X35S) and double (2X35S) CaMV 35S promoter and super promoter, respectively, than in seedlings expressing control vectors (Figure 9A). *CHS* mRNA level was increased 4- and 3-fold in 2X35S::*MYB75* and super::*MYB75* seedlings, respectively (Figure 9B). However, *CHS* expression was not increased in 1X35S::*MYB75* seedlings despite its 60-fold higher *MYB75* expression, which suggests a threshold expression level or the requirement of other MYB75-modulated co-activators for *CHS* activation. Importantly, consistent with increased *CHS* expression, high level of anthocyanin (purple coloration) was readily detectable in cotyledons of 2X35S::*MYB75* and super::*MYB75* seedlings but not in seedlings infected with a vector control, 1X35S::*MYB75*, or super::*gusA-intron* (Figure 9C). No increase of anthocyanin accumulation from super::*gusA-intron* seedlings indicated that the specificity of the observed anthocyanin phenotype was due to the transient expression of *MYB75* rather than a secondary effect from the infection or the overexpression of any foreign protein. Importantly, AGROBEST also enables the transient expression of *MYB75* to monitor its downstream *CHS* expression and anthocyanin accumulation in Col-0 seedlings. We show that transient expression of *MYB75* driven

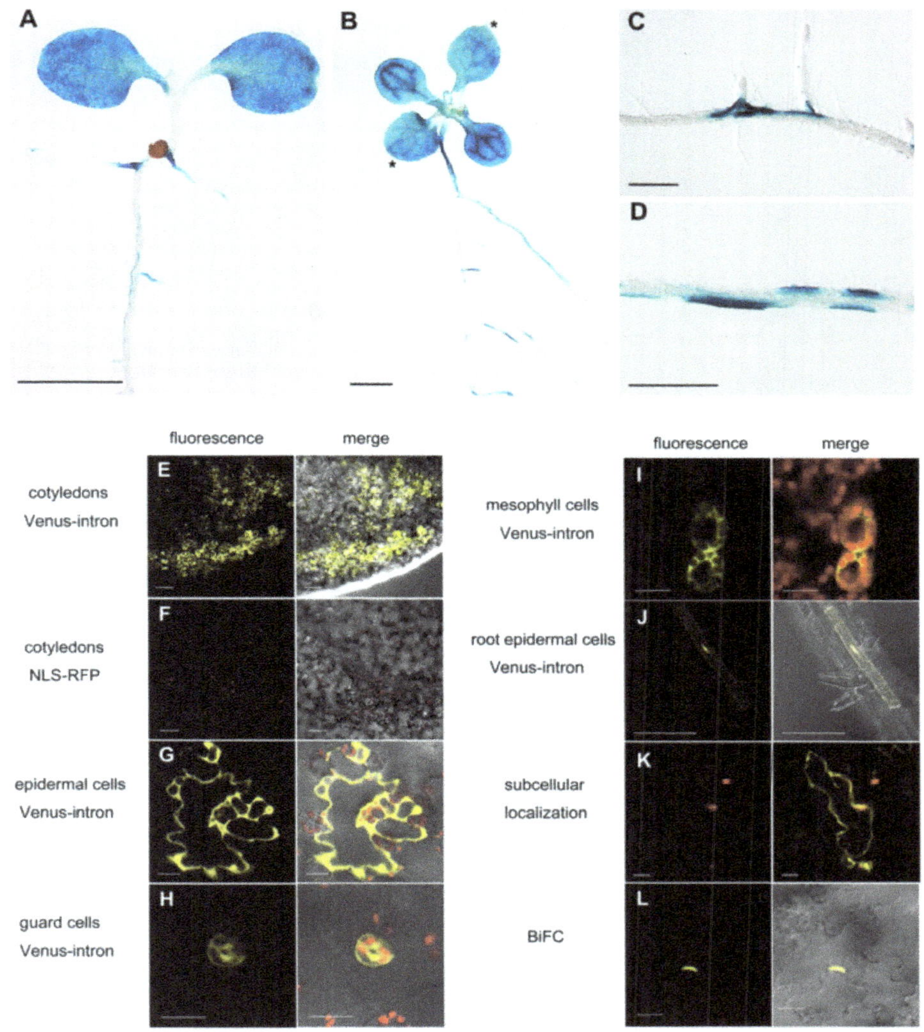

Figure 7 Transient transformation events in different organs and cell types. **(A-D)** Four-day-old **(A and C-D)** or 7-d-old **(B)** *Arabidopsis efr-1* seedlings were infected with C58C1(pTiB6S3ΔT)^{H} carrying pBISN1 by the AGROBEST method and analyzed for GUS staining. GUS staining was detected in true leaves (**B**, indicated by asterisk), cotyledons **(A and B)**, main roots near lateral initiation site **(C)**, and elongation zone **(D)**. **(E-L)** Confocal microscopy of 4-day-old *Arabidopsis efr-1* seedlings infected with C58C1(pTiB6S3ΔT)^{H} carrying various vectors for transient expression of indicated fluorescent proteins by the ABM200 method. Fluorescence signals for 35S::Venus-intron or 35S::NLS-RFP were detected in cotyledons **(E and F)**. Venus-intron signals were detected in different types of cells, including epidermal cells **(G)**, guard cells **(H)**, mesophyll cells **(I)** of cotyledon, and root epidermal cells **(J)**. **(K)** Subcellular localization of Venus-intron and NLS-RFP by co-infection of 2 *Agrobacterium* strains expressing 35S::Venus-intron or 35S::NLS-RFP. **(L)** Protein–protein interaction by BiFC of nYFP-ASK1 and TIR1-cYFP. Images show fluorescence alone **(K)** and/or merged with bright field **(E, F, J and L)** or chloroplast fluorescence **(G-I)**. Scale bars are 2 mm **(A and B)**, 0.5 mm **(C and D)**, 100 μm **(E, F and J)**, 50 μm **(L)** and 20 μm **(G-I and K)**. BiFC, bimolecular fluorescence complementation.

by the 2X35S promoter results in significantly higher of *MYB75* mRNA levels than vector control in Col-0 seedlings (Figure 10A). Remarkably, *CHS* mRNA levels were also upregulated in 2X35S::*MYB75* seedlings (Figure 10B), in which a moderate increase in anthocyanin accumulation was also detected (Figure 10C). This result strongly suggested the broad application of AGROBEST for gain-of-function studies not limited to the immune-compromised mutant.

Discussion

AGROBEST enables high transient transformation and expression efficiency in intact *Arabidopsis* young seedlings

In this study, we developed a simple, fast, reliable, and robust transient expression system named AGROBEST and uncovered the key factors enabling 100% of infected seedlings with high transgene expression efficiency in *Arabidopsis* seedlings. Remarkably, AGROBEST appears to achieve the highest transient expression efficiency in

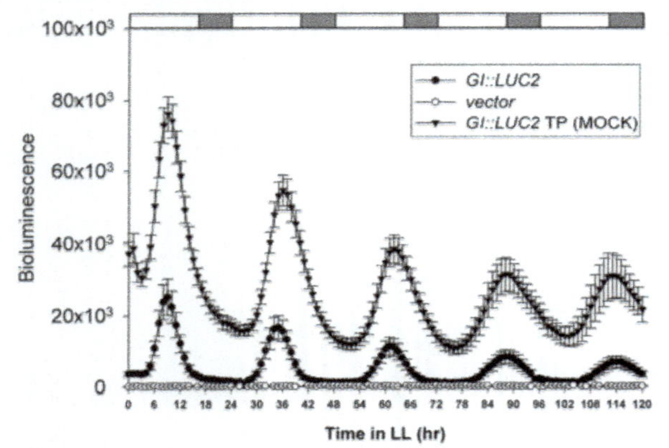

Figure 8 Monitoring *Arabidopsis* circadian rhythm by transient expression of *GIGANTEA*::luciferase (*GI::LUC2*). Four-day-old *Arabidopsis efr-1* seedlings were infected with *Agrobacterium* C58C1(pTiB6S3ΔT)^H carrying a vector (pCAMBIA1390) or p1390-*GI-LUC2* by the AGROBEST method for 3 days in a 16-h/8-h light/dark cycle (75 μmol m^{-2} s^{-1}), then transferred to 1/2 MS liquid medium in the presence of 100 μM Timentin and 0.5 mM luciferin and grown under continuous light at 40 μmol m^{-2} s^{-1} for up to 5 days. The *GI::LUC2* transgenic *Arabidopsis* plant (TP) cultured in identical conditions without *Agrobacterium* infection was a positive control. Real-time bioluminescence signals were photographed and the luciferase intensity is shown as mean ± SEM from 12 seedlings expressing *GI::LUC2*. Similar results were obtained from at least 3 independent experiments. The white and gray regions indicate subjective light and dark periods, respectively.

the EF-Tu receptor *efr-1* mutant as compared to the wild-type Col-0, flagellin receptor mutant *fls2*, and DEX-inducible AvrPto transgenic line. This result is consistent with a previous finding in agroinfiltrated *Arabidopsis* adult leaves showing increased transient GUS expression efficiency in *efr-1* [24]. Because of no detectable phenotype impairing the growth and development in the *efr* mutant [24], the use of the *efr* mutant has an advantage over DEX-inducible AvrPto in seedling stages. Thus, more selected elimination of specific PRRs such as EFR with minimal effects on hormonal signaling, cell death and seedling growth may be a preferred system for *Agrobacterium*-mediated high transient expression efficiency. Interestingly, *N. benthamiana* leaves, which are commonly used for *Agrobacerium*-mediated transient transformation, also lack the EFR receptor [45].

Unexpectedly, we discovered that AGROBEST also enables high transient expression efficiency in wild-type Col-0 seedlings. The significantly higher transient expression activity by AGROBEST than the FAST and Marion et al. methods likely accounts for the success of our gain-of-function experiments, which have not been shown previously [26,27]. Of note, *efr-1* seedlings remained poorly transformed by FAST method as compared with the significantly increased transient expression in *efr-1* by AGROBEST or the Marion et al. method. The reason

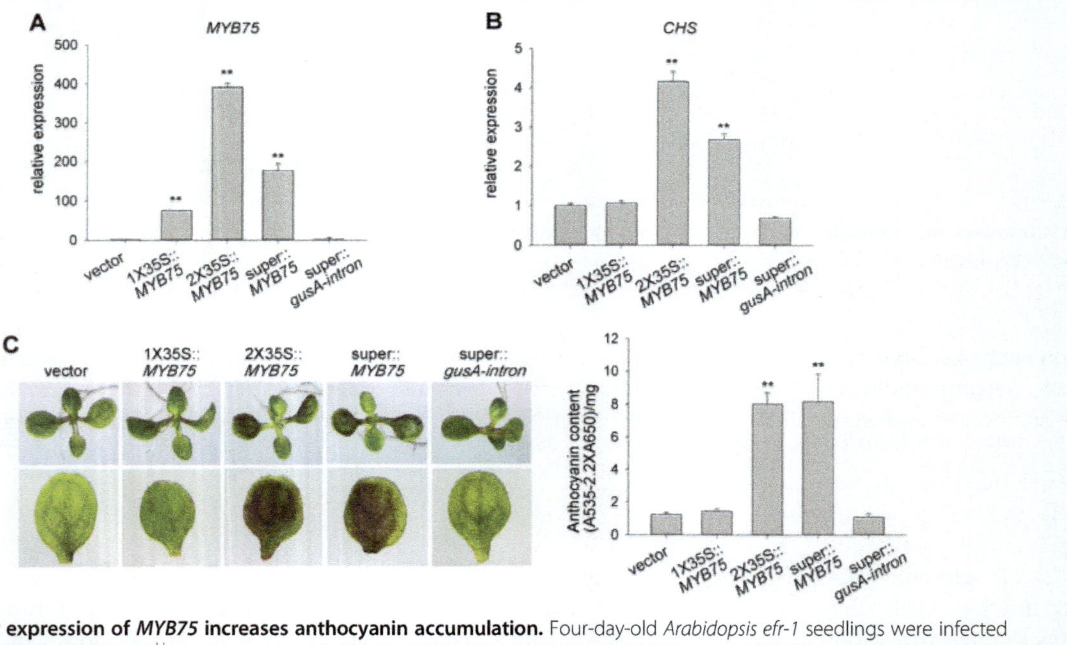

Figure 9 Transient expression of *MYB75* increases anthocyanin accumulation. Four-day-old *Arabidopsis efr-1* seedlings were infected with *Agrobacterium* C58C1(pTiB6S3ΔT)^H carrying a vector (pCAMBIA1390), 35S::*MYB75*, 2X35S::*MYB75*, super::*MYB75*, or super::*gusA-intron* by the AGROBEST method. At 3 dpi, co-cultivation medium was replaced with MS medium containing 100 μM Timemtin for additional incubation for 3 days. qRT-PCR of relative expression of *MYB75* **(A)** and *CHS* **(B)** with representative data shown with mean ± SD from 3 technical repeats. Similar results were obtained from three independent experiments. Zeiss inverted microscopy of anthocyanin accumulation in seedlings (upper panels) and cotyledons (lower panels) and quantification **(C)**. Data for anthocyanin content are mean ± SD from 4 repeats (2 biological repeats from each of 2 independent experiments, 20–30 seedlings for each biological repeat), Values significantly different from that obtained with vector are denoted (**$P < 0.01$, by Student's *t* test).

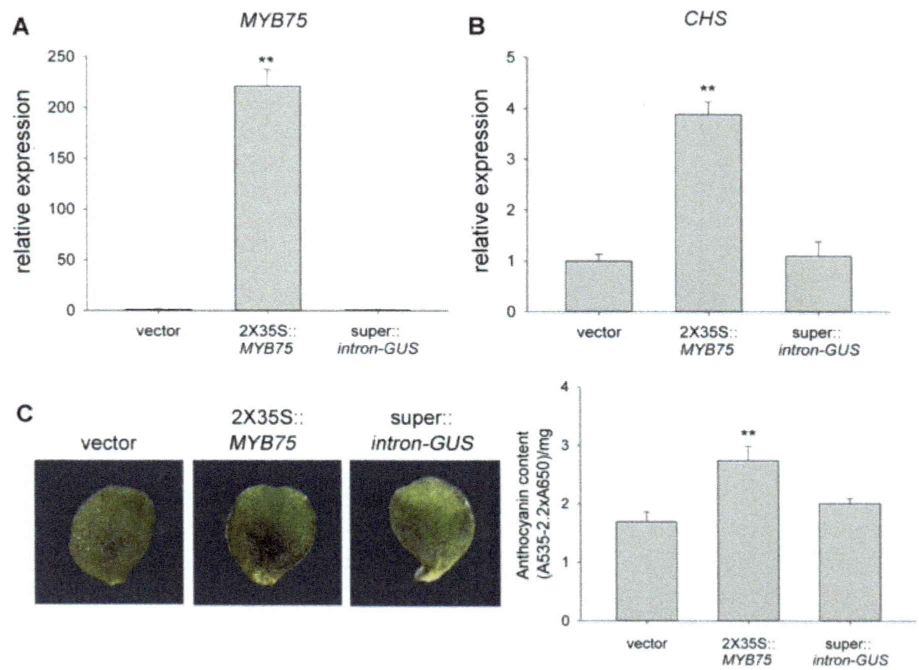

Figure 10 Transient expression of *MYB75* increases anthocyanin accumulation in Col-0 seedlings. Four-day-old *Arabidopsis* Col-0 seedlings were infected with *Agrobacterium* C58C1(pTiB6S3ΔT)^H carrying a vector (pCAMBIA1390), 2X35S::*MYB75*, or super::*gusA-intron* by the AGROBEST method. At 3 dpi, co-cultivation medium was replaced with MS medium containing 100 μM timemtin for additional incubation for 3 days. qRT-PCR of relative expression of *MYB75* **(A)** and *CHS* **(B)** with representative data shown with mean ± SD from 3 technical repeats. Similar results were obtained from three incependent experiments. Zeiss inverted microscopy of anthocyanin accumulation in seedlings (upper panels) and cotyledons (lower panels) and quantification **(C)**. Data for anthocyanin content are mean ± SD from 3 independent experiments (20–30 seedlings for each biological repeat, 3 biological repeats for each independent experiment). Values significantly different from those obtained with vector are denoted (**$P < 0.01$, by Student's *t* test).

underlying this discrepancy is unknown, but the yellowish and retarded-growth seedlings after co-cultivation with *Agrobacterium* in the dark for 2 days by the FAST method may contribute to the observed phenotype. Our AGROBEST method, applying a lower density of *Agrobacterium* cells (OD_{600} 0.02 as opposed to OD_{600} 0.5 for the FAST method and OD_{600} 2 for the Marion et al. method) for co-cultivation with seedlings without any mechanical treatment (e.g., vacuum infiltration) or chemical treatment (e.g., the addition of surfactant Silwet L-77) offers advantages to maintain infected seedlings with normal growth and a physiological state without injury. The success of transiently expressing the circadian rhythm reporter in *Arabidopsis* seedlings may open a new platform to rapidly test the circadian behaviors of *Arabidopsis* mutants, bypassing the process of introducing a circadian reporter gene into the mutants by crossing. Most remarkably, AGROBEST allows for high transient expression of the *MYB75* transcription factor and subsequently upregulates the expression of its downstream gene *CHS* in both *efr-1* and Col-0 seedlings. This result suggested the broad application of AGROBEST to study transcription factor action.

Widespread and differential transient transformation events in different organs and cell types

AGROBEST has a breakthrough performance by enabling high and homogeneous transient GUS expression efficiency in shoots of 100% infected Col-0 or *efr-1* seedlings. The successful transient expression in roots, although with less efficient transformation events (~70% of seedlings with GUS staining in roots), is also remarkable and not previously detected [26,27]. Interestingly, preferential transformation events occurring at the initiation sites of lateral roots or the root elongation zone of infected intact seedlings were also previously detected in wounded *Arabidopsis* roots [46]. High transformation of *Arabidopsis* roots may require further loosening or opening of cell walls or wounding, which was not included in our infection conditions. Because we observed similar transient expression levels and transformation efficiency in roots of Col-0 and *efr-1* seedlings (Additional file 1: Table S1), EFR may play no or little role in seedling root transformation efficiency under our infection conditions. Consistently, *EFR* is expressed at low levels in Col-0 seedling roots [47], which were not responsive to the EF-Tu peptide elf26, as evidenced by limited induction of immune marker genes and callose deposition in the roots of Col-0 seedlings [48]. Because the flg22 peptide derived from *Agrobacterium* flagellin is inactive in *Arabidopsis* [24,34,35] and the flagellin receptor mutant exhibited similar transformation efficiency as Col-0 in our seedling assays, the flagellin receptor FLS2 may not be involved in *Agrobacterium*-triggered plant innate immune responses and therefore did not compromise *Agrobacterium*-mediated transient gene expression. Future investigations could examine whether the absence of the peptidoglycan receptor [49] or yet-to-be identified receptors in recognizing additional MAMPs such as polysaccharides [50] could increase the transformation efficiency in seedling roots.

Key factors for high transient transformation/expression efficiency

During this course of our method development, we also uncovered new factors critical for the high transient transformation/expression efficiency in *Arabidopsis* seedlings. One factor is the addition of AB salts in MS medium buffered with acidic pH 5.5 during *Agrobacterium* infection, which allows for significantly higher transient expression efficiency than in MS medium alone. Another breakthrough is the use of the disarmed *A. tumefaciens* strain C58C1(pTiB6S3ΔT)^H, which offers the highest transient expression efficiency with the least adverse impact on plant growth over other tested strains. Root growth was severely inhibited on infection with other tested *A. tumefaciens* strains including the transfer-incompetent Δ*virB2*. These data indicate that the transport of T-DNA and T4SS effectors into plant cells by a virulent C58 strain may not suppress host immune responses like that observed in T3SS effectors from *Pseudomonas syringae* [51]. We observed that C58C1(pTiB6S3ΔT)^H achieved higher transient expression efficiency in both Col-0 and *efr-1* seedlings than other *A. tumefaciens* strains tested. The agent also had little impact on root growth inhibition of infected seedlings by the ABM50 method (Figure 3). The results suggested that the *A. tumefaciens* strain C58C1 (pTiB6S3ΔT)^H is the main factor affecting the root growth difference. EFR may play a minor role in root growth inhibition because we observed slightly stronger root growth inhibition in Col-0 than *efr-1* seedlings infected with C58C1(pTiB6S3ΔT)^H. This finding is consistent with limited root growth inhibition detected in Col-0 seedlings in response to EF-Tu peptide elf18 as compared with strong root growth inhibition induced by flg22 [47]. The observed inverse association of root growth inhibition and transient expression efficiency suggested that C58C1(pTiB6S3ΔT)^H may circumvent a plant defense barrier to enable high transient expression levels in *Arabidopsis* seedlings. However, interestingly, root length was significantly lower in Col-0 and *efr-1* seedlings with C58C1(pTiB6S3ΔT)^H infection than in uninfected seedlings (MOCK), despite the significantly higher transient expression efficiency with the AGROBEST than the ABM50 method (Figure 3). Thus, although C58C1(pTiB6S3ΔT)^H remains a strain causing the least inhibition in seedling root growth as compared to other *A. tumefaciens* strains, whether the observed root growth inhibition results from PTI contributing to reduce transient expression efficiency requires future investigation. Other factors in addition to PTI may contribute to the enhanced transient expression efficiency by AGROBEST.

C58C1(pTiB6S3ΔT)^H has been known to achieve high transformation efficiency in several plant species including *Arabidopsis*, but the underlying mechanism is not known. The nomenclature of *Agrobacterium* strains used in plant transformation experiments is often simplified, which causes confusion and could sometimes be misleading. C58C1(pTiB6S3ΔT)^H is often simplified as C58C1 in the plant community. C58C1 is in fact named after curing pTiC58 from the wild-type virulent strain C58, and rifampicin (Rif)-resistant strains are selected from C58C1 for convenient use to acquire various disarmed Ti plasmids transferred from different *Agrobacterium* strains [52,53]. Therefore, C58C1(pTiB6S3ΔT)^H is a Rif-resistant C58C1 harboring the octopine-type Ti plasmid pTiB6S3 with the removal of the T-DNA region [31] and containing a pCH32 helper plasmid with increased expression of virulence genes *virG* and *virE2* [32]. GV3101(pMP90) is a C58-derived disarmed strain, in which pMP90 is a nopaline-type Ti plasmid, pTiC58, with the removal of T-DNA [38]. Therefore, in theory, C58C1(pTiB6S3ΔT)^H should share the same chromosomal background with GV3101(pMP90) and only differ in the use of different Ti plasmids and the presence of the helper plasmid pCH32. Future work to determine which genetic factor(s) contribute to increased transient expression efficiency with less growth inhibition by C58C1(pTiB6S3ΔT)^H will shed light on understanding the molecular mechanisms underlying the observed high transient transformation and expression efficiency.

Conclusions

In this study, we developed a valuable and novel method, named AGROBEST, and uncovered the key factors enabling this unprecedented high transient transformation and reporter gene expression efficiency in the immune receptor mutant *efr-1* and in wild-type Col-0 *Arabidopsis* seedlings. The applicability for transient expression of *MYB75* in activating downstream gene expression in a Col-0 background further suggested that AGROBEST may be a feasible method to use in examining transcription factor actions or gain-of-function studies in different *Arabidopsis* ecotypes/genotypes. Because most plants do not harbor *EFR*, which is only present in *Brassicaceae*

[24], the established method may be applicable in other plant species. This fast, sensitive, and quantitative assay was routinely used with culture plates, which are easily scaled up for quick and systematic screens. Importantly, this method nicely compliments the commonly used *Arabidopsis* mesophyll-protoplast transfection [15,16] and *Agrobacterium*-mediated leaf infiltration in *N. benthamiana* [17] for gene functional studies and provides advantages for its high reproducibility without advanced skills. Furthermore, AGROBEST may be an alternative method for evaluating *Agrobacterium* virulence and discovering and dissecting gene functions involved in various steps of *Agrobacterium*-mediated DNA transfer. The method may help unravel the mechanisms underlying *Agrobacterium* infection in unwounded cells/tissues.

Methods

Materials and growth condition

Strains, plasmids, and primer sequences used in this study are in Additional file 2: Table S2 and Additional file 3: Table S3. The bacterial growth conditions and procedures for plasmid and mutant constructions are described in Additional file 4: Methods S1. *Arabidopsis thaliana* plants included ecotype Columbia-0 (Col-0), Wassilewskija (Ws-2), T-DNA insertion mutants *efr*-1 (SALK_044334) and *fls2* (SALK_093905) and the DEX-inducible AvrPto transgenic line generated in a Col-0 background were obtained from the Arabidopsis Biological Resource Center (Ohio). Seeds were sterilized in 50% bleach (v/v) containing 0.05% Triton X-100 (v/v) for 10 min, rinsed 5 times with sterile water, and incubated at 4°C for 3 days. For germination, 10 seeds were transferred to 1 ml 1/2 MS liquid medium (1/2 MS salt supplemented with 0.5% sucrose (w/v), pH 5.5 [pH adjusted to 5.7 by KOH but pH 5.5 after autoclaving], in each well of a 6-well plate. Germination and growth took place in a growth room at 22°C under a 16-hr/8-hr light–dark cycle (75 μmol m^{-2} s^{-1}).

Agrobacterium infection in *Arabidopsis* seedlings

For AGROBEST infection assay, *A. tumefaciens* was freshly streaked out from -80°C glycerol stock onto a 523 agar plate for 2-day incubation at 28°C. A fresh single colony from the plate was used to inoculate 5 ml of 523 liquid medium containing appropriate antibiotics for shaking (220 rpm) at 28°C for 20–24 hr. For pre-induction of *A. tumefaciens vir* gene expression, *A. tumefaciens* cells were pelleted and re-suspended to OD_{600} 0.2 in various liquid media including LB, LB-MES (LB with 10 mM MES, pH 5.7) [53,54] or AB-MES (17.2 mM K_2HPO_4, 8.3 mM NaH_2PO_4, 18.7 mM NH_4Cl, 2 mM KCl, 1.25 mM $MgSO_4$, 100 μM $CaCl_2$, 10 μM $FeSO_4$, 50 mM MES, 2% glucose (w/v), pH 5.5) [37] with different concentrations of acetosyringone (AS; 0, 50 or 200 μM) without antibiotics, then shaken (220 rpm) at 28°C for 12–16 hr. Before infection, *A. tumefaciens* cells were pelleted and re-suspended in desired co-cultivation liquid media to OD_{600} 0.02. The growth medium of *Arabidopsis* seedlings was replaced with 1 ml *A. tumefaciens* cells freshly prepared above and incubated in the same growth room until further analysis. Three-day-old seedlings were treated with 10 μM DEX for 1 day before infection for the following 3 days. When the removal of *Agrobacterium* cells was required, co-cultivation medium was removed after the chosen infection time and replaced with 1 ml freshly prepared MS medium containing 100 μM Timentin and incubated for additional days before analysis. The procedures for the seedling transient transformation assay using the method optimized by Marion et al. and FAST Method developed by Li et al. were performed [26,27] and described in Additional file 4: Methods S1. Unless indicated, 10 seedlings grown in each well were infected and 3 biological repeats were performed in each independent experiment.

Plant RNA extraction and quantitative RT-PCR

RNA was extracted from *Arabidopsis* seedlings as described [55]. An amount of 4 μg total RNA was used to synthesize first-strand cDNA with SuperScript III Reverse Transcriptase (Invitrogen) and oligo dT primer. Quantitative PCR involved the Applied Biosystems QuantStudio 12 K Flex Real Time PCR machine and Power SYBRR Green PCR Master Mix (Invitrogen). *Arabidopsis ACTIN 2* (At3g18780) or *UBC21* (At5g25760) was an internal control.

GUS staining and activity assays

Seedlings were stained with 5-bromo-4-chloro-3-indolyl glucuronide (X-Gluc) at 37°C for 6 hr unless indicated or quantified with a fluorescence substrate (4-methylumbelliferyl-β-D-glucuronide [MUG]) as described [56]. For MUG assay, fluorescence was determined using a 96 microtiter-plate reader (Bio-Tek Synergy Mx, 356 nm excitation 455 nm emission with ±20 nm filter) and calculation of specific GUS enzyme activity was based on the standard curve of 0.5–500 pmole (0.5, 5, 50 and 500 pmole) 4-MU standards obtained from the same microtiter plate. For relative GUS activity, the fluorescence signal value was normalized by an equal amount of proteins with subtraction of the background fluorescence signal detected by the mock control.

Confocal microscopy

Fluorescence signals were observed by use of a Zeiss LSM 510 Meta Confocal microscope. Venus signals were observed at 488-nm excitation with an HFT 488/514-nm filter and emission with NTF 515- and BP 505- to 530-nm filters. RFP signals were observed at 488-nm excitation with an HFT 405/488-nm filter and emission with NFT 545 and LP 650 filters.

Transient expression of *MYB75* and anthocyanin content assay

Four-day-old seedlings were infected with *A. tumefaciens* strain C58C1(pTiB6S3ΔT)^H carrying the control or *MYB75*-expressing binary vector in ABM-MS liquid medium for 3 days. The co-cultivation medium was then replaced with 1 ml fresh MS medium (1/2 MS, 2% sucrose (w/v), pH adjusted to 5.7 by KOH but pH 5.5 after autoclaving) containing 100 μM Timemtin and then incubated for 3 days. For anthocyanin content assay, seedlings were blot-dried briefly, weighed, ground into powder with liquid nitrogen and mixed with 1 ml extraction buffer (0.12 M HCl, 18% isopropanol (v/v)). The mixture was boiled for 90 sec and centrifuged at 16000 × g for 15 min. The supernatant was collected and measured at OD_{535} (A535) and OD_{650} (A650). Anthocyanin content was calculated as A535 - (2.2 × A650)/fresh weight (g) [57].

Transient expression of *GI::LUC2* and bioluminescence measurement

Four-day-old seedlings were infected with *A. tumefaciens* strain C58C1(pTiB6S3ΔT)^H carrying p1390-*GI::LUC2* or empty vector (pCAMBIA1390) in ABM-MS co-cultivation medium. At 3 dpi, each seedling was transferred to MS medium (1/2 MS, pH adjusted to 5.7 by KOH but pH 5.5 after autoclaving) containing 100 μM Timentin and 0.5 mM luciferin in a black 96-well plate. Bioluminescence activity was measured and analyzed as described [43].

Luciferase activity assay

Arabidopsis seedlings after infection were surface sterilized with 1% bleach (0.05% sodium hypochlorite) for 5–10 min and washed with sterile water 3 times to remove bacteria before assay. The washing step is essential to minimize the background signals expressed in bacteria because of the use of intron-less *LUC2* reporter. For photography, 10 seedlings infected by each method were placed in a clean 15-cm square Petri dish and covered with 100 μl 1 mM luciferin. Luciferase intensity was imaged by use of the XENOGEN IVIS lumina system with 5-sec exposure time. Bioluminescence assay involved the luciferase assay system (Promega). Briefly, 10–15 seedlings after a washing were blot-dried with tissue paper before being frozen with liquid nitrogen and stored at -80°C. Seedlings were ground into fine powder by liquid nitrogen, mixed with 300 μl cell-culture lysis reagent (Promega), and centrifuged at 16000 × g for 10 min at 4°C. Supernatant was 100× diluted with cell-culture lysis reagent. In total, 20 μl cell lysate was mixed with 100 μl Luciferase Assay Reagent and the signal was detected by use of lumat LB 9507 (Berthold Technologies). The bioluminescence signal was normalized to the protein amount of each sample quantified by the Bradford protein assay (Bio-Rad).

Additional files

Additional file 1: Table S1. Transient transformation efficiency of shoots and roots of *Arabidopsis* Col-0 and *efr-1* seedlings.

Additional file 2: Table S2. Bacterial strains and plasmids.

Additional file 3: Table S3. Primer information.

Additional file 4: Methods S1. Methods for bacterial strains and plasmids and infection method by Li et al. (FAST method) and by Marion et al.

Competing interests

The authors declare that they have no competing interests.

Authors' contributions

HYW participated the experimental designs, performed most of the experiments, analyzed data, and drafted the manuscript. KHL and WLC participated in method optimization. YCW and JFW performed experiments. CYC participated in data analysis. SHW and JS participated in experimental designs and helped drafting the manuscript. EML conceived of the study, participated in method optimization and experimental designs, coordinated the project, and wrote the manuscript. All authors read and approved the final manuscript.

Acknowledgements

The authors thank Hau-Hsuan Hwang, Lay-Sun Ma, Jer-Sheng Lin, and Po-Yuan Shih for discussion and critical reading of the manuscript; and Yajie Niu and Hoosun Chung for preliminary transient expression tests on different seedling ages. We thank Dr. Inhwan Hwang for NLS-RFP; Drs. Stanton Gelvin and Lan-Ying Lee for pBISN1, Venus-intron and BiFC vectors; and Ms. Mei-Jane Fang from the Cell Biology Core Laboratory at the Institute of Plant and Microbial Biology, Academia Sinica, for excellent technical support on confocal microscopy. This work was supported by research grants from the National Science Council of Taiwan (NSC 99-2918-I-001-005 and NSC 101-2321-B-001-033 to E. M. Lai, NSC100-2311-B-001-028-MY3 to S. H. Wu) and the US National Institutes of Health (R01GM60493 and R01GM70567 to J. Sheen).

Author details

[1]Institute of Plant and Microbial Biology, Academia Sinica, Taipei 11529, Taiwan. [2]Department of Plant Pathology and Microbiology, National Taiwan University, Taipei 10617, Taiwan. [3]Department of Molecular Biology and Center for Computational and Integrative Biology, Massachusetts General Hospital, Boston, MA 02114, USA. [4]Department of Genetics, Harvard Medical School, Boston, MA 02114, USA. [5]Center for the Study of Biological Complexity, Virginia Commonwealth University, Richmond, VA 23284, USA.

References

1. Dandekar AM, Fisk HJ: **Plant transformation: *Agrobacterium*-mediated gene transfer.** *Methods Mol Biol* 2005, **286:**35–46.
2. Gelvin SB: **Agricultural biotechnology: gene exchange by design.** *Nature* 2005, **433:**583–584.
3. Tzfira T, Citovsky V: ***Agrobacterium*-mediated genetic transformation of plants: biology and biotechnology.** *Curr Opin Biotechnol* 2006, **17:**147–154.
4. Gelvin SB: ***Agrobacterium*-mediated plant transformation: the biology behind the "gene-jockeying" tool.** *Microbiol Mol Biol Rev* 2003, **67:**16–37.
5. Gelvin SB: **Plant proteins involved in *Agrobacterium*-mediated genetic transformation.** *Annu Rev Phytopathol* 2010, **48:**45–68.
6. Gelvin SB: **Traversing the cell: *Agrobacterium* T-DNA's journey to the host genome.** *Front Plant Sci* 2012, **3:**52.
7. McCullen CA, Binns AN: ***Agrobacterium tumefaciens* and plant cell interactions and activities required for interkingdom macromolecular transfer.** *Annu Rev Cell Dev Biol* 2006, **22:**101–127.
8. Fronzes R, Christie PJ, Waksman G: **The structural biology of type IV secretion systems.** *Nat Rev Microbiol* 2009, **7:**703–714.

9. Brencic A, Angert ER, Winans SC: **Unwounded plants elicit *Agrobacterium vir* gene induction and T-DNA transfer: transformed plant cells produce opines yet are tumour free.** *Mol Microbiol* 2005, **57:**1522–1531.
10. Clough SJ, Bent AF: **Floral dip: a simplified method for *Agrobacterium*-mediated transformation of *Arabidopsis thaliana*.** *Plant J* 1998, **16:**735–743.
11. Escudero J, Hohn B: **Transfer and integration of T-DNA without cell injury in the host plant.** *Plant Cell* 1997, **9:**2135–2142.
12. Narasimhulu SB, Deng XB, Sarria R, Gelvin SB: **Early transcription of *Agrobacterium* T-DNA genes in tobacco and maize.** *Plant Cell* 1996, **8:**873–886.
13. Ryu CM, Anand A, Kang L, Mysore KS: **Agrodrench: a novel and effective agroinoculation method for virus-induced gene silencing in roots and diverse *Solanaceous* species.** *Plant J* 2004, **40:**322–331.
14. Mysore KS, Kumar CT, Gelvin SB: ***Arabidopsis* ecotypes and mutants that are recalcitrant to *Agrobacterium* root transformation are susceptible to germ-line transformation.** *Plant J* 2000, **21:**9–16.
15. Sheen J: **Signal transduction in maize and *Arabidopsis* mesophyll protoplasts.** *Plant Physiol* 2001, **127:**1466–1475.
16. Yoo SD, Cho YH, Sheen J: ***Arabidopsis* mesophyll protoplasts: a versatile cell system for transient gene expression analysis.** *Nat Protoc* 2007, **2:**1565–1572.
17. Vaghchhipawala Z, Rojas CM, Senthil-Kumar M, Mysore KS: **Agroinoculation and agroinfiltration: simple tools for complex gene function analyses.** *Methods Mol Biol* 2011, **678:**65–76.
18. Kim J, Somers DE: **Rapid assessment of gene function in the circadian clock using artificial microRNA in *Arabidopsis* mesophyll protoplasts.** *Plant Physiol* 2010, **154:**611–621.
19. Li JF, Chung HS, Niu Y, Bush J, McCormack M, Sheen J: **Comprehensive protein-based artificial microRNA screens for effective gene silencing in plants.** *Plant Cell* 2013, **25:**1507–1522.
20. Kapila J, De Rycke R, Van Montagu M, Angenon G: **An *Agrobacterium*-mediated transient gene expression system for intact leaves.** *Plant Sci* 1997, **122:**101–108.
21. Sheludko YV, Sindarovska YR, Gerasymenko IM, Bannikova MA, Kuchuk NV: **Comparison of several *Nicotiana* species as hosts for high-scale *Agrobacterium*-mediated transient expression.** *Biotechnol Bioeng* 2007, **96:**608–614.
22. Tsuda K, Qi Y, Nguyen LV, Bethke G, Tsuda Y, Glazebrook J, Katagiri F: **An efficient *Agrobacterium*-mediated transient transformation of *Arabidopsis*.** *Plant J* 2012, **69:**713–719.
23. Wroblewski T, Tomczak A, Michelmore R: **Optimization of *Agrobacterium*-mediated transient assays of gene expression in lettuce, tomato and *Arabidopsis*.** *Plant Biotechnol J* 2005, **3:**259–273.
24. Zipfel C, Kunze G, Chinchilla D, Caniard A, Jones JD, Boller T, Felix G: **Perception of the bacterial PAMP EF-Tu by the receptor EFR restricts *Agrobacterium*-mediated transformation.** *Cell* 2006, **125:**749–760.
25. Kim MJ, Baek K, Park CM: **Optimization of conditions for transient *Agrobacterium*-mediated gene expression assays in *Arabidopsis*.** *Plant Cell Rep* 2009, **28:**1159–1167.
26. Li JF, Park E, von Arnim AG, Nebenfuhr A: **The FAST technique: a simplified *Agrobacterium*-based transformation method for transient gene expression analysis in seedlings of *Arabidopsis* and other plant species.** *Plant Methods* 2009, **5:**6.
27. Marion J, Bach L, Bellec Y, Meyer C, Gissot L, Faure JD: **Systematic analysis of protein subcellular localization and interaction using high-throughput transient transformation of *Arabidopsis* seedlings.** *Plant J* 2008, **56:**169–179.
28. Boller T, Felix G: **A renaissance of elicitors: perception of microbe-associated molecular patterns and danger signals by pattern-recognition receptors.** *Annu Rev Plant Biol* 2009, **60:**379–406.
29. Tena G, Boudsocq M, Sheen J: **Protein kinase signaling networks in plant innate immunity.** *Curr Opin Plant Biol* 2011, **14:**519–529.
30. Segonzac C, Zipfel C: **Activation of plant pattern-recognition receptors by bacteria.** *Curr Opin Microbiol* 2011, **14:**54–61.
31. Deblaere R, Bytebier B, De Greve H, Deboeck F, Schell J, Van Montagu M, Leemans J: **Efficient octopine Ti plasmid-derived vectors for *Agrobacterium*-mediated gene transfer to plants.** *Nucleic Acids Res* 1985, **13:**4777–4788.
32. Hamilton CM: **A binary-BAC system for plant transformation with high-molecular-weight DNA.** *Gene* 1997, **200:**107–116.
33. Zipfel C, Robatzek S, Navarro L, Oakeley EJ, Jones JD, Felix G, Boller T: **Bacterial disease resistance in *Arabidopsis* through flagellin perception.** *Nature* 2004, **428:**764–767.
34. Bauer Z, Gomez-Gomez L, Boller T, Felix G: **Sensitivity of different ecotypes and mutants of *Arabidopsis thaliana* toward the bacterial elicitor flagellin correlates with the presence of receptor-binding sites.** *J Biol Chem* 2001, **276:**45669–45676.
35. Felix G, Duran JD, Volko S, Boller T: **Plants have a sensitive perception system for the most conserved domain of bacterial flagellin.** *Plant J* 1999, **18:**265–276.
36. Gelvin SB: ***Agrobacterium* virulence gene induction.** *Methods Mol Biol* 2006, **343:**77–84.
37. Lai EM, Kado CI: **Processed VirB2 is the major subunit of the promiscuous pilus of *Agrobacterium tumefaciens*.** *J Bacteriol* 1998, **180:**2711–2717.
38. Koncz C, Schell J: **The promoter of TL-DNA gene 5 controls the tissue-specific expression of chimaeric genes carried by a novel type of *Agrobacterium* binary vector.** *Mol Genet Genomics* 1986, **204:**383–396.
39. Berger BR, Christie PJ: **The *Agrobacterium tumefaciens virB4* gene product is an essential virulence protein requiring an intact nucleoside triphosphate-binding domain.** *J Bacteriol* 1993, **175:**1723–1734.
40. Shan L, He P, Li J, Heese A, Peck SC, Nurnberger T, Martin GB, Sheen J: **Bacterial effectors target the common signaling partner BAK1 to disrupt multiple MAMP receptor-signaling complexes and impede plant immunity.** *Cell Host Microbe* 2008, **4:**17–27.
41. Chen CC, Liang CS, Kao AL, Yang CC: **HHP1, a novel signalling component in the cross-talk between the cold and osmotic signalling pathways in *Arabidopsis*.** *J Exp Bot* 2010, **61:**3305–3320.
42. Michael TP, Mockler TC, Breton G, McEntee C, Byer A, Trout JD, Hazen SP, Shen R, Priest HD, Sullivan CM, Givan SA, Yanovsky M, Hong F, Kay SA, Chory J: **Network discovery pipeline elucidates conserved time-of-day-specific cis-regulatory modules.** *PLoS Genet* 2008, **4:**e14.
43. Wang Y, Wu JF, Nakamichi N, Sakakibara H, Nam HG, Wu SH: **LIGHT-REGULATED WD1 and PSEUDO-RESPONSE REGULATOR9 form a positive feedback regulatory loop in the *Arabidopsis* circadian clock.** *Plant Cell* 2011, **23:**486–498.
44. Borevitz JO, Xia Y, Blount J, Dixon RA, Lamb C: **Activation tagging identifies a conserved MYB regulator of phenylpropanoid biosynthesis.** *Plant Cell* 2000, **12:**2383–2394.
45. Lacombe S, Rougon-Cardoso A, Sherwood E, Peeters N, Dahlbeck D, van Esse HP, Smoker M, Rallapalli G, Thomma BP, Staskawicz B, Jones JD, Zipfel C: **Interfamily transfer of a plant pattern-recognition receptor confers broad-spectrum bacterial resistance.** *Nat Biotechnol* 2010, **28:**365–369.
46. Yi H, Mysore KS, Gelvin SB: **Expression of the *Arabidopsis* histone *H2A-1* gene correlates with susceptibility to *Agrobacterium transformation*.** *Plant J* 2002, **32:**285–298.
47. Ranf S, Eschen-Lippold L, Pecher P, Lee J, Scheel D: **Interplay between calcium signalling and early signalling elements during defence responses to microbe- or damage-associated molecular patterns.** *Plant J* 2011, **68:**100–113.
48. Millet YA, Danna CH, Clay NK, Songnuan W, Simon MD, Werck-Reichhart D, Ausubel FM: **Innate immune responses activated in *Arabidopsis* roots by microbe-associated molecular patterns.** *Plant Cell* 2010, **22:**973–990.
49. Willmann R, Lajunen HM, Erbs G, Newman M-A, Kolb D, Tsuda K, Katagiri F, Fliegmann J, Bono J-J, Cullimore JV, Jehle AK, Götz F, Kulik A, Molinaro A, Lipka V, Gust AA, Nürnberger T: ***Arabidopsis* lysin-motif proteins LYM1 LYM3 CERK1 mediate bacterial peptidoglycan sensing and immunity to bacterial infection.** *Proc Natl Acad Sci USA* 2011, **108:**19824–19829.
50. Dow M, Newman MA, von Roepenack E: **The induction and modulation of plant defense responses by bacterial lipopolysaccharides.** *Annu Rev Phytopathol* 2000, **38:**241–261.
51. He P, Shan L, Lin NC, Martin GB, Kemmerling B, Nurnberger T, Sheen J: **Specific bacterial suppressors of MAMP signaling upstream of MAPKKK in *Arabidopsis* innate immunity.** *Cell* 2006, **125:**563–575.
52. Hellens R, Mullineaux P, Klee H: **Technical focus:a guide to *Agrobacterium* binary Ti vectors.** *Trends Plant Sci* 2000, **5:**446–451.
53. Van Larebeke N, Engler G, Holsters M, Van den Elsacker S, Zaenen I, Schilperoort RA, Schell J: **Large plasmid in *Agrobacterium tumefaciens* essential for crown gall-inducing ability.** *Nature* 1974, **252:**169–170.
54. Burch-Smith TM, Schiff M, Liu Y, Dinesh-Kumar SP: **Efficient virus-induced gene silencing in *Arabidopsis*.** *Plant Physiol* 2006, **142:**21–27.
55. Jackson AO, Larkins BA: **Influence of ionic strength, pH, and chelation of divalent metals on isolation of polyribosomes from tobacco leaves.** *Plant Physiol* 1976, **57:**5–10.

56. Kim K-W, Franceschi V, Davin L, Lewis N: **β-Glucuronidase as reporter gene.** In *Methods in Molecular Biology, Arabidopsis Protocols. Volume 323.* Edited by Salinas J, Sanchez-Serrano J. Totowa, New Jersey, USA: Humana Press; 2006:263–273.
57. Lange H, Shropshire W, Mohr H: **An analysis of phytochrome-mediated anthocyanin synthesis.** *Plant Physiol* 1971, **47:**649–655.

7

Protocol: a beginner's guide to the analysis of RNA-directed DNA methylation in plants

Huiming Zhang[1*], Kai Tang[1], Bangshing Wang[1], Cheng-Guo Duan[1], Zhaobo Lang[1] and Jian-Kang Zhu[1,2*]

Abstract

Background: DNA methylation is a conserved epigenetic mark that controls genome stability, development and environmental responses in many eukaryotes. DNA methylation can be guided by non-coding RNAs that include small interfering RNAs and scaffold RNAs. Although measurement of DNA methylation and regulatory non-coding RNAs is desirable for many biologists who are interested in exploring epigenetic regulation in their areas, conventional methods have limitations and are technically challenging. For instance, traditional siRNA detection through RNA hybridization requires relatively large amount of small RNAs and involves radioactive isotopes. An alternative approach is RT-qPCR that employs stem loop primers during reverse transcription; however, it requires a prerequisite that the exact sequences of siRNAs should be known.

Results: By using the model organism Arabidopsis thaliana, we developed an easy-to-follow, integrative procedure for time-efficient, quantitative measurement of DNA methylation, small interfering RNAs, and scaffold RNAs. Starting with simplified nucleic acid manipulation, we examined DNA methylation levels by using Chop PCR (methylation-sensitive enzyme digestion followed by PCR), which allowed for fast screening for DNA methylation mutants without the need of transgenic reporters. We deployed a simple bioinformatics method for mining published small RNA databases, in order to obtain the nucleotide (nt) sequences of individual 24nt siRNAs within the regions of interest. The protocol of commercial TaqMan Small RNA Assay was subsequently optimized for reliable quantitative detection of individual siRNAs. We used nested qPCR to quantify scaffold RNAs that are of low abundance and without Poly-A tails. In addition, nuclei fraction enables separation of chromatin-associated scaffold RNAs from their cognate non-scaffold transcripts that have been released from chromatin.

Conclusions: We have developed a procedure for quantitative investigations on nucleic acids that are core components of RNA-directed DNA methylation. Our results not only demonstrated the efficacy of this procedure, but also provide lists of methylation-sensitive restriction enzymes, novel DNA methylation marker loci, and related siRNA sequences, all of which can be valuable for future epigenetic studies. Importantly, step-by-step protocols are provided in details such that the approaches can be easily followed by biologists with little experience in epigenetics.

Keywords: Epigenetics, RNA-directed DNA methylation, siRNA, Scaffold RNA, Pol IV, Pol V, Nuclei fractionation

Background

DNA methylation is a major epigenetic mark that can confer transcriptional silencing of genes and transposable elements. RNA-directed DNA methylation (RdDM), which involves complimentary pairing between small RNAs [including small interfering RNAs (siRNAs) and piwi-associated RNAs (piRNAs)] and scaffold RNAs, has been established in plants and mammals [1-5]. DNA methylation is subject to dynamic regulation during establishment, maintenance, and removal in response to developmental and environmental cues [6]. In order to understand epigenetic regulation, it is important to characterize the histone modification patterns and to determine the levels of DNA methylation and small RNAs, as well as of scaffold RNAs that are chromatin-associated long non-coding RNAs (Figure 1A). This study focuses on assays for DNA methylation, small RNAs and scaffold RNAs, since methods for characterizing histone modification patterns such as chromatin immunoprecipitation (ChIP) have been covered

* Correspondence: zhang731@purdue.edu; jkzhu@purdue.edu
[1]Department of Horticulture and Landscape Architecture, Purdue University, West Lafayette, IN 47907, USA
[2]Shanghai Center for Plant Stress Biology, Shanghai Institute of Biological Sciences, Chinese Academy of Sciences, Beijing, Shijingshan, China

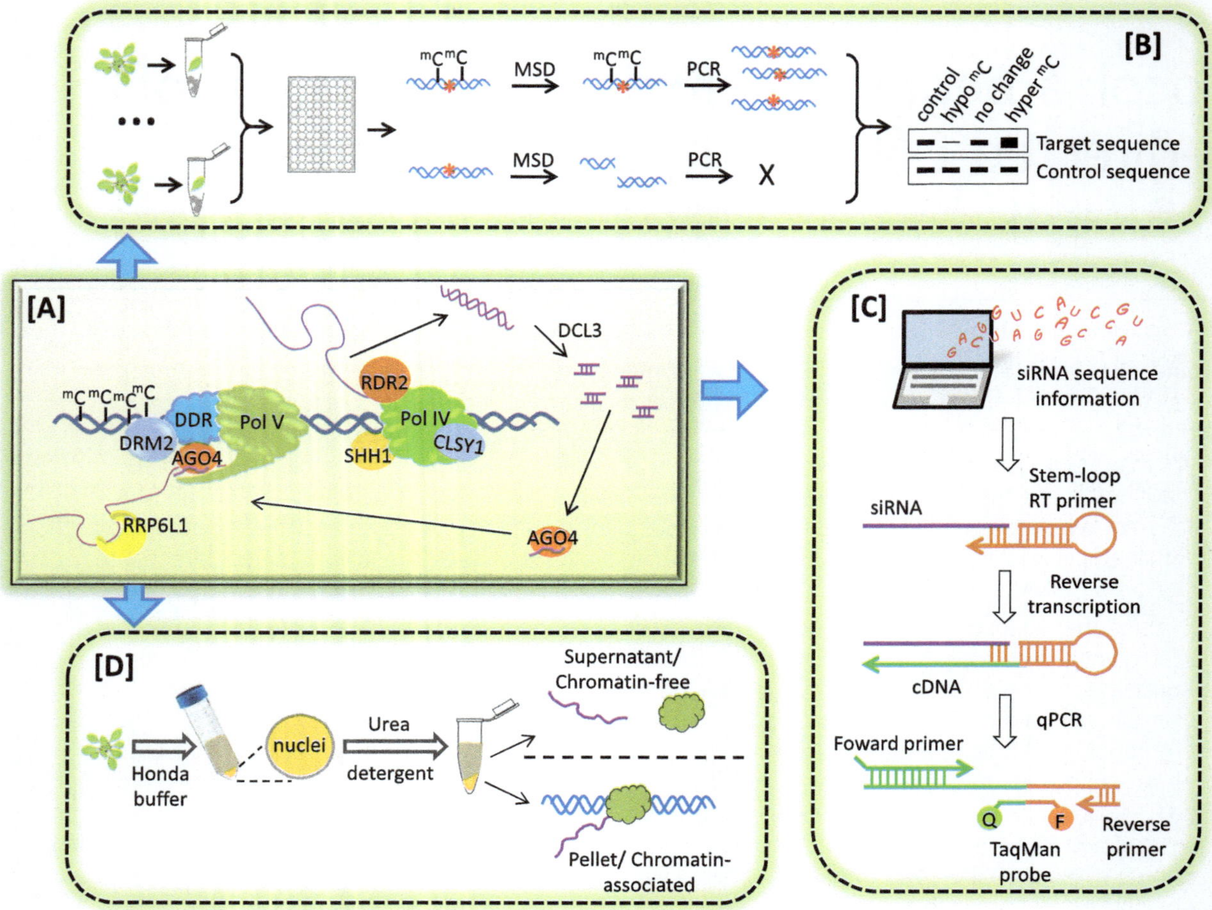

Figure 1 An integrative procedure for studying nucleic acids in RdDM. [A] A model of RdDM in Arabidopsis. Briefly, Pol IV initiates siRNA production by producing single-stranded non-coding RNAs, while Pol V synthesizes scaffold RNAs that recruits Argonaute-associated siRNAs. The complimentary pairing between the two types of non-coding RNAs, as well as physical interactions between proteins in the methylation complex, results in targeting of the methyltransferase DRM2 to RdDM loci. For simplicity, not all known RdDM components are shown. **[B]** Chop PCR-based screen for mutants showing abnormal DNA methylation patterns. Red asterisks represent enzyme restriction sites. Cytosine methylation overlapping restriction sites will protect DNA from methylation-sensitive digestion (MSD), while enzymatic cleavage results in failure of subsequent PCR amplification of the target sequences. Note that the enzyme McrBc is special in that it cleaves methylated- but not unmethylated DNA sequences (see also Additional file 2: Table S1). **[C]** RT-qPCR detection of siRNAs by using stem-loop primers during reverse transcription. **[D]** Schematic procedure of nuclei fractionation, which separates chromatin-associated RNAs from RNAs that are not associated with chromatin. See Methods for details.

extensively [7,8]. Genome-wide DNA methylation and siRNA accumulation patterns can be precisely profiled by next-generation-sequencing. At individual loci, conventional methods for obtaining data on DNA methylation and non-coding RNAs include bisulfite sequencing for DNA methylation, RNA hybridization for siRNAs, and RT-PCR (reverse transcription followed by polymerase chain reaction) for scaffold RNAs [9,10]. Although commonly used, these methods have limitations. Bisulfite sequencing is costly and time-consuming, and the assay requires considerable experience in several important steps of the experimental procedure including designing suitable primers. Northern blotting of siRNA requires large amounts of small RNA input and involves radioactive isotopes. RT-PCR is semi-quantitative and, if applied alone, does not distinguish scaffold RNAs from those transcripts that have been released from chromatin. To overcome these limitations, we used the model organism *Arabidopsis thaliana* to develop PCR- or qPCR-based methods for studying nucleic acids in RdDM. These optimized approaches form an integrative procedure that can be followed by biologists with little experience in epigenetic studies (Figure 1).

Results and discussion

The first part of this procedure is to detect alteration in DNA methylation levels. Alternative to bisulfite sequencing, DNA methylation levels in different samples can be compared by Chop PCR, in which methylation-sensitive enzyme digestion of DNA is followed by PCR amplification of the target sequence (Figure 1B). By using Chop PCR, we developed a method for epigenetic mutant screen that does not need a transgenic reporter gene. A large population of Arabidopsis mutants is commercially available from seed stock sources such as ABRC (Arabidopsis Biological Resource Center) and can be used for Chop PCR-based mutant screen. Genomic DNA samples were prepared by a simple home-made recipe that is suitable for simultaneous DNA extraction from multiple samples. We used the methylation-sensitive restriction enzyme Hae III and examined DNA methylation at *AtSN1*, a known RdDM target locus. Hundreds of Arabidopsis mutants were screened, resulting in discovery of several DNA hypomethylation mutants including *atrrp6l1* alleles that were recently characterized [11]. In addition to Hae III, we have examined some other methylation-sensitive restriction enzymes and correspondingly demonstrated the efficacy of these enzymes in examining several RdDM marker loci in Arabidopsis (Additional file 1: Figure S1; Additional file 2: Table S1). Different restriction enzymes can be used to assay methyaltion in the different sequence contexts, i.e., CG, CHG, and CHH (H represents A, T, or C). Although Chop PCR examines less cytosine than bisulfite sequencing, observation of DNA hypomethylation by Chop PCR in multiple marker loci is sufficient to demonstrate an epigenetic function. Quantification of the difference in DNA methylation levels can be achieved by Chop qPCR (Additional file 1: Figure S1B). Parallel amplification of the marker locus from non-digested DNA samples is commonly used to show loading controls. However, we recommend amplifying DNA sequences without restriction sites, from the same digested DNA samples, as internal controls, because experimental accuracy can be increased while the amount of required DNA can be reduced. Compared to traditional forward genetic screens that involve transgenic reporter genes, Chop PCR-based screen does not require generation of a transgenic reporter line and the subsequent mutated population, neither required is the process of map-based cloning that can be time-consuming.

The second part of this procedure is quantification of siRNAs (Figure 1C). The levels of siRNAs can be semi-quantitatively detected by RNA hybridization, which typically uses ≥20 μg RNAs that are enriched in sizes smaller than 200 bp [12]. The whole process takes several days from RNA extraction to final results. Alternative to RNA hybridization, small RNAs can be detected by RT-qPCR that employs stem loop primers during reverse transcription [13]. This approach takes only a few hours starting from RNA extraction and involves the patented TaqMan Small RNA Assays. However, the protocol of TaqMan Small RNA Assays, which was initially implemented to detect miRNAs, does not readily enable detection of lower abundance small RNAs such as 24 nt siRNAs that mediate RdDM. In addition, these PCR-based quantitative measurements require, as a prerequisite, knowing the exact sequences of small RNAs; meanwhile mining published small RNA datasets for such information typically requires skills in bioinformatics. Therefore we developed a simple bioinformatics method for selecting siRNAs to quantify and optimized the TaqMan Small RNA Assays protocol for detecting 24 nt siRNAs.

To obtain nucleotide sequences for 24 nt siRNAs from the regions of interest, we searched small RNA sequencing databases. This can be done by BLAST searches of published small RNA databases, such as Zhang et al. [14], for siRNAs that match the regions of interest. However, such analysis requires skills in bioinformatics. We thus provide an alternative analysis approach that can be applied in any laboratory, by using Arabidopsis *AtSN1* locus as an example. *AtSN1* locates within 1 kb downstream of the gene At3g44006 [9]. Browsing *At3g44006* and its downstream region in IGB (Integrated Genome Brower) identified *AtSN1* as *At3TE63860*, which is a transposable element that starts at 15794808 and ends at 15794619. Arabidopsis small RNA nucleotide sequence information with genomic coordinates is available in the UDelaware Small RNA database [15]. Since 24 nt siRNAs are proposed to pair with scaffold RNAs [1-4], we screened *AtSN1* small RNAs for 24 nt siRNAs that are annotated as Waston strand, i.e., template DNA strand. This means that these siRNAs have the same sequences as the template DNA strand (despite the U/T difference) and thus can potentially pair with scaffold RNAs. To increase the chance of detecting individual siRNAs, these 24 nt siRNAs were subjected to a second screen for those that showed abundance of ≥ 1 hit per million reads, according to the UDelaware Small RNA database, in the wild type *Col-0* ecotype. Among these candidates, one siRNA exhibited sequence homology to a 43 nt probe that had been used for detecting *AtSN1* siRNAs by Northern blots [10]. Thus we chose this siRNA for customer TaqMan Small RNA Assay design.

The manufacture-recommended protocol of TaqMan Small RNA Assay suggests using 1-10 ng total RNA per reaction. We were able to observe stress-induced regulation of miRNAs by using ≥ 50 ng total RNA as input (Additional file 1: Figure S2). However, most 24 nt siRNAs that we examined were undetectable under similar condition. We found that 400 ng small RNA (the ≤200 bp fraction of total RNA), instead of total RNA, per RT reaction is sufficient for consistently detecting all 24 siRNAs that we tested, including the *AtSN1* siRNA (Figure 2A). We used *snoR101*, a small nucleolar RNA, as the internal control to ensure equal

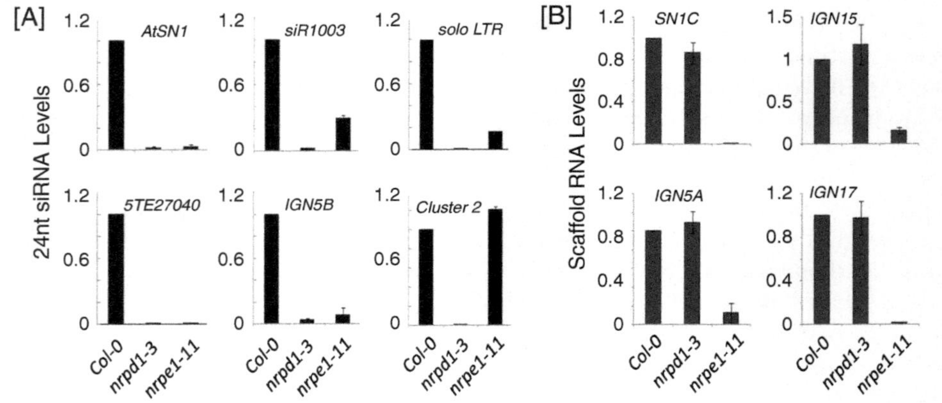

Figure 2 Quantitative measurement of 24nt siRNAs and scaffold RNAs in Arabidopsis. [A] Quantification of individual 24nt siRNAs by TaqMan small RNA assays. The levels of siRNAs were normalized by using *snoR101* as internal control. **[B]** Quantification of Pol V-dependent scaffold RNAs by gene/scaffold RNA-specific RT-qPCR. *Actin2* was used as internal control. Error bars indicate SD, n = 3 technical replicates. All experiments were performed with at least two biological repeats. Representative images are shown.

RNA input. During the RT reactions, we mixed RNA with 2 μL 20X (manufacture-defined concentration) siRNA RT primer together with 2 μL *snoR101* RT primer, instead of using 3 μL siRNA RT primer alone as suggested by the manual. Subsequently, the levels of siRNA and *snoR101* were independently measured by quantitative PCR. Because siRNA and *snoR101* are simultaneously reverse transcribed in the same tube, the level of *snoR101* reflects RT efficiency of each sample, allowing for elimination of false results that may be caused by different RT efficiency between samples.

The effectiveness of using TaqMan Assays to quantify 24nt siRNA levels was confirmed by using the Arabidopsis mutants *nrpd1-3* that is defective in RNA polymerase Pol IV, which is required for biogenesis of almost all 24nt siRNAs [14], and *nrpe1-11* that is defective in Pol V, of which mutation decreases the levels of some Pol IV-dependent siRNAs [16]. Compared to wild type plants (*Col-0*), *nrpd1-3* exhibited depletion of all examined siRNAs including *AtSN1, siR1003, solo LTR, 5TE27040, IGN5B,* and *Cluster 2* (Figure 2A). The *nrpe1-11* mutant also showed substantial reduction in the levels of the examined siRNAs except *Cluster 2* (Figure 2A), which is known to be Pol IV-dependent and Pol V-independent [17]. The patterns observed in our RT-qPCR measurements are consistent with published RNA hybridization results [10,17] and whole genome small RNA sequencing results [18]. Sequences of these individual 24nt siRNAs are listed in Additional file 2: Table S2. Designing siRNA probes requires only submitting individual siRNA sequences to the manufacturer website of the patented TaqMan Assays.

The third part of studying RdDM nucleic acids is to quantitatively measure chromatin-associated scaffold RNAs. In Arabidopsis, Pol V transcription generates scaffold RNAs that are of low abundance and are devoid of poly-A tail [9]. Therefore we used scaffold RNA-specific primers during reverse transcription for the first-strand cDNA synthesis, and subsequently performed nested qPCR. Such an approach successfully measures Pol V-transcribed scaffold RNAs, as demonstrated by comparison between *Col-0* and *nrpe1-11* (Figure 2B). Although the levels of scaffold RNAs are in proportion to the pool of total Pol V transcripts, sometimes nuclei fractionation is needed to distinguish chromatin-associated scaffold RNAs from those Pol V transcripts that have been released from chromatin (Figure 1D), as was the case in our recent study of AtRRP6L1 that functions in retention of Pol V transcripts at chromatin [11]. Nuclei can be isolated with Honda buffer and subsequently lysed in the presence of urea and detergent, followed by centrifugation that sediments chromatin-associated RNAs in the pellet (chromatin-associated fraction), while leaving released RNAs in the supernatant (chromatin-free fraction) (Figure 1D). As an indication of successful nuclei fractionation, genomic DNA should be nearly absent in the chromatin-free fractions (Additional file 1: Figure S3).

To provide an easy-to-follow procedure, experimental details are described step by step in the Methods.

Conclusions

To summarize, we have developed an integrative procedure for studying nucleic acids that are central players in RNA-directed DNA methylation and other epigenetic regulation (Figure 1). The results not only demonstrated the efficacy of this procedure (Figure 2; Additional file 1: Figure S1-S3), but also provide a list of novel RdDM marker loci (Additional file 2: Table S1) and the related methylation-sensitive restriction enzymes (Additional file 2: Table S1), as well as the nucleotide sequences of a list of individual 24nt siRNA (Additional file 2: Table S2), which can be valuable for future epigenetic studies. Moreover, this procedure requires little previous experimental experience in epigenetics or bioinformatics, and thus can broadly benefit the scientific community.

Methods

I. Chop-PCR/qPCR

<I-1> Preparation of genomic DNA samples

1. Prepare 1.5 mL eppendorf tubes, each containing 3 Chrome steel beads (BioSpec Products, Inc, Cat.# 11079132C).
   * Note: Steps 1 to 3 are optimized for preparing multiple samples. Start from Step 4 if samples are homogenized by other methods.
2. Take one or two mature leaves and put into labeled tubes. Make sure the tubes are tightly capped.
3. Freeze tubes in liquid nitrogen shortly, then shake tubes vigorously in a box to shatter the samples.
   *Note: Be aware that a capped tube with liquid nitrogen in it will pop open or even explode under room temperature.
4. Add 500 µL 2% CTAB (Cetyltrimethyl ammonium bromide) solution to each tube (or per ~100 mg sample if homogenized by other methods), then incubate at 65°C for 30-60 min.
   * Note: To make 100 mL CTAB solution (modified from [19]): 2g CTAB, 10 mL 1 M Tris-HCl (pH = 7.5), 28 mL 5 M NaCl, 4 mL 0.5 M EDTA (pH = 8.0), H_2O.
5. Add 500 µL chloroform, then votex vigorously and centrifuge at ≥ 12000 g at room temperature for 10 min.
6. Transfer aqueous phase into new tubes, then mix with equal volume isopropanol.
7. Store at -20°C for ≥ 30 min.
8. Centrifuge at ≥ 12000 g at room temperature for 10 min.
9. Wash the pellet with 0.5 mL 70% ethanol.
10. Air dry pellets and dissolve in 50 µL nuclease-free water (we use autoclaved double-distilled water).
11. Proceed to steps in < I-2 > for Chop-PCR if doing mutant screening. Alternatively, for Chop-qPCR, continue with DNA purification as described below.
    * Note: Although DNA samples extracted by this way have RNA contamination, they can be used for rough estimations of DNA methylation levels during initial mutant screen. RNase treatments should be performed for Chop qPCR and for semi-quantitative Chop PCR that is used for the purposes other than rough estimation.
12. Add 2 µL RNase A (10mg/mL) per 500 µL DNA solution, mix well and incubate at 37°C for 1 hr.
    * Note 1: Concentration of DNA solutions can vary depending on the yield of crude extraction using CTAB solution.
    * Note 2: If making RNase A solution from powders, boiling the solution for 15–20 minutes usually eliminate most of its DNases.
13. Add equal volume *phenol: chloroform: isoamyl alcohol* (24:1:1, pH 8.0), vortex vigorously for 5 sec, then centrifuge at ≥ 12000 g at room temperature for 10 min.
14. (Optional) Transfer the aqueous phase to a new tube, repeat Step 2.
15. Transfer the aqueous phase to a new tube, add equal volume *phenol: chloroform*, vortex vigorously for 5 sec, then centrifuge at ≥ 12000 g at room temperature for 8 min.
16. Transfer the aqueous phase to a new tube, add 2 volume ethanol, 0.1 volume NaOAc (pH 5.2, 3 M), mix well.
17. Store at -20°C for ≥ 30 min.
18. Centrifuge at ≥ 12000g at room temperature for 10 min.
19. Wash the pellet 2X with 0.5 mL 70% ethanol.
20. Air dry pellets and dissolve in nuclease-free water.

<I-2 > Enzymatic DNA digestion and PCR/qPCR

1. Measure genomic DNA concentration using Nanodrop.
2. Digest equal amount (typically 0.5 to 1 µg) of genomic DNA with a methylation-sensitive enzyme in a 20 µl reaction mixture.
   * Note: While following manufacture protocols to set up reactions for each enzyme, the digestion time can be extended to overnight.
3. After digestion, check DNA methylation levels at the enzyme-targeted marker locus by qPCR or PCR. For semi-quantitative PCR,use 1 µL of the digested DNA as template in a 20 µl reaction mixture that contains 0.15 unit of commercial Taq DNA polymerase. Run a DNA gel (1.2% agarose) to compare the PCR results from different samples. For qPCR, prepare 20 µl reaction mixture following manufacture protocols and use 1 µl of the digested DNA as template per reaction.
   * Note 1: DNA input amounts in different samples should be normalized by amplifying an internal control sequence that does not contain the restriction site of the enzyme used. Alternatively, non-digested DNA can be used as PCR/qPCR templates for input normalization.
   * Note 2: In semi-quantitative PCR, too many PCR cycles will probably yield in saturated amplification and thereby mask small differences in template inputs. For optimized observation of differences,

PCR can be repeated with increased or decreased cycle numbers after initial assessment. A positive control, e.g., the Pol IV mutant *nrpd1-3*, should be examined in parallel to the wild type.

Optional: For Chop-qPCR, follow steps 4 to 10 to analyze the results

4. Export qPCR results to Excel file.
5. Calculate average Ct value for the non-digested control in each sample.
6. Calculate "ΔCt = Ct of the examined amplicons – average Ct of *the non-digested control*".
7. Calculate average ΔCt value of one sample (e.g., wild type plants).
8. Calculate the comparative Ct values ($\Delta\Delta$Ct) for other samples (e.g., mutant plants). $\Delta\Delta Ct = \Delta Ct_{(mutant)} - \text{average } \Delta Ct_{(wild\ type)}$.
9. Calculate relative abundance of the amplicons by using this function: "= 2^(-ΔΔCt)".
10. Calculate average of relative abundance of the amplicons; and calculate standard deviation by using the function "=STDEV(values to be calculated)".

II. siRNA RT-qPCR

<II-1> Isolation of small RNAs

1. Homogenize ≤100 mg plant tissues in liquid nitrogen by using mortar and pestle. Transfer the powder into a 1.5 mL eppendorf tube (pre-chilled in liquid nitrogen) and add 1 mL RNAzol® RT (Molecular Research Center Inc)
2. Add 0.4 mL water per 1 mL of RNAzol® RT. Mix well and store at room temperature for 5-15 min.
   * Note: Autoclaved distilled water is good enough for this step.
3. Centrifuge the mixture ≥12,000 g at room temperature for 15 min.
   * Note: this step precipitates DNA, protein, and polysaccharide.
4. Transfer the supernatant to a new tube and add 0.4 mL 75% ethanol per 1 mL of supernatant. Mix well and store at room temperature for 10 min.
5. Centrifuge ≥12,000 g for 8 min at room temperature or at 4°C.
   * Note: this step pellets RNA > 200 bp.
6. Transfer the supernatant to a new tube and add 0.8 volume of isopropanol. Mix well and store at 4°C for 30 min.
   * Note: the mixture can be stored at -20°C for longer time.
7. Centrifuge ≥12,000g for 15 min at room temperature or at 4°C.
8. Wash the pellet with 70% isopropanol twice.
   * Note: After removing the bulk isopropanol solution, centrifuge briefly and use pipette to remove the residual liquid. By doing this, RNA pellets dissolved without drying perform well in the downstream RT-PCR reactions.
9. Dissolve the pellet in 20-40 μL RNase-free water, without drying, in RNase-free water. Keep on ice and proceed to reverse transcription, or store at -80°C.

<II-2> Quantitative measurement of siRNAs by RT-qPCR

1. Search small RNA databases to determine sequences of siRNAs to be examined (see main text for an example).
   * Note: Additional file 2: Table S2 lists some Arabidopsis 24nt siRNAs that have been detected by RT-qPCR in this study.
2. Submit siRNA sequences to Life Technologies for TaqMan Small RNA Assay design by following the online instructions.
3. Place the order after the designing has been completed.
   * Note: we suggest downloading the TaqMan Small RNA Assay protocol at this point, as well as to order the TaqMan MicroRNA Reverse Transcription Kit and the TaqMan Universal PCR Master Mix as recommended in the protocol.
4. Use TaqMan MicroRNA Reverse Transcription Kit to prepare reverse transcription (RT) master mixture and keep on ice. The following recipe is for one reaction:
   100 mM dNTPs 0.15 μL
   MultiScribe TM Reverse Transcriptase (50 unit/μL) 1.0 μL
   10X Reverse Transcription Buffer 1.5 μL
   RNaseOut (40 unit/μL) 0.9 μL
   RNase-free Water 4.26 μL
   Total Volume: 7 μL
5. In an RNase-free PCR tube, add 400 ng small RNA in a volume of 5 μL. Then add 2 μL *snoR101* RT primer and 2 μL siRNA RT primer. Mix well gently.
   400ng small RNA is sufficient for consistently detecting all 24 nt siRNAs that we have examined. For certain abundant siRNAs such as siR1003 that originates from 5S rDNA repeats, 100 ng small RNA is sufficient.
6. Incubate the RNA-primers mixture at 85°C for 5 min, followed by 60°C for 5 min. Then put the mixture on ice immediately and for at least 1 min.
7. Add 7 μL RT mixture to each RNA-primers mixture. Mix well gently. Keep in ice for 5 min.
8. Perform RT reaction using the following conditions consecutively: 16°C for 30 min, 42°C for 30 min, 85°C for 5 min. Keep the samples on ice after RT reaction.

* Note: Because TaqMan stem-loop primer recognizes a specific short sequence and the qPCR reverse primer is complementary to the stem-loop RT primers (see Figure 1C in main text), genomic DNA contamination, if any, will not be a problem for this experiment. Thus no-RT control is unnecessary. In fact, we did not detect qPCR signals even using genomic DNA.

9. Prepare the qPCR mixture as the following:
   TaqMan Small RNA Assay (20×) 1.0 μL
   cDNA from RT reaction (Step 8) 1.2 μL
   TaqMan Universal PCR Master Mix II (2×) 10.0 μL
   Nuclease-free water 7.8 μL
   Total Volume: 20 μL
   * Note 1: run qPCR individually for *snoR101* and the siRNA and perform 3 technical repeats.
   * Note 2: The patented TaqMan Small RNA Assays provide qPCR primers in the tubes labeled as "TaqMan Small RNA Assay (20×)" without giving primers information.
10. Perform qPCR in a real-time PCR instrument using the following program:
    Step I: 50°C (optional) 2 min
    Step II: 95°C (Enzyme activation) 10 min
    Step III: 95°C (Template denaturing) 15 sec
    Step IV: 60°C (Anneal/extend) 1 min
    Step V: go to Step III, 39 times (totally 40 cycles)
    Step VI: End
    * Note: TaqMan Small RNA Assay PCR signals will be detected in the FAM channel by the qPCR instrument.
11. Export results to Excel file.
12. Calculate average Ct value for snoR101 in different replicates.
13. Calculate "ΔCt = Ct of siRNA − average Ct of *snoR101*" from the same RT reaction.
14. Calculate average ΔCt value of one RNA sample (e.g., wild type plants).
15. Calculate the comparative Ct values (ΔΔCt) for other samples (e.g., mutant plants). $\Delta\Delta Ct = \Delta Ct_{(mutant)} - \text{average } \Delta Ct_{(wild\ type)}$.
16. Calculate relative siRNA abundance by using this function: "= 2^(-ΔΔCt)".
17. Calculate average of relative siRNA abundance; and calculate standard deviation by using the function "=STDEV(values to be calculated)".
    * Note: Mean value and standard deviation can also be calculated based on multiple biological replicates. The expression patterns, e.g., release of transcriptional silencing, should always be the same for different biological replicates, although the standard deviation between different biological replicates is usually greater than that of technical repeats from one single biological replicate.

III. Scaffold RNA RT-qPCR

<III-1> Quantitative measurement of scaffold RNAs without nuclei fractionation

1. Extract total RNA by using Trizol (Invitrogen) following the manufacturer's protocol.
   * Note: Scaffold RNAs may be a few hundred nucleotides, so for this experiment we do not recommend using those commercial RNA extraction kits that preferentially extract RNA > 200 bp.
2. Remove DNA contamination in the RNA samples by using TURBO DNA-free™ (Ambion).
   * Note: We typically treat 10 μg RNA with TURBO DNase in a 20 μL reaction, and use maximum amount of RNA during RT reaction while following the manufacturer RT protocol.
3. Perform RT reaction by using SuperScript™ III First-Strand Synthesis SuperMix (Invitrogen). In each reaction, use mixed gene/scaffold RNA-specific primers (the scaffold RNA and the internal control *ACTIN2*). For a 10 μL reaction, use 0.5 μL pre-mixed primers (2 mM each).
   * Note 1: We have tried using some other RT kits and found that, in our hands, SuperScript works best for low-abundance non-coding RNA detection.
   * Note 2: We recommend using the same amount of RNA for each RT reaction. We typically use 1 μg total RNA for a reaction in 10 μL.
   * Note 3: No-RT control reactions can be performed by making RT mixtures as describe above but without reverse transcriptase. This will ensure no false qPCR signals from DNA contamination in RNA samples.
4. Perform qPCR by using SYBR® Green Master Mix. PCR for scaffold RNA and *ACTIN2* should be separated.
   Step I: 95°C 3 min
   Step II: 95°C 10 sec
   Step III: Tm 30 sec
   Step IV: go to Step II 39 times + plate read (totally 40 cycles)
   Step V: increase from 65°C to 95°C, increment 0.5°C for 5 sec, plus plate read
   Step VI: End
   * Note 1: Because cDNAs are synthesized by using gene-/scaffold RNA-specific primers, PCR amplification that follows has to be nested PCR, i.e., positions of the qPCR primers on the DNA coding strand should be upstream of the gene-/scaffold RNA-specific RT primers.
   * Note 2: A single peak in the melt curve generated by Step V should indicate the specificity

of the primers. This can also be examined by DNA gel electrophoresis.

5. Use *ACTIN2* as internal control to analyze the results (Refer to Steps *V-11 to V-17* for details).
   * Note: By starting with equal amount of RNA, different samples should show similar Ct value for *ACTIN2* in qPCR reactions. Alternatively, another housekeeping gene can be examined to ensure similar RNA quality.

<III-2> Quantitative measurement of scaffold RNAs with nuclei fractionation

1. Prepare Honda buffer (0.4 M Sucrose, 2.5% Ficoll, 5% Dextran T40, 25 mM Tris-HCl, pH 7.4, 10 mM MgCl2, 0.5% Triton X-100, 0.5 mM PMSF, 0.1% pepstatin A, 0.1% aprotinin, 10 mM β-mercaptoethanol, 8 unit/mL RNaseOUT) (modified from [9]).
   * Note: add PMSF, pepstatin A, aprotinin, β-mercaptoethanol, and RNaseOUT before use.
2. Grind 3 gram ~12-day-old plants with liquid nitrogen.
   * Note: We recommend using the same amount of plant materials for each sample.
3. Transfer the power to a 50 mL Falcon tube prefilled with 15 mL cold Honda buffer. Homogenize by vortex vigorously. Keep the tubes in ice for 15-30 min.
4. Filter the homogenate through 4 layers of Miracloth, followed by centrifugation at 1500 g at 4°C for 5 min.
5. Re-suspend the pellet in 1 mL Honda buffer, then transfer to a new 1.5 mL eppendorf tube, followed by centrifugation at 1800 g at 4°C for 5 min to pellet the nuclei.
   * Note: Save the supernatant for cytosol analysis if wished.
6. Gently rinse the nuclei pellet with ice-cold wash buffer (1X PBS, 1 mM EDTA, 8 unit/mL RNaseOUT).
7. Re-suspend the pellet in 0.5 mL prechilled glycerol buffer (20 mM Tris-HCl, pH 7.9, 75 mM NaCl, 0.5 mM EDTA, 0.85 mM DTT, 50% glycerol, 0.125 mM PMSF, 0.1% pepstatin A, 0.1% aprotinin, 10 mM β-mercaptoethanol, 160 unit/mL RNaseOUT) by gently flicking the tube.
8. Add 0.5 mL cold nuclei lysis buffer (10 mM HEPES, pH 7.6, 1mM DTT, 7.5 mM MgCl2, 0.2 mM EDTA, 0.3 M NaCl, 1 M Urea, 1% NP-40, 0.5 mM PMSF, Proteinase inhibitor cocktail, 0.1% pepstatin A, 0.1% aprotinin, 10 mM β-mercaptoethanol, 160 unit/mL RNaseOUT).
9. Gently vortex for 2 × 2 sec, incubate on ice for 2 min, centrifuge at 14,000 rpm at 4°C for 2 min to separate the chromatin fraction and the nucleoplasmic fraction.
10. Treat the supernatant (nucleoplasmic fraction) with proteinase K at 37°C for 1 hr, followed by traditional acidic phenol/chloroform extraction and ethanol precipitation for RNA.
11. Gently rinse the chromatin pellet twice with 0.5 mL cold wash buffer and then dissolve in 1 mL Trizol for RNA extraction.
12. Refer to *Section III-1* for RT-qPCR of scaffold RNAs.

Additional files

Additional file 1: Figure S1. Examples of methylation-sensitive Chop PCR and Chop qPCR. DNA methylation at a group of transposon loci were examined by Chop PCR or Chop qPCR. Mutants were compared with the wild type Arabidopsis (*Col-0*). For BsmF I, BstBI, or Hae III digestion, a DNA sequence from *SKP1* without the restriction sites was examined as loading control. For HpyCH4 IV, 5TE27040 was amplified as non-digestion control. **[A]** Chop PCR. Restriction enzymes BsmF I, HpyCH4 IV, BstB I, and Hae III are indicted on the right. The examined loci are annotated on the left. **[B]** Chop qPCR. Hae III was used for digestion. Methylation levels, as indicated by qPCR signals, in the mutants were relative to those in *Col-0*. *SKP1* was used as loading control. Error bars indicate SD, n ≥ 3. Primers are listed in Table S3. **Figure S2.** Detection of Arabidopsis miRNAs by using TaqMan Small RNA Assay. Aerial portions of plants grown with (P deficient) or without (control) phosphate deficiency stress were examined. Each RT reaction used 80 ng total RNA and *snoR101* was used as internal control. Error bars indicate SD, n = 3. **Figure S3.** Quality control of nuclei fractionation. qPCR detection of genomic DNA (gDNA) in the chromatin-associated fraction (pellet) and the chromatin-free fraction (supernatant) after nuclei fractionation. *SKP1* gDNA levels in the chromatin-free fraction were presented relative to those in chromatin-associated fraction. Error bars indicate SD, n = 3. Primers are listed in Table S3. (PDF 363 kb)

Additional file 2: Table S1. Methylation-sensitive restriction enzymes for Chop PCR. **Table S2.** Arabidopsis 24nt siRNAs detected by using TaqMan Small RNA Assays. **Table S3.** Primers used for PCR/qPCR in this report.

Competing interests

The authors declare that they have no competing interests.

Authors' contributions

HZ performed the experiments. KT participated in siRNA sequence analysis. BW participated in Chop PCR and siRNA RT-qPCR. C-GD participated in Chop PCR. ZL participated in Chop PCR. J-KZ conceived of the study and wrote the manuscript with HZ. All authors read and approved the final manuscript.

Acknowledgements

This work was supported by US National Institutes of Health grants to J-K Zhu. We thank Rebecca Stevenson and the reviewers for critical reading of the manuscript.

References

1. Matzke M, Kanno T, Daxinger L, Huettel B, Matzke AJM: **RNA-mediated chromatin-based silencing in plants.** *Curr Opin Cell Biol* 2009, **21**:367–376.
2. Law JA, Jacobsen SE: **Establishing, maintaining and modifying DNA methylation patterns in plants and animals.** *Nat Rev Genet* 2010, **11**:204–220.
3. Zhang H, Zhu JK: **RNA-directed DNA methylation.** *Curr Opin Plant Biol* 2011, **14**:142–147.
4. Pikaard CS, Haag JR, Pontes OMF, Blevins T, Cocklin R: **A transcription fork model for Pol IV and Pol V-dependent RNA-directed DNA methylation.** *Cold Spring Harb Symp Quant Biol* 2012, **77**:205–212.
5. Watanabe T, Tomizawa SI, Mitsuya K, Totoki Y, Yamamoto Y, Kuramochi-Miyagawa S, Iida N, Hoki Y, Murphy PJ, Toyoda A, Gotoh K, Hiura H, Arima T, Fujiyama A, Sado T, Shibata T, Nakano T, Lin H, Ichiyanagi K, Soloway PD, Sasaki H: **Role for piRNAs and noncoding RNA in de novo DNA**

methylation of the imprinted mouse Rasgrf1 locus. *Science* 2011, **332**:848–852.

6. Zhang H, Zhu JK: Active DNA demethylation in plants and animals. *Cold Spring Harb Symp Quant Biol* 2012, **77**:161–173.
7. Saleh A, Alvarez-Venegas RUL, Avramova Z: An efficient chromatin immunoprecipitation (ChIP) protocol for studying histone modifications in Arabidopsis plants. *Nat Protoc* 2008, **3**:1018–1025.
8. Kasinathan S, Orsi GA, Zentner GE, Ahmad K, Henikoff S: High-resolution mapping of transcription factor binding sites on native chromatin. *Nat Methods* 2014, **11**:203–209.
9. Wierzbicki AT, Haag JR, Pikaard CS: Noncoding transcription by RNA polymerase Pol IVb/Pol V mediates transcriptional silencing of overlapping and adjacent genes. *Cell* 2008, **135**:635–648.
10. He XJ, Hsu YF, Pontes O, Zhu J, Lu J, Bressan RA, Pikaard C, Wang CS, Zhu JK: NRPD4, a protein related to the RPB4 subunit of RNA polymerase II, is a component of RNA polymerases IV and V and is required for RNA-directed DNA methylation. *Genes Dev* 2009, **23**:318–330.
11. Zhang H, Tang K, Qian W, Duan C, Wang B, Zhang H, Wang P, Zhu X, Lang Z, Yang Y, Zhu JK: An Rrp6-like protein positively regulates non-coding RNA levels and DNA methylation in Arabidopsis. *Mol Cell.* in press.
12. Xie Z, Johansen LK, Gustafson AM, Kasschau KD, Lellis AD, Zilberman D, Jacobsen SE, Carrington JC: Genetic and functional diversification of small RNA pathways in plants. *PLoS Biol* 2004, **2**:E104.
13. Chen C, Ridzon DA, Broomer AJ, Zhou Z, Lee DH, Nguyen JT, Barbisin M, Xu NL, Mahuvakar VR, Andersen MR, Lao KQ, Livak KJ, Guegler KJ: Real-time quantification of microRNAs by stem-loop RT-PCR. *Nucleic Acids Res* 2005, **33**:e179–e179.
14. Zhang X, Henderson IR, Lu C, Green PJ, Jacobsen SE: Role of RNA polymerase IV in plant small RNA metabolism. *Proc Natl Acad Sci U S A* 2007, **104**:4536–4541.
15. Nakano M, Nobuta K, Vemaraju K, Tej SS, Skogen JW, Meyers BC: Plant MPSS databases: signature-based transcriptional resources for analyses of mRNA and small RNA. *Nucleic Acids Res* 2006, **34**(Database issue):D731–5.
16. Mosher RA, Schwach F, Studholme D, Baulcombe DC: PolIVb influences RNA-directed DNA methylation independently of its role in siRNA biogenesis. *Proc Natl Acad Sci U S A* 2008, **105**:3145–3150.
17. Zheng B, Wang Z, Li S, Yu B, Liu JY, Chen X: Intergenic transcription by RNA polymerase II coordinates Pol IV and Pol V in siRNA-directed transcriptional gene silencing in Arabidopsis. *Genes Dev* 2009, **23**:2850–2860.
18. Zhang H, Ma ZY, Zeng L, Tanaka K, Zhang CJ, Ma J, Bai G, Wang P, Zhang SW, Liu ZW, Cai T, Tang K, Liu R, Shi X, He XJ, Zhu JK: DTF1 is a core component of RNA-directed DNA methylation and may assist in the recruitment of Pol IV. *Proc Natl Acad Sci U S A* 2013, **110**:8290–8295.
19. Clarke JD: Cetyltrimethyl Ammonium Bromide (CTAB) DNA miniprep for plant DNA isolation. *Cold Spring Harb Protoc* 2009, doi: 10.1101/pdb.prot5177.

8

A virus-induced gene silencing (VIGS) system for functional genomics in the parasitic plant *Striga hermonthica*

Dinah Kirigia[1*], Steven Runo[2] and Amos Alakonya[1]

Abstract

Background: *Striga hermonthica* is a hemiparasitic weed that infects cereals in Sub Sahara Africa (SSA) resulting in up to 100% grain yield loss. This significant loss in grain yields is a major contributor to food insecurity and poverty in the region. Current strategies to control the parasite are costly, unavailable and remain unpracticed by small-scale farmers, underscoring the need for more economical and sustainable control strategies. Development of resistant germplasm is the most sustainable strategy in the control of *S. hermonthica*, but is constrained by paucity of resistance genes for introduction into crop germplasm. RNA interference (RNAi) has potential for developing host-derived resistance against *S. hermonthica* by transformation of host crops with RNAi sequences targeted at critical Striga genes. The application of RNAi in management of *S. hermonthica* is however constrained by lack of efficient high throughput screening protocols for the candidate genes for silencing, as well as sub optimal delivery of siRNAs into the parasite. In comparison to stable transformation, viral induced gene silencing (VIGS) is a rapid and powerful tool for plant functional genomics and provides an easy and effective strategy in screening for putative candidate genes to target through RNAi. In addition, VIGS allows for a secondary amplification of the RNAi signal increasing the siRNA threshold and facilitates siRNA transport through viral movement proteins. We tested the efficiency of the Tobacco *rattle virus* (TRV1 and TRV2) VIGS vectors in silencing *S. hermonthica phytoene desaturase* (PDS) gene through agrodrench and agro-infiltration.

Results: We report the validation of VIGS in *S. hermonthica* using a silencing cassette generated from *TRV* with a *PDS* gene insert. Agro-infiltrated and agro-drenched *S. hermonthica* leaves showed photo-bleaching phenotypes typical for PDS silencing within 7 and 14 days post infection respectively. In both cases *S. hermonthica* plants recovered from photo-bleaching effects within 28 days post inoculation. The transformation efficiency of the VIGS protocol in *S. hermonthica* was (60 ± 2.9)%.

Conclusion: These results demonstrate that the TRV-VIGS system work in *S. hermonthica* and can be used for candidate gene validation for their role in the parasite development and parasitism, with the ultimate goal of developing resistant transgenic maize.

Keywords: *Striga hermonthica*, Viral induced gene silencing, Agro-drench, Agro-infiltration, Tobacco rattle virus, *Phytoene desaturase*

* Correspondence: dinahkarimi@gmail.com
[1]Jomo Kenyatta University of Agriculture and Technology, Institute of Biotechnology Research, P. O. Box 62000-00200, Nairobi, Kenya
Full list of author information is available at the end of the article

Background information

Maize is an important staple food crop for majority of people in Sub-Sahara Africa (SSA) [1,2]. Maize grain yields are below the demand hence leading to food insecurity and poverty in the region. The low maize yields result from various biotic and abiotic factors that combined cause cereal grain loss worth of US$3 billion annual [3-5]. The most devastating biotic constraint to maize production in SSA is *Striga hermonthica* (Del.) Bentha, a root hemi-parasitic weed of maize which causes up to 100% grain loss annually [3,6,7]. The life cycle of *S. hermonthica* is intimately synchronized with that of its host, and the seeds of the parasite only germinate in response to chemical signals present in root exudates of the host [8]. *Striga hermonthica* infects maize by forming haustoria connections with the host vasculature resulting in syphoning of water and nutrients [8,9]. Although some *S. hermonthica* control strategies have been proposed and practiced, Striga seed bank in soils has continued to build up and the parasite has continued to spread to previously non-infected arable land [10,11].

Genetic engineering through cross species RNA interference (RNAi) technology offers great promise in parasitic plant management [12-19]. However, its applicability in *S. hermonthica* management has been constrained by lack of methods to deliver the silencing molecules and the lack of candidate genes to target [20]. The recent report on horizontal gene transfer from *Sorghum bicolor* to *S. hermonthica* has increased prospects of delivering the silencing RNA molecules from cereal hosts to *S. hermonthica* [21]. This suggests the possibility of RNAi in some monocots which have been reported to be recalcitrant to transformation. Host-derived resistance using RNAi dependent on efficient delivery of siRNAs from the host to the parasite in order to determine if the gene causes an alteration in the parasites' phenotype, reviewed in [15]. An alternative approach to determine if a gene has a function on the parasite would be to develop a high throughput genetic transformation protocol for Striga. These two approaches present challenges, as transformation of grasses (especially rice, maize wheat and sorghum) is recalcitrant [22-26] and no protocols exist for Striga transformation yet.

Viral induced gene silencing (VIGS) is a technique that employs recombinant viruses to specifically reduce endogenous gene activity through plant innate silencing mechanisms called Post-Transcriptional Gene Silencing (PTGS) [27]. The VIGS vectors are usually standard binary Ti-plasmids that contain a viral genome and a fragment of the host plant's target gene. The vectors are introduced in the plants via *Agrobacterium tumefaciens* infection that results in the transfer of the T-DNA containing the viral genome into the host genome of at least one cell, where it is transcribed, and translated [28,29] . This leads to the production of double-stranded RNAs (dsRNAs) due to self-assembly of viral ssRNA into hairpins or complementary sequences derived from sense and antisense viral ssRNA strands [30]. Dicer-like proteins cleave these viral dsRNAs into short interfering RNAs (siRNA) duplexes of 21–24 nucleotides (nt) in length [27-29]. These siRNAs are incorporated into a RNA-induced Silencing Complex (RISC) that guide and cleave complementary RNAs [31,32]. The virus-derived silencing signal is amplified and spreads systemically throughout the plant [33]. Amplification of VIGS results in down-regulation of target gene [27,34]. VIGS is not a stable transformation strategy but works transiently and therefore could be used as a powerful and rapid tool in gene validation for loss-of-function.

We report efficient Tobacco *rattle virus* (TRV-(1&2) VIGS vectors in silencing *S. hermonthica Phytoene desaturase (PDS)* gene. These findings have far reaching application in designing RNAi strategies based on host derived resistance.

Results

VIGS induced RNAi on *S. hermonthica* PDS causes photo-bleaching

Viral induced technique through agro-drench methods was effective in *S. hermonthica* plants. This was evidenced by down regulation of the PDS gene resulting into photo-bleached phenotypes on the leaves of *S. hermonthica* plants. The bleaching appeared on plants agro-drenched with the *Agrobacterium* strain GV3101 habouring the TRV1 and mixed with GV3101 having TRV2 vector containing the PDS insert (Figure 1a, b, c, d, e). The photo-bleaching effects appeared on the 14th day after agro-drench but the plants recovered on the 28th day post infection (Figure 1c) and (Figure 1e) respectively. There was no photo-bleaching on the control agro-drenched plants leaves (Figure 1f-t). Similarly agro-infiltrated *S. hermonthica* plants developed photo-bleached phenotypes on the leaves of PDS targeted *S. hermonthica* plants. The effects appeared on the 7th day after agro-infiltration (Figure 2b) and the plants recovered on the 28th day post agro-infiltration (Figure 2d). The negative control agro-infiltrated plants did not show photo-bleaching symptoms (Figure 2f-o).

VIGS induced RNAi is because of down-regulation of *S. hermonthica* PDS

To confirm if silencing of the PDS had occurred in *S. hermonthica* due to infiltration and agro-drench, RT-PCR analysis was done using PDS and TRV gene specific primers. The PDS primers were designed to prime outside the region of homology between the VIGS vector and target mRNA. The RT-PCR of the cDNA from

Figure 1 Viral induced gene silencing (VIGS) via Agro-drench method on *S. hermonthica*. a, **b**, **c**, **d**, and **e** are plants agro-drenched with *Agrobacterium* strain GV3101 with TRV2- PDS (GV3101/TRV2-PDS), and mixed in 1:1 ratio with GV3101 containing TRV1 empty vectors (GV3101/TRV2), in days 1, 7, 14, 21 and 28 post agro-drench respectively; **f**, **g**, **h**, **i** and **j** were agro-drenched with GV3101 vectors containingTRV1 empty (GV3101/TRV1) in days 1,7,14,21 and 28 respectively. **k**, **l**, **m**, **n**, and **o** plants were agro-drenched with GV3101 only, while **p**, **q**, **r**, **s** and **t** were only watered in days 1, 7, 14, 21 and 28 after agro-drench respectively. All scale bars represent 5 cm.

photo-bleached *S. hermonthica* leaves, using PDS primers amplified a 250 bp fragment as expected (Figure 3a). However, fragments were extremely faint due to down regulation of the PDS gene in these plants. The negative control plants treated with GV3101/TRV1, GV3101 empty and those which were only watered had bands of higher intensity indicating the PDS was highly expressed in these plants because there was no down-regulation (Figure 3b, c, d, respectively). Additionally, the cDNA amplification with TRV primers indicated the success of VIGS by amplifying a 400 bp fragment on plants treated with GV3101/TRV2 with the PDS insert (Figure 3e). Amplification of the cDNA from plants treated with; GV3101/TRV1 empty, GV3101 empty and water only did not have the 400 bp fragment with TRV primers (Figure 3f, g, h respectively). The *S. hermonthica* DNA amplification with primers for Actin gene showed the 426 bp fragment indicating the quality of the cDNA of plants from all the four treatments (Figure 3i, j, k, l).

VIGS efficiency on *S. hermonthica*

Statistical analysis using student's *t*-test revealed that the results of VIGS were successful on plants targeted for PDS silencing (treated with GV3101/TRV2/PDS + GV3101/TRV1) in *S. hermonthica*. There was indication of photo-bleaching by 60.2 ± 2.9 percentage of *S. hermonthica* plants targeted for PDS plants silencing in Agro-infiltration method, while in agro-drench only 10.3 ± 1.5. None of the negative control treatments indicated photo-bleaching effects in both methods (Table 1).

Discussion

Plants induce homology dependent defense mechanisms in response to attack by virus, therefore engineering a virus into a plant to target a gene of interest results in silencing of the gene through PTGS [29]. The PTGS mechanisms are similar to those of RNA interference in plants [27]. Our experiments demonstrate that VIGS using TRV vectors with the PDS gene resulted in inhibition of carotenoid biosynthesis in *S. hermonthica.* This was evidenced by down-regulation of the *PDS* gene in *S. hermonthica* plants resulting in photo bleaching phenotypes at 7 and 14 days post inoculation in agro-infiltrated and agro-drenched plants respectively. The early appearance of silencing in agro-infiltrated *S. hermonthica* plants as compared to the agro-drenched ones could be ascribed to high efficiency in the delivery of the signals that initiate the innate silencing machinery of PGS in the *S. hermonthica* leaves. In our case the VIGS vector was introduced in *S. hermonthica* via *A. tumefaciens* infection. It is also possible that *A. tumefaciens* was more efficient at transferring the T-DNA containing the viral genome and

Figure 2 Viral induced gene silencing (VIGS) via agro-infiltration method on *S. hermonthica*. a, b, c, d, and **e** are plants agro-infiltrated with *Agrobacterium* strain GV3101 habouring TRV2-PDS (GV3101/TRV2-PDS), and mixed in 1:1 ratio with GV3101 containing TRV1 empty vectors (GV3101/TRV2), in days 1, 7, 14, 21 and 28 after agro-infiltration respectively; **f, g, h, i** and **j** are plants agro-infiltrated with GV3101 vectors containing TRV1 empty (GV3101/TRV1) in days 1, 7, 14, 21 and 28 after infiltration respectively; **k, l, m, n,** and **o** are plants agro-infiltrated with GV3101 only, in days 1, 7, 14, 21 and 28 after agro-infiltration respectively. All scale bars represent 5 cm.

PDS into the host genome in the cells of the leaves than in the root and stem cells, this could have therefore delayed all the downstream silencing steps in agro-drenched plants [30]. Further these results show that a better mechanism in spreading the silencing signal after introduction exists in *S. hermonthica* leaf tissue than in the root or stem tissue. The translocation of PTGS silencing factor may utilize both short-range cell-to-cell movements through plasmodesmata as well as phloem-associated long-range transport mechanisms [35,36]. The RNA-dependent RNA Polymerase6 (RDR6) is required for long-range transport, possibly by amplifying the silencing signal [33]. From our experiments we are not able to verify if RDR6 in leaf, root and stem cells of *S. hermonthica* is responsible for the difference in the efficiency. The silencing efficiency however has been reported to be proportional to the number of silencing molecules in the cells [37].

Phytoene desaturase (PDS) is a key enzyme involved in carotenoid biosynthesis pathway [28]. It's known that reduced levels of photo-protective carotenoids leads to rapid destruction of chlorophyll by photo-oxidation that results to white or bleached phenotypes [38]. The recovery of the photo-bleached plants at 28 days post inoculation is attributed to the transient nature by which VIGS is expressed in cells. The negative control plants treated with GV3101/TRV1, GV3101 empty and water did not show bleaching characteristic of the PDS silencing effects, because for any silencing to occur, the PDS insert must be contained in TRV2 expression vector that encode the virus coat protein genes responsible for viral replication [39]. The control plants could not therefore initiate the PTGs silencing machinery. The silencing of the PDS was therefore as a result of the infected *S. hermonthica* plants employing the innate PTGS as defense mechanism against the TRV. The PTGS as a response has been widely reported in plants [27,30,34].

This study has established a VIGS protocol that can be used for reverse genetics or functional genomics studies in *S. hermonthica*. This approach also ensures that the gene validation can proceed without laboring with stable transformation of the injurious parasite or its recalcitrant monocot hosts. Although there is limited evidence that genetic material can be exchanged between *S. hermonthica* and its hosts [20,21], the developed tools could be independently used without having to worry about delivery of enough of the silencing molecules through from the host to parasite via the haustoria. In fact once the factors that enable *S. hermonthica* to uncontrollably

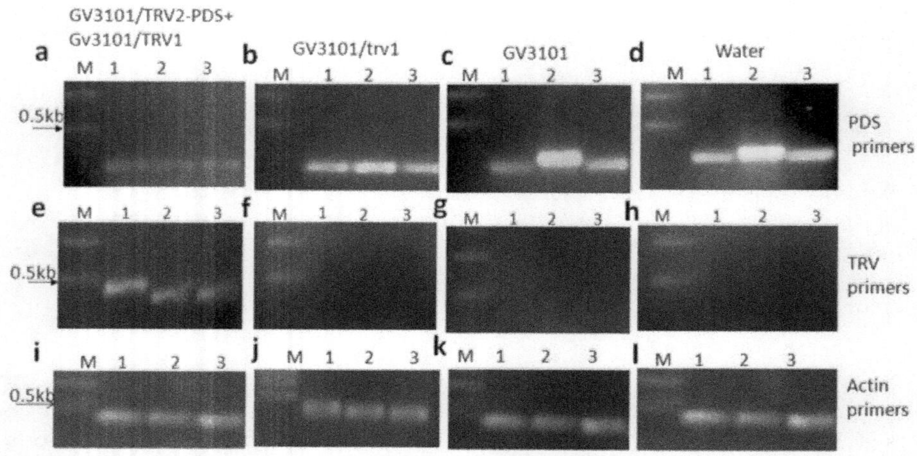

Figure 3 RT-PCR to confirm silencing of *PDS* gene on *S. hermonthica*. The PDS and TRV gene specific primers were used. **(a)** *S. hermonthica* treated with GV3101 with the TRV2 vector containing the PDS gene insert (GV3101/TRV2-PDS) + GV3101/TRV1) and amplified with PDS primers. Lane (M) represents 1 kb ladder while 1, 2, and 3 are the replicates. **(b)** The cDNA plants treated with GV3101 containing TRV1 empty vectors (GV3101/TRV1) with three replicates (1, 2 and 3) and amplified with PDS primers. **(c)** Three replicates (1, 2, 3) of plants treated with GV3101 empty and amplified with PDS primers. **(d)** Represents plants treated only with water and amplified with PDS primers. **(e)** *S. hermonthica* cDNA amplified with TRV primers from plants treated with GV3101/TRV2-PDS + GV3101/TRV1 vectors with three replicates (1, 2, 3). **(f)** *S. hermonthica* cDNA from Plants treated with GV3101/TRV1 empty vectors, three replicates (1, 2, and 3) amplified with TRV primers. **(g)** Plants treated with GV3101 empty, three replicates (1, 2, 3) amplified with TRV primers. **(h)** Plants treated with water only, three replicates (1,2,3), cDNA amplified with TRV primers. Images (**i**, **j**, **k** and **l**) are the internal control using Actin primers for the *S. hermonthica* cDNA from plants treated with GV3101/TRV2-PDS + GV3101/TRV1, Plants treated with GV3101/TRV1 only, Plants treated with GV3101 only and plants treated with water only, respectively.

colonize its host are identified, the parasite could be basted by boasting host defense mechanisms through gene over-expression techniques along the identified pathways. In such a case delivery of resistance molecules will not be in question, as the host will directly exhibit resistance to *S. hermonthica* on attachment.

Conclusion

We have demonstrated that TRV VIGS vectors could be used in functional genomics in the parasitic weed *S. hermonthica.* Although VIGS was more efficient through agro infiltration than agro-drench, using the PDS gene obviously results in an above ground leaf phenotype, it remains to be seen what will be observed during validation in *S. hermonthica* parasitism genes where most of the promising phenotypes are expected to occur in the roots where haustoria colonizes. Finally, with the available genetic resources at the Parasitic Plant Genome Project, and the developed tools will aid in the validation and identification of genes responsible for unabated *S. hermonthica* parasitism. The identification could lead to a variety of transgenic approaches that could lead to development of *S. hermonthica* resistant germplasm not only in maize but in other cereal hosts as well.

Materials and methods

VIGS plasmids

Tobacco rattle virus (TRV)-derived vectors were provided by Prof. Dinesh Kumar from the University of California-Davis. The Tobacco rattle virus contains bipartite positive-sense RNA genome (RNA1 and RNA2). The TRV1 vector represented RNA1 which encodes two viral replication proteins, a movement protein and a seed transmission factor. The provided TRV2 represented the RNA2 and encodes the coat protein and a nematode transmission plant kingdom [40]. The two

Table 1 VIGS efficiency in *S. hermonthica*

Treatment	No of transformed plants	% of PDS transformed plants (photo-bleached)	% of PDS negative plants (not photo-bleached)
GV3101/TRV2-PDS + GV3101/TRV1			
Agro-infiltration	12.9 ± 2.9	60.2 ± 2.9*	39.8 ± 5
Agro-drench	12.2 ± 1.5	10.3 ± 1.5*	89.7 ± 3
GV3101/TRV1	8.0 ± 1.5	0.0	100.0*
GV3101	8.0 ± 1.0	0.0	100.0*

*Significant values; $P \leq 0.05$.

binary vectors were separately transformed in the *A. tumefaciens* Strain GV3101 which was the delivery vehicle for the viral vectors into plants through agro-inoculation and agro-drench. *Agrobacterium* colonies carrying TRV1, and TRV2 vector with the PDS insert (Figure 4) were grown separately in Luria bertani (LB) liquid media containing Kanamycin 50 mg/l and rifampicin 1 mg/l overnight. From the overnight culture, 1 ml was picked from each tube and again sub-cultured separately for 2 hours in 10 ml LB media containing Kanamycin 50 mg/L and rifampicicn 1 mg/L and 150 μM of acetosyringone to induce the virulence genes. The separate 10 mL liquid cultures were grown at 28°C in darkness until they attained an optical density (OD) of 0.6. The two cultures were then spinned for 15 minutes at 10000 revolutions per minute (RPM) and resuspended in an induction buffer containing 150 μM acetosyringone, 10 mM of MES and 10 mM of $MgCl_2$, adjusted to PH 5.6 and grown again for 2 hours. *Agrobacterium* strains GV3101 containing TRV1 and TRV2-PDS were then mixed in a 1:1 ratio and used in the agro-inoculation experiments (Figure 4).

Agro-drench

Agro-drench involved applying the mixture of GV3101/TRV2PDS and GV3101/TRV1 (1:1) ratio directly onto the soil adjacent to the crown part of 3–4 week old *S. hermonthica* plants as described by [39] with slight modifications. The experiments involved six plants and were repeated three times. The negative controls were six plants treated with GV3101/TRV1, GV3101, and water separately. All the negative controls were replicated thrice as well. Pictures were taken after every seven days and the numbers of plants showing photo-bleaching effects were recorded.

Agro-infiltration

For leaf agro-infiltration in *S. hermonthica*, all the young leaves on the upper part of the plant were infiltrated by pricking the lower side of the leaves with a wire brush. Using cotton wool the GV3101/TRV2-PDS and GV3101/TRV1 mixture was gently applied on the pricked leaves until they became fully wet. Eight *S. hermonthica* plants were used and the experiment was repeated three times. The control *S. hermonthica* plants were separately infiltrated with GV3101 empty, GV3101/TRV1 and water only. Plants were photographed every seven days and the number of plants showing photo-bleaching effects recorded.

Screening for silencing of *phytoene desaturase* gene through reverse transcriptase-polymerase chain reaction (RT-PCR)

Leaf tissues from three of plants from all the treatments were collected and ground in liquid nitrogen using a pestle and mortar. Approximately 20 mg of ground tissues was used for total RNA extraction as per the instructions of the RNeasy® mini kit (Qiagen, Cat no 74104, Valencia. U.S.A). The total RNA was subjected to DNAse treatment and incubated at 37°C for 15 minutes. The total RNA extracted was then reverse transcribed to cDNA synthesis using Superscript™ III first stand synthesis system (Invitrogen, CAT 18080–051, Carlsbad, U.S.A). The first strand cDNA synthesis reactions were primed using random heximers. The synthesized cDNA was then amplified using PDS primers and TRV primers. The PDS

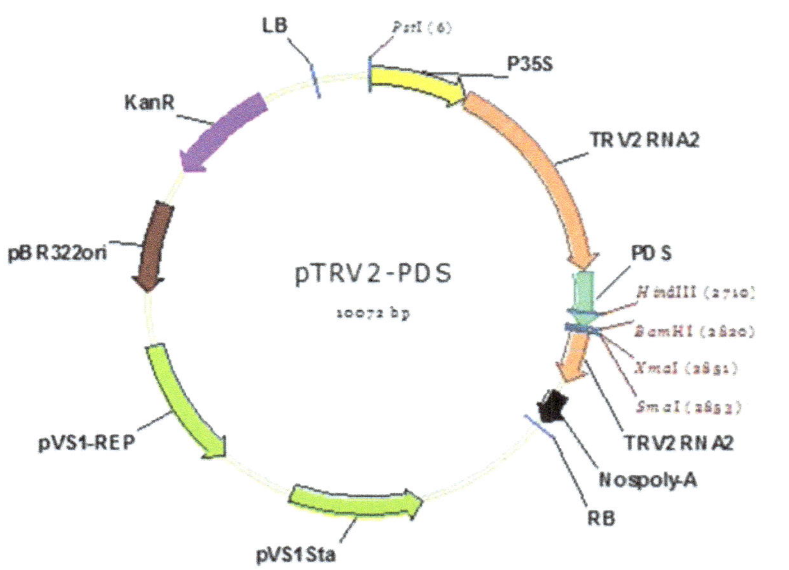

Figure 4 TRV2 vector with the PDS insert.using the TRV2 vector details provided in the ABRC database [41], the map was constructed using Vector NTI software, version 11.5.2.

primers were (Forward primer 5′-GAGAAACATGGTTC AAAAATGG-3′and reverse primer 5′-AACACAAAAG CATCTCCCTC-3′). The PDS primers were designed to prime outside the region of homology between the VIGS vector and the target mRNA. The TRV primers were (Forward 5′-ACTCACGGGCTAACAGTGCT-3′ and reverse primer 5′-GACGTATCGGACCTCCACTC-3′. The PCR was set with 94°C denaturation temperature, 55°C annealing temperature and 74°C extension temperature for 40 cycles. Gel electrophoresis was performed at 100 volts using 1% of agarose loaded with 5 μl of each sample reaction. Gel pictures were taken under a Ultraviolet light illuminator after a 30 minutes run.

Abbreviations

SSA: Sub Saharan Africa; PDS: *Phytoene desaturase*; Bp: Base pairs; Kb: Kilobytes; RPM: Revolutions per minute; TRV: Tobacco rattle virus; PTGS: Post transcription gene silencing; OD: Optical density; RNAi: Ribonucleic acid interference; VIGS: Viral induced gene silencing; siRNA: Small interfering ribonucleic acid; T-DNA: Transfer DNA of the tumor-inducing (Ti) plasmid in Agrobacterium tumefaciens; dsRNA: Double stranded ribonucleic acid; ssRNA: Single stranded ribonucleic acid; RDR6: RNA-dependent RNA Polymerase6; LB: Luria bertani; cDNA: Complementary deoxibonucleic acid; nt: Nucleotide.

Competing interest

The authors declare that they have no competing interests.

Authors' contribution

DK performed the experiments and participated in manuscript preparation, SR planned the experiments and participated in manuscript preparation, AA planned the experiments and participated in manuscript preparation. All authors have read and approved the final manuscript.

Acknowledgments

The authors wish to acknowledge the International Foundation for Science (IFS) for funding the project and the Plant transformation Laboratory at Kenyatta University where this work was conducted.

Author details

[1]Jomo Kenyatta University of Agriculture and Technology, Institute of Biotechnology Research, P. O. Box 62000-00200, Nairobi, Kenya. [2]Department of Biochemistry and Biotechnology, Kenyatta University, P.O. Box 43844, 00100 Nairobi, Kenya.

References

1. Food and Agriculture Organization: **The state of food insecurity in the world: addressing food insecurity in protracted crises.** 2010, [https://www.fao.org/docrep/013/i1683e/i1683e]
2. Erenstein O, Kassie G-T, Langyintuo A, Mwangi W: *Characterization of Maize Producing Household in Drought Prone Regions of Eastern Africa.* CIMMYT; 2011 [http://repository.cimmyt.org/xmlui/bitstream/handle/10883/1306/95930.pdf?sequence=3]
3. AATF: *Empowering African Farmers to Eradicate S. Hermonthica from Maize Crop Lands.* Nairobi, Kenya: The African Agricultural Technology Foundation; 2006 [http://www.aatf-africa.org/userfiles/Empowering_African_farmers]
4. Ejeta G: **Breeding for *S. hermonthica* resistance in Sorghum: exploitation of intricate host parasite biology.** *Crop Sci* 2007, **47:**216–227.
5. M'mboyi F, Mugo S, Murenga M, Ambani L: *Maize Production and Improvement in Sub-Saharan Africa.* Nairobi, Kenya: African Biotechnology Stakeholders Forum(ABSF); 2010.
6. Rich PJ, Ejeta G: **Towards effective resistance to *S. hermonthica* in African maize.** *Plant Signal Behav* 2008, **3:**618–621.
7. Parker C: **Observations on the current status of Orobanche and Striga problems worldwide.** *Pest Manag Sci* 2009, **65:**453–459.
8. Yoder J: **Parasitic plant responses to host plant signals: a model for subterranean plant-plant interactions.** *Curr Opin Plant Biol* 1999, **2:**65–70.
9. Westwood JH, Yoder JI, Timko MP, DePamphilis CW: **The evolution of parasitism in plants.** *Trends Plant Sci* 2010, **15:**227–235.
10. Khan Z, Pickett J, Wadhams L, Hassanali A, Midega C: **Combined control of *S. hermonthica* and stemborers by maize-*Desmodium spp* intercrops.** *Crop Prot* 2006, **25:**989–995.
11. De Groote H, Wangare L, Kanampiu F, Odendo M, Diallo A, Karaya H, Friesen D: **The potential of herbicide Tolerant maize technology for *S. hermonthica* control in Africa.** *Agr Syst* 2008, **97:**83–94.
12. Tomilov A, Tomilova N, Wroblewski T, Michelmore R, Yoder J: **Trans-specific gene silencing between host and parasitic plants.** *Plant J* 2008, **56:**389–397.
13. Yoder J, Gunathilake P, Wu B, Tomilova N, Tomilov AA: **Engineering host resistance against parasitic weeds with RNA interferance.** *Pest Manag Sci* 2009, **65:**460–466.
14. Aly R, Cholakh H, Joel DM, Leibman D, Steinitz B, Zelcer A, Naglis A, Yarden O, Gal-On A: **Gene silencing of mannose 6-phosphate reductase in the parasitic weed *Orobanche aegyptiaca* through the production of homologous dsRNA sequences in the host plant.** *Plant Biotechnol* 2009, **7:**487–498.
15. Runo S, Alakonya A, Machuka J, Sinha N: **RNA interference as a resistance mechanism against crop parasites in Africa: a 'Trojan horse' approach.** *Pest Manag Sci* 2011, **67:**129–136.
16. Runo S: **Engineering host-derived resistance against plant parasites through RNA interferance: challanges and opportunities.** *Bioeng Bugs* 2011, **2:**208–213.
17. Yoder J, Scholes JD: **Host plant resistance to parasitic weeds;recent progress and bottlenecks.** *Curr Opin Plant Biol* 2010, **13:**478–484.
18. Alakonya A, Kumar R, Koenig D, Kimura S, Townsley B, Runo S, Garces HM, Kang J, Yanez A, David-Schwartz R, Rakefet, Machuka J, Sinha N: **Interspecific RNA interference of SHOOT MERISTEMLESS-like disrupts *Cuscuta pentagona* plant parasitism.** *Plant Cell* 2012, **24:**3153–3166.
19. Bandaranayake P, Yoder J: **Trans-specific gene silencing of acetyl-CoA carboxylase in a root-parasitic plant.** *Mol Plant Microbe Interact* 2013, **26:**575–584.
20. De Framond A, Rich PJ, McMillan J, Ejeta G: **Effects of Striga Parasitism of Transgenic Maize Armed With RNAi Constructs tar-Geting Essential S. Asiatica Genes.** In *Integrating New Technologies for Striga Control.* Edited by Ejeta G, Gressel J. Singapore: World Scientific Publishing Co; 2007:185–196.
21. Yoshida S, Maruyama S, Nozaki H, Shirasu K: **Horizontal gene transfer by the parasitic plant *Striga hermonthica*.** *Science* 2010, **328**(5982):1128.
22. Hiei Y, Ohta S, Komari T, Kumashiro T: **Efficient transformation of rice (*Oryza sativa L.*) mediated by *Agrobacterium* and sequences-analysis of the boundaries of the T-DNA.** *Plant J* 1994, **6:**271–282.
23. Ishida Y, Saito S, Ohta S, Hiei Y, Komari T, Kumashiro T: **High efficiency transformation of maize (*Zea mays L.*) mediated by *Agrobacterium tumefaciens*.** *Nat Biotechnol* 1996, **14:**745–750.
24. Zhao ZY, Gu WN, Cai TS, Tagliani L, Hondrd D, Bond D, Schroeder S, Rudert M, Pierce D: **High throughput genetic transformation mediated by *Agrobacterium tumefaciens* in maize.** *Mol Breed* 2002, **8:**323–333.
25. Cheng M, Hu T, Layton J, Liu CN, Fry JE: **Desiccation of plant tissue post-*agrobacterium* infection enhances T-DNA delivery and increase stable transformation efficiency in wheat.** *In Vitro Cell Dev Biol Plant* 2003, **39:**595–604.
26. Zhao ZY, Cai TS, Tagliani L, Miller M, Wang N, Pang H: **Agrobacterium-mediated sorghum transformation.** *Plant Mol Biol* 2000, **44:**789–798.
27. Voinnet O: **RNAi silencing as a plant immune system against viruses.** *Trends Genet Engin* 2001, **17:**449–459.
28. Lu R, Martin-Hernandez AM, Peart JR, Malcuit I, Baulcombe DC: **Virus- induced gene silencing.** *Plants Methods* 2003, **30:**296–303.
29. Robertson D: **VIGS vectors for gene silencing: many targets, many tools.** *Ann Rev Plant Biol* 2004, **55:**495–519.
30. Donaire L, Wang Y, Gonzalez-Ibeas D, Mayer KF, Aranda MA, Lave CL: **Deep-sequencing of plant viral small RNAs reveals effective and widespread targeting of viral genomes.** *Virology* 2009, **392:**203–214.
31. Waterhouse PM, Fusaro AF: **Plant science. Viruses face a double defense by plant small RNAs.** *Science* 2006, **313:**54–55.
32. Ding SW, Voinnet O: **Antiviral immunity directed by small RNAs.** *Cell* 2007, **130:**413–426.
33. Kalantidis K, Schumacher HT, Alexiadis T, Helm JM: **RNA silencing movement in plants.** *Biol Cell* 2008, **100:**13–26.

34. Baulcombe DC: **RNA silencing.** *Curr Biol* 2002, **21**:82–84.
35. Himber C, Dunoyer P, Moissiard G, Ritzenthaler C, Voinnet O: **Transitivity-dependent and -independent cell-to-cell movement of RNA silencing.** *EMBO J* 2003, **22**:4523–4533.
36. Voinnet O, Baulcombe DC: **Systemic signaling in gene silencing.** *Nature* 1997, **389**:553–555.
37. Dunoyer P, Himber C, Voinnet O: **DICER-LIKE 4 is required for RNA interference and produces the 21-nucleotide small interfering RNA component of the plant cell-to-cell silencing signal.** *Nat Genet* 2005, **37**:1356–1360.
38. Becker A, Lange I: **VIGS -genomics goes functiona.** *Trends Plant Sci* 2009, **15**:1–4.
39. Choong-Min R, Anand A, Kang L, Mysore K: **Agrodrench: a novel and effective agroinoculation method forvirus-induced gene silencing in roots and diverseSolanaceous species.** *Plant J* 2004, **40**:322–331.
40. Verchot-Lubicz: **Soil borne viruses: advances in cell-to-cell movement, gene silencing, and pathogen-derived resistance.** *J Plant Pathol* 2002, **62**:55–63.
41. **Arabidopsis biological resource center data base.** http://abrc.osu.edu.

9

AFLPsim: an R package to simulate and detect dominant markers under selection in hybridizing populations

Francisco Balao[1,2*] and Juan Luis García-Castaño[1]

Abstract

Background: In spite of a large diversity of approaches to investigate loci under selection from a population genetic perspective, very few programs have been specifically designed to date to test selection in hybrids using dominant markers. In addition, simulators of dominant markers are very scarce and they do not usually take into account hybridization.

Results: Here, we present a new, multifunctional, R package for dominant genetic markers, *AFLPsim*. This package can simulate dominant markers in hybridizing populations and implements genome scan methods for detecting outlier dominant loci in hybrids. In addition, it includes tools for further manipulating the results, plotting them and other tasks. We describe and tabulate the major functions implemented in *AFLPsim*. In addition, we provide some demonstration of its use and we perform a comparative study with other software. Finally, we conclude by briefly describing the input and output formats.

Conclusions: The R package *AFLPsim* application provides several useful tools in the context of hybridization studies. It can simulate dominant markers in hybridizing populations and predict their demographic evolution. In addition, we implement a new genome scan method for detecting outlier dominant loci in hybrids, which shows a rather high sensitivity and is very conservative in comparison with Gagnaire *et al.*'s, BAYESCAN and *introgress*. The application is downloadable at http://cran.r-project.org/web/packages/AFLPsim/.

Keywords: Demographic simulation, Dominant markers, Genome scan, Hybridization, Outlier loci, R package

Background

The study of natural hybridization has been the focus of much attention in Evolutionary Biology [1,2]. Hybridization has recently been perceived as a catalyst not only for speciation but also for major evolutionary innovations [3]. Hybrid zones offer a "window on the evolutionary process" involving divergence at many loci by a balance between dispersal and selection [4]. Cline theory has provided a conceptual framework to understand the forces maintaining hybrid zones and to help infer the relevant evolutionary parameters describing the introgression of traits across hybrid zones [5]. However, in non-stable hybrid zones, selection in early generations has a central role in the establishment and fate of hybrids and progenitors [1,4]. In these early generations, several hybrid categories can be easily distinguished (first generation hybrids –F_1–, outcrosses between F_1 individuals –F_2–, or backcrosses to parental "A" –BxA– and backcrosses to parental "B" –BxB–, for instance). Hence, identifying markers under selection on these early-generated hybrids can provide tremendous knowledge about the stability of hybrid zones.

Although next generation sequencing (NGS) has transformed our ability to identify the genes underpinning selection/adaptation [6], a complementary approach with potentially neutral markers such as Amplified Fragment Length Polymorphisms (AFLPs) allows a cost-effective screening of the genomes of a large number of individuals [7]. AFLP has been very successfully used in the identification of hybrids [8,9] and outlier loci presumably under selection [10-12]. However, in spite of a large diversity of approaches to investigate loci under selection from a

* Correspondence: fbalao@us.es
[1]Departamento de Biología Vegetal y Ecología, Universidad de Sevilla, Ap-1095, 41080 Sevilla, Spain
[2]Department of Systematic and Evolutionary Botany, University of Vienna, Rennweg 14, Vienna 1030, Austria

population genetic perspective [13-15], no program has been specifically designed to date to test selection in hybrids using dominant markers. Nevertheless, genomic clines have been used to identify molecular markers with patterns of introgression inconsistent with neutrality (e.g. *introgress* [16,17]). However, detecting outliers in a hybridizing framework using dominant markers can be a real challenge. Long periods of time are needed to identify loci that have experienced a history of weak selection, as cumulative effect is necessary to produce a detectable signal in the DNA polymorphism of the underlying loci [18]. Therefore, for hybrids in early generations, loci under weak selection would remain undetectable. In addition, dominance imposes difficulties to estimate allelic frequencies, especially when the frequency of the presence-allele is high [19]. Scoring errors and a low sample size can also affect the correct estimate of the allelic frequencies. Furthermore, previous methods used to detect dominant loci under selection have shown a substantial proportion of false positives among the detected outliers [20]. Hence, genome scan in hybrid zones should be capable to correctly estimate expected allelic frequencies under neutrality in hybrids overcoming these problems. Moreover, it should be sensitive enough to detect loci under moderate selection as well as keeping the false positive rate close to null.

Additionally, being able to simulate dominant markers in a hybrid zone is important to obtain a better knowledge about expected patterns of hybridization. When experimental treatments are not feasible, in-silico simulations have been widely used to test population genetic hypotheses [21]. Furthermore, genetic simulations have been also used to understand the statistical efficiency of several genetic methods and to compare different approaches [20]. Finally, simulations are of practical use when analysing data from a real system, as they can compare observed genetic distributions with the theoretically expected ones. In spite of its importance, simulators of dominant markers are very scarce [22] and they do not usually take into account hybridization, but see [23]. Again, dominance of markers is one of the major problems in the simulation process, as allele frequencies have to be estimated from phenotypic data.

Here, we describe *AFLPsim*, a software package designed to overcome these limitations by implementing a dominant marker simulator of hybridization and two genome scan algorithms (Gagnaire *et al.*'s [11] and a new method called bal&gar-ca) specifically designed to detect outlier markers in recent generated hybrids (F_1, F_2, BxA and BxB). The software is written in a statistical, open source, scripting R language [24], and released under the GPL license to guarantee the continuing availability of the source code.

Implementation

A list of the major functions in the *AFLPsim* library is shown in Table 1. These functions cover methods in simulation, genome scan, and manipulation and visualization of results, which are detailed below. More details can be found in the software manual, which is available at https://github.com/fbalao/AFLPsim.

Simulating hybridization and demographic evolution

Our software generates diploid hybrid genotypes, under the hypothesis of the Hardy-Weinberg equilibrium, Mendelian inheritance of markers and not linkage disequilibrium, by calculating observed allele frequencies in parental populations and the expected frequencies in the different hybrid classes (F_1, F_2 and backcrosses in both directions). Observed frequencies can be calculated from simulated parental populations with the 'hybridsim' function (following a beta distribution – [25]) or from two user-specified parental phenotypes ('hybridize' function),

Table 1 Functions of the *AFLPsim* package

Function name	Description
Simulations	
demosimhybrid	conducts demographic analysis in hybrid populations.
hybridize	generates multilocus dominant hybrids individuals from parental profiles.
hybridsim	generates multilocus dominant parental and hybrid individuals.
Genome scan	
bayescan	calls BAYESCAN 2.1 program [12] from R to a set of populations.
gscan	conducts genome scan on F_1 and backcross individuals [10].
hybridindex	estimates the hybrid index calling the *introgress* package.
Plotting	
plot.hybridsim	plots phenotypic frequencies of hybrids on a neutral hybridization model.
plot.demosimhybrid	plots results of 'demosimhybrid' function.
Data manipulation	
sim2adegenet	converts simulation to the *genind* format [25].
sim2arlequin	writes the input file for ARLEQUIN [26] from the simulation results.
sim2bayescan	writes the input file for BAYESCAN [12] from the simulation results.
sim2popgene	writes the input file for POPGENE [27] from the simulation results.
sim2introgress	converts simulation to the *introgress* format [16].
sim2newhybrids	writes the input file for NEWHYBRIDS [8] from the simulation results.
sim2structure	writes the input file for STRUCTURE [28] from the simulation results.

being the allele frequencies of dominant markers calculated using a square-root procedure [19]. For the F_1 hybrids the expected phenotypic frequency of each band is

$$E(f_{F_1}) = p_A + p_B - p_A p_B \quad (1)$$

where p_A is the frequency of the presence-allele in the parental population A:

$$p_A = 1 - \sqrt{1 - \textit{frequency of the band}} \quad (2)$$

and p_B is the frequency of the presence-allele in the parental population B ($p_B = 1 - \sqrt{1 - \textit{frequency of the band}}$).

For the other hybrid classes, we proceed in the same way. For example, for backcrosses with parental A, the expected frequency, based again on the parental allele frequencies, is

$$E(f_{BxA}) = \frac{3p_A + p_B - p_A^2 - p_A p_B}{2} \quad (3)$$

In addition, we implement phenotypic directional selection on the dominant allele, i.e. we modify the frequency of those individuals bearing a selected fragment regardless they are homozygous or heterozygous. Phenotypic selection on a specific marker is simulated with a conceptually simple variable (s), which ranges from -1 to $+\infty$. This coefficient is 0 when there is no selection and it varies following negative and positive directional selection (negative and positive values, respectively). The expected frequency is calculated with the formula

$$E(f_{F_1}) = \frac{w(p_A + p_B - p_A p_B)}{w(p_A + p_B - p_A p_B) + (1 - p_A)(1 - p_B)} \quad (4)$$

where w is the 'fitness', which relates to the selection coefficient (s) through the equation

$$w = 1 + s \quad (5)$$

Users can choose both the intensity of this coefficient (s) and the number of markers under selection.

The interplay of genetic and ecological processes often has important effects on the fate of the hybrid zones. Models can be powerful tools for investigating different control scenarios before undertaking expensive field trials. Plenty of mathematical models describing the dynamics and genetics of hybridization have been widely used. Briefly, these are categorised as ecological or genetic models. The advantages and disadvantages of these diverse approaches have been discussed in detail [26] and the utility of each model depends on the scenario and the supporting data. We implement a modified version of the genetic model of demographic evolution in hybrid zones ('demosimhybrid') developed by Epifanio & Philipp [27]. This heuristic model simulates the proportion of parentals, F_1, F_x and backcrosses (with both parentals) individuals for each generation. The contribution of each taxon following admixture and hybridization depends on three independent variables: (1) the initial proportion of parental taxa; (2) the fitness gradient among parental and hybrid taxa; and, (3) the assortative mating between these taxa. Composition at any time (t, in generations) is calculated by multiplying its initial abundance by its relative fitness, and then, by the probability of mating, using the general equations 6 & 7.

$$S_G = \frac{\phi_t \times \omega_G}{\sum \phi_t \times \omega_G} \quad (6)$$

where S_G is the proportion of a taxon G surviving to reproduction, ϕ_t is the frequency, before selection, of the taxon at the beginning of the generation t, and ω_G is its fitness. The expected contribution of a taxon to the subsequent generation is determined by the equation

$$\phi_{t+1} = S_G \times M \quad (7)$$

where M is the assortative mating matrix.

Epifanio & Philipp's model suffers from several pitfalls, e.g. migration is not taken into account and it is not spatially explicit [26,28]. However, it has been successfully used to explain the extinction of progenitors in several hybrids zones [28,29].

Genome scan

AFLPsim performs two approaches for statistically seeking outlier loci ('gscan') in different hybrid classes (F_1 and backcrosses). The first one, called 'gagnaire' [11] is based on a binomial test to assess any significant deviation between the observed and the expected frequencies for each marker. Briefly, Gagnaire *et al.*'s method estimates the frequency of the presence-allele based on one minus the square-root of the absence (null homozygote) frequency (Eq. 2). Therefore, using these parental frequencies (f_A and f_B), expected band presence frequencies are then calculated for each hybrid category with the Eq. 1 (for F_1 individuals) and the derived ones (e.g., Eq. 3). A binomial test is then performed to test for significant deviation between observed and expected frequencies of band presence at each locus in each hybrid category. For example, for F_1 individuals:

$$P(f_{OF_1} = f_{F_1}) \sim B(f_{OF_1} \times n_{F_1}, n_{F_1}, f_{F_1})$$

where f_{OF_1} is the observed frequency of band presence at one locus, n_{F_1} is the number of F_1 hybrids and f_{F_1} is the expected frequency of band presence at that locus.

Using this method, some fragments could appear to be under selection when in fact they are not (i.e. false positives), because the observed parental frequencies (f_A and f_B) can be biased by sampling error. For this reason, we implement a more conservative method (bal&gar-ca) calculating parental frequencies for each marker through $\sqrt{(1-\alpha)}$ confidence intervals (in this case, $\alpha = 0.05$) by the Clopper-Pearson 'exact' procedure, which is based on a beta distribution [30]. Every combination of one interval end from one parental and one interval end from the other parental lead to an expected value within the neutral expectation surface, and the four values delimit a $1-\alpha$ probability portion of it.

$$\textit{Lower endpoint "Parental A"} : LEA = Beta\left(\frac{1-\sqrt{(1-\alpha)}}{2}; f_A \times n_A\ ,\ n_A - f_A \times n_A + 1\right)$$

$$\textit{Upper endpoint "Parental A"} : UEA = Beta\left(1-\frac{1-\sqrt{(1-\alpha)}}{2}; f_A \times n_A + 1,\ n_A - f_A \times n_A\right)$$

$$\textit{Lower endpoint "Parental B"} : LEB = Beta\left(\frac{1-\sqrt{(1-\alpha)}}{2}; f_B \times n_B,\ n_B - f_B \times n_B + 1\right)$$

$$\textit{Upper endpoint "Parental B"} : UEB = Beta\left(1-\frac{1-\sqrt{(1-\alpha)}}{2}; f_B \times n_B + 1,\ n_B - f_B \times n_B\right)$$

Applying Eq. 2 to these values we obtain the frequency intervals of the presence-allele in each parental (i.e. p_{LEA} , p_{UEA}, p_{LEB}, p_{UEB}), and applying the Eq. 1, we finally obtain the four estimated values of the expected frequency of the presence-allele for F_1 under neutrality (fLL_{F_1}, fUL_{F_1}, fLU_{F_1}, fUU_{F_1}). For instance:

$$E(fLL_{F_1}) = p_{LEA} + p_{LEB} - p_{LEA} \times p_{LEB}$$

To test if a specific locus behaves as an outlier, the average real offspring value $\left(f_{OF_1}\right)$ is confronted against these four estimated values, considering the two following possibilities:

(i) If it is within the two most extreme values, we conclude the fragment is not under selection (i.e. $P(f_{OF_1} = f_{F_1}) = 1$).

$$\text{If } f_{min} < f_{OF_1} < f_{max} \text{ then } P(f_{OF_1} = f_{F_1}) = 1$$
$$\text{where } f_{max} = max\{fLL_{F_1}, fUL_{F_1}, fLU_{F_1}, fUU_{F_1}\}$$
$$\text{and } f_{min} = min\{fLL_{F_1}, fUL_{F_1}, fLU_{F_1}, fUU_{F_1}\}$$

(ii) If not, we choose the closest of the four frequencies to be the expected value of the binomial test.

$$\text{If } f_{OF_1} > f_{max} \text{ then } P(f_{OF_1} = f_{F_1}) \sim B(f_{OF_1} \times n_{F_1}, n_{F_1}, f_{max})$$

$$\text{If } f_{OF_1} < f_{min} \text{ then } P(f_{OF_1} = f_{F_1}) \sim B(f_{OF_1} \times n_{F_1}, n_{F_1}, f_{min})$$

In both methods, the False Discovery Rate (FDR) correction is used to counteract for multiple comparisons and control for the expected proportion of the incorrectly rejected null hypotheses.

In addition, we also include a function that calls the BAYESCAN program [13] from R to perform a Bayesian estimation of selection; in this way, we facilitate the analysis and plotting of results of this efficient software. Moreover, the hybrid index (i.e. the genome-wide admixture) for the simulated hybrid individuals is calculated with the 'hybridindex' function. The maximum likelihood estimates (together with the 95% confidence intervals) of this hybrid index can be obtained with this function, which is a wrapper for the 'est.h' function of the package *introgress* [17].

Data manipulation and visualisation

AFLPsim functions do not require external input files out of the R environment (Table 1). However, for the simulation of hybrids from user-specified parental data, these should be loaded to R as a *matrix* or a *data.frame*. Simulation results can be readily used by multivariate and phylogenetic methods of other R packages (e.g. *ade4* [31]; *adegenet* [32]). Our package is also able to export ARLEQUIN formatted data [33] and POPGENE [34] to estimate summary statistics (e.g. *F*-statistics, Shannon index or polymorphic loci) from the data set. Our package is also able to export data formatted for several popular population genetic computer programs such as STRUCTURE [35] and NEWHYBRIDS [9]. In addition, *AFLPsim* contains functions that produce graphics for visualising the expected frequencies under neutrality for loci under selection across the different hybrid classes. Finally, our package includes a function that plots the results of the demographic evolution model in a hybrid zone.

Results and discussion

To demonstrate the capacities of *AFLPsim*, we assessed some comparisons on the behaviour of our genome scan method vs. that of other software. We have also created several illustrative examples, which can be easily reproduced. An example of the application of *AFLPsim* for investigating introgression patterns (bal&gar-ca method) and demographic dynamics (Epifanio & Philipp's model) in a hybrid zone can be found in [28].

Genome scan comparisons

We investigated the performance of our method (bal&gar-ca) under different scenarios using a simulation study and its efficiency was compared with that of Gagnaire *et al.*'s, BAYESCAN and *introgress*. Although BAYESCAN is not specifically designed for hybrids, we used it for comparison, as it is one of the most popular genome scan software. In this case, we used both parentals and F_1 hybrids as three independent populations. *introgress* has been used to explore introgression between genomes through the genomic cline method. This method is able to identify those molecular markers with introgression patterns inconsistent with neutrality and, therefore, detecting possible loci under selection.

We performed three different simulation experiments to compare the outlier detection efficiency of different genome scan methods, different selection coefficients, different sample sizes and different percentage of selected loci in the genome. We carried out all the simulations on the University of Oslo Bioportal (*www.bioportal.uio.no*) using the 'hybridsim' function.

1) Comparison of different genome scan methods

We compared the results obtained with four methods (bal&gar-ca, Gagnaire *et al.*'s, *introgress* and BAYESCAN), with three selection coefficients ($s = 2.162$, $\log(s+1) = 0.5$, weak selection; $s = 99$, $\log(s+1) = 2$, strong selection; $s = 999$, $\log(s+1) = 3$, very strong selection). 100 F_1 hybrid individuals were simulated with 'hybridsim' from two parental populations (A and B) of 100 individuals each one. We simulated 1000 independent loci (i.e. based on linkage disequilibrium), where 100 were introgressed under directional selection in the F_1. Every scenario (i.e. every selection coefficient) was replicated 100 times. Genome scans were performed on simulations with our method (bal&gar-ca), that by Gagnaire *et al.*, *introgress* and BAYESCAN 2.1, assessing the ability of each one for detecting selection of different intensities as well as the proportion of failures as false positives.

Results showed, on the one hand, that our method was very conservative and that its sensitivity (i.e., the rate of true positives) was lower than that by Gagnaire *et al.* [11] and *introgress*; on the other hand, BAYESCAN almost invariably failed to detect outliers for the simulated datasets (Table 2). However, our method had a null false positive rate unlike Gagnaire *et al.*'s and *introgress* (~15% and ~13% false outlier detection, respectively).

2) Impact of different scenarios on outlier detection

In the second approach we compared specifically our method with Gagnaire *et al.*'s and we tested the effect of the sampling bias, selection coefficient and percentage of loci under selection on the detection of outliers. A range of possible scenarios were run regarding: (i) different parental sampling sizes (3, 5, 10, 30, 50, 100, 300, 500 and 1000 individuals, from the original populations of 1000 individuals each) for a $s = 99$ [$\log(s+1) = 2$]; (ii) different values of the selection coefficient (-0.999, -0.990, -0.900, -0.684, 0.000, 2.162, 9.000, 99.000 and 999.000; equivalent to $\log(s+1)$ values of -3.00, -2.00, -1.00, -0.50, -0.25, -0.10, -0.05, 0.00, 0.05, 0.10, 0.25, 0.50, 1.00, 2.00 and 3.00, respectively) with a parental sampling of 100 individuals each (from the original populations of 1000 individuals each); (iii) different proportions of selected loci from the 1000 simulated ones (0.0%, 0.5%, 1.0%, 5.0% and 10.0%). Every scenario was replicated 100 times.

Results (Figure 1) showed, on the one hand, that the bal&gar-ca method was very conservative and that its sensitivity (i.e., the rate of true positives) was lower than Gagnaire *et al.*'s. However, the bal&gar-ca method had a null false positive rate unlike Gagnaire *et al.*'s (~15% false outlier detection). Regarding the sampling bias (affecting the estimates of parental frequency), although the sensitivity (Figure 1a) of the bal&gar-ca method is specially lower than Gagnaire *et al.*'s at low sample sizes (i.e. with a high deviation in the parental frequency estimates), in these cases Gagnaire *et al.*'s suffers from a much higher false positive rate (Figure 1b). Both methods behave in a similar way in relation to varying selection coefficients and percentage of selected loci (Figure 1c–f). In our simulation, the sensitivity of both methods decreases at low values of the selection coefficient. Although detection of loci under selection becomes null in the bal&gar-ca method, its rate of false positives is null as well. Otherwise, the type I error rate of Gagnaire *et al.*'s method remains about 15% regardless of the selection coefficient values. Lastly, the mean rates of true and positive values obtained by both methods are not affected by the percentage of selected loci (Figure 1e–f). However, when the percentage of selected loci increases, the standard deviation of the sensitivity values decreases.

Table 2 Summary of sensitivity (true positive rate ± SD) and type I error rate (false positive rate ± SD) for outlier methods tested with 100 simulated data for three regimes of divergent selection

Method	Weak selection ($s = 2.162$)		Strong selection ($s = 99$)		Very strong selection ($s = 999$)	
	Sensitivity	Type I error	Sensitivity	Type I error	Sensitivity	Type I error
bal&gar-ca	0.124 ± 0.039	0.000 ± 0.000	0.479 ± 0.050	0.000 ± 0.000	0.487 ± 0.049	0.000 ± 0.000
Gagnaire *et al.*'s	0.572 ± 0.052	0.147 ± 0.012	0.621 ± 0.048	0. 149 ± 0.012	0.619 ± 0.049	0.147 ± 0.013
introgress	0.454 ± 0.034	0.127 ± 0.015	0.717 ± 0.061	0.126 ± 0.010	0.723 ± 0.052	0.128 ± 0.014
BAYESCAN	0.025 ± 0.021	0.029 ± 0.001	0.035 ± 0.019	0.012 ± 0.004	0.080 ± 0.019	0.037 ± 0.005

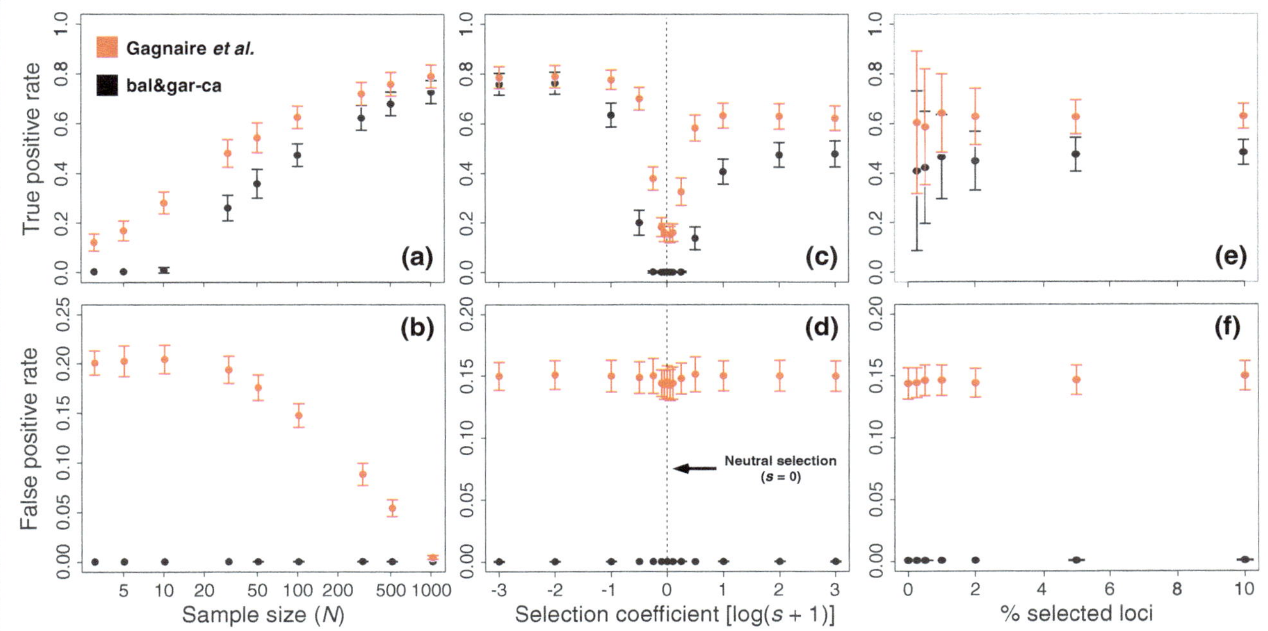

Figure 1 Effect of the parental sample size (*N*) on (a) the detection of true outlier loci and (b) rate of false positives, for the two methods. Influence of the selection coefficient (*s*) on **(c)** the detection of true outlier loci and **(d)** rate of false positives. Effect of the percentage of selected loci on **(e)** the detection of true outlier loci and **(f)** rate of false positives. Simulations (*N* =100) were performed for 900 neutrally introgressed loci and 100 loci under selection for Gagnaire *et al.*'s (in red) and bal&gar-ca (in black) methods. Bars for ± standard deviation values.

3) Effect of the frequency estimation method

The original formulation of Gagnaire *et al.*'s and bal&gar-ca methods use the square-root method to estimate the null-allele frequency. As the allele frequency estimation is difficult in dominant markers, we investigated the effect on the outlier detection of using a Bayesian estimator of the null-allele frequency, $\hat{q}$, with non-uniform priors [36]. For the bal&gar-ca method, we also modified the calculation of the frequencies of the parental allele confidence intervals, by using a quantile-based 95% probability interval based on the Bayesian estimator of the squared standard error, $s^2_{\hat{q}}$ [36]. To compare this Bayesian method and the previous approach for a particular common scenario, 900 independent loci plus 100 selected loci ($s = 99$) were simulated in 100 F_1 hybrid individuals from two parental populations (A and B) with 100 individuals each. One hundred simulations were carried out to compare the results of both methods (Gagnaire *et al.*'s and bal&gar-ca) with both different allelic frequencies estimation (square-root and Bayesian).

Figure 2 shows the results for different allelic frequency estimates in the two genome scan methods. Bayesian estimation has strong effect in both methods. For Gagnaire *et al.*'s method, using Bayesian estimations slightly decreased its sensitivity, but it highly decreased its rate of false positives (from 15% to 4%). For the bal&gar-ca method, using Bayesian estimations improved its sensitivity, but its rate of false positives also increased (reaching almost 3% –which, nevertheless, was not very high).

In stable hybrid zones with several generations of hybrids, *introgress* seems to be a compromise solution. However, in transitory recent hybrid zones, bal&gar-ca and Gagnaire *et al.*'s methods are better options, depending on the size of the hybrid zone, our sampling strategy and the scope of the genome scan. Gagnaire *et al.*'s is useful when we have a good population sampling. However, when the sample sizes are modest and we want to avoid any possible false positive, bal&gar-ca is the advised method. Although Bayesian estimation of frequency allele has been proved to be useful in many cases, it is not advised for the bal&gar-ca method as it increases its type I error rate. Finally, none of these methods is advisable when the selection coefficient shows low values, which is not strange as, in these cases, long periods of time would be needed to accumulate detectable signal in the DNA [18].

Simulating hybridization with selection and genome scan for F_1 individuals

In this example, we carry out the simulation of two parental populations of 100 individuals and 100 F_1 hybrids for a total of 300 markers using the 'hybridsim' function. Positive selection was simulated with $s = 10$ for 15 out of 300 markers.

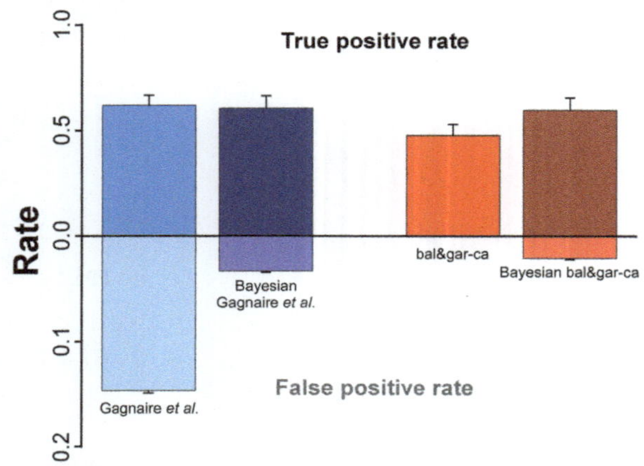

Figure 2 Effect of the allele frequency estimation method (square-root and Bayesian estimations –non-labelled and labelled columns, respectively) on the detection of true outlier loci and rate of false positives for Gagnaire *et al.*'s and bal&gar-ca methods. Bars for the standard deviation.

Firstly, we need to load *AFLPsim* and set the random seed number (arbitrarily to 123) for reproducibility.

```
> require(AFLPsim)
> set.seed(123)
> f1hybrid<-hybridsim(Nmarker=300, Na=100,
Nb=100, Nf1=100, +type='selection',
hybrid='F1', S=10, Nsel=15)
```

This is an object 'hybridsim' that contains the presence-absence matrices for the parentals (PA and PB) and for the hybrid classes (in this case, F_1). In addition, this object contains the loci under selection (`f1hybrid$SelMarkers`) and the selection coefficient used in the simulation (`f1hybrid$S`).

```
> f1hybrid$F1
$F1
M1 M2 M3 ...
F1_1 0 1 0 ...
F1_2 1 1 0 ...
.
.
.
> f1hybrid$SelMarkers
[1] 13 26 46 55 69 78 149 158 161 177 183
229 230 289 290
> f1hybrid$S
[1] 10
```

Then we perform a genome scan with the bal&gar-ca method setting the type parameter to the correct hybrid class (i.e. F_1 hybrids). The results of the genome scan analysis are saved in a data object (outlier), which includes the *P*-values of the binomial test after FDR (`outlier$fdrf1`), and the loci identity of those markers with $P < 0.05$ (`outlier$Outliers`). Moreover, this object is used as the input for the 'plot.hybridsim' function to generate a plot of the outlier markers and the expected frequencies under neutrality (Figure 3).

```
> outlier<-gscan (f1hybrid, type='F1',
method='bal&gar-ca')
> str (outlier)
List of 2
$ Pvalues :'data.frame': 300 obs. of 1
variable:
..$ fdrf1: num [1:300] 1 1 1 1 1 1 1 1 1 1 ...
$ Outliers: num [1:10] 26 46 55 69 149 161
177 229 230 289
> plot.hybridsim(f1hybrid,hybrid='F1', +
markers=outlier$Outliers)
```

In this example, we detected 10 out of 15 loci under selection (67% sensitivity) and we did not obtain any false positive.

Simulating demographical evolution under hybridization

Finally, we simulated hybridization on one area and evaluated its demographical consequences. For the initial frequencies we created a vector with the frequencies of Parental A, Parental B, F_1, Backcross to Parental A, Backcross to

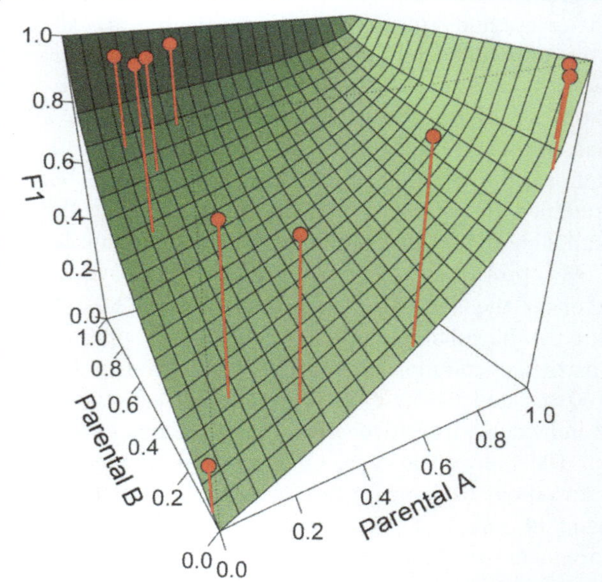

Figure 3 Three-dimensional scatter plots showing significant outlier loci detected by the 'gscan' function for the simulated F_1 hybrids. The green-coloured surface shows the theoretical probability of observing a dominant marker as a function of the band presence frequency in each parental species. The difference between the observed and the theoretical band frequency is represented with a vertical line joining both values.

Parental B and F_x. In our case, we fix Parental A and Parental B initial frequencies to 0.5.

```
> freqinit<-c (0.5,0.5,0,0,0,0)
```

Then, we create a matrix of assortative mating using the matrix function, and allow crosses between all taxa with the same probability.

```
> matingmat<-matrix (1,ncol=6,nrow=6)
```

In this example, parentals have similar fitness but F_1 individuals' is lower than parentals'. Here, we want to force asymmetrical introgression and breakdown occurs after F_1 hybrids, with posterior hybrid generations (F_x) and backcrosses to Parental B being sterile, whereas backcrosses to Parental A (BxA) would have a similar fitness to F_1 individuals. Hence, fitness would be modified as following:

```
> fitness<-c (1,1,0.5,0.5,0,0)
```

We obtain a matrix with the frequency of each taxon in eight generations. Parental A dominates the hybrid zone after eight generations, and displaces the other parental and the hybrids. We used the 'plot.demosim' function to visualise this demographic evolution (Figure 4):

```
> set.seed(123)
```

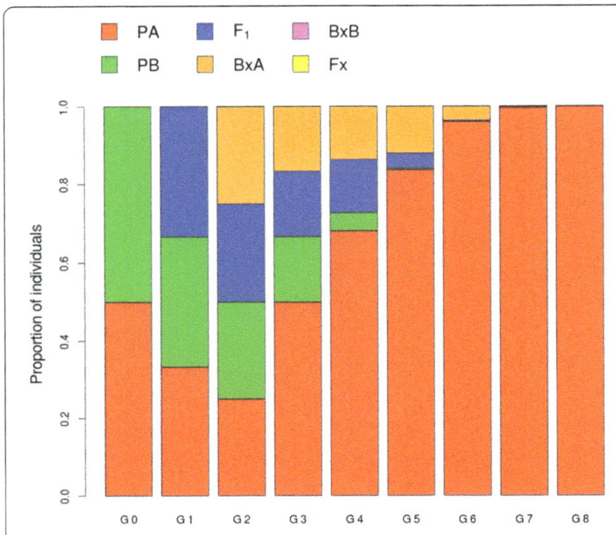

Figure 4 Simulated demographic evolution of a hybrid zone under similar initial proportions of the parentals, using a modified version of Epifanio & Philipp's model. Each bar represents the relative proportion of each parental and hybrid category (see legend) in the area over 8 generations (G0-G8), until Parental A (PA) dominates the area.

```
> results<-demosimhybrid(freqinit,
matingmat, fitness)
> results
     PA    PB    F1    BPA BPB Fx
G0 0.500 0.500 0.000 0.000 0  0
G1 0.333 0.333 0.333 0.000 0  0
G2 0.250 0.250 0.250 0.250 0  0
G3 0.500 0.167 0.167 0.167 0  0
G4 0.682 0.045 0.136 0.136 0  0
G5 0.838 0.003 0.040 0.120 0  0
G6 0.962 0.000 0.002 0.036 0  0
G7 0.998 0.000 0.000 0.002 0  0
G8 1.000 0.000 0.000 0.000 0  0
attr(,"class")
[1] "demosim.hybrid"
> plot.demosimhybrid (results)
```

Conclusions

This simulation study showed the interest of performing comparative studies in hybridization analytical software. Here we fill an important gap in this kind of software, as the R package *AFLPsim* application provides several useful tools in the context of hybridization studies. This is true specifically in relation to dominant markers with low sample sizes, in order to obtain markers under selection with a low rate of false positives and with a rather high sensitivity. None of the methods used is advisable when the selection coefficient shows low values. Moreover, *AFLPsim* provides a demographic method to study evolution in this context. Finally, we hold an on-going project to implement the bal&gar-ca method with biallelic codominant markers as well as single nucleotide polymorphisms (SNPs).

Availability and requirements

Project name: *AFLPsim*
Project home page: http://cran.r-project.org/web/packages/AFLPsim/
Operating system(s): Windows, Mac OS, Linux
Programming language: R
Other requirements: R version 2.15 or higher
License: GPL-2 | GPL-3
Any restrictions to use by non-academics: None

Competing interests

The authors declare that they have no competing interests.

Authors' contributions

FB and JLG-C developed and coded the R package *AFLPsim*. FB drafted the manuscript and JLG-C assisted in drafting the manuscript. Both authors read and approved the final manuscript.

Acknowledgements

We thank Dr. M. Escudero for help in the 'demosimhybrid' code, for testing the software and providing feedback. Dr. O. Paun and two anonymous reviewers provided invaluable comments on a previous version of the manuscript. While working on this package, FB was partly funded by a Marie

Curie Intra-European post-doctoral fellowship ('TRANSADAPTATION') at the Department of Systematic and Evolutionary Botany (University of Vienna) and Junta de Andalucía, Proyecto de Excelencia (2005/RNM484) and Ministerio de Educación y Ciencia, *Flora iberica* VIII (CGL2009-08178) and *Flora iberica* IX (CGL2012-32914), which were financed by FEDER funds.

References

1. Burke JM, Arnold ML: **Genetics and the fitness of hybrids.** *Annu Rev Genet* 2001, **35:**31–52.
2. Soltis PS, Soltis DE: **The role of hybridization in plant speciation.** *Annu Rev Plant Biol* 2009, **60:**561–588.
3. Mallet J: **Hybrid speciation.** *Nature* 2007, **446:**279–283.
4. Barton NH: **The role of hybridization in evolution.** *Mol Ecol* 2001, **10:**551–568.
5. Barton NH, Hewitt GM: **Analysis of hybrid zones.** *Annu Rev Ecol Syst* 1985, **16:**113–148.
6. Stapley J, Reger J, Feulner PGD, Smadja C, Galindo J, Ekblom R, Bennison C, Ball AD, Beckerman AP, Slate J: **Adaptation genomics: the next generation.** *Trends Ecol Evol* 2010, **25:**705–712.
7. Meudt HM, Clarke AC: **Almost forgotten or latest practice? AFLP applications, analyses and advances.** *Trends Plant Sci* 2007, **12:**106–117.
8. O'Hanlon P, Peakall R, Briese D: **Amplified fragment length polymorphism (AFLP) reveals introgression in weedy *Onopordum* thistles: hybridization and invasion.** *Mol Ecol* 1999, **8:**1239–1246.
9. Anderson EC: **Bayesian inference of species hybrids using multilocus dominant genetic markers.** *Philos Trans R Soc Lond B Biol Sci* 2008, **363:**2841–2850.
10. Liu Z, Nichols A, Li P, Dunham RA: **Inheritance and usefulness of AFLP markers in channel catfish (*Ictalurus punctatus*), blue catfish (*I. furcatus*), and their F1, F2, and backcross hybrids.** *Mol Gen Genet* 1998, **258:**260–268.
11. Gagnaire PA, Albert V, Jónsson B, Bernatchez L: **Natural selection influences AFLP intraspecific genetic variability and introgression patterns in Atlantic eels.** *Mol Ecol* 2009, **18:**1678–1691.
12. Zha H, Milne RI, Sun H: **Asymmetric hybridization in *Rhododendron agastum*: a hybrid taxon comprising mainly F_1s in Yunnan, China.** *Ann Bot* 2010, **105:**89–100.
13. Foll M, Gaggiotti O: **A genome-scan method to identify selected loci appropriate for both dominant and codominant markers: a Bayesian perspective.** *Genetics* 2008, **180:**977–993.
14. Bazin E, Dawson KJ, Beaumont MA: **Likelihood-free inference of population structure and local adaptation in a Bayesian hierarchical model.** *Genetics* 2010, **185:**587–602.
15. Antao T, Beaumont MA: **Mcheza: a workbench to detect selection using dominant markers.** *Bioinformatics* 2011, **27:**1717–1718.
16. Gompert Z, Buerkle CA: **A powerful regression-based method for admixture mapping of isolation across the genome of hybrids.** *Mol Ecol* 2009, **18:**1207–1224.
17. Gompert Z, Buerkle CA: ***introgress*: a software package for mapping components of isolation in hybrids.** *Mol Ecol Resour* 2010, **10:**378–384.
18. Storz JF: **Using genome scans of DNA polymorphism to infer adaptive population divergence.** *Mol Ecol* 2005, **14:**671–688.
19. Bonin A, Ehrich D, Manel S: **Statistical analysis of amplified fragment length polymorphism data: a toolbox for molecular ecologists and evolutionists.** *Mol Ecol* 2007, **16:**3737–3758.
20. Pérez-Figueroa A, García-Pereira MJ, Saura M, Rolán-Alvarez E, Caballero A: **Comparing three different methods to detect selective loci using dominant markers.** *J Evol Biol* 2010, **23:**2267–2276.
21. Caballero A, Quesada H, Rolán-Alvarez E: **Impact of Amplified Fragment Length Polymorphism size homoplasy on the estimation of population genetic diversity and the detection of selective loci.** *Genetics* 2008, **179:**539–554.
22. Laval G, Excoffier L: **SIMCOAL 2.0: a program to simulate genomic diversity over large recombining regions in a subdivided population with a complex history.** *Bioinformatics* 2004, **20:**2485–2487.
23. Nielsen EE, Bach LA, Kotlicki P: **Hybridlab (Version 1.0): a program for generating simulated hybrids from population samples.** *Mol Ecol Notes* 2006, **6:**971–973.
24. R Development Core Team: **R: a language and environment for statistical computing.** 2014, http://www.R-project.org/.
25. Wright S: **Evolution in Mendelian populations.** *Genetics* 1931, **16:**97–159.
26. Hall R, Ayres DR: **What can mathematical modeling tell us about hybrid invasions?** *Biol Invasions* 2009, **11:**1217–1224.
27. Epifanio J, Philipp D: **Simulating the extinction of parental lineages from introgressive hybridization: the effects of fitness, initial proportions of parental taxa, and mate choice.** *Rev Fish Biol Fish* 2000, **10:**339–354.
28. Balao F, Casimiro-Soriguer R, García-Castaño JL, Terrab A, Talavera S: **Big thistle eats the little thistle: Does unidirectional introgressive hybridization endanger the conservation of *Onopordum hinojense*?** *New Phytol in press* doi:10.1111/nph.13156.
29. Bettles C, Docker M, Dufour B, Heath DD: **Hybridization dynamics between sympatric species of trout: loss of reproductive isolation.** *J Evol Biol* 2005, **18:**1220–1233.
30. Clopper CJ, Pearson ES: **The use of confidence or fiducial limits illustrated in the case of the binomial.** *Biometrika* 1934, **26:**404–413.
31. Dray S, Dufour AB: **The *ade4* package: implementing the duality diagram for ecologists.** *J Stat Softw* 2007, **22:**1–20.
32. Jombart T: ***adegenet*: a R package for the multivariate analysis of genetic markers.** *Bioinformatics* 2008, **24:**1403–1405.
33. Excoffier L, Laval G, Schneider S: **Arlequin ver. 3.0: An integrated software package for population genetics data analysis.** *Evol Bioinform Online* 2005, **1:**47–50.
34. Yeh FC, Boyle TJB: **Population genetic analysis of co-dominant and dominant markers and quantitative traits.** *Belgian J Bot* 1997, **129:**157.
35. Falush D, Stephens M, Pritchard JK: **Inference of population structure using multilocus genotype data: dominant markers and null alleles.** *Mol Ecol Notes* 2007, **7:**574–578.
36. Zhivotovsky LA: **Estimating population structure in diploids with multilocus dominant DNA markers.** *Mol Ecol* 1999, **8:**907–913.

10

Protocol: a fast and simple *in situ* PCR method for localising gene expression in plant tissue

Asmini Athman[1,3†], Sandra K Tanz[2†], Vanessa M Conn[1,3], Charlotte Jordans[1,3], Gwenda M Mayo[3,4], Weng W Ng[1,3], Rachel A Burton[5], Simon J Conn[3,6] and Matthew Gilliham[1,3*]

Abstract

Background: An important step in characterising the function of a gene is identifying the cells in which it is expressed. Traditional methods to determine this include *in situ* hybridisation, gene promoter-reporter fusions or cell isolation/purification techniques followed by quantitative PCR. These methods, although frequently used, can have limitations including their time-consuming nature, limited specificity, reliance upon well-annotated promoters, high cost, and the need for specialized equipment. *In situ* PCR is a relatively simple and rapid method that involves the amplification of specific mRNA directly within plant tissue whilst incorporating labelled nucleotides that are subsequently detected by immunohistochemistry. Another notable advantage of this technique is that it can be used on plants that are not easily genetically transformed.

Results: An optimised workflow for in-tube and on-slide *in situ* PCR is presented that has been evaluated using multiple plant species and tissue types. The protocol includes optimised methods for: (i) fixing, embedding, and sectioning of plant tissue; (ii) DNase treatment; (iii) *in situ* RT-PCR with the incorporation of DIG-labelled nucleotides; (iv) signal detection using colourimetric alkaline phosphatase substrates; and (v) mounting and microscopy. We also provide advice on troubleshooting and the limitations of using fluorescence as an alternative detection method. Using our protocol, reliable results can be obtained within two days from harvesting plant material. This method requires limited specialized equipment and can be adopted by any laboratory with a vibratome (vibrating blade microtome), a standard thermocycler, and a microscope. We show that the technique can be used to localise gene expression with cell-specific resolution.

Conclusions: The *in situ* PCR method presented here is highly sensitive and specific. It reliably identifies the cellular expression pattern of even highly homologous and low abundance transcripts within target tissues, and can be completed within two days of harvesting tissue. As such, it has considerable advantages over other methods, especially in terms of time and cost. We recommend its adoption as the standard laboratory technique of choice for demonstrating the cellular expression pattern of a gene of interest.

Keywords: *In situ* PCR, RT-PCR, Cell-specific localisation, Plant tissue, Immunohistochemistry

Introduction

The function of only a small fraction of genes from any species has been experimentally verified. For the majority of plant genes, their function are either unknown or have been predicted through homology with a known gene from another species [1]. With the advent of next generation sequencing, this disparity between annotated genes and those that are functionally characterized is likely to increase.

An important step toward understanding the role of a gene product is to identify the tissue and cell type within which it is expressed. However, transcript abundance is predominantly analysed using whole plants or organs. Despite the fact that few transcripts are expressed exclusively within a single cell type, many genes are preferentially expressed in only a few cell types [2]. Analysing the transcript abundance of such genes in whole plants or

* Correspondence: matthew.gilliham@adelaide.edu.au
†Equal contributors
1 ARC Centre of Excellence in Plant Energy Biology, University of Adelaide, Glen Osmond, SA, Australia
3 Waite Research Institute & School of Agriculture, Food and Wine, University of Adelaide, PMB1, Glen Osmond, SA 5064, Australia
Full list of author information is available at the end of the article

organs may be misleading and will reflect the net expression for those particular genes in all cells rather than in the cell types in which they are actually expressed. This is of particular relevance for genes expressed predominantly in cell types that comprise only a small percentage of the total analysed tissue. Even if transcripts of genes of interest are highly abundant in one or few cell types the total number of the transcripts of interest may be diluted to below the detection limits within the whole tissue RNA pool. Such a phenomenon is not uncommon as there are a number of specific cell-types that appear to have a profound influence on the performance of the plant and act as 'gatekeepers' for a particular physiological process [3,4]. These cell-types contain proteins and signalling cascades that are primed for a particular physiological role so have unique transcriptional profiles. Such cell-types include, but are not limited to, guard cells (that undergo large turgor changes to control whole plant gas exchange), xylem parenchyma (which control net xylem content so have a major influence on long-distance nutrient and water transport), and pericycle (which initiate lateral root development so have a major role in root architecture). The transcriptional profiling of whole tissues can therefore be extremely misleading when ascribing a function to a gene and caution should be observed when interpreting expression data from tissue samples containing more than one cell-type. Awareness of such issues has led to the development of techniques that have allowed researchers to examine the expression of specific genes in specific cell types.

Common assays for detecting where transcripts are expressed can be classified in several ways (Table 1). Ordinarily they rely on one of the following: 1) the generation of transgenic plants with indicator proteins or nucleic acid/ribosomal binding proteins; 2) the isolation of single cells or single cell contents; 3) the hybridisation of RNA/cDNA directly in tissue; or 4) a combination of 1 and 2. These techniques can be classified into two further groups: A) those that first isolate RNA from a tissue or specific cell-type and then identify a transcript of interest in that RNA pool or, B) those that search directly for the transcript of interest in living or fixed plant tissue. We have summarised the advantages and disadvantages of these methodologies in Table 1.

Many of these methods have long lead times as they require the generation of transgenic plants so in these instances can be unsuitable for species that cannot be transformed or for the analysis of multiple genes. Also,

Table 1 Techniques available for *in planta* cell level expression analysis

Technique	Type	Comments	References
Specific gene promoter: indicator protein fusions [i.e. beta-glucuronidase (GUS) or fluorescent proteins]	1, B	Long lead times of >2 months to get stably expressed genes, homozygous gene expression, sequences other than the promoter may control gene expression, no guarantee that promoter fragment chosen is correct, not suitable for plants that cannot be transformed.	[5]
Laser Capture Microdissection (LCM) and Single Cell Sampling (SiCSA)	2, A	Difficulty in isolating certain cell-types (e.g. vascular cells using SiCSA), long tissue prep (~2 weeks) for LCM, need for specialised equipment	[6,7]
in situ PCR	3, B	No specialised equipment required apart from vibratome, simple and fast method. No detailed protocol for plants until this manuscript, especially for agarose- embedded vibratome- sectioned in tube PCR. Can be performed on plants that are difficult to, or cannot be transformed. We do not recommend fluorescence detection of *in situ* PCR products due to interference with autofluorescence from plant tissue. As it is possible to do separate PCRs on adjoining tissue sections or replicate tissues from other plants we see no great advantage in multiplexing *in situ* PCR as it causes multiple complications to what is a robust and relatively simple technique. We do not recommend multiplexing *in situ* PCR due to the differential abundance of transcripts in the same cell and consequently the saturation of products, generation of non-specific products and problems with signal separation.	
in situ hybridisation (ISH)	3, B	High detection threshold (10–20 copies per cell for ISH vs 1–2 copies per cell for *in situ* PCR), need to design a specific probe that hybridises to RNA while *in situ* PCR uses the same primers as qPCR, much cheaper. The View RNA Assay (Affymetrix) enables automated multiple transcript detection using fluorescence by employing the principles of ISH to detect nucleic acid targets within specific cells/cell-types. Its use is limited to abundant transcripts and laboratories equipped with and/or experienced in FFPE (formalin-fixed paraffin embedded) and frozen tissue preparation.	[8]
Protoplasting of fluorescently labelled cells and single cell sorting (FACS)	4, A	Can assay multiple transcripts at the same time. Potential damage responses of tissue. Needs specialised sorting flow cytometer. Limited to analysis of cells that are fluorescently labelled. Cannot give detail of all cells in which a particular gene is expressed.	[9,10]
Nuclear sorting, INTACT (isolation of nuclei tagged in specific cell types) or ribosomal binding techniques	4, A	When combined with microarrays or RNAseq can assay multiple transcripts at the same time. Limited to analysis of cells that are fluorescently labelled. Cannot give detail of all cells in which a particular gene is expressed. Not easily replicated.	[11-13]
In situ RNA sequencing	3, B	Offers the prospect of obtaining whole transcriptomes and more from single cells in tissue sections. Not yet optimised in any tissue. Has not been performed for plant tissue.	[14]

many of these methods are difficult to adopt in a laboratory not set up for such techniques without a significant outlay as they utilize expensive and specialized equipment and expertise. Of the non-transgenic techniques, *in situ* hybridization and *in situ* PCR are methods that allow gene expression analysis at a single cell resolution. Two disadvantages of *in situ* hybridization revolve around sensitivity. Firstly, low-copy mRNAs often result in a false negative signal and secondly, highly homologous members of the same gene family are difficult to distinguish. Both of these issues are addressed with *in situ* PCR because a specific region of the gene transcript is amplified during multiple cycles of the PCR resulting in a highly sensitive technique.

In situ PCR is a semi-quantitative method that has been used to detect target RNA using colourimetry or fluorescence. Therefore it is useful to detect in which cell types genes are expressed. To precisely quantitate transcript levels in tissues or single cells, *in situ* PCR should be used in conjunction with a quantitative transcript detection technique (such as Single cell sampling and analysis (SiCSA) in combination with RNASeq, qPCR, or microarrays (Table 1).

In situ PCR has been described and performed extensively in animals and a detailed protocol has been made available [15]. Reports of *in situ* PCR on plant material either include lengthy tissue preparations, thermal cycling of slide-mounted specimens, or a detailed protocol was not supplied [16-20]. Here, we have further developed the *in situ* PCR technique, allowing improved resolution of transcript localisation. We also include detailed recommendations for the implementation of this technique in a variety of plant tissues to improve its utility (Figure 1). This includes both an in-tube protocol, which uses a standard thermocycler block for routine samples and an on-slide methodology, which uses a slide thermal cycler for use with fragile tissues. Although this alternative is not necessary for most applications of *in situ* PCR, it has advantages for very thin sections (<15 μm), or for delicate dissected tissues such as epidermal peels. We have used this method in published articles for the localisation of transcripts in wheat, barley and soybean [21-24], but due to the constraints of most journals have not had the opportunity to publish the methodology in the detail required for others to follow. Here, to outline the technique we include a full protocol and we show the preferential localisation of transcripts in various plant tissues and cell types, including guard cells, mesophyll, xylem parenchyma, pericycle, and different reproductive tissues.

Protocol

Overview

To preserve the morphology of the plant tissue and to anchor the transcripts to their cellular origin, freshly harvested plant tissue is immediately fixed in an ethanol/acetic acid/formaldehyde solution with the penetration of the fixative enhanced through vacuum infiltration (Figure 1). The plant tissue is then embedded in agarose for subsequent sectioning on a vibratome (Figure 1). Sections are collected into a microfuge tube or placed on a slide. DNase treatment, reverse transcription and *in situ* PCR can all be carried out in the microfuge tube using a standard thermal cycler or on the slide using an *in situ* block. DNase treatment is carried out to remove genomic DNA (Figure 1), which would otherwise lead to erroneous detection of signal. The DNase-treated RNA in the tissue is converted to cDNA by reverse transcription (Figure 1). The resulting cDNA is amplified by standard gene-specific PCR integrating Digoxigenin (DIG) labeled nucleotides (Figure 1). The sections are incubated with an anti-DIG antibody conjugated to alkaline phosphatase, which binds to the DIG-labeled PCR products (Figure 1). Adding specific substrates for alkaline phosphatase allows the colourimetric detection of the DIG-labeled PCR products, delimited to the site of RNA synthesis. At this point, if the protocol was carried out in a microfuge tube, the sections are transferred onto a microscope slide (Figure 1). After mounting the slides, the processed tissue sections can be visualized by bright field microscopy (Figure 2).

Plant growth and tissue sampling

The protocol described here concentrates on the amplification of RNA/cDNA from leaf and root tissues across a range of plant species (Table 2). By optimizing various aspects of the conditions, this method can be adapted easily to tissue types other than leaf or root (Table 2). The plant-growth method is inconsequential for the success of this technique, but we recommend hydroponics for root tissue samples to simplify harvesting of tissue [25]. Roots of plants grown in soil or agar should be immersed repeatedly in sterile water using a soft paintbrush to remove particulate matter while minimising damage to the roots.

Prior to performing *in situ* PCR, it is important to consider the time of day that the tissue will be harvested, the growth stage of the plant, and the tissue type, as these factors may influence gene expression. As a general rule, we sampled at relative midday with respect to the photoperiod. For leaves, we use young and undamaged leaves, harvesting material with the midrib intact as this provides extra stability to the sections. Excising the leaf tip, outer margins and base (>2 mm from petiole) simplifies vibratome sectioning, handling and visualization of sections. We generally avoid older root sections with a large number of lateral root hairs as the root hairs tend to stick to the blade during vibratome sectioning, and this may result in whole root samples being pulled out of the agarose block. The time taken to prepare the material for fixation

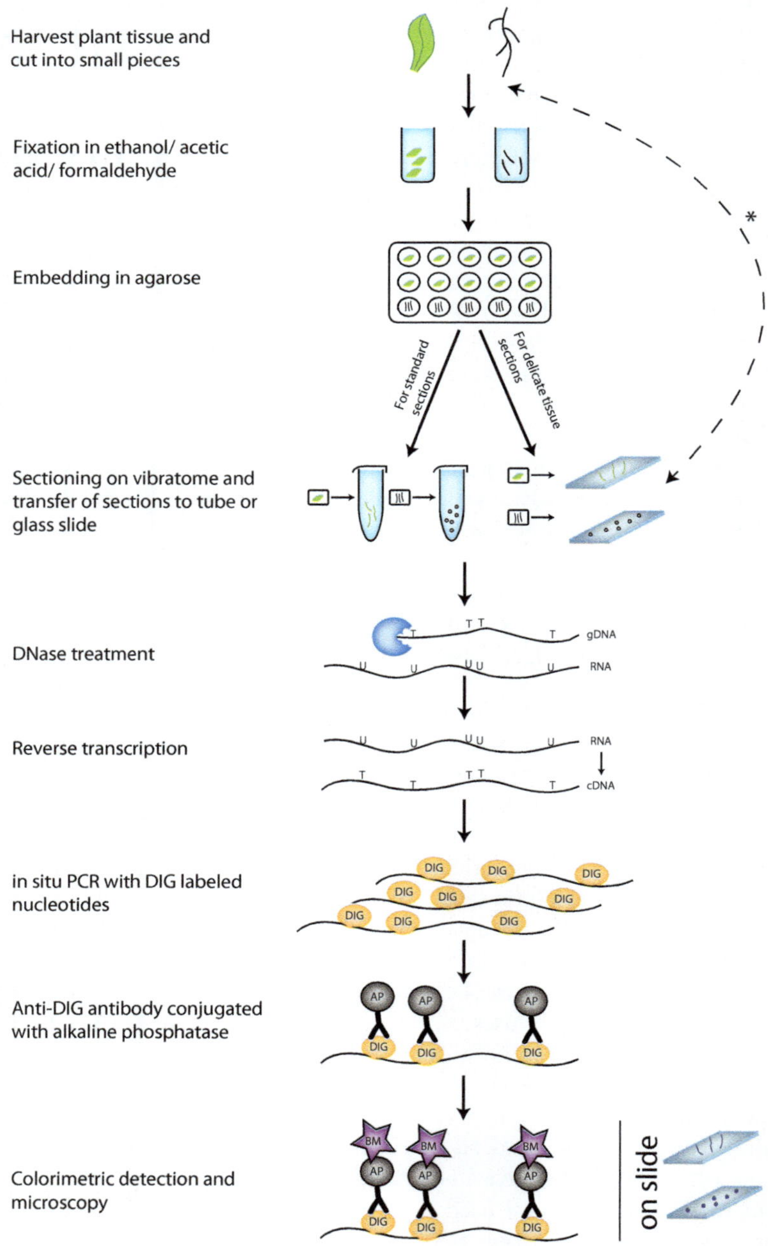

Figure 1 Schematic representation of the *in situ* PCR pipeline. Plant tissue is fixed in an ethanol/acetic acid/formaldehyde solution, followed by embedding in agarose and sectioning. Sections are collected into a microfuge tube in which DNase treatment, reverse transcription and *in situ* PCR are carried out using a thermocycler. During *in situ* PCR, DIG labeled nucleotides are incorporated into the PCR products. An anti-DIG antibody conjugated with alkaline phosphatase and an alkaline phosphatase substrate are used for the detection of the DIG labeled PCR products. These are visualized under a microscope. Thin fragile sections are placed onto a glass slide after sectioning, while *non-embedded, non-sectioned samples (i.e. epidermal peels) are placed directly onto slides for all processes from fixation to the final PCR step (dotted arrow).

should be kept to a minimum as RNA may begin to degrade and tissue begins to wilt. In our experience we can prepare tissue in under 1 minute. This will not avoid very rapid changes in expression caused by handling; such changes are unavoidable in most applications. The number of leaves or roots to be harvested depends on the number of transcripts to be tested by *in situ* RT-PCR, with at least 3 reproducible biological replicates per tissue per condition being required to confirm expression patterns – this means that more than three tissue replicates will need to be harvested and fixed to account for damage or loss of tissue etc.

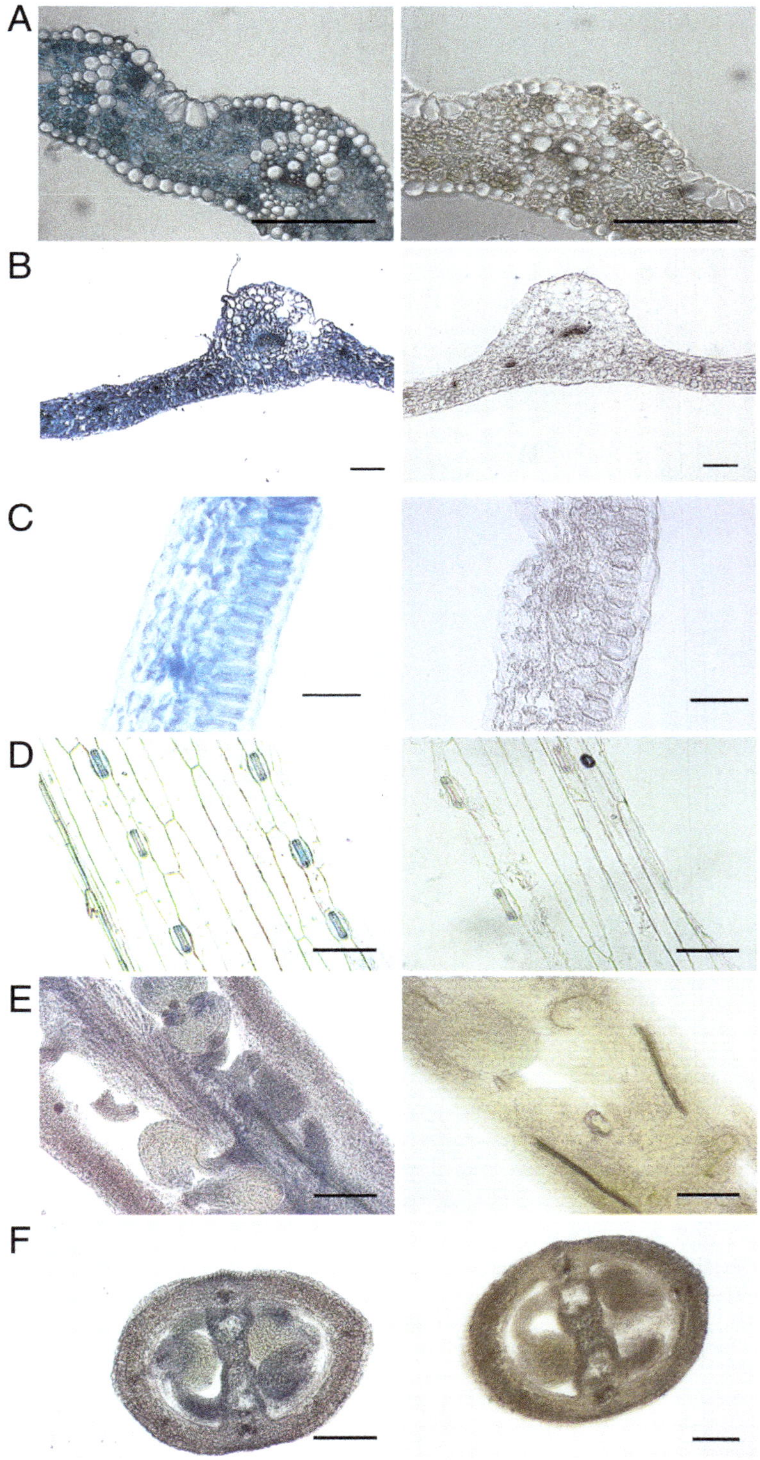

Figure 2 ***In situ*** **PCR for various transcripts in barley and Arabidopsis.** Blue staining demonstrates presence of cDNA (left panel) while a brown colour indicates absence of amplified target cDNA within cells (right panel). **(A, left)** 18S ribosomal RNA positive control to show staining in all cell-types of barley leaf, **(B, left)** 18S ribosomal RNA positive control to show staining in all cell-types of Arabidopsis leaf, **(C, left)** *AtCAX1* expression predominantly in the mesophyll of Arabidopsis leaf, **(D, left)** guard cell staining of a barley Aluminum-activated malate transporter *(ALMT)* *(SL1251)* from an *in situ* PCR performed on epidermal peels. **(E and F, left)** detection of an Arabidopsis *ALMT14* in the septum **(E)** and embryo **(F)** of developing flowers/siliques. The same primers and conditions were used for the corresponding negative controls (A-F, right panels, Table 6). Scale bars represent 100 μm.

Table 2 Various alkaline phosphatase (AP) substrates used with this *in situ* PCR protocol on different plant species and tissues

Plant species	Tissue	AP substrate
Arabidopsis thaliana	Leaf	BM Purple
		Elf97
		Vector Blue
Arabidopsis thaliana	Flower	BM Purple
Cleome gynandra	Leaf	BM purple
Cleome hassleriana	Leaf	BM purple
Glycine max	Root	BM purple
Hordeum vulgare	Leaf	BM purple
		Elf97
Hordeum vulgare	Epidermal peel	BM Purple
		Elf97^{++}
Oryza sativa	Leaf	BM Purple
		Vector Blue
Oryza sativa	Root	BM Purple
Solanum lycopersicum	Leaf	Vector Blue
Triticum aestivum	Root	BM Purple
Triticum durum	Leaf	Elf97
Triticum durum	Root	BM Purple
		NBT/BCIP^{X}
		Vector Blue^{++}
Vitis vinifera	Root	Vector Blue
Vitis vinifera	Leaf	BM Purple
Zea mays	Leaf	BM Purple

Most AP substrates were used successfully for specific detection of the transcript in question, with some exceptions as indicated. $^{++}$indicates detection with significant background staining. ^{X}indicates no detectable signal.

Tissue preparation

The method of tissue preparation used in this protocol involves formaldehyde/acetic acid/ethanol fixation and agarose-embedding, a fast and simple method that preserves the morphology of the plant tissue and keeps the RNA molecules intact [26]. Formaldehyde is a small and highly permeable molecule [27], which accurately preserves cellular morphology, condensing fixation times of most samples to two hours. It is reported that fixation in excess of four hours for leaves and 12 hours for roots in formaldehyde can result in excessive cross-linking of proteins to nucleic acids [28,29], which can interfere with the subsequent reverse transcription resulting in no detectable staining or in significant background staining of the sample.

Following fixation, the tissue should be embedded in molten agarose and allowed to cool for at least 1.5 hours prior to sectioning. This is significantly shorter than the time required for paraffin embedding (2–7 days). The RNA in the agarose-embedded tissue is stable for at least three weeks and can be detected on plant tissue sections cut as thin as 20 μm using a vibratome. Better image clarity is observed with thinner sections but these are more easily damaged than thicker sections during the wash steps. We recommend 50 μm thick sections for roots and 60–70 μm sections for leaves. While the sections may not be as thin as those that can be cut from frozen or paraffin-embedded tissues (4–15 μm), cell morphology is well preserved and transcripts are easily detected following *in situ* PCR in tissue sections of up to 75 μm thick. Furthermore, the absence of the pre- and post-treatment steps that are required with paraffin embedding and cryosectioning, make agarose embedding a technique suitable for any standard laboratory without the need for specialized equipment.

Generally, we embed three leaf samples or eight root samples per agarose block. The sectioning of one of these agarose blocks should yield enough tissue sections to perform *in situ* PCRs on at least three transcripts. From our experience, to be confident in the localisation of a transcript we suggest that at least ten tissue sections are visualised per transcript per biological replicate to check that the expression patterns are consistent across tissues in at least three biological replicates. To account for loss or damage to samples during processing we recommend starting with three times the amount of sections during vibratome sectioning and at least one additional biological replicate (so examine tissues from four plants). During sectioning the tissue usually separates from the agarose, so the number of sections can be counted onto the slide or tube for processing.

In some cases (such as with epidermal peels or the seed aleurone), tissue can be dissected from the plant without the need for sectioning on a vibratome. The processing of these non-embedded tissues should be carried out entirely on glass slides from fixation through to the final PCR and staining steps. In addition, if thin (<50 μm) or particularly fragile sections are needed for PCR, then it is again best to avoid in-tube processing of these samples; instead, these tissues are best processed on slides. To do this, when sections are taken post-fixation and embedding, they should be collected directly onto the slide for the DNAse treatment step and through to the final PCR and staining steps. These variations in the sample processing are highlighted in the protocol below.

Primer design

Primers should be designed to the cDNA sequence of a plant species. To test if the primer sequences align to genomic DNA (gDNA), the primer sequence can be compared against available gDNA sequences using BLAST (http://blast.ncbi.nlm.nih.gov/). The primers should amplify a 150–300 bp product, since this size product simplifies amplification within the cellular matrix and provides sufficient DIG incorporation (and thus signal) per amplicon. Oligonucleotides should be 18–24 bases long, have an annealing temperature of 55°C and if possible, span an exon-exon junction to prevent the amplification of gDNA. If the latter is not possible, the primers should be designed to span an intron sequence (i.e. the two primers are on two different exons) so that possible amplification from gDNA is eliminated by using a short extension time during the *in situ* PCR. The following standard primer design features should also be observed: they should amplify a single product; contain no internal secondary structure (hairpin, palindromes, self-dimers, repeats and runs in excess of four bases); contain 40–60% G/C; and, should not be self-complementary to avoid primer dimer formation. To assist with these guidelines, free online primer design and primer analysis tools can be used, such as Primer3 (http://simgene.com/Primer3) and NetPrimer (http://www.premierbiosoft.com/netprimer/).

Optimization of PCR conditions

Before conducting *in situ* PCR on any plant tissue, the amplification conditions must be optimised in solution, such as the concentration of magnesium and dNTPs, the type of polymerase used, the annealing temperature, extension time and cycle number. Optimised parameters invariably transfer successfully to the *in situ* PCR protocol. To determine the number of cycles that should be used for the *in situ* PCR, it is critical to choose the minimum cycle number that gives an identifiable product to maximise the chance of differentiating abundance between cell types, albeit semi-quantitatively, for the gene of interest. The cycle number used for *in situ* PCRs may be different for each gene to be tested. When choosing a cycle number for the *in situ* PCR, it needs to be considered that the PCRs are not likely to be as efficient for cDNA contained within plant tissue as compared to cDNA in solution, and the incorporation of DIG-dUTP during the *in situ* PCR may have an inhibitory effect on the DNA polymerase. Considering the above points, adding 1–2 cycles to the cycle number obtained from the in-solution PCR is recommended when conducting the *in situ* PCR.

RT-PCR

Since *in situ* PCR is primarily used for detecting a particular mRNA, an *in situ* reverse transcription (RT) step precedes the *in situ* PCR. During the RT, mRNAs are converted to cDNAs, either by a random primed or a specific primed RT reaction. Random primed RT reactions use random hexamers that are six bases long and convert all RNA molecules (mRNAs, rRNA and ncRNAs) into cDNA. We generally use the specific primed RT reaction, where a specific primer designed to the 3′ end of the transcript of interest is used to reverse transcribe only the required transcript from the mRNA. A fragment of this transcript is then amplified in the following PCR using either a pair of nested gene-specific primers or the RT-primer plus a gene-specific primer 5′ of this. For the design of the RT primer, the standard primer design guidelines apply as described above under 'Primer design'.

Amplicon labeling and detection

Amplicons can be labeled directly with a fluorescent label, which allows their subsequent detection by microscopy, or indirectly using either digoxigenin (DIG) or biotin. DIG-labeled PCR products can be detected using enzyme-based colour detection systems such as alkaline-phosphatase (AP) or peroxidase, using an anti-DIG-enzyme conjugate that binds to the DIG label on the amplicon. Biotin-labeled amplicons can be recognized using a streptavidin-enzyme conjugate for colourimetric detection in a similar fashion. We had most success with the colourimetric detection of an anti-DIG-AP conjugate that recognizes DIG-labeled-dUTP incorporated into the amplicon during the *in situ* PCR. We have tested different AP substrates with varying success for different plant tissues and species (Table 2). This table includes fluorescent substrates but we found these problematic due to significant detection of autofluorescence from the tissue interfering with our substrate fluorescence (see Results and discussion).

Validation and controls

To validate the expression pattern of a gene, we routinely perform *in situ* PCR on replicate tissues from four different plants. In addition, as mentioned above, multiple technical replicates (~10) are useful per biological replicate to validate the site of amplification and detection. A negative control that tests for the presence of contaminating gDNA should be included alongside each transcript analyzed by *in situ* PCR. This consists of tissue samples, prepared as for the test samples but with the omission of the reverse transcriptase enzyme during the RT step. A positive control should also be included for each tissue sample tested. This should be a gene that is uniformly expressed in all cells; if the effect of a treatment is being tested in your experiment, this gene should remain stably expressed across different treatments. The ribosomal 18S transcript, which is expressed across the majority of plant species and is ubiquitously expressed in all cells, is recommended.

Materials

Reagents

- RNase Zap (Sigma, cat. no. R2020).
- Fixative (see REAGENT SETUP).

- Ethanol (Chem-supply, cat no. 64-17-5, 100% Analytical Reagent).
- Acetic acid (Southern Cross Science, cat no. AA009-2.5LP).
- Formaldehyde (Sigma, cat. no. 252549-4 L).
- Wash Buffer 1 (see REAGENT SETUP).
- 1× PBS (see REAGENT SETUP).
- Agarose (Bioline, cat. no. BIO-41025) (see REAGENT SETUP).
- Ultra-low gelling temperature agarose (Sigma, cat. no. A2576) (see REAGENT SETUP).
- 1 L sterile water (ice-cold).
- 40 U $\mu L^{-}1$ RNaseOUT (Invitrogen, cat. no. 10777019).
- 10× Turbo DNA-free Buffer from Turbo DNA-free kit (Ambion, cat. no. AM1907). Only use the buffer TURBO DNaseI did not work well on plant sections.
- RNase-Free DNase Set (1500 U, lyophilised; Qiagen, cat. no. 79254) (see REAGENT SETUP).
- 0.5 mM EDTA, pH 8.0 (Chem-Supply, cat no. 6381-92-6) (see REAGENT SETUP).
- dNTPs (10 mM each; Bioline, cat. no. BIO-39025).
- Thermoscript RT kit (Invitrogen, cat. no. 11146–024).
- RT reaction solution (see REAGENT SETUP).
- HiFiPhusion polymerase (Thermo Scientific, 2 U μL^{-1}, cat. no. F-530S).
- Digoxigenin-11-dUTP, alkali stable (DIG-11-dUTP, 1 mM; Roche, cat. no. 11093088910).
- Gene-specific forward and reverse primers.
- PCR reaction solution (see REAGENT SETUP).
- 1× Block solution (see REAGENT SETUP).
- Anti-Digoxigenin-AP, Fab fragments (Anti-DIG antibody; Roche, cat. no. 11093274910) (see REAGENT SETUP).
- Wash Buffer 2 (see REAGENT SETUP).
- AP substrates: BM Purple AP Substrate (Roche, cat. no. 11442074001) or Vector Blue Substrate Kit (Abacus ALS, cat. no. VESK5300) or NBT-BCIP solution (Sigma-Aldrich, cat. no 72091) or ELF97 Endogenous Phosphatase Detection Kit (Life Technologies, cat. no. E-6601).
- 40% glycerol (VWR, cat. no. 24397.296) (see REAGENT SETUP).
- Clear nail varnish.

Equipment

- Micro scissors (Wescott spring scissors; Proscitech, cat. no. TS1084).
- Soft paintbrushes
- Tweezers (ProSciTech, cat no. T045-212).
- Sterile scalpel (ProSciTech, cat.no. METP2325-10-CE).
- Sterile 150 mm petri dishes (Sarstedt, cat.no. 82.1184.001).
- 2 mL microfuge tubes (VWR, cat. no. 211–2120).
- Vacuum infiltrator (EYELA Aspirator, cat. no. A-3S).
- 3 MM Whatmann filter paper (diameter 90 mm; VWR, cat. no. WHAT1001-090). Sterilize by autoclaving.
- Sterile 12-well plates (Adelab, cat. no. CNG3513).
- 0.2 mL PCR tubes (Adelab, cat. no. AXYPCR-02-C).
- Vibratome Leica VT1200 with magnetic specimen plate, buffer tray and ice bath (see EQUIPMENT SETUP).
- Superglue (Selleys Quick Fix).
- Double-edge blade (ProSciTech, cat. no. L056).
- Single-edge blade (ProSciTech, cat. no. L055C).
- Thermal cycler (Geneworks, cat. no. GS00001).
- *In-situ* block (Geneworks, cat. no. GSBOSITU).
- Microscope slides (ProSciTech, cat. no. G300B) or StarfrostSilane-prep slides (Sigma, cat. no. S4651-72EA).
- Frame-seal slide chambers (BioRad, cat. no. SLF0601).
- Coverslips (ProSciTech, cat.no. GCC2222).
- Glass pipette.
- For brightfield imaging: Leica, AS LMD microscope.
- For fluorescence detection: Zeiss Axiophot microscope: DAPI, BP 365/12, FT 395, LP 397.
- OR, Nikon A1R laser scanning confocal microscope: UV excitation using DAPI laser line (405 nm ex).

Reagent setup

Fixative 63% Ethanol, 5% acetic acid, 2% formaldehyde. Prepare fresh and use within 3 h. Keep on ice.

Caution

Hazardous chemicals should be handled in a fume cupboard using appropriate personal protective equipment.

Wash Buffer 1 63% Ethanol, 5% glacial acetic acid. Prepare fresh and keep on ice.

Caution

Hazardous chemicals should be handled in a fume hood using appropriate personal protective equipment

10× PBS stock solution, pH 7.5 0.1 M Na_2HPO_4, 1.3 M NaCl, pH 7.5. Dissolve 14.19 g Na_2HPO_4 in 800 mL of sterile water. Adjust to pH 7.5 by adding 37% HCL. Add 75.9 g of NaCl and add sterile water to a total volume of 1 L. This solution can be kept for 1 year at room temperature.

1× PBS, pH 7.5 Dilute the stock 10× PBS at 1:10 ratio to give a final concentration of 0.01 M Na_2HPO_4 and 0.13 M NaCl. This solution can be kept up for 1 year at room temperature.

5% Agarose or **5% Ultra-low gelling agarose** Dissolve 2.5 Agarose (leaves) or ultra-low gelling agarose (roots) in 50 mL 1× PBS. Can be kept at 4°C for 6 months.

Critical step

Ultra-low gelling agarose is recommended for embedding root tissue, while regular agarose works best on leaves. On the day of use, melt the agarose by heating in a water bath in a microwave oven under medium power and keep in a water bath/incubator at a temperature higher than the gelling temperature. We incubate at 50°C for 5% regular agarose and 37°C for 5% ultra-low gelling agarose.

RNase-Free DNase Set (lyophilised) Add 1.5 mL of RNase-free water (from kit) to lyophilized DNase I to give final concentration of 1 U μL^{-1}.

Critical step

Buffer of this kit is optimized for on-column DNA digestion; do not use it. It is recommend to use 10× Turbo DNA-free Buffer from Turbo DNA-free kit (Ambion) (see REAGENTS).

0.5 mM EDTA, pH 8.0 Add 18.6 g EDTA to ~80 mL of water. Adjust pH to 8.0 with NaOH. The disodium salt of EDTA will not go into solution until the pH is approximately 8.0. Add water to a total volume of 100 mL and sterilize by autoclaving.

RT reaction solution Use Thermoscript RT kit (see REAGENTS) and make up the first master mix according to the manufacturer's instructions and the number of reactions. This initial master mix will contain reverse primer, dNTPs, water and tissue sections. An aliquot of the second master mix (Table 3) is added to the tube or slide containing the first master mix after denaturation of the first master mix.

PCR reaction solution Use HiFi Phusion polymerase kit (see REAGENTS) and make up a master mix on ice according to the number of reactions (Table 4).

10× Block stock solution Mix 10 mg BSA in 1 mL 10x PBS. May be stored at −20°C for 1 year.

1× Block solution Just before use, dilute 10× Block stock solution 1:10 with sterile water to give 1× Block solution with a final concentration of 1% BSA in 1× PBS.

Critical step

Ensure you make enough 1× Block solution for blocking step and dilution of the Anti-DIG antibody. Need approximately 150 μL 1× Block solution per sample.

Anti-Digoxigenin-AP Fab fragment This is the Anti-DIG antibody conjugated with alkaline phosphatase (AP). Dilute Anti-DIG-AP Fab fragments 1:500 using 1× Block solution. For 10 samples prepare 550 μL: dilute 1.1 μL Anti-DIG-AP in 548.9 μL 1× Block.

Table 3 RT reaction solution

Component	Stock conc.	Final conc./ amount	1× (μL, in-tube)	1× (μL, on-slide)
cDNA synthesis buffer (kit)	5×	1×	4	20
DTT (kit)	0.1 M	5 mM	1	5
RNaseOUT (kit)	40 U μL^{-1}	40 U	1	1
Thermoscript RT (kit)	15 U μL^{-1}	15 U	1	1

Table 4 PCR reaction solution

Component	Stock conc.	Final conc./ amount	1× (μL, in-tube)	1× (μL, on-slide)
HiFi Phusion buffer	5×	1×	10	20
dNTPs	10 mM	0.2 mM	1	2
DIG-11-dUTP	1 mM	4 μM	0.2	0.4
HiFi Phusion polymerase	2 U μL^{-1}	1 U	0.5	0.5
Sterile water			33.3	67.1

Wash Buffer 2 0.1 M Tris-Cl, 0.15 M NaCl, pH 9.5. Can be kept at 4°C for 1 year.

40% glycerol Dilute 4 mL of 100% glycerol with water to give a final volume of 10 mL.

Equipment setup

Vibratome setup The Leica VT1200 vibrating microtome comes with a magnetic specimen plate, buffer tray and ice bath. Assemble the vibratome according to the manufacturer's instructions. Add sterile ice-cold water to the buffer tray. The water provides a flotation medium for the sections. To maintain the cold temperature, fill the integrated ice bath with crushed ice.

Frame-seal slide chambers If performing the *in-situ* PCR directly on slides the frame-seal should be adhered to the slide following the manufacturer's instructions. Label all slides with species, gene name, whether negative or positive control, and the date.

Thermal cyclers If performing in-tube PCR, any standard thermal cycler is suitable. If performing reactions on slides, a thermal cycler with a block that can hold slides is necessary (see EQUIPMENT).

Procedure

Critical step

General guidelines for working with RNA apply to avoid contamination with RNases. All solutions need to be RNase-free. Clean all surfaces and utensils with Ethanol and RNase Zap (see *Reagents*) or equivalent. Use filter tips or new sterile tips for RNA work only.

Sample preparation and fixation

Timing ~3.5 h (leaves), ~20 h (roots), ~1.5 h (non-embedded, non-sectioned samples)

1 Prepare fixative fresh and before harvesting plant material (see REAGENT SETUP). Place on ice.

Caution

Hazardous chemicals should be handled in a fume hood, and using appropriate personal protective equipment.

For non-embedded, non-sectioned samples: If the material to be tested is less than 50 μm thick (i.e. only 1–2 cell layers as for epidermal peels or aleurone layers), the agarose embedding and vibratome sectioning steps (Figure 1) are omitted from the protocol. All processes from fixation through to *in situ* PCR and staining are performed directly on a microscope slide.

2 Aliquot 1.8 mL fresh fixative into 2 mL microfuge tubes (1 per species and tissue) and place on ice.

For non-embedded, non-sectioned samples: Aliquot 100 μL fresh fixative onto the rectangular frame seal slide chamber on a glass slide and keep on a cold flat surface.

3 Harvest plant tissue using micro-scissors and place onto a sterile surface (i.e. petri dish). Using a sterile scalpel, cut leaves into 3 × 5 mm pieces and roots 5 mm long making sure each cut is perpendicular to the long axis of the tissue. For 10 *in situ* PCRs it is recommended to prepare 12 leaf pieces (4 agarose blocks with 3 samples each) and 32 root pieces (4 agarose blocks with 8 samples each).

For non-embedded, non-sectioned samples: The samples are placed directly onto the glass slides containing the fixative and incubated three times for 15 min on a cold flat surface, changing the fixative every time. Proceed to step 7 (washes).

Critical step

Samples should be limited in size to a maximum of 5 mm in width or length, as the fixative can only penetrate up to 5 mm through the tissue. Include the middle vein in leaf pieces as this assists in maintaining the integrity of the leaf sections during the procedure.

4 Using a soft paintbrush, transfer leaf or root pieces immediately into 1.8 mL ice-cold fixative in 2 mL tubes. Carefully invert tubes three times and ensure samples are fully submerged and move unrestricted. This is done to ensure fixative is saturating and can easily access all surfaces of the samples. If the samples do not float freely, transfer some of the samples to a second tube with cold, fresh fixative.

Critical step

The time between harvesting the plant tissue and submerging it in fixative should be as short as possible.

5 Vacuum infiltrate (400 mm Hg) twice for 1.5 min (2 mL tubes open lid). Between infiltrations, cap tubes and gently mix samples.

6 Incubate leaf samples for 3–4 h in fixative on ice. Incubate roots for at least 12 h in fixative in the dark at 4°C. For leaf samples only: During the 3–4 h incubation time change the fixative every hour taking care not to damage the tissue. Keep tubes on ice in fume hood.

Washes

Timing 1 h

7 Wash three times for 10 min in 1.8 mL of ice-cold Wash Buffer 1 (see REAGENT SETUP). After last wash, remove as much of the wash buffer as possible.

For non-embedded, non-sectioned samples: Use 100 μL of Wash Buffer 1.

8 Wash three times for 3 min in 1.8 mL of ice-cold 1x PBS (see REAGENT SETUP). After last wash, remove 600 μL of PBS and discard.

For non-embedded, non-sectioned samples: Wash three times in 100 μL of 1×PBS, add 85 μL sterile water and 2.5 μL of 40 U μL^{-1} RNaseOUT (=100 U) (see REAGENT SETUP) to each slide to give a final volume of 87.5 μL then proceed directly to step 29 (DNase treatment).

Embedding

Timing ~1 h

9 Place three sterile 3MM Whatman filter papers into the lid of a round petri dish (1 per sample).

Steps 10 and 11, process one sample at a time:

10 Invert contents of a 2 mL tube over the filter paper in the petri dish lid. Use a soft paintbrush to gently transfer any remaining samples onto filter paper.

11 In a sterile 12-well plate, half-fill a well with the molten 5% agarose (leaves) or ultra-low gelling agarose (roots) (see REAGENT SETUP). Immediately transfer tissue pieces into the molten agarose using a soft paintbrush. Ensure tissue pieces are fully submerged in the agarose. After orienting the tissue pieces, place 12-well plate on ice for agarose to solidify.

Critical step

Leaves: Place three leaf pieces in each well. Using two paintbrushes, orient the leaf pieces with flat faces parallel to each other and each epidermis perpendicular to the bottom of the plate (so the leaf pieces are stacked like the leaves of a book). Repeat at least four times in four different wells.

Roots: On filter paper, gently roll 8 root pieces parallel to each other to form a tight bundle. Using two paintbrushes or gloved hands transfer one bundle (8 root pieces) into an agarose-filled well, orienting the long axis (epidermis) perpendicular to the bottom of the plate. Repeat at least four times in four different wells.

12 Seal plates using parafilm to prevent agarose drying out and store samples at 4°C for at least 3 h prior to sectioning to ensure the agarose is solidified thoroughly for sectioning on a vibratome. RNA should remain stable for at least three weeks.

PAUSE POINT.

Sectioning

Timing ~3.5 h

For thin fragile sections: If sections thinner than 50 μm are desired from agarose embedded samples, we recommend proceeding on silanised glass slides [15] from this point to minimise tissue disturbance. Tissue sections are held in place between glass slide and plastic seal of frame-sealed chambers preventing the sections from coming in contact with the pipette tip, which would otherwise cause tissue damage.

13 Prepare 0.2 mL PCR tubes and keep on ice during vibratome sectioning (1 tube per transcript to be tested, including controls):

For both non-embedded, non-sectioned samples AND thin fragile sections: Prepare frame-seal chambers on silane glass slides [15] and pipette water and RNaseOUT onto the slides according to 'leaf tissue' volumes below.

For leaf tissue, add 85 μL sterile water and 2.5 μL of 40 U μL^{-1} RNaseOUT (=100 U) (see REAGENT SETUP) to each tube to give a final volume of 87.5 μL. Leaf sections will be placed into this solution.

For root tissue, add 5 μL sterile water and 2.5 μL of 40 U μL^{-1} RNaseOUT (=100 U) (see REAGENT SETUP) to each tube. Four times 20 μL sterile water containing root sections will be added to this solution to give a final volume of 87.5 μL.

14 Using a sterile scalpel, cut a square agarose block from a well in the 12-well plate, taking care not to cut the embedded tissue. Remove agarose edges. Use tweezers to lift the square agarose block onto a sterile petri dish. Alternatively, flip the 12-well plate upside down and then place the block on a petri dish. Using a sterile scalpel cut off excess agarose closer to the tissue to reduce the size of the block, taking care not to cut the embedded tissue.

15 Place superglue onto the magnetic vibratome specimen plate slightly off centre (so agarose block can be orientated later). The amount should be enough to cover the base of the block. Immediately place agarose block onto superglue using tweezers or gloved hands.

16 Place the specimen plate into its buffer tray, cover the tray with its lid and place on ice for at least 5 min for the superglue to dry.

17 Fill the buffer tray with ice-cold sterile water until the agarose block is just covered. The cooled water provides a flotation medium for the sections.

18 Place the buffer tray into the ice bath and fill the ice bath with crushed ice to maintain the cold temperature of the water.

19 Assemble ice bath/buffer tray/specimen plate onto vibratome.

20 Place a double-edge blade into the holder, tighten and turn into position for sectioning. The blade should sit just above the agarose block and be in contact with the water.

21 If required, orient the agarose block by turning the specimen plate. Section leaves lengthwise along the leaf. This helps to keep the leaf sections intact and prevents leaf pieces to be pulled out of the agarose block.

22 Before setting section thickness for cutting (raising the stage), move the blade back until ~0.5 cm behind the agarose block. Set thickness of sections for preliminary trimming at ~200 μm to cut away any excess agarose until the tissue is reached. Set velocity (0.4 mm sec^{-1}) and amplitude (0.4 mm).

23 Press 'Run' to cut section and again to stop. Remember to reverse blade before raising stage for next cut.

24 Once the blade is in contact with the tissue, continue sectioning at 60–70 μm (leaves) or 50 μm (roots). Repeat 6–7 times. These first few sections will be damaged from processing and can be discarded. Collect the sections from the next 3 cuttings onto a microscope slide.

Critical step

Generally, the sections will split from the agarose and float into the cold water. Use a soft paintbrush to collect leaf sections onto the glass slide. To collect root sections use a 200 μL micropipettor with a cut-off tip and pipette water with root sections onto the glass slide.

25 Check the integrity of the collected sections under a bright field microscope (using 4×, 10× and 20× objectives). If the sections look 'blurred', the tissue is not cut at a perpendicular angle and the adjustable stage needs to be tilted/rotated in order to correct the orientation of the block, ensuring the long axis of the tissue is perpendicular to the blade.

26 If the sections are intact and appear 'in focus', start collecting fresh sections into 0.2 mL PCR tubes containing sterile water with 100 U RNaseOUT (prepared at step 13). Keep tubes on ice.

For thin fragile sections: Sections are transferred to the silane glass slide with water and RNaseOUT (prepared at step 13). Keep the glass slide on a cold, dry surface.

Critical step

Leaves: Use a soft paintbrush to collect at least 15 leaf sections into each PCR tube containing sterile water and 100 U RNaseOUT (87.5 μL). This takes about 30–45 min per tube. Take care to collect sections to the tip of the paintbrush, as it can be difficult to remove sections from the paintbrush after transfer. If sections are not separating fully from the agarose, gently tease away the agarose using a paintbrush or tweezers.

Roots: Collect root sections using a 200 μL pipette with a cut-off tip, pipette up water with root sections from buffer tray and transfer into an empty tube. Once roots settle to the bottom of the tube, remove excess water until 20 μL remains in the tube. Alternatively, remove all water and add 20 μL water to the tube. Transfer 20 μL of water with root sections into PCR tubes containing sterile water and 100 U RNaseOUT. Repeat this process four times in total to collect at least 35 root sections in a final volume of 87.5 μL.

If multiple transcripts are to be tested, the collection of sections should be done concurrently rather than consecutively to ensure sections across different *in situ* PCRs are similarly sized and are from the same region of the tissue (i.e. collect sections from the first cut into PCR tube 1, sections from the second cut into PCR tube 2, etc.). This is important as leaf and root morphology and possibly expression of genes may vary along the tissue.

Take care to frequently change the blade or the section of the blade that is cutting the tissue, because the cut from a blunt blade tends to be jagged and uneven.

If root sections do not separate from ultra-low gelling temperature agarose after cutting, transfer them into a tube with sterile water using a paintbrush. Heat at 65°C for 10 min to melt the agarose. Wash three times with sterile 65°C warm water before proceeding.

27 Once enough sections have been collected, disassemble the ice bath/buffer tray/specimen plate from the vibratome. Cut around the bottom of the agarose block using a single-edge blade, before slicing underneath the block to lift it off the specimen plate. If there is a substantial amount of plant material left in the block, use tweezers or paintbrushes to transfer the agarose block back into the 12-well plate. Viable tissue samples can be kept at 4°C for at least three weeks.

28 Clean the superglue from the specimen plate by carefully scraping a single-edge blade over it, taking care not to damage the specimen plate.

DNase treatment

Timing ~1 h

Critical step

Take great care when pipetting solutions into and out of 0.2 mL PCR tubes or from the glass slide containing the sections. They are very easily damaged and may stick to the pipette tip. Gently pipette up and down to mix the sample and if this not possible, flick tubes using finger. Do not use a centrifuge. Sections should never be allowed to dry; prepare mastermixes for next step during the current incubation such that as soon as solution 1 is removed, solution 2 can be added.

29 On ice, to the 0.2 mL PCR tubes containing 87.5 μL sterile water, RNaseOUT and tissue sections, add 10 μL of 10× Turbo DNA-free buffer (Ambion) and 2.5 μL of 1 U μL^{-1} DNase I (Qiagen) (see REAGENT SETUP) to give a final volume of 100 μL.

For both non-embedded, non-sectioned samples AND thin fragile sections: Add the above DNase reaction mixture directly onto the tissue samples on the glass slide containing water and RNaseOUT.

30 Incubate at 37°C for 45 min in a thermocycler with either a 0.2 ml PCR tube block or a slide block.

31 Add 3.3 μL of 0.5 M EDTA, pH 8.0 (see REAGENT SETUP) to give a final concentration of 15 mM. Heat inactivate the DNase I enzyme at 70°C for 15 min. Place on ice.

Critical step

For non-embedded, non-sectioned samples AND thin fragile sections: In our experience the glass slide is prone to cracking due to expansion and contraction of the metal PCR block. To avoid this we recommend removing the slides prior to the final cooling step. For instance, if the thermal cycler is set to cool down to 4°C after the 70°C EDTA incubation in the DNase step, remove the slide in the final 5 seconds of the 70°C incubation and place on a cold (4°C) flat surface. This also applies to subsequent incubations (see steps 37 and 39).

32 Using a 200 μL pipette, carefully remove the DNase solution and discard.

33 Wash twice for 1 min in 150–200 μL ice-cold sterile water.

34 Using a 200 μL pipette and 10 μL pipette gently remove as much of the water as possible.

Reverse transcription

Timing ~1.5 h

Critical step

For the negative controls, add all reagents except Thermoscript RT. Alternatively exclude negative controls from the RT and keep samples at 4°C in 1× PBS during steps 35–39.

35 On ice, to each 0.2 mL PCR tube containing the sections, add 10 μL of sterile water and 2 μL of 10 mM dNTPs.

For non-embedded, non-sectioned samples AND thin fragile sections: Add 60.5 μL of sterile water and 10 μL of 10 mM dNTPs.

36 Add 1 μL of 10 μM gene-specific reverse primer.

For non-embedded, non-sectioned samples AND thin fragile sections: Add 2.5 μL of 10 μM gene-specific reverse primer.

Critical step

Reverse primers that have previously been used in qPCR generally work well as primer in RT and in the subsequent *in situ* PCR. Alternatively, design a new reverse primer that is 5′ of the gene-specific reverse primer used in the *in situ* PCR (see "Primer design" and "RT-PCR" for more details).

37 In a thermal cycler, incubate samples for 5 min at 65°C and hold at 4°C for at least 1 min. Place on ice.

For non-embedded, non-sectioned samples AND thin fragile sections: See critical note under step 31.

38 Add 7 μL of RT reaction solution (see REAGENT SETUP, Table 3).

For non-embedded, non-sectioned samples AND thin fragile sections: Add 27 μL of RT reaction solution.

39 In a thermal cycler, incubate for 1 h at 50°C, 5 min at 85°C, and then hold at 4°C.

For non-embedded, non-sectioned samples AND thin fragile sections: See critical note under step 31.

PAUSE POINT.

In situ PCR

Timing 1–2.5 h

40 Place tubes/glass slides into a rack on ice. Using a 200 μL pipette carefully remove the RT solution.

41 Wash twice for 1 min in 150-200 μL ice-cold sterile water.

42 Using a 200 μL and 10 μL pipette remove as much of the water as possible.

43 On ice, to each 0.2 mL PCR tube containing the sections add 45 μL of PCR reaction solution (see REAGENT SETUP).

For non-embedded, non-sectioned samples AND thin fragile sections: Add 90 μL of PCR reaction solution.

44 Add 2.5 μL of each 10 μM gene-specific forward and reverse primers to give final reaction volume of 50 μL.

For non-embedded, non-sectioned samples AND thin fragile sections: Add 5 μL of each 10 μM reverse and forward primers to give a final reaction volume of 100 μL.

45 Cycling conditions will vary according to primer Tm, product size, and the abundance of the transcript of interest. A sample protocol is given below for primers designed with an annealing temperature of 55°C and a product size of approximately 200 bp.

Initial denaturation:

98°C 30"

Cycling. Repeat 20-35× (depending on results from the Test PCR):

98°C 10"

55°C 25"

72°C 5"

Final elongation:

72°C 7'

10°C hold

For non-embedded, non-sectioned samples AND thin fragile sections: See critical note under step 31.

Colourimetric detection of DIG labeled PCR products

Timing ~3-4 h

For non-embedded, non-sectioned samples AND thin fragile sections: For on-slide *in situ* PCR, remove the frame-seal chamber from the slide completely using a scalpel and complete all remaining steps on the slide. During incubations, place the slides within a small and sealable plastic box to minimize evaporation.

46 Place 0.2 mL PCR tubes into a rack on ice. Using a 10 μL pipette, carefully pipette off the rest of the PCR solution.

47 Wash tissue sections twice for 5 min in 150–200 μL 1× PBS (see REAGENT SETUP).

48 Using a 200 μL and 10 μL pipette remove as much of the 1× PBS as possible.

49 Gently add 100 μL of 1× Block solution (see REAGENT SETUP) to the sections and incubate for 30 min on ice.

50 During incubation, dilute Anti-DIG-AP antibody 1:500 in 1× Block solution (see REAGENT SETUP). Keep on ice.

51 Using a 200 μL pipette and 10 μL pipette, carefully pipette off as much of the 1× Block solution as possible.

52 Add 50 μL of diluted Anti-DIG-AP antibody to the tissue sections and incubate at room temperature for 1 h.

53 Using a 200 μL pipette, remove the antibody solution.

54 Wash tissue sections twice for 15 min at room temperature in Wash Buffer 2. Keep sections in Wash Buffer 2 after second wash.

55 For detection move to a Bright Field microscope equipped with a camera.

56 Label silanised microscope slides. Label should include species, gene name, whether negative or positive control, and the date.

57 Using a glass pipette big enough to pipette leaf sections or a 200 μL pipette with cut-off tip for root sections, transfer the sections along with Wash Buffer 2 onto a silanised microscope slide. A paintbrush should be used if any sections stick to the wall of the tube.

58 Using a 200 μL pipette, remove any excess Wash Buffer from the slide.

Critical step

Hold pipette tip perpendicularly to the slide surface and remove liquid very slowly.

59 Add 50 μL BM purple or any substrate for the AP enzyme to each sample on the glass slides. Some substrates, such as BM purple are light sensitive. Only take an aliquot from the stock and keep in the dark and on ice. Once added to the slides, cover the slides to protect from light and let develop in the dark until a purple-blue signal appears.

60 Initially incubate for 10 min at room temperature. Check staining under the microscope (4× and 10× objectives, without cover slide) every 15–30 min. Keep slides in the dark.

Critical step

The time of staining may vary anywhere from 10 min to over 2 h depending on species, the abundance of the transcript, and the PCR efficiency. If no signal is detected after 1 h, pipette off the AP substrate and add fresh 50 μL BM purple. Do not leave sections to develop overnight, as they will become saturated.

Mounting and microscopy

Timing ~30 min

61 Once the stain has developed and the DIG-labeled PCR products are detected, use a 200 μL pipette remove the AP substrate from the slide.

62 Add 100 μL Wash Buffer 2 and wash three times 5 min to remove residual stain and any debris.

63 Wash once with 100 μL sterile water. Using a 200 μL pipette, remove all the water.

64 Using a 200 μL pipette, mount sections in 40–150 μL of 40% glycerol. The amount of glycerol will depend on the number of sections on your slide. Take care all the sections are covered but do not use an excessive amount of glycerol.

65 Place coverslip on top and seal corners with nail varnish. If the amount of glycerol is too much, such that the coverslip floats, pipette off some glycerol then seal.

Critical step

Do not place the coverslip by sliding down from one edge (as usually practiced), as all the sections will float to one side. Simply drop the coverslip from above.

66 Sections may be visualized under a bright field or fluorescence microscope for the next 3 weeks. Store slides at room temperature.

Timing overview

Steps 1–6, fixation: 1 h procedure plus 2.5 h incubation (leaves) or 19 h incubation (roots).

Steps 7 + 8, washes: 1 h.

Steps 9–12, embedding: 1 h procedure plus minimum 3 h incubation or overnight.

Steps 13–28, sectioning: 3.5 h.

Steps 29–34, DNase treatment: ~1 h.

Steps 35–39, reverse transcription: ~30 min procedure plus 1 h incubation.

Steps 40–45, *in situ* PCR: up to 2.5 h.

Steps 46–60, detection of DIG labeled PCR products: up to 4 h.

Steps 61–66, mounting: 30 min.

Troubleshooting

Advice on troubleshooting is listed in Table 5.

Results and discussion

We use *in situ* PCR as a core technique for gene functional characterisation studies. All the evidence we have gathered so far suggests that it reliably localises the target transcript to the cell-types in which it is expressed. We see distinct patterns of staining when examining the localisation of different transcripts, and these are consistent across technical and biological replicates, and when available, with the results of published studies for the same gene using different transcript localization techniques. For instance, in Figure 2A, B and 3C using the primers listed in Table 6, we detect the presence of 18S ribosomal RNA in all cells, as would be expected. We have previously seen the preferential expression of *AtCAX1* in the mesophyll using single cell sampling and laser capture microdissection [6,7,30] and we show the same using *in situ* PCR (Figure 2C). Figure 3 shows the localisation of a transcript that encodes a sodium transporter in near isogenic lines (NIL) of *Triticum durum* (durum wheat) cultivar Tamaroi [21]. The near isogenic lines of durum wheat were made after introgression of a genomic fragment from *Triticum monococcum* in *Triticum durum* by conventional crossing and backcrossing with the durum wheat parent [21]. The + *Nax2* NIL contains *Triticum monococcum TmHKT1;5-A* (Figure 3B) whereas the – Nax2 NIL lacks *TmHKT1;5-A* (Figure 3B). Homologues of this gene are expressed in the stele of bread wheat [24], rice [31], Arabidopsis [32] and other plants [33], and we see a very similar expression pattern in durum wheat (Figure 3B). Here, we also present examples of using *in situ* PCR on delicate tissues for two genes encoding putative organic acid transporters. One is present in the guard cells in epidermal peels from barley leaves (Figure 2D), a common localization for this gene family [34-36] and the other localizes to the septum and embryo of Arabidopsis flower/siliques (Figure 2E and 2F).

Unfortunately, as with every PCR based technique, the experiment may sometimes fail for an unforeseen (and seemingly inexplicable reason), but, in our hands, the vast majority of experiments have yielded usable results, and if care is taken then it will consistently work. However, in the course of optimizing this technique we have encountered several examples of what may go wrong – this includes a total absence of detectible transcript, over-staining, and even erroneous detection of transcript (see also Table 5 for troubleshooting tips). Incomplete fixation of RNA and clearing of tissue can result in detection of the target transcripts in fewer cells than in which it is actually expressed. This is exemplified by the lack of detection of 18S RNA in all cell types either through poor penetration of fixative (Figure 4A) or poor clearing of chlorophyll (Figure 4B). This is the reason why it is essential to detect a 'house-keeping' transcript, which is equally expressed in all cell-types, at the same time as your target transcript to make sure you have complete and equal fixation and clearing. Performing sectioning on very delicate tissues, including older root tissues, is difficult as they can often fall apart during processing (Figure 4C) so these types of samples are best processed on slides. Figure 4D shows an example of what can happen if the primers detect genomic DNA; the nucleus stains and shows a pattern quite distinctive from cytoplasmic cDNA staining. This scenario can be avoided by optimal primer design.

We attempted to use fluorescence detection of gene expression in tissue sections as we hypothesized we would achieve an improvement over what is possible with colourimetric detection in terms of signal intensity and contrast with the background image. However, due to the inherent autofluorescence in our no RT control sections (of chlorophyll in leaf tissue, and vascular and cuticular regions in leaf and root tissue control samples (Figure 4E–H)) the use of fluorescence was less satisfactory as a detection method for gene expression in our

Table 5 Troubleshooting suggestions

Step	Problem	Possible reason	Solution
3	Inconsistent staining in the positive control sample (as in Figure 4A and B)	i) RNA is degraded	i) Reduce the time between harvesting plant tissue and placing it in fixative. Follow general rules for working with RNA to prevent contamination with RNases (see also start of protocol).
		ii) Over-fixation	ii) Reduce overall fixation time, but increase pressure/time of vacuum infiltration to ensure fixative penetrates.
		iii) Under-fixation (tissue may be too large for fixative to penetrate).	iii) Pieces must be kept at a maximum of 5x5 mm to ensure fixative can penetrate entire tissue. Increase pressure/time of vacuum infiltration to ensure fixative penetrates.
11, 20	Poor morphology (as in Figure 4C)	i) Long axis of sample is not perpendicular to blade on vibratome	i) Be quick when orienting the tissue pieces in the molten agarose and place the sample on ice for the agarose to solidify. Examine the first few sections cut on the vibratome and if they appear smeared, adjust the orientation of the block by either tilting the adjustable stage or removing the agarose block and making sure the side stuck on the stage is perfectly flat.
		ii) Sections have been damaged during processing	ii) Be careful not to damage the tissue sections during processing (use a paintbrush for manipulation, do not vortex or centrifuge tubes, prevent pipette tip from contacting sections during multiple rounds of pipetting, consider performing experiments on-slide).
29	Nuclear staining throughout the sample (as in Figure 4D)	Amplification of gDNA	Design primers that are split across an exon-exon junction to eliminate the possibility of amplifying gDNA. Alternatively, design primers that have 1 or more large introns between them such that reducing the elongation time during PCR only allows amplification of the smaller cDNA product.
			Increase the incubation time of the DNase treatment or change the DNase enzyme.
45, 60	Staining of sections appears very dark	Saturated staining	Reduce the number of cycles in the PCR and/or reduce the overall staining time.
60	Positive result in the negative "no RT" control or weak non-specific staining in the positive control	Background staining caused by presence of endogenous alkaline phosphatase enzymes	Add levamisole to the substrate to inhibit endogenous alkaline phosphatases.

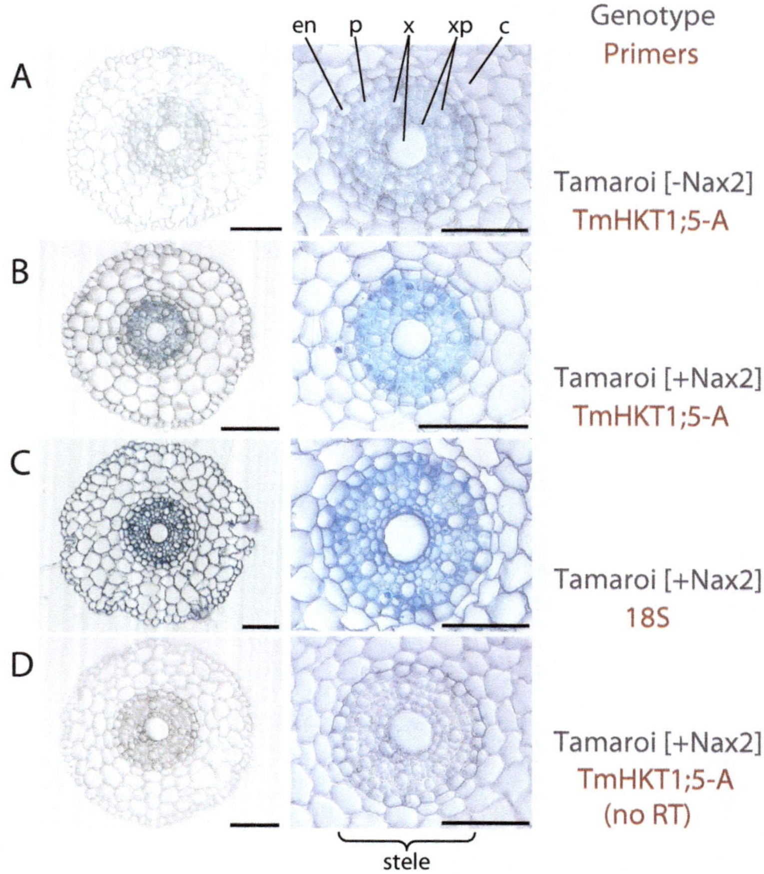

Figure 3 ***In situ*** **PCR demonstrating vasculature-specific expression of a sodium transporter (*****TmHKT1;5-A)*** **in durum wheat (*****Triticum durum*****) root sections.** The blue stain indicates the presence of transcripts while the images in the right panel are magnifications of figures in the left panel. **(A)** shows that *TmHKT1;5-A* is undetectable in Tamaroi [–*Nax2*] (the near-isogenic line without *Nax2/TmHKT1;5-A*). **(B)** depicts the stele-specific expression of *TmHKT1;5-A* in Tamaroi [+*Nax2*]. In **(C)** the expression of 18S ribosomal RNA is seen in all cell-types (positive control). **(D)** is the negative control where the reverse transcription (RT) step was omitted in Tamaroi [+*Nax2*]. c, cortex; en, endodermis; p, pericycle; x, xylem; xp, xylem parenchyma. Scale bars represent 100 μm. This figure was originally published in [21].

Table 6 List of primers used during *in situ* PCR

Gene name	Accession number	Primer sequence (5′ to 3′)
Hv18S	AK251731.1	**qF:** GGTAATTCCAGCTCCAAT
		qR: GTTTATGGTTGAGACTAG
TmHKT1; 5-A	DQ646339.2	**qF:** GACCACAAAAGGATAACAAGCA
		qR: AGAACATGACAGCAATGAGAGC
GmPIP1; 2b	XM_003532769.2	**qF:** TGTTTTTGTATGTGCTTGCTTG
		qR: TCCATTCAGAGTGTCACAAATACA
AtCAX1	NM_201901.3	**qF:** AGTTGCGTTAGGCTCTGC
		qR: TTGATGTCCCAAGTGAATG
HvSL1251	AK371960.1	**qF:** GGTCACAACCACGGCTATTT
		qR:GTCTTGAATGAGGGCAGAGC
AtALMT14	NM_001125913.1	**qF:** CGGTAGACATAACCCCAACG
		qR: TCAACACCACAATCCTGCTC

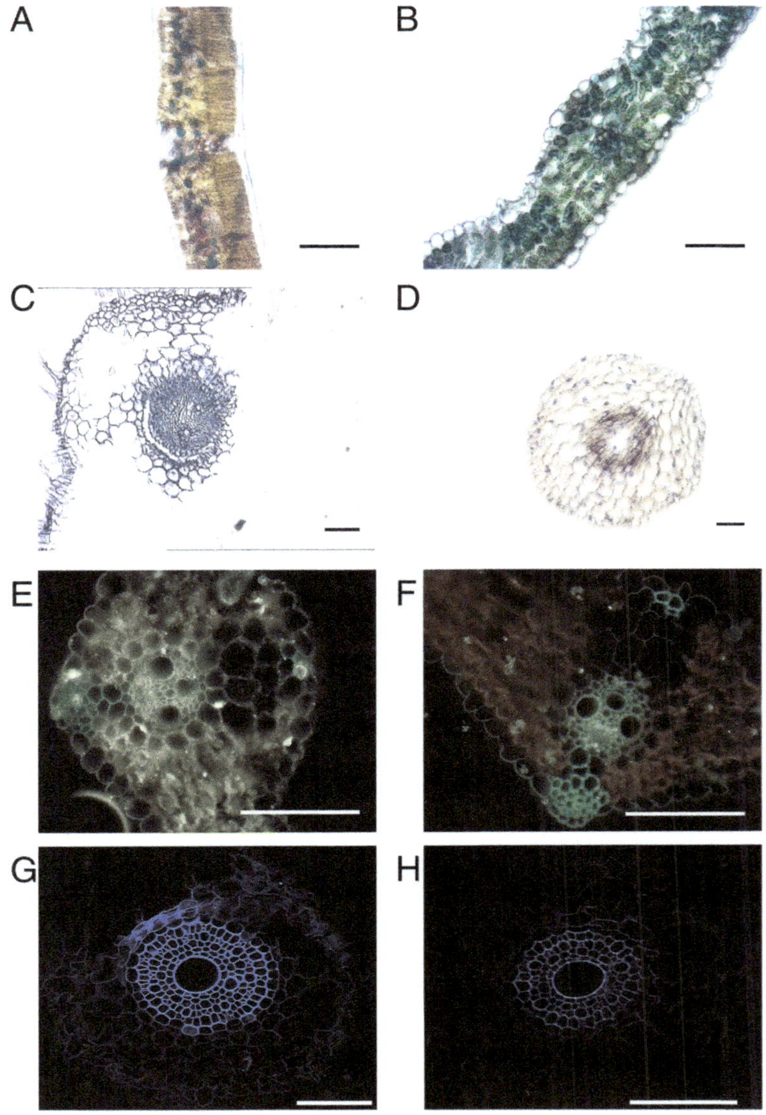

Figure 4 Examples of tissue sections where *in situ* PCR was unsuccessful. Inconsistent, non-uniform expression of 18S rRNA in *Vitis vinifera* leaf **(A)** and *Hordeum vulgare* leaf **(B)**. Severely damaged *Glycine max* root cells **(C)**. Distinct staining of cell nuclei (genomic DNA) in *Triticum durum* root section **(D)**. Fluorescent staining using Elf97 in *Hordeum vulgare* leaf (**E**, *Hv18S* rRNA and **F**, *Hv18S* rRNA negative control; Zeiss stereofluorescence microscope) and *Triticum durum* root sections (**G**, *TmHKT1; 5-A* and **H**, *TmHKT1;5-A* negative control; Nikon confocal microscope). Please see Table 5 for troubleshooting tips. Scale bars represent 100 μm.

hands. For instance, when we used fluorescence to investigate the expression of *TmHKT1;5-A* in the + *Nax2* NIL we detected a similar signal in our no RT negative control and our sample (Figure 4G and H); this contrasts with no signal in our no RT control using colourimetric detection (Figure 3D) or in the NIL –*Nax2* (Figure 3A). Despite the lower signal of the autofluorescence we could not eliminate it using lower gain or spectral unmixing when using confocal microscopy. The use of fluorescence also has additional complications compared to the more reliable and straightforward colourimetric detection (such as cost of fluorescent substrates or primers, stability of fluorescent compounds and the cost and access to fluorescence/confocal microscopes). Pesquet et al. [18] used fluorescence to detect *in situ* PCR products in plant tissues, however the protocol was not outlined and the technique has not been used very often; problems with autofluorescence interference when using fluorescent detection of *in situ* PCR products have since been noted [37].

Conclusion

We advocate the use of colourimetric detection of *in situ* PCR products for gene expression localization in plant tissue. Once the primers have been validated to be specific to the required target, this technique is a relatively quick and powerful tool to spatially define the expression profile of any gene of interest. We recommend the adoption of this technique as a standard method for the localization of genes in plant tissue.

Competing interests

The authors declare that they have no competing interest.

Authors' contributions

MG, SKT and AA wrote the manuscript. RAB and SJC initiated the development of the *in situ* technique on plant tissue, GMM provided advice and training on microtomy, AA conducted most of the development (with CJ) and performed the wheat and soybean *in situ* PCR, CJ performed Arabidopsis CAX experiments, VMC performed and developed the technique for on-slide PCR, and performed the experiments on barley, epidermal peels and flowers. WWN performed additional experiments on barley. All authors commented on the manuscript.

Acknowledgments

This work was supported by Australian Research Council [DP130104205, FT130100709 and CE140100008 to MG, and DE120100307 to SKT] and a University of Adelaide, Faculty of Sciences equipment grant to SJC, RAB and MG.

Author details

[1]ARC Centre of Excellence in Plant Energy Biology, University of Adelaide, Glen Osmond, SA, Australia. [2]ARC Centre of Excellence in Plant Energy Biology, The University of Western Australia, Perth, WA, Australia. [3]Waite Research Institute & School of Agriculture, Food and Wine, University of Adelaide, PMB1, Glen Osmond, SA 5064, Australia. [4]Adelaide Microscopy Waite Facility, University of Adelaide, Glen Osmond, SA, Australia. [5]ARC Centre of Excellence in Plant Cell Walls, University of Adelaide, Glen Osmond, SA, Australia. [6]Centre for Cancer Biology, Division Immunology, Level 3, Frome Road, Adelaide, SA, Australia.

References

1. Rhee SY, Mutwil M: **Towards revealing the functions of all genes in plants.** *Trends Plant Sci* 2014, **19**:212–221.
2. Li S, Pandey S, Gookin TE, Zhao Z, Wilson L, Assmann SM: **Gene-sharing networks reveal organizing principles of transcriptomes in Arabidopsis and other multicellular organisms.** *Plant Cell* 2012, **24**:1362–1378.
3. Chaumont F, Tyerman SD: **Aquaporins: highly regulated channels controlling plant water relations.** *Plant Physiol* 2014, **164**:1600–1618.
4. Henderson S, Gilliham M: **The 'gatekeeper' concept: cell-type specific molecular mechanisms of plant adaptation to abiotic stress.** In *Mechanisms of Plant Adaptation.* Edited by Laitinen R. New Jersey: Wiley; 2015. in press.
5. Taylor CB: **Promoter fusion analysis: an insufficient measure of gene expression.** *Plant Cell* 1997, **9**:273–275.
6. Conn SJ, Gilliham M, Athman A, Schreiber AW, Baumann U, Moller I, Cheng N-H, Stancombe MA, Hirschi KD, Webb AAR, Burton R, Kaiser BN, Tyerman SD, Leigh RA: **Cell-specific vacuolar calcium storage mediated by CAX1 regulates apoplastic calcium concentration, gas exchange, and plant productivity in Arabidopsis.** *Plant Cell* 2011, **23**:240–257.
7. Roy SJ, Conn SJ, Mayo GM, Athman A, Gilliham M: **Transcriptomics on small samples.** In *Methods in Molecular Biology. Volume 913.* Edited by Shabala S, Cuin TA. Humana Press: New Jersey; 2012:335–350.
8. Nuovo GJ: **In situ localization of PCR-amplified DNA and cDNA.** In *Methods in Molecular Biology. Volume 123.* Edited by Darby IA. New Jersey: Humana Press; 2000:217–238.
9. Birnbaum K, Jung JW, Wang JY, Lambert GM, Hirst JA, Galbraith DW, Benfey PN: **Cell type-specific expression profiting in plants via cell sorting of protoplasts from fluorescent reporter lines.** *Nat Methods* 2005, **2**:615–619.
10. Brady SM, Orlando DA, Lee J-Y, Wang JY, Koch J, Dinneny JR, Mace D, Ohler U, Benfey PN: **A high-resolution root spatiotemporal map reveals dominant expression patterns.** *Science* 2007, **318**:801–806.
11. Deal RB, Henikoff S: **A simple method for gene expression and chromatin profiling of individual cell types within a tissue.** *Dev Cell* 2010, **18**:1030–1040.
12. Zhang C, Barthelson RA, Lambert GM, Galbraith DW: **Global characterization of cell-specific gene expression through fluorescence-activated sorting of nuclei.** *Plant Physiol* 2008, **147**:30–40.
13. Grindberg RV, Yee-Greenbaum JL, McConnell MJ, Novotny M, O'Shaughnessy AL, Lambert GM, Arauzo-Bravo MJ, Lee J, Fishman M, Robbins GE, Lin X, Venepally P, Badger JH, Galbraith DW, Gage FH, Lasken RS: **RNA-sequencing from single nuclei.** *Proc Natl Acad Sci U S A* 2013, **110**:19802–19807.
14. Ginart P, Raj A: **RNA sequencing *in situ*.** *Nat Biotechnol* 2014, **32**:543–544.
15. Bagasra O: **Protocols for the in situ PCR-amplification and detection of mRNA and DNA sequences.** *Nat Protoc* 2007, **2**:2782–2795.
16. Johansen B: **In situ PCR on plant material with sub-cellular resolution.** *Ann Bot* 1997, **80**:697–700.
17. Koltai H, Bird DM: **High throughput cellular localization of specific plant mRNAs by liquid-phase in situ reverse transcription-polymerase chain reaction of tissue sections.** *Plant Physiol* 2000, **123**:1203–1212.
18. Pesquet E, Barbier O, Ranocha P, Jauneau A, Goffner D: **Multiple gene detection by *in situ* RT-PCR in isolated plant cells and tissues.** *Plant J* 2004, **39**:947–959.
19. Lee YH, Tegeder M: **Selective expression of a novel high-affinity transport system for acidic and neutral amino acids in the tapetum cells of Arabidopsis flowers.** *Plant J* 2004, **40**:60–74.
20. Przybecki Z, Siedlecka E, Filipecki M, Urbanczyk-Wochniak E: **In situ reverse transcription PCR on plant tissues.** In *Methods in Molecular Biology. Volume 334.* Edited by Pellestor F. Humana Press: New Jersey; 2006:181–198.
21. Munns R, James RA, Xu B, Athman A, Conn SJ, Jordans C, Byrt CS, Hare RA, Tyerman SD, Tester M, Gilliham M: **Wheat grain yield on saline soils is improved by an ancestral Na^+ transporter gene.** *Nat Biotechnol* 2012, **30**:360–U173.
22. Vandeleur RK, Sullivan W, Athman A, Jordans C, Gilliham M, Kaiser BN, Tyerman SD: **Rapid shoot-to-root signalling regulates root hydraulic conductance via aquaporins.** *Plant Cell Environ* 2014, **37**(2):520–538.
23. Tiong J, McDonald GK, Genc Y, Pedas P, Hayes JE, Toubia J, Langridge P, Huang CY: **HvZIP7 mediates zinc accumulation in barley (*Hordeum vulgare*) at moderately high zinc supply.** *New Phytol* 2014, **201**:131–143.
24. Byrt C, Xu B, Krishnan M, Lightfoot D, Athman A, Jacobs A, Watson-Haigh N, Munns R, Tester M, Gilliham M: **The Na^+ transporter, TaHKT1;5-D, limits shoot Na^+ accumulation in bread wheat.** *Plant J* 2014, doi:10.1111/tpj.12651.
25. Conn SJ, Hocking B, Dayod M, Xu B, Athman A, Henderson S, Aukett L, Conn V, Shearer MK, Fuentes S, Tyerman SD, Gilliham M: **Protocol: optimising hydroponic growth systems for nutritional and physiological analysis of Arabidopsis thaliana and other plants.** *Plant Methods* 2013, **9**:4.
26. Fox CH, Johnson FB, Whiting J, Roller PP: **Formaldehyde fixation.** *J Histochem Cytochem* 1985, **33**:845–853.
27. Thavarajah R, Mudimbaimannar VK, Elizabeth J, Rao UK, Ranganathan K: **Chemical and physical basics of routine formaldehyde fixation.** *J Oral Maxillofac Pathol* 2012, **16**:400–405.
28. Dapson RW: **Macromolecular changes caused by formalin fixation and antigen retrieval.** *Biotech Histochem* 2007, **82**:133–140.
29. Puchtler H, Meloan SN: **On the chemistry of formaldehyde fixation and its effects on immunohistochemical reactions.** *Histochemistry* 1985, **82**:201–204.
30. Gilliham M, Athman A, Tyerman SD, Conn SJ: **Cell-specific compartmentation of mineral nutrients is an essential mechanism for optimal plant productivity - another role for TPC1?** *Plant Signal Behav* 2011, **6**:1656–1661.
31. Ren ZH, Gao JP, Li LG, Cai XL, Huang W, Chao DY, Zhu MZ, Wang ZY, Luan S, Lin HX: **A rice quantitative trait locus for salt tolerance encodes a sodium transporter.** *Nat Genet* 2005, **37**:1141–1146.
32. Sunarpi Horie T, Motoda J, Kubo M, Yang H, Yoda K, Horie R, Chan WY, Leung HY, Hattori K, Konomi M, Osumi M, Yamagami M, Schroeder JI, Uozomi N: **Enhanced salt tolerance mediated by AtHKT1 transporter-induced Na^+ unloading from xylem vessels to xylem parenchyma cells.** *Plant J* 2005, **44**:928–938.
33. Waters S, Gilliham M, Hrmova M: **Plant high-affinity potassium (HKT) transporters involved in salinity tolerance: structural insights to probe differences in ion selectivity.** *Int J Mol Sci* 2013, **14**:7660–7680.

34. Gruber BD, Ryan PR, Richardson AE, Tyerman SD, Ramesh S, Hebb DM, Howitt SM, Delhaize E: **HvALMT1 from barley is involved in the transport of organic anions.** *J Exp Bot* 2010, **61:**1455–1467.
35. Meyer S, Mumm P, Imes D, Endler A, Weder B, Al-Rasheid KAS, Geiger D, Marten I, Martinoia E, Hedrich R: **AtALMT12 represents an R-type anion channel required for stomatal movement in Arabidopsis guard cells.** *Plant J* 2010, **63:**1054–1062.
36. De Angeli A, Zhang J, Meyer S, Martinoia E: **AtALMT9 is a malate-activated vacuolar chloride channel required for stomatal opening in Arabidopsis.** *Nat Commun* 2013, **4:**1084.
37. Boubourakas IN, Voloudakis AE, Fasseas K, Resnick N, Koltai H, Kyriakopoulou PE: **Cellular localization of peach latent mosaic viroid in peach sections by liquid phase *in situ* RT-PCR.** *Plant Pathol* 2011, **60:**468–473.

11

Rhizoslides: paper-based growth system for non-destructive, high throughput phenotyping of root development by means of image analysis

Chantal Le Marié*, Norbert Kirchgessner, Daniela Marschall, Achim Walter and Andreas Hund

Abstract

Background: A quantitative characterization of root system architecture is currently being attempted for various reasons. Non-destructive, rapid analyses of root system architecture are difficult to perform due to the hidden nature of the root. Hence, improved methods to measure root architecture are necessary to support knowledge-based plant breeding and to analyse root growth responses to environmental changes. Here, we report on the development of a novel method to reveal growth and architecture of maize root systems.

Results: The method is based on the cultivation of different root types within several layers of two-dimensional, large (50 × 60 cm) plates (rhizoslides). A central plexiglass screen stabilizes the system and is covered on both sides with germination paper providing water and nutrients for the developing root, followed by a transparent cover foil to prevent the roots from falling dry and to stabilize the system. The embryonic roots grow hidden between a Plexiglas surface and paper, whereas crown roots grow visible between paper and the transparent cover. Long cultivation with good image quality up to 20 days (four fully developed leaves) was enhanced by suppressing fungi with a fungicide. Based on hyperspectral microscopy imaging, the quality of different germination papers was tested and three provided sufficient contrast to distinguish between roots and background (segmentation). Illumination, image acquisition and segmentation were optimised to facilitate efficient root image analysis. Several software packages were evaluated with regard to their precision and the time investment needed to measure root system architecture. The software 'Smart Root' allowed precise evaluation of root development but needed substantial user interference. 'GiaRoots' provided the best segmentation method for batch processing in combination with a good analysis of global root characteristics but overestimated root length due to thinning artefacts. 'WhinRhizo' offered the most rapid and precise evaluation of root lengths in diameter classes, but had weaknesses with respect to image segmentation and analysis of root system architecture.

Conclusion: A new technique has been established for non-destructive root growth studies and quantification of architectural traits beyond seedlings stages. However, automation of the scanning process and appropriate software remains the bottleneck for high throughput analysis.

Keywords: Root system architecture, Maize, High-throughput, Imaging, Rhizoslide, Breeding, Agriculture

Background

The direct selection of efficient root systems is an important aim for a second green revolution enabling to increase yield in low input agriculture [1]. As costs of fertilizers rise and some fertilizers, especially phosphorus, become limited, there is an increasing interest in understanding the genetic control of RSA traits. Still, the ability to phenotype roots with sufficient throughput remains the bottleneck. Throughput is needed to close the phenotype-to-genotype gap, either by classical mapping of quantitative trait loci (QTL) or by association mapping [2]. In practice at least 100 – 500 individuals are needed for a QTL or association study [3]. Here we focus on the root system of cereal roots, especially maize. The root systems of cereals consist of three different below ground root types, the embryonic primary and seminal roots and the shoot borne crown

* Correspondence: chantal.lemarie@usys.ethz.ch
Institute of Agricultural Sciences, ETH Zurich, Universitätstrasse 2, 8092 Zurich, Switzerland

roots [4,5]. The crown roots are separated from the embryonic roots by the mesocotyl, which elongates to place the shoot base close to the soil surface. While the embryonic root system can be studied directly after germination, crown roots in the example of maize develop around the two leaf stage [6].

Various phenotyping platforms were developed to monitor root growth non-invasively in soil and with high throughput in hydroponics, aeroponics, agar and on germination paper [7-10]. However, these platforms are limited to the assessment of the embryonic root system, disregarding, that the crown roots dominate the root system of a mature plant [11]. There is a certain risk that seedling root traits are of little relevance for the development beyond the seedling stage. For maize, there is strong evidence that the embryonic roots, especially the primary root, behave differently compared to the crown roots. All three root types (primary, seminal and crown roots) are under different genetic control [12] and the early development of embryonic roots is affected strongly by seed size, seed quality, timing of germination and other processes as discussed by Hund et al. [5].

There is a lack of high throughput phenotyping methodologies enabling to study crown roots and their response to environmental stimuli. The reason for this is that cultivating plants with accessible crown root systems is far from being trivial: Containers need to be larger, require more space, and the access and measurement of the root systems requires more time. Different approaches were taken to access root systems of larger plants. Soil based systems such as rhizotrons and containers [13,14] or systems observing roots in undisturbed soil in pots (e.g. via computed tomography; [15] or magnetic resonance imaging; [16]). Usually these systems offer high precision but have limited throughput or require major investments for automation as in the case of GROWSCREEN Rhizo [17]. Otherwise, soil free systems are preferred for large-scale genetic screens, as they can be assessed more rapidly and enable for a sufficient number of replicates [7,9,18-21]. One widely used soil free cultivation method is to grow plants on germination paper. Such paper is not only used for routine germination testing but also to assess root traits as it is easy to handle, can be kept free from pathogens and enables to manage a high number of replicates on a limited space. Moreover, the access to the root system is simple and coloured paper, unlike soil, provides favourable optical contrast between background and roots, thereby making automatic digital image processing possible [10]. On germination paper a wide range of experiments were performed. Investigations focused e.g. on the interaction between roots and rhizobacteria [22] as well as on the effects of temperature [23], low water potential induced by polyethylene glycol [24,25], nutrient deficiencies [26,27] and aluminium toxicity [28] on root growth.

The paper-based systems developed so far have the above mentioned disadvantage, that only the early embryonic root system can be assessed. Several factors complicate the enlargement of such systems in order to assess the development of crown roots. The most critical point is the increasing overlap and parallel growth of roots on the two dimensional paper surface. To circumvent this, different root types may be grown in a layered sandwich of paper as we will outline in Methods. Here we describe the biological basis of this approach: The mesocotyl, situated between the scutellar node bearing seminal roots and the first node bearing crown roots, elongates to place the shoot base at the soil surface. Mesocotyl elongation is stimulated by darkness [29]. If seeds are placed in the dark between two papers, the embryonic roots are growing between the papers, while the mesocotyl elongates and places the crown roots on top of the papers.

For a sufficient throughput, not only the cultivation method, but also image acquisition and image analysis are of high importance. A sophisticated image analysis is essential for fast and meaningful RSA analysis [30]. The process of image acquisition needs to be optimized with respect to an optimal contrast between roots and background and with respect to a sufficient spatial resolution to clearly visualize and quantify also finer roots with small diameter. Past studies showed that hyperspectral data can be used to elucidate differences between soil and roots or to identify plants infected with root rot [31,32]. In this context, it is important to identify wavelengths with an optimal contrast between root and background.

A wide range of literature is available dealing with software that enables for image based root system analysis [33-45] and there is an online database comparing the different software packages that are already available [46]. Yet, in the context of our study, it is crucial to explain how the optimal software should be chosen for the purpose of parameter extraction on rhizoslides, to clarify under which circumstances such a software performs best and to outline putative pitfalls. Available software for RSA analysing ranges from completely automated analysis delivering global root data via semi-automated systems to hand measurements enabling detailed measurement of a wide range of traits [33,34,36-40,42-45,47].

In summary, growth pouches as described by Hund et al. [10] have the disadvantages that i) only the early, embryonic root system may be studied and ii) an intense user interference is required to mount the pouches on the imaging station and open opaque foil covering the roots. The aim of this project was to develop a paper-based root observation system, so called rhizoslides, that enable for i) a characterization of post-embryonic cereal root systems and ii) automated or semi-automated image acquisition and processing.

Results

Root slides enable the separation of crown roots

The root slides enabled to grow maize seedlings to three fully developed leaves before the first embryonic root reached the bottom edge of the paper. The nodal root system was observed until the four-leaf stage. The plants needed 10 and 20 days to reach the three and four leaf stage, respectively. A separation between embryonic and postembryonic roots was achieved by physically separating roots into the different layers of the rhizoslide sandwich construction. Embryonic roots were growing in the invisible layer between the plexiglass sheet and the germination paper (Figures 1A, B and C), whereas the crown roots grew in the outermost, visible layer, on top of the germination paper covered with a transparent PE foil (Figure 1C). 90% (= 41 roots of 9 plants) of the crown roots grew on top of the germination paper and only 10% (=4 roots of 9 plants) between the paper and the plexiglass (See Additional file 1). The embryonic roots grown under the germination paper could be visualized using backlighting (Figure 1D). Tested alternatives to the separation of embryonic and postembryonic roots were to grow them not separated but either on both sides of the plexiglass sheet on the germination paper or on one site of the plexiglass sheet. The advantage of this method is the opportunity to monitor all root types at the same time without the usage of backlighting. The disadvantage is that space is very limited and roots start to grow parallel and cross each other. With increasing

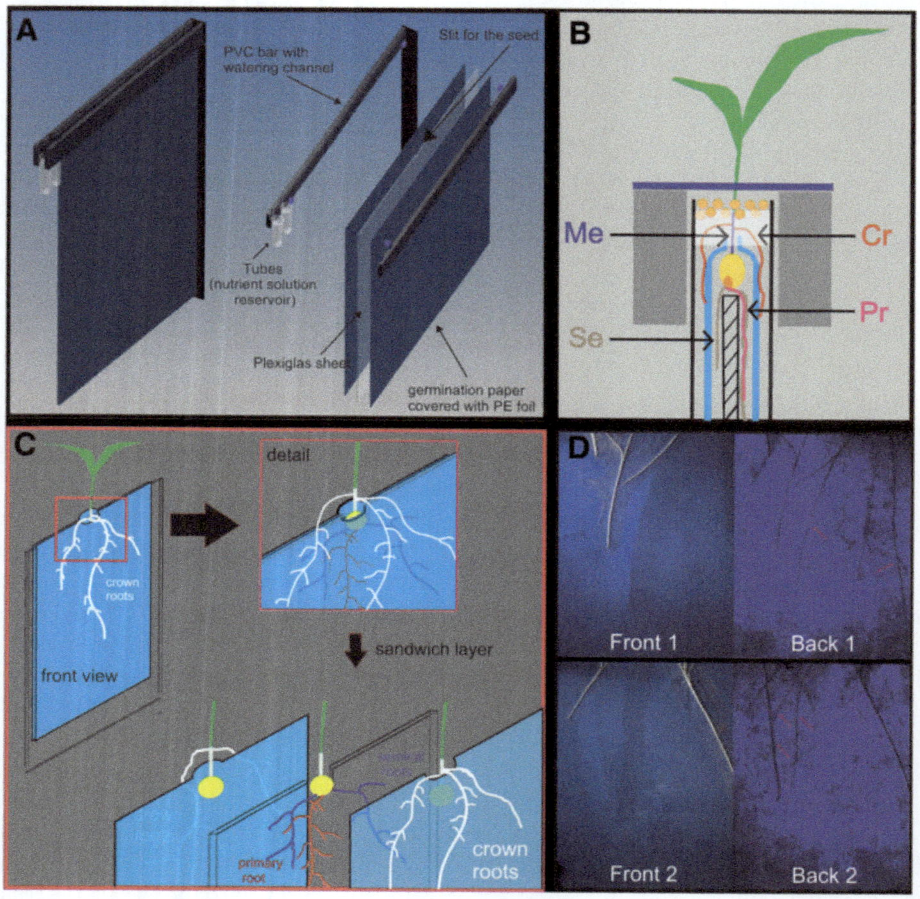

Figure 1 Construction of the rhizoslides. A: Root slides consistent of a plexiglass sheet covered with germination paper and a transparent PE foil belt with PVC bars with watering channels. Tubes on the site serve as nutrient solution reservoir. **B**: Cross section of the rhizoslide, illustrating the seed placement and separation of embryonic and crown roots. The seed is placed between the germination paper and the space between the PVC bars is filled with a layer of Potassium polycarbonate and a granulate substrate. Me: Mesocotyl; Cr: Crown root; Se: Seminal root; Pr: Primary root. **C**: Schematic figure of the separation of the embryonic and shoot-born crown roots: Embryonic roots are growing hidden below the germination paper whereas crown roots are growing visible on the top of the germination paper. **D**: Images taken of one slide with front- or backlight. Front 1 and Back 1 (complementary Front 2 and Back 2) are images of the same side taken with either front or backlight. Front images show crown roots whereas the transmitted light allows detection of seminal roots as well.

age, root tracing became demanding. Furthermore, fungal growth was successfully suppressed by moistening the germination paper with the fungicide Captan and by adding Captan to the nutrient solution. The nine replications of the control slides (no fungicide) were all colonised by fungi whereas both Captan concentrations (2.5 g L^{-1} and 5 g L^{-1}) reduced the colonization to one out of nine slides (See Additional file 2). Most fungi on the non-treated paper were *Chromelosporium fulvum* (See Additional file 3). Total root length did not differ between the treated and non-treated plants (data not shown), but plant development was delayed compared to the control plants (See Additional file 2).

Reflections are overcome using polarization filters and a staggered flash

We aimed to optimize image acquisition to enable imaging through the transparent cover foil with a minimal disturbance or reflectance of light, haze or droplets on the surface of the foil.

The minimum tonal value method, i.e. combining the left and right image by keeping only the minimum tonal value present in either image, resulted in a lower amount of reflections of the bends in the surface of the covering transparent foil (Figures 2A and B; upper blue circle) and a reduction of reflections by droplets (Figures 2A and B; lower blue circle). It also increased the contrast

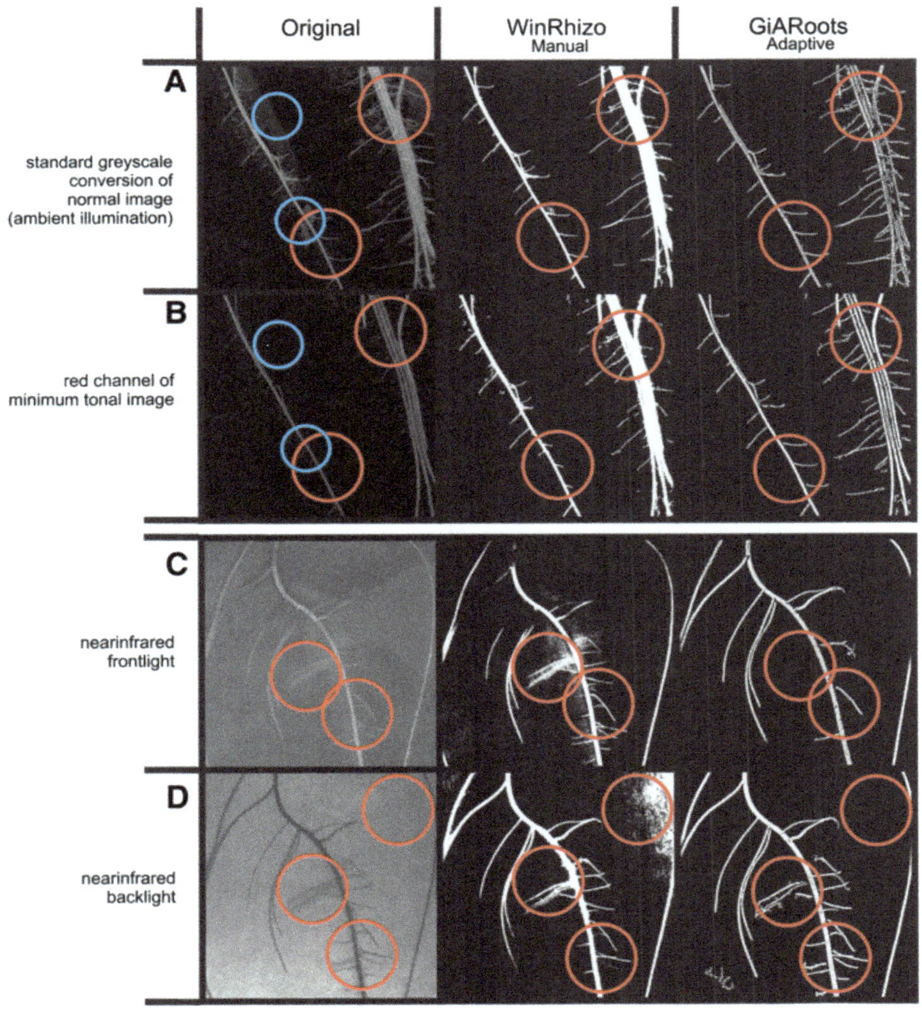

Figure 2 Imaging and thresholding methods. Images of roots grown on either Anchor blue (A + B) or Sebio grey (C + D). Red circles highlight regions for which the different thresholding methods yielded contrasting results (Lateral roots vanished, parallel growing roots became one root or the intensity of background noise). Blue circles indicate the removal of droplets and reflections. **A**: Image taken with diffuse lighting. **B**: Two images taken with flash light (right/left side) including polarization filters on flash and camera lens and combined to a minimum tonal image. The red channel was used for the conversion into greyscale. **C**: Image taken with near-infrared front lighting. **D**: Image taken with near-infrared backlighting. Thresholding was done using the WinRhizo or GiARoots routines. Only the routines resulting in the best separation between root and background are shown.

between roots and background compared to ambient illumination (Figures 2A and B). The higher contrast presumably resulted from the shadows from left and right illumination, which were retained in the combined image. A further advantage of the shadows was a better distinction between roots growing in parallel (Figures 2A and B; upper right red circle). A slight disadvantage was that the detection of the origin of lateral roots became more difficult as they emerged in the shadowed region (Figures 2A and B; lower left red circle).

Red light created the strongest contrast

We used spectral reflectance to elucidate at which wavelengths the contrast between roots and paper background is maximized. Based on this information we aimed to identify which colour channel of the available camera would be best suited to segment between roots and paper background. The reflection of germination paper behaved differently depending on colour and/or texture and there were differences in reflectance between the root and the papers (Figure 3). The root reflected in the entire range between 400 and 1000 nm with small differences in reflection intensity. A similar pattern was observed for the white Whatman paper, where the intensity was two times higher than for the root. Also the light blue Whatman paper showed a higher reflection compared to the root and reflection maxima were situated in the blue and infrared range. All strongly blue and grey colored papers (steel blue (Anchor), Whatman blue, Sebio grey) showed a similar pattern of a high reflection in the blue range, a decrease in green and red and an increase in the near-infrared range. With the exception of the steel blue paper, the intense reflection of these papers in the blue range was lower than the reflection of the root. Between 560 and 720 nm the reflection of all three blue papers was two times lower compared to the reflection of the root (Figure 3). Depending on these results, images were taken and colour channels tested. The best distinction between root and background was obtained with the red channel for steel blue, Whatman blue and Sebio grey (See Additional file 4A). Whatman white showed similar poor results for all three channels.

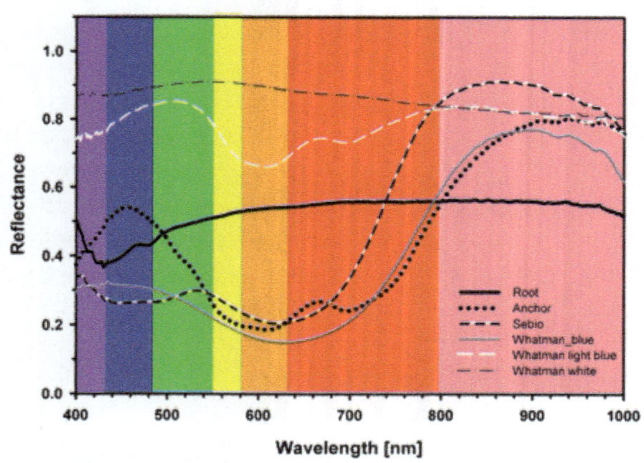

Figure 3 Hyperspectral reflectance of root and paper. Reflectance of the root and five tested germination papers (Anchor, Whatman blue, light blue, and white and Sebio grey) in the spectrum from 400–1000 nm. Colouring indicates the spectral range of blue, green and red light. Slightly red coloured is the near-infrared range (790–1000 nm).

Near infrared backlight enables root growth studies

Images taken in the near-infrared range (940 nm) confirm the observation of a slight contrast between root and paper due to a high reflectance in the near-infrared range of the papers. Best results were obtained using Sebio grey paper (see Additional file 4B), but the low contrast and noise due to reflections resulted in a loss of lateral roots (Figure 2C). However, using near-infrared back lighting, segmentation between root and background was good and only negligible losses of root structures occurred (Figure 2D). However, both thresholding methods still had problems to handle the background noise resulting from the paper texture. Compared to other papers, the Sebio paper had the advantage of a fine texture and thinness, which produced only slight background noise in the backlight image (Figure 2D). In contrast, the texture noise of the steel blue germination paper (Figure 1D) was too high for segmentation methods applicable to date.

Image analysis software

We conducted a literature search to identify software with the ability to perform a RSA analysis of complex root systems grown on two dimensional images. As a result, eight potentially suitable software packages were chosen (See Additional file 5) and three of them could be successfully installed and tested (all software was downloaded in October 2012). Two of them, WinRhizo (WinRhizo Pro 2009b, Régent Instruments Inc.) and GiARoots [37] offer batch processing with the opportunity to perform manual adjustments. The third software, SmartRoot [40] is semi-automated. We were not able to test the remaining five software packages for different reasons: DigiRoot [48] and RootReader2D [7] could be successful installed, but did not accurate work with the supplied material (incomplete/wrong marking of the roots). EZ-Rhizo [33] could be successful installed, but the software stopped working immediately after starting the analysis. We tested whether it would work with different image formats or resolution and requests assistance from the developer, which remained unanswered. RootTrace [42] could not be successful installed in spite of intense support by the developer. The software DART [39] could be installed, but not opened. Our help request was answered with the comment that the software was not developed to a stage where it would stably run on any system.

GiARoots facilitates best automatic thresholding routine

In addition to improve image quality, we evaluated different methods for image segmentation. The thresholding methods of WinRhizo and GiARoots were tested and compared. These two software packages were chosen because they make an automated conversion into binary images possible and worked well with the supplied material. SmartRoot was not included in this part of the analysis, as it works on the basis of greyscale images. In WinRhizo, the automatic routine did not eliminate all background noise; in contrast the manual adjustment facilitated a better reduction of the background noise without a loss of information (See Additional file 6). In WinRhizo, it was not possible to distinguish roots growing in parallel to each other (See Additional file 6, blue circles). However, the software GiARoots enabled to some extent distinguishing between roots growing in parallel, but the images of the segmented roots were incomplete, making it difficult to identify individual roots (See Additional file 6, blue circles). Both routines (adaptive and double-adaptive threshold) enabled a good separation between root and background and no big difference was observed between them (See Additional file 6). Lateral roots with a weaker contrast were clearly visible after adaptive thresholding in GiARoots and manual adjustment of the threshold in WinRhizo, but they were neither detectable using the global threshold value in WinRhizo, nor the double adaptive threshold in GiARoots (See Additional file 6; red circles). Based on these results, images taken on either steel blue germination paper, Whatman blue, white or Sebio grey were transferred into greyscale images using the red channel and were segmented using the adaptive threshold of GiARoots (Figure 4). The best results, regarding the ratio (reduction of background noise)/(loss of lateral roots), were obtained on steel blue germination paper and Sebio grey. For both papers a good separation between root and background was obtained with a minor loss of lateral roots.

High repeatability of SmartRoot

To determine the influence of the user on the results using semi-automated software, a test for repeatability was done using SmartRoot.

In SmartRoot, the variance of root length detected in ten different images was related to the overall variance created by different users and the interaction between users and image content. Image processing was highly repeatable with respect to the overall length detected for the lateral roots (repeatability of 0.99) and axile roots (repeatability of 0.97). Total measured length of lateral roots ranged from 152 to 164 cm; total length of axile roots varied from 162 to 165 cm, depending on the user.

Good correlations for SmartRoot and WinRhizo

As SmartRoot enables for a user-defined, controlled tracing of the whole root system we considered the output of this software package as the one that represents best the real root system length. Compared to SmartRoot, WinRhizo underestimated the total root length due to the fact that it could not detect lateral roots that only showed a small contrast between root and background (Figures 5B and C). In contrast to this, GiARoots rendered much higher root lengths. Hereupon, we evaluated the images showing the thinned objects and observed thinning artefacts (Figure 5D). We anticipated that these differences were a result of artefacts caused by root hairs, reflections and other effects leading to a ragged edge between root object and background. Therefore, the output would be comparable under optimal image conditions, defined by smooth edges of the roots and by a perfect contrast between root and background. To test this hypothesis, we reconstructed the root systems traced in the ten images using the data of the vectorized root system supplied by SmartRoot. This resulted in ten images of "artificial roots" with an optimal contrast and a known length. Indeed, visually all programs detected the roots without losses or false tracing (Figure 5E-H). However, there were still differences in total root length. The values obtained with SmartRoot differed 0–5% compared to the original pixel length, WinRhizo differed 0–4% and GiARoots differed 2–22%. Even more important than the absolute values are the correlations between the results obtained with the three programs. For the artificial root images, the correlations between all programs were satisfying with r^2 values between 0.91 and 0.97 (See Additional file 7B), but for the original images, the correlations were much lower (0.33 GiARoots-WinRhizo; 0.54 WinRhizo-SmartRoot; 0.67 GiARoots-SmartRoot) (See Additional file 7A). In addition to total root length measurements, all three programs enable to study further traits of root system architecture. As the measured traits and methods differed strongly among the software packages, they could not be used for software comparison. Total root length was chosen as a common trait for differentiating the performance of the software packages. Other traits such as lateral root number, angle between roots etc. were less good indicators of the performance of a software package. As already shown, software packages often underestimate the length of a certain lateral root; hence, their applicability increases with increasing length of each lateral root, and, therefore with total root length. Similar examples are traits based on diameter calculations as surface area (cm^2) or volume (cm^3). Correlations for the root diameter were low for all three programs (between 0.025 and 0.51) for

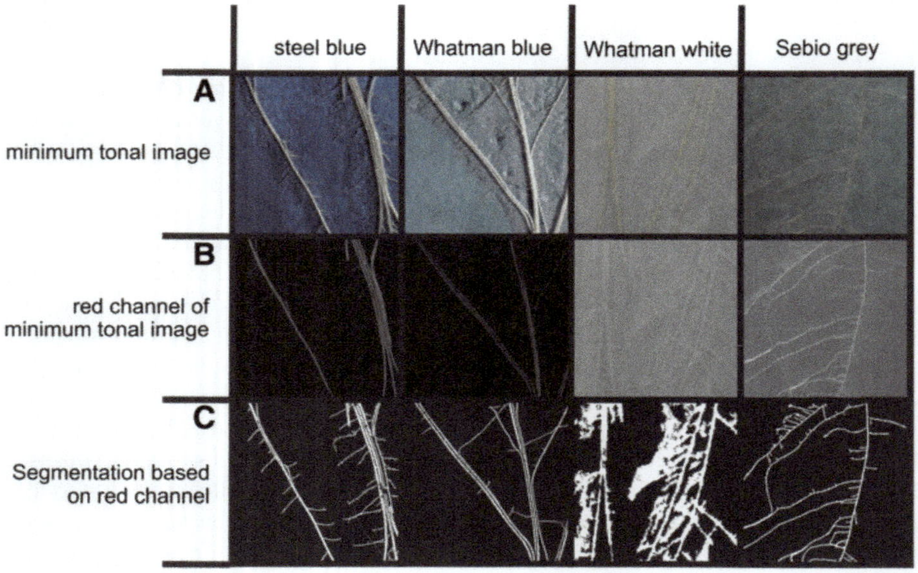

Figure 4 Application of optimal image processing. The optimized image processing protocol as described in Figure 6, was applied to the four most promising papers identified based on spectral imaging (Figure 3). **A**: Images of roots grown on Anchor, Whatman blue and white or Sebio paper. Two images were taken with flash light from the right/left side and combined to one image. **B**: Conversion into greyscale using the red channel for conversion. **C**: Segmentation of the root system using the adaptive threshold of GiARoots.

the original images and consequently for calculated traits based on the diameter as well (See Additional file 7A). Diameters measured by SmartRoot and WinRhizo correlated for the artificial roots (0.96), but for GiARoots the correlations remained low with both programs (0.07; 0.08) (See Additional file 7B). As WinRhizo also enables for topology analysis, the time investment for a topology analysis in WinRhizo as well as for SmartRoot was investigated. The images were taken from root systems of plants with two fully developed leaves grown in small pouches (21 × 29.5 cm). These data set has been described previously [10]. The analysis was divided into four steps and time investment for each step was recorded. For both programs the last step (lateral root tracing/assignment of ranks) was most time consuming (See Additional file 8). The analysis of the images used in our study took between 8 and 40 minutes depending strongly on the number of lateral roots. A correlation between number of lateral roots and the

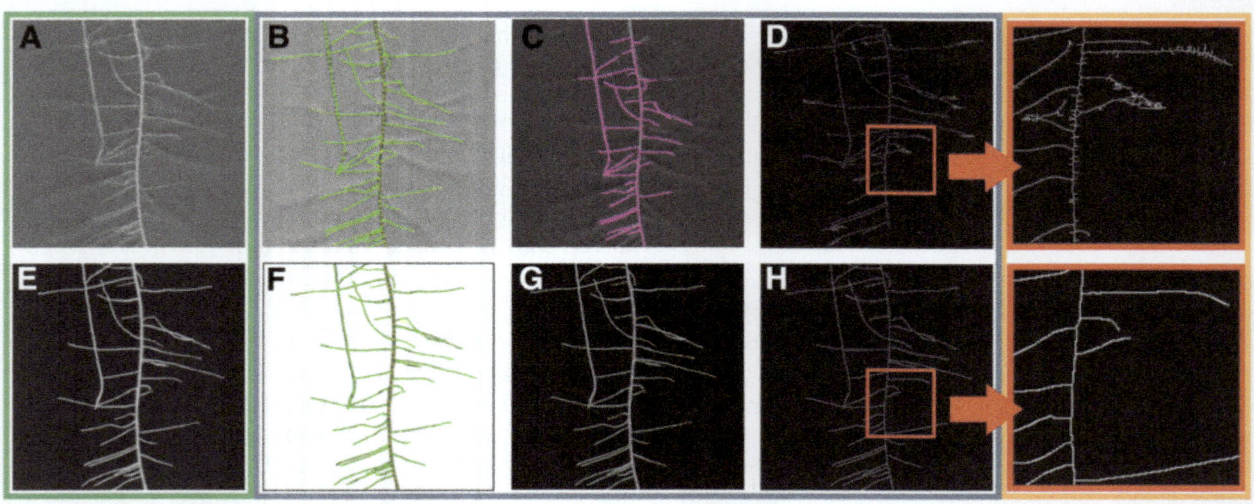

Figure 5 Tracking of the root system by the three software packages. **A**: Greyscale image of a root system section. **B-D**: Tracking of the roots in SmartRoot **(B)**, WinRhizo **(C)** or GiARoots **(C)** using the image shown in **A**. **E**: Artificial root of **A** derived from vectorization in SmartRoot. **F-H**: Tracking of the roots in SmartRoot **(F)**, WinRhizo **(G)** or GiARoots **(H)** using the binary image shown in **E**.

required time for analysis could be observed for WinRhizo ($R^2 = 0.76$) and SmartRoot ($R^2 = 0.44$). Although both programs enable topology analysis, the obtained traits are not identical and not all traits are immediately extractable from the output but must be calculated (See Additional file 9). An advantage of SmartRoot compared to WinRhizo is the clear arrangement of the output file. This facilitates to obtain information for every single root immediately (See Additional file 10B). In contrast, in the WinRhizo output every measured section (link) is listed and the data needed to be summarized by the user first, before meaningful traits can be extracted (See Additional file 10A). This data reorganization is very time intensive. GiARoots does not enable for a manual topology analysis as all images are batch processed.

Discussion

The aim of this work was to create a growth system that enables for non-destructive and potentially high-throughput quantification of root system architecture traits. Ultimately, this system should be applicable for genome mapping of crown root characteristics. The "sandwich" composition of the paper layers enabled distinguishing between embryonic and postembryonic roots as they grew in different layers. Postembryonic crown roots are a major focus, as they account for the major part of the adult root system [11], and the ability to study their response to stresses offers a major advantage.

The rhizoslide sandwich system is a good compromise between the space-saving and handling capacities of a 2D-system and the advantage of 3D-systems enabling for an unlimited spread of roots in three dimensions. Single layered systems, even if up scaled to larger paper size, have the disadvantage that roots will increasingly overlap and crown roots will be difficult to measure. Three dimensional systems based on agar, aero- or hydroponics circumvent these problems [7-9]. Agar has got the advantage that roots stay in place and do not overlap. However, space is usually limited and keeping the agar free of pathogens is laborious. Therefore, studies are preferably performed over a short time period. Hydro- and aeroponic cultivation make it possible to studying a high number of individuals over a long time period, but roots change their position and this complicates the image analysis.

A big advantage of the rhizoslides is that they facilitate studying the response of particular root types, especially crown roots, to changes of the root environment in space and time. For example, different concentrations of nutrients can be applied to the paper on the left and right side of each slide. In split-root setups responses to stimuli such as altered nutrient availability can be studied [49-51]. Besides root system architecture traits, the plasticity and the dynamic alteration of root growth to changing environments can be observed. Previous studies showed that root morphology and growth can change in acclimation processes to nutrient availability as short as well as long term response [52-55]. Rhizoslides offer the potential to study such response on a large number of plants. In the past, studies were done on monocot species [10,56-58] as well as on dicot species [59] using the so called paper-roll setup or growth pouches. For example, Watt et al. grew wheat seedlings in a paper-roll setup and found positive correlations of root length between seedlings grown in the paper-roll setup or in the field, but not with the reproductive stage [56]. These findings emphasize the importance to work with later developmental stages. Potentially the paper-sandwich is perfectly suited to study the fibrous root system of monocot species e.g. rice, wheat, or barley. The mesocotyl elongation is used to separate embryonic from crown roots. Eventually, small adjustments (e.g. a smaller slit in the plexiglass sheet) have to be made to keep the smaller seeds in space and to ensure an elongation of the mesocotyl above the paper edge. For dicot species, which form one tap root undergoing secondary thickening, the sandwich system is less suited. Still such roots can be studied on one site of the Plexiglass sheet. This has the already mentioned disadvantage that space is very limited and roots start to grow parallel and cross each other. Furthermore, in the current rhizoslide version, the Plexiglass plate bends, as the adhesive power of the nutrient solution connects it tightly with the paper and the covering foil. Apparently, each material has a different coefficient of expansion. Therefore, it is advantageous to grow either two plants on one plate (each on one side) or to enable root growth on both sides of the plate.

Furthermore, the adjustment of parameters of the rhizosphere, such as pH, CO_2 or O_2 can be analysed via Optodes in complementation to growth analysis [60]. Rhizoslides are not only an opportunity to perform high throughput screening for RSA traits, as usually done for QTL mapping, but allow to do more precise effect studies on a small scale with high temporal resolution. The lack of automation is currently the only bottleneck to achieve high temporal resolution.

We optimized the imaging system to enable automation. The necessity to remove the foil covering the roots in order to avoid reflections of the cover itself and of droplets on the inside of the cover was a major bottleneck hampering automation. These reflections could be successfully minimized by using polarization filters in combination with the combined images with left and right illumination, respectively. Polarization filters are commonly used to reduce noise due to reflections and were already successfully used by Clark et al. [7] in a hydroponic system. We took this approach further, by combining two images, each illuminated from a different angle. By retaining only the lower (darker) pixel value of each image in the final minimum tonal image, we utilized the optical properties of the three dimensional structures in the image: bright reflections on round droplets or on folds in the covering foil appeared at different locations depending on the illumination. These artefacts were minimized. Similarly, the cast shadows of the roots appeared either on the left or the right side and were maximized. Accordingly, the minimum tonal image had a strong local contrast between roots and background combined with reduced noise due to reflections. The possibility to take images without removal of a cover is a major advantage compared to the pouch system described by Hund et al. [10] which had to be opened manually. However, background noise by condensed water could not be completely removed by image combination, but by usage of backlighting.

Maximal contrast between roots and paper background can be achieved by using the red channel of the RGB images. This conclusion is based on our analysis of spectra of the root compared to those of various paper backgrounds. Hund et al. (2009) reported the saturation channel to be best suited for image segmentation. However, differences between the red and the saturation channel in the earlier study were very small. Several studies showed that illumination within the visible range affects root growth [29,61,62]. A frequently used alternative is near-infrared (NIR) light as so far no negative effect on root morphology could be observed [63,64]. NIR illumination was used to differentiate between roots and soil background [32] and to illuminate roots grown in aeroponics (personal communication, Draye, X.). In Rhizoslides only NIR backlighting resulted in a sufficient contrast and, with a double layer of thick steel blue germination paper, this contrast was not sufficient for segmentation methods available to date. This makes NIR unsuitable as light source in paper-based rhizoslides. Furthermore, in previous studies with growth pouches, the influence of the scanning light on root growth during imaging was negligible (Hund et. al 2009). It remains to be tested, whether an increased frequency of illumination in high-throughput screening approaches would have systematic effects on root morphology.

The resolution of the camera was high enough to detect first-order lateral roots of maize. Given the dimension of the imaged area of the slide of 490 mm widths in combination with the 21 mega pixel camera, a pixel size of 0.13 mm was achieved. A minimum of three pixel are required to detect roots by means of an image processing software. This three-pixel diameter of 0.39 mm is in the range of the lateral root diameter of maize. Hund et al. (2004) reported lateral root diameters of maize in the range of 0.26 to 0.47 mm in plants grown in sand substrate under chilling conditions. The diameters of lateral roots in pouches are usually below a threshold value of around 0.5 mm [10,24] with average diameters ranging between 0.25 and 0.306 mm [23]. However, Mac Cully et al. (1987) reported lateral roots as thin as 0.07 mm which would be below the threshold detected by the current setup. Accordingly, it will not be possible to distinguish between lateral roots diameters and to detect very fine root. The solution would be higher resolution. The four times smaller A4-size growth pouches in combination with a 28 megapixel scanner [10], yield an almost tenfold resolution of 0.042 mm px^{-1}. By stitching multiple images or zooming into particular regions of interest, resolution on rhizoslides can be increased to a point where even monitoring of root hairs may be possible.

Suitable software remains a bottleneck. The three software packages, offered different strengths but had also severe weaknesses. Dependent on the research question WinRhizo and GiARoots offer the advantage of simple batch processing without additional user interference. The thresholding algorithm of GiARoots is more advantageous compared to WinRhizo when it comes to the elucidation of inhomogeneities in the root system. Accordingly, GiARoots provided a much better global segmentation. However, we did not test the color analysis in WinRhizo as an option for enhanced segmentation. After segmentation, GiARoots delivers basic characteristics of a root system with the lowest time investment and without influence of the user. A negative point for GiARoots is that it needs images with a good contrast to avoid false tracing. These artefacts may lead to a serious overestimation of total root length. The images derived from our rhizoslides did not provide sufficient contrast to avoid such artefacts. The automatic routines in WinRhizo provide root lengths that can be grouped in user-defined diameter classes. Using this root-length in diameter class distribution, roots may be classified in large-diameter axile roots and small diameter lateral roots [10,65]. This approach was efficiently used for high throughput image analysis in genome mapping studies [25,66,67] and it may be applicable for rhizoslides. WinRhizo and Smart Root offer the possibility to perform an in-depth topology analysis. For such an analysis an intense user interaction is needed to allocate lateral roots to their parental origin. For both WinRhizo and SmartRoot manual tagging of root for topology analysis is time intensive and ranges from 8 to 40 min for a root system grown on a small 21 × 29 cm paper. In the four times larger rhizoslides, a much higher time investment is needed unless the focus is on individual, representative roots. Furthermore, the user might bias the results as a high degree of user interaction is required. Although we could not detect strong bias among the three different test persons we recommend controlling potential systematic differences among users by an appropriate experimental design. A difficult part of the topology analysis in WinRhizo is the extraction of information from the generated output, as it is not intuitive and traits such as root length of single roots must be calculated. Furthermore, WinRhizo's topology analysis does not allow for a simplified tracking of roots though multiple images of a time series. These difficulties could be the reason, why it was not used for genome mapping in crops to date and only in a small number of topology studies [68,69], as far as we know. SmartRoot offers both, a topology analysis with an intuitive output of the summary statistic and the possibility to tag and track roots through successive images of a time series [46]. The software was developed to enable more complex analysis of the RSA to do QTL analysis amongst others.

For many research questions, global root traits or quantitative traits as generated with automatic routines in WinRhizo or GiARoots are sufficient. A study with hundreds of individuals as needed in QTL analysis, for example, would require massive investment of time for manual root tracing. However, SmartRoot may prove suitable if only parts of the root system, e.g. some representative crown roots are to be measured in more detail. Such an approach would be feasible, even for quantitative genetic studies. For example, Trachsel et al. [66] measured the length of the primary axile root on more than 1000 plants using the ruler tool of Adobe Photoshop. As SmartRoot allows tagging and tracing of individual roots in image series in a convenient manner, it is particularly suited to monitor temporal changes in growth rates. Furthermore, there are research questions with a smaller number or repetitions that need to study e.g. single root scale. For these cases a program such as SmartRoot is optimal. To sum up, so far there is no optimal software solution for every setup. Instead the most suitable method must be chosen depending on the research question and the maintainable time investment.

Conclusions

A new technique has been established for high-throughput non-destructive root growth studies and quantification of architectural traits beyond seedlings stages. The method allows studying root growth of crown roots and seminal roots independently under heterogeneous environmental conditions. Transparent foil sheets covering both sides of the sandwich construction allow for a rapid screening of the maize root system growing within the rhizoslide. In future, the usability for other crop species should be tested and necessary adaptations identified. The reflections of the foil could be successfully eliminated by a newly developed imaging setup and image processing. In future, rhizoslides can be used to study a wide range of research questions on a small scale as well as with a high number of replicates necessary e.g. for QTL analysis. A future challenge will be the establishment of a system allowing the automation of the imaging process to increase the screening speed of huge sets of genotypes. Of the tested software packages, each offered specific strengths. Specifically, we identified the segmentation algorithms of GiARoots to be optimal, we found the most precise automated measurement of root length using WhinRhizo and we saw a user friendly topology analysis combined with the ability to trace roots in successive images as the major advantages of SmartRoot. Improved next generation software solutions should ideally combine these strengths.

Methods

Plant material

All experiments were carried out with the maize hybrid Bonfire supplied by Delley seeds and plants Ltd (DSP Ltd), Switzerland.

Materials

The rhizoslides (version 2.7) consist of two PVC bars (600 × 60 × 10 mm) and an acrylic sheet (530 × 650 × 4 mm) fixed with two screws between the bars (Figure 1A). Between acrylic sheet and bar, an 8 mm flat washer was placed to obtain a slit for the roots (see Additional file 11). On one side of the bars, 25 mL PE tubes (Semadeni AG, Ostermundigen, Switzerland) were placed to act as water/nutrient solution reservoirs (Figure 1A). On the inner side of each bar, a channel was mortised to hold a watering system. The watering system consisted of two glass fibre wicks (Ø = 2 mm) (Suter-Kunststoffe AG, Fraubrunnen, Switzerland), each surrounded with a PVC tube (outer diameter 5 mm; inner diameter 3 mm) (GVZ-Gossart AG, Otelfingen, Switzerland). The wick system allowed the transport of the nutrient solution via capillary force from the two reservoirs to the right and the left sides on the germination paper, respectively. The acrylic sheet was covered with wet germination paper (490 × 610 mm) on both sides serving as substrate. These were in turn covered by a transparent oriented polypropylene (OPP) foil with micro holes of 70 μm to allow for gas exchange (Maag, GmBH, Iserlohn, Germany). The foil is widely used in the packaging industry for cooled, fresh food to allow for gas exchange and to avoid droplets and fog on the transparent cover. Steel blue germination paper (Anchor Steel Blue Seed Germination Blotter, Anchor Papers Co, USA) (Anchor) proved useful in several studies evaluating root growth and development in growth pouches [10,20,23-25,28,70,71]. Unless mentioned otherwise, this paper was used for all standard tests. In addition, we tested four alternative germination papers with respect to their optical contrast to the root objects, i.e. light blue (FP3621), blue (FP3644), and white (FP5703) germination paper by Whatman (GE Healthcare Life

Science, Glattbrugg, Switzerland) and Sebio grey (FP3236, Albet-Hahnemuehle S.L, Dassel, Germany).

Cultivation conditions

For sterilization, the germination paper was heated in three cycles from room temperature to 80°C and kept at this temperature for at least 120 min. Between the heating periods the paper was kept for 20–22 h in an oven at 37°C and 50% relative humidity [72]. Maize seeds were surface sterilized with sodium hypochlorite for 15 min. and rinsed with deionized water for 5 min. Subsequently, seeds were kept for 48 h at 26°C in the dark for germination and were then transferred into the rhizoslides. To prevent fungi growth, a method described by Bohn et al. [73] was used. The germination paper was moistened with water containing 2.5 g L^{-1} Malvin (Syngenta Agro AG, Dielsdorf, Switzerland) containing the active component Captan. Plants were cultivated in a climate chamber (PGW36, Conviron, Winnipeg, MB, Canada) refurbished with new control unit, compressor and slight ceiling (Kälte 3000, Landquart, Switzerland). Environmental settings were a day period of 14 h light, at a temperature of 26/18°C (day/night) at seed level, 60% humidity and a light intensity of 230 μmol photosynthetically active radiation m^{-2} s^{-1} at plant canopy level supplied with a mixture of 2/3 Cool White (Philips TL5 HO 54 W/865; Philips, Zurich, Switzerland) and 1/3 GRO LUX (Sylvania F36W/GRO, Sylvania, Germany, Munich) light tubes.

Root type separation into rhizoslide layers using mesocotyl elongation

The placement of the seed was done immediately after germination to avoid damaging the primary root. The seed was placed between the germination papers into a slit on the upper edge of the plexiglass sheet. The paper edge was placed at the plexiglass edge, surrounding the seed (Figure 1B). Paper clips or paper fastener stuck the two papers on the site of the seed. For mesocotyl elongation, the seed was kept in the dark. To keep the seed and mesocotyl wet Potassium polycarbonate moistened with deionized water containing 2.5 g L^{-1} Captan was injected between the PVC bars on top of the seed. To avoid incidence of light, a granulated substrate was placed on top of the Potassium polycarbonate and a cover was placed on top of the bars (See Additional file 12). To test whether a separation of embryonic and crown roots was possible in the described setup, nine plants were grown in the rhizoslides and the whole root system was harvested layer by layer. The number of roots per root type in each of the four layers was recorded.

Verification of fungi reduction

One consequence of the longer cultivation period was fungal infection. The most prominent fungus was *Chromelosporium fulvum*. To verify that the fungicide treatment could successfully reduce the number of fungal infections without an influence on root morphology, a preliminary experiment was done using a modified pouch setup based on the method described by Hund et al. [10]. Every pouch was supplied with nutrient solution over a wick that was hanging in a single 50 mL tube (Greiner, Frickenhausen, Germany). The tube was filled with sterile nutrient solution containing either 0 g L^{-1}, 2.5 g L^{-1} or 5 g L^{-1} Malvin (Syngenta Agro AG, Dielsdorf, Switzerland) (n = 9). Tubes were refilled every 48 h. After ten days, images of the root systems were taken, infection rated and plants were harvested for biomass measurements.

Image acquisition and pre-processing

For standard imaging, images were taken either with a 21 mega pixel full-frame digital single-lens reflex camera (EOS 5D Mark II, Canon, Tokyo, Japan) equipped with a 50 mm lens (compact macro 50 mm f/2.5, Canon, Tokyo, Japan). The resolution of the images was around 0.13 mm/pixel. The camera was equipped with a circular polarization filter (Hama, Augsburg, Germany) and was placed in 1 m distance parallel to the pouch surface. Two studio flash lights (Walimex pro VC 400, Burgheim, Germany) were used to illuminate the rhizoslides. The lights were positioned at an angle of 30° and a distance of 1 m to the left and right in front of the slide, respectively. The front of the lights were equipped with linear polarization filters (Foto Mayr, Dietzenbach, Germany). For each side of the rhizoslide, two images were taken: one illuminated from the right, the other one illuminated from the left. To trigger these staggered left/right flashes, a microcontroller was built and programmed in Arduino 1.0 (http://arduino.cc/en/) to activate a different flash each time the camera trigger was released. The microcontroller connected the camera and the flashes with a computer and was triggered by CanonEOSUtility Software (V2.1 Canon Inc. 2011) (See Additional file 13). Colour 24 bit RGB images were taken and directly stored on the hard drive by the CanonEOSUtility Software. A backlight was used to evaluate the possibility to measure the embryonic roots, covered by the germination paper, by means of their reduced transmission of light compared to the paper. Roots were backlighted with a continuous spotlight and images were taken from the front (f/4; 1/6 s). In case of infrared images, a monochrome CCD camera (Scorpion SCOR-20SO; Point Grey Research, Vancouver, BC, Canada) equipped with a standard lens (25 mm; Cosmicar/Pentax, The Imaging Source, Bremen, Germany) and an infrared interference filter (940 nm; Edmund Optics, Karlsruhe, Germany) was used. For lighting, a LED panel (880/940 nm) or infrared diode-fields (940 nm) were used. The camera had a resolution of 0.22 mm/pixel and the display detail was approximately 10 × 10 cm.

Hyperspectral microscopy

To identify the wavelength with the maximal contrast between root and background, a Darkfield transmission optical microscope (CytoViva Hyperspectral Imaging System (HSI), Auburn (AL) USA) was used. The reflection of Steel blue germination paper (Anchor Steel Blue Seed Germination Blotter, Anchor Papers Co, USA), light blue (FP3621), blue (FP3644), and white (FP5703) germination paper from Whatman (GE Healthcare Life Science, Glattbrugg, Switzerland) and Sebio grey (FP3236) (Albet-Hahnemuehle S.L, Dassel, Germany) were recorded in the range of 400 to 1000 nm. Measurements were done using Environment for Visualization software (ENVI 4.8, Exelis Visual Information Solutions, Inc., Boulder, USA) that can extract complete spectral information from single or multiple pixels. The reflection spectrum of a white reflectance standard with spectralon served as reference (WS-1-SL, Ocean Optics, Ostfildern, Germany). Spectra resulted from average values of 13 081 pixels per spectrum. Exposure times were 0.21 s for the spectralon, 0.4 s for Steel blue germination paper, Sebio grey and Whatman blue, 0.3 s for Whatman light blue and 0.25 s for Whatman white.

To correct for differences in exposure time, the intensity of the reflection signal at each wavelength was corrected using the following formula:

$$corrected\ intensity = \frac{Intensity\ value}{exposure\ time\ (spectralon) \cdot exposure\ time\ (sample)}$$

In the following, the intensity at each individual reflection signal wavelength was normalized by dividing through the reflection intensity of the spectralon.

Image analysis

Images, pre-processed in Matlab, were successively analyzed by digital image analysis software developed for root image analysis. The utilized software was WinRhizo (Régent Instruments, Québec, Canada, 2003a) GiARoots or SmartRoot [37,40]. The images were combined to one 24 bit RGB image using Matlab (Version 7.12 The Mathworks, Natick, MA, USA) by keeping only the minimum tonal value present in either image (minimum tonal image). Each of the three RGB colour channels was tested to figure out which channel delivered the best distinction between root and background. A Matlab (Matlab Version 7.12) routine was used to i) balance inhomogeneous illumination, ii) combine the images taken with right or left flash light to one image by using the pixel with the lower tonal value iii) extract the color channel with the highest contrast between roots and background iv) identify each individual by reading the label in the image and v) rename the image with the label content (Figure 6).

Optical differentiation between root and background (called segmentation or thresholding) was done in WinRhizo or GiARoots. In WinRhizo the automated threshold or a manual adaption of the threshold was done by choosing the tonal value with the best noise to root relation. All pixels above this value are assumed as background while all pixels below this value are considered as root. The threshold value of WinRhizo is used for the segmentation of the whole image. In GiARoots an adaptive thresholding or double adaptive thresholding was done. Using the adaptive threshold, the entire image is broken up into smaller square arrays of a certain block size. Within each block, the mean pixel intensity is calculated and all pixels with the same intensity +/− a selectable proportion are considered to be part of the root network, all others are considered to be part of the background. The double adaptive threshold looks at the behavior of the mean intensity as a function of the neighborhood size and classifies the pixel as foreground if a sufficiently large decrease/increase in its values is achieved within a specified range of neighborhood sizes (for details see Galkowskyi et al. 2012).

Software comparison

To compare the performance of the programs, a dataset of ten root sytems scanned on steel blue germination paper were analysed using WinRhizo, GiARoots or SmartRoot. The images were part of a previous study published in 2009 [10]. In contrast to WinRhizo and GiARoots, Smart root allows for user interference. Therefore, to determine the effect of user interference on root detection by the software, ten images of two contrasting genotypes were measured repeatedly with SmartRoot by three different persons. Furthermore, the time investment performing a topology analysis using WinRhizo or SmartRoot was investigated. The analysis was divided in four steps and time was recorded for every step separately. The classification into steps is not identical for both software packages as their procedures were different, but as close as possible. WinRhizo: 1. Step: Automatic analysis of the image and setting of the segmentation threshold, 2. Step: Excluding non-volitional regions, 3. Step: combining and cutting of root fragments and 4. Step: Allocation of the root order. SmartRoot: 1. Step: Automatic labeling of seminal roots, 2. Step: Manual correction of seminal roots, 3. Step: Automatic labeling of lateral roots and 4. Step: Manual correction of lateral roots.

Artifical roots

SmartRoot delivers xml files with the position of every node used to analyze the root system. These data, created for every analyzed image, could be used to generate artificial root images by a Matlab script as follows: First points and diameters were read from xml-files written by Smartroot. The points were interpolated by splines to get the complete root line of each single root. Gaussian

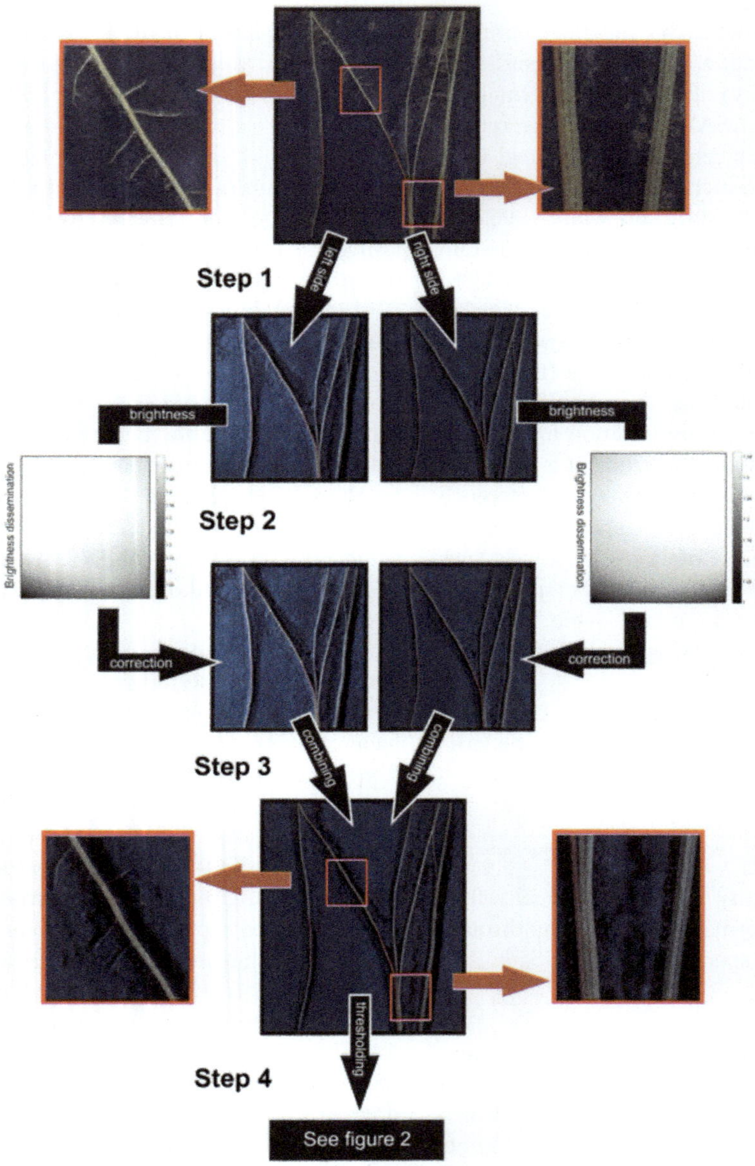

Figure 6 Work flow of the image processing. Images in the center row illustrate the workflow and images on right and left the effect of noise reduction. Center row: Step 1: Two images are taken, one with illumination from the right and one with illumination from the left side. Step 2: A correction for inhomogeneous brightness is done. The diagrams illustrate the inhomogeneous brightness for which need to be corrected on the right/left image. Step 3: Images taken with either right or left illumination were combined using minimal tonal value of each pixel. Step 4: A conversion into greyscale using the red channel is done followed by thresholding. For further details see Figure 2. Left and right row: Images show the effect of noise reduction (due to droplets on the inside of the covering foil) (left and right) and an enhanced differentiation between parallel growing roots (right) after the corrections and combination of the two images.

distributions were positioned along the longitudinal axis of the artificial root in a way that matched full width at half maximum of the distribution with the root diameter at each position. These artificial root systems were used to compare software performance under optimal contrast between root and background.

Statistics

The variance component of the user interaction experiments were estimated with ASREML-R [74] by setting the factors "user" and "image" as random in a model containing no fixed factor. To estimate the repeatability, we divided the variance of the determined axile and lateral root length within the sampled images (σ^2_{image}) with the overall variance due to image variance plus image-by-user interaction variance (σ^2_{error}).

$$R2 = var_{image} / \left(var_{image} + var_{error}\right)$$

The experiment determining the effect of the fungicide on fungal infection and plant growth was a complete randomized block design with 9 replications. Each experimental unit consisted of one rhizoslide containing one plant. A mixed linear model was calculated in ASREML-R as

$$Y_{ij} = f_i + r_j + \varepsilon_{ij}$$

where Y_{ij} is the ith plants in three leaf stage, number of infected plants or plant biomass in the jth replication, f_i is the fungicide concentration (i = no fungicide, 2.5 g/L or 5 g/l Captan), r_j is the replication (j = 1, ..., 9), and εij is the residual error. The factor replication was set as random.

Additional files

Additional file 1: Separation of embryonic and crown roots based on the paper sandwich method (Figure 1C and D). Black bars indicate crown roots growing on top of the paper surface and grey bars indicate crown roots growing under the paper. A replicate represents one rhizoslide planted with one plant.

Additional file 2: Fungicide effect. Vigor traits measured for plants treated with Captan (2.5 g L^{-1} or 5 g L^{-1}) and plants with no fungicide (control). N = 9. Significance level $p < 0.001$(***); $p < 0.01$(**); $p < 0.05$(*); $p < 0.1$(.).

Additional file 3: Images of Chromelosporium fulvum. A: Image taken with a consumer camera, **B:** magnifying glass 33 times magnified, **C:** microscope 1000 times magnified.

Additional file 4: Channel separation and NIR lighting. A: Comparison of the conversion into greyscale images using either the blue, green or red channel or all three channels (grey). **B**: Images taken of roots growing on steel blue, Whatman blue, Whatman white or Sebio grey germination paper using NIR front or backlighting.

Additional file 5: List of programs suitable for 2D root system analysis. List includes relevant root traits accessible with these programs.

Additional file 6: Thresholding done with four different thresholding routines. In WinRhizo either an automatic selection or a manual adaptation of the tonal value was chosen. In GiARoots the adaptive threshold and the double adaptive threshold were tested. Red circles indicate the loss of lateral roots due to segmentation and blue circles the difficulties to separate parallel growing roots.

Additional file 7: Correlation of total root length, surface and diameter between WinRhizo, SmartRoot and GiARoots. A: Based on original images of the roots. **B**: Based on images of the artificial roots.

Additional file 8: Time costs for a topology analysis using WinRhizo or SmartRoot. The analysis was divided into four steps and after every step the time was recorded. The classification into steps is not identical for both software packages as their procedures were different, but as close as possible. WinRhizo: 1. Step: Automatic analysis of the image and setting of the segmentation threshold, 2. Step: Excluding non-volitional regions, 3. Step: Combining and Cutting of root fragments and 4. Step: Allocation of the root order. SmartRoot: 1. Step: Automatic labeling of seminal roots, 2. Step: Manual correction of seminal roots, 3. Step: Automatic labeling of lateral roots and 4.Step: Manual correction of lateral roots. Data points are mean values (n = 5) ± standard deviation).

Additional file 9: Root traits measured with WinRhizo, SmartRoot and GiARoot listed for one exemplary root. Root traits were either calculated by the program (direct) or calculated by the user based on the output (indirect). In WinRhizo the length of a single root must be calculated. Therefore, all traits based on single root measurements could not be directly extracted from the output file. All these traits are marked with "*".

Additional file 10: Modified output tables of WinRhizo (A) and SmartRoot (B). A: WinRhizo divides the root system in so called axis and links. A link is a segment on a root 0th order or a 1st order root. An axis is a group of connected links. The yellow highlighted parts are informations about the links and the orange regions about the axis. Green highlighted is the summarizing section. In the second column appears operator (axis, link or summary (DEV-) and in the following columns traits describing this segments are listed. **B**: SmartRoot organizes the data based on roots. In the second column appears the root notation chosen by the user and in the following columns traits describing this root are listed. Values in **A** and **B** are measured for an exemplary root system.

Additional file 11: Constructional drawing of the rhizoslides.

Additional file 12: Sandwich method to separate embryonic and crown roots. A: Granulate substrate to avoid incidence of light. **B**: Potassium polycarbonate on top of the seed to moisten the seed and the mesocotyl. **C**: Crown roots emerging on top of the germination paper edge surrounded by Potassium polycarbonate.

Additional file 13: Diagram of circuit connected to Arduino. Camera focus and release are controlled by an optocoupler as well as right and left illumination. The illumination is connected to a relay to switch the higher current of the LEDs.

Abbreviations
RSA: Root system architecture; QTL: Quantitative trait loci; NIR: Near-infrared reflectance.

Competing interests
The authors declare that they have no competing interests.

Authors' contributions
The method was conceived by CLM and AH. The development was supported by AW and NK. DM did the software evaluation, analysis and tests. All scripts necessary for imaging and image processing were written by NK. Experiments and data analysis were performed by CLM. Technical equipment was developed and constructed by AH, NK and CLM. All authors read and approved the final manuscript.

Acknowledgements
The authors thank Ueli Merz and Adrian Leuchtmann (ETH Zurich) for their help identifying the fungi colonizing our system; Alexander Gogos (Agroscope, Zurich) for his support with the hyperspectral microscopy; Pascal Schaad (Sebio GmbH, Germany) for supplying custom sized samples of germination paper; Alexandra Metzger (ETH Zurich) for technical assistance; Xavier Draye (Université catholique Louvain, Belgium) for his support with NIR lighting; Guillaume Lobet, (Université Liège, Belgium) for his support with the analysis in SmartRoot and Patrick Flütsch (ETH Zurich) for manufacturing the rhizoslides.
This research received funding from the European Community Seventh Framework Programme FP7-KBBE-2011-5 under grant agreement no.289300.

References
1. Lynch JP: **Turner Review No. 14 Roots of the second green revolution.** *Aust J Bot* 2007, **55:**493–512.
2. Springer: *High-Throughput Phenotyping in Plants.* Heidelberg: Humana Press; 2012.
3. Rafalski JA: **Association genetics in crop improvement.** *Curr Opin Plant Biol* 2010, **13:**174–180.

4. Hochholdinger F, Park WJ, Sauer M, Woll K: **From weeds to crops: genetic analysis of root development in cereals.** *Trends Plant Sci* 2004, **9:**42–48.
5. Hund A, Reimer R, Messmer R: **A consensus map of QTLs controlling the root length of maize.** *Plant Soil* 2011, **344:**143–158.
6. Singh V, van Oosterom EJ, Jordan DR, Messina CD, Cooper M, Hammer GL: **Morphological and architectural development of root systems in sorghum and maize.** *Plant Soil* 2010, **333:**287–299.
7. Clark RT, Famoso AN, Zhao K, Shaff JE, Craft EJ, Bustamante CD, McCouch SR, Aneshansley DJ, Kochian LV: **High-throughput 2D root system phenotyping platform facilitates genetic analysis of root growth and development.** *Plant Cell Environ* 2013, **36:**454–466.
8. Babé A, Lavigne T, Séverin JP, Nagel KA, Walter A, Chaumont F, Batoko H, Beeckman T, Draye X: **Repression of early lateral root initiation events by transient water deficit in barley and maize.** *Philos Trans R Soc Lond B Biol Sci* 2012, **367:**1534–1541.
9. Ingram PA, Zhu J, Shariff A, Davis IW, Benfey PN, Elich T: **High-throughput imaging and analysis of root system architecture in Brachypodium distachyon under differential nutrient availability.** *Philos Trans R Soc Lond B Biol Sci* 2012, **367:**1559–1569.
10. Hund A, Trachsel S, Stamp P: **Growth of axile and lateral roots of maize: I development of a phenotying platform.** *Plant Soil* 2009, **325:**335–349.
11. Smith S, De Smet I: **Root system architecture: insights from Arabidopsis and cereal crops.** *Philos Trans R Soc Lond B Biol Sci* 2012, **367:**1441–1452.
12. Hochholdinger F, Woll K, Sauer M, Dembinsky D: **Genetic dissection of root formation in maize (Zea mays) reveals root-type specific developmental programmes.** *Ann Bot-London* 2004, **93:**359–368.
13. Smit AL, Groenwold J: **Root characteristics of selected field crops: data from the Wageningen Rhizolab (1990–2002).** *Plant Soil* 2005, **272:**365–384.
14. Kuchenbuch RO, Ingram KT, Buczko U: **Effects of decreasing soil water content on seminal and lateral roots of young maize plants.** *Journal of Plant Nutrition and Soil Science-Zeitschrift Fur Pflanzenernahrung Und Bodenkunde* 2006, **169:**841–848.
15. Mooney SJ, Pridmore TP, Helliwell J, Bennett MJ: **Developing X-ray computed tomography to non-invasively image 3-D root systems architecture in soil.** *Plant Soil* 2012, **352:**1–22.
16. Jahnke S, Menzel MI, van Dusschoten D, Roeb GW, Buhler J, Minwuyelet S, Blumler P, Temperton VM, Hombach T, Streun M, Jahnke S, Menzel MI, van Dusschoten D, Roeb GW, Buhler J, Minwuyelet S, Blumler P, Temperton VM, Hombach T, Streun M, Beer S, Khodaverdi M, Ziemons K, Coenen HH, Schurr U: **Combined MRI-PET dissects dynamic changes in plant structures and functions.** *Plant J* 2009, **59:**634–644.
17. Nagel KA, Putz A, Gilmer F, Heinz K, Fischbach A, Pfeifer J, Faget M, Blossfeld S, Ernst M, Dimaki C: **GROWSCREEN-Rhizo is a novel phenotyping robot enabling simultaneous measurements of root and shoot growth for plants grown in soil-filled rhizotrons.** *Funct Plant Biol* 2012, **39:**891–904.
18. Famoso AN, Clark RT, Shaff JE, Craft E, McCouch SR, Kochian LV: **Development of a novel aluminum tolerance phenotyping platform used for comparisons of cereal aluminum tolerance and investigations into rice aluminum tolerance mechanisms.** *Plant Physiol* 2010, **153:**1678–1691.
19. Nagel KA, Kastenholz B, Jahnke S, Van Dusschoten D, Aach T, Muehlich M, Truhn D, Scharr H, Terjung S, Walter A, Schurr U: **Temperature responses of roots: impact on growth, root system architecture and implications for phenotyping.** *Funct Plant Biol* 2009, **36:**947–959.
20. Hund A, Reimer R, Stamp P, Walter A: **Can we improve heterosis for root growth of maize by selecting parental inbred lines with different temperature behaviour?** *Philos Trans R Soc Lond B Biol Sci* 2012, **367:**1580–1588.
21. Chen YL, Dunbabin VM, Diggle AJ, Siddique KHM, Rengel Z: **Development of a novel semi-hydroponic phenotyping system for studying root architecture.** *Funct Plant Biol* 2011, **38:**355–363.
22. Planchamp C, Balmer D, Hund A, Mauch-Mani B: **A soil-free root observation system for the study of root-microorganism interactions in maize.** *Plant Soil* 2013, **367:**605–614.
23. Reimer R, Stich B, Melchinger AE, Schrag TA, Sørensen AP, Stamp P, Hund A: **Root response to temperature extremes: association mapping of temperate maize (Zea mays L).** *Maydica* 2013, **58:**156–168.
24. Ruta N, Liedgens M, Fracheboud Y, Stamp P, Hund A: **QTLs for the elongation of axile and lateral roots of maize in response to low water potential.** *Theor Appl Genet* 2009, **120:**621–631.
25. Ruta N, Stamp P, Liedgens M, Fracheboud Y, Hund A: **Collocations of QTLs for seedling traits and yield components of tropical maize under water stress conditions.** *Crop Sci* 2010, **50:**1385–1392.
26. Liao H, Rubio G, Yan X, Cao A, Brown KM, Lynch JP: **Effect of phosphorus availability on basal root shallowness in common bean.** *Plant Soil* 2001, **232:**69–79.
27. Zhu J, Kaeppler SM, Lynch JP: **Mapping of QTLs for lateral root branching and length in maize (Zea mays L.) under differential phosphorus supply.** *Theor Appl Genet* 2005, **111:**688–695.
28. Trachsel S, Stamp P, Hund A: **Effect of high temperatures, drought and aluminum toxicity on root growth of tropical maize (*Zea Mays L.*) seedlings.** *Maydica* 2010, **55:**249–260.
29. Flint LH: **Light and the elongation of the mesocotyl in corn.** *Plant Physiol* 1944, **19:**537–543.
30. Zobel RW: **Hardware and software efficacy in assessment of fine root diameter distributions.** *Comput Electron Agr* 2008, **60:**178–189.
31. Yang C, Everitt JH, Fernandez CJ: **Comparison of airborne multispectral and hyperspectral imagery for mapping cotton root rot.** *Biosyst Eng* 2010, **107:**131–139.
32. Nakaji T, Noguchi K, Oguma H: **Classification of rhizosphere components using visible–near infrared spectral images.** *Plant Soil* 2008, **310:**245–261.
33. Armengaud P, Zambaux K, Hills A, Sulpice R, Pattison RJ, Blatt MR, Amtmann A: **EZ-Rhizo: integrated software for the fast and accurate measurement of root system architecture.** *Plant J* 2009, **57:**945–956.
34. Burton AL, Williams M, Lynch JP, Brown KM: RS: **Software for high-throughput analysis of root anatomical traits.** *Plant Soil* 2012, **357:**189–203.
35. Clark RT, MacCurdy RB, Jung JK, Shaff JE, McCouch SR, Aneshansley DJ, Kochian LV: **Three-dimensional root phenotyping with a novel imaging and software platform.** *Plant Physiol* 2011, **156:**455–465.
36. French A, Ubeda-Tomás S, Holman TJ, Bennett MJ, Pridmore T: **High-throughput quantification of root growth using a novel image-analysis tool.** *Plant Physiol* 2009, **150:**1784–1795.
37. Galkovskyi T, Mileyko Y, Bucksch A, Moore B, Symonova O, Price C, Topp C, Iyer-Pascuzzi A, Zurek P, Fang S: **GiA Roots: software for the high throughput analysis of plant root system architecture.** *BMC Plant Biol* 2012, **12:**116.
38. Regent Instruments Inc: *WinRhizo.* Canada: Regent Instruments Inc; 2009.
39. Le Bot J, Serra V, Fabre J, Draye X, Adamowicz S, Pages L: **DART: a software to analyse root system architecture and development from captured images.** *Plant Soil* 2010, **326:**261–273.
40. Lobet G, Pages L, Draye X: **A novel image-analysis toolbox enabling quantitative analysis of root system architecture.** *Plant Physiol* 2011, **157:**29–39.
41. Mairhofer S, Zappala S, Tracy SR, Sturrock C, Bennett M, Mooney SJ, Pridmore T: **RooTrak: automated recovery of three-dimensional plant root architecture in soil from X-ray microcomputed tomography images using visual tracking.** *Plant Physiol* 2012, **158:**561–569.
42. Naeem A, French AP, Wells DM, Pridmore TP: **High-throughput feature counting and measurement of roots.** *Bioinformatics* 2011, **27:**1337–1338.
43. Pierret A, Gonkhamdee S, Jourdan C, Maeght J-L: **IJ_Rhizo: an open-source software to measure scanned images of root samples.** *Plant Soil* 2013, **373:**531–539.
44. Pound MP, French AP, Atkinson J, Wells DM, Bennett MJ, Pridmore TP: **RootNav: navigating images of complex root architectures.** *Plant Physiol* 2013, **162:**1802–1814.
45. Stefanelli D, Fridman Y, Perry RL: **DigiRoot (TM): new software for root studies.** *Eur J Hortic Sci* 2009, **74:**169–174.
46. Lobet G, Draye X, Périlleux C: **An online database for plant image analysis software tools.** *Plant Methods* 2013, **9:**38.
47. Clark RT, Shaff J, Aneshansley DJ, Kochian L: **Advances to Whole Root System Quantification Tools and Techniques.** In *The Proceedings of the International Plant Nutrition Colloquium XVI.* UC Davis: Department of Plant Sciences; 2009. Retrieved from: http://www.escholarship.org/uc/item/40v4x53m.
48. Stefanelli D, Fridman Y, Perry RL: **DigiRoot (TM): new software for root studies.** *European Journal of Horticultural Science* 2009, **74:**169–174.
49. Yu P, Li X, Yuan L, Li C: **A novel morphological response of maize (Zea mays) adult roots to heterogeneous nitrate supply revealed by a split-root experiment.** *Physiol Plantarum* 2014, **150:**133–144.
50. Pfeifer J, Faget M, Walter A, Blossfeld S, Fiorani F, Schurr U, Nagel K: **Spring barley shows dynamic compensatory root and shoot growth**

responses when exposed to localized soil compaction and fertilization. *Funct Plant Biol* 2014, **41**:581–597.
51. Drew M, Saker L: **Nutrient supply and the growth of the seminal root system in barley II. localized, compensatory increases in lateral root growth and rates of nitrate uptake when nitrate supply is restricted to only part of the root system.** *J Exp Bot* 1975, **26**:79–90.
52. Walter A, Silk WK, Schurr U: **Environmental effects on spatial and temporal patterns of leaf and root growth.** *Annu Rev Plant Biol* 2009, **60**:279–304.
53. Zhu J, Kaeppler SM, Lynch JP: **Mapping of QTL controlling root hair length in maize (Zea mays L.) under phosphorus deficiency.** *Plant Soil* 2005, **270**:299–310.
54. Mollier A, Pellerin S: **Maize root system growth and development as influenced by phosphorus deficiency.** *J Exp Bot* 1999, **50**:487.
55. Liu J, Li J, Chen F, Zhang F, Ren T, Zhuang Z, Mi G: **Mapping QTLs for root traits under different nitrate levels at the seedling stage in maize (Zea mays L.).** *Plant Soil* 2008, **305**:253–265.
56. Watt M, Moosavi S, Cunningham SC, Kirkegaard JA, Rebetzke GJ, Richards RA: **A rapid, controlled-environment seedling root screen for wheat correlates well with rooting depths at vegetative, but not reproductive, stages at two field sites.** *Ann Bot-London* 2013, **112**:447–455.
57. Rahnama A, Munns R, Poustini K, Watt M: **A screening method to identify genetic variation in root growth response to a salinity gradient.** *J Exp Bot* 2011, **62**:69–77.
58. Almgren I, Gustafsson M, AS F, ält, Lindgren H, Liljeroth E: **Interaction between root and leaf disease development in barley cultivars after inoculation with different isolates of Bipolaris sorokiniana.** *J Phytopathol* 1999, **147**:331–337.
59. Bonser AM, Lynch J, Snapp S: **Effect of phosphorus deficiency on growth angle of basal roots in Phaseolus vulgaris.** *New Phytol* 1996, **132**:281–288.
60. Blossfeld S, Schreiber CM, Liebsch G, Kuhn AJ, Hinsinger P: **Quantitative imaging of rhizosphere pH and CO2 dynamics with planar optodes.** *Ann Bot-London* 2013, **112**:267–276.
61. Björn L, Suzuki Y, Nilsson J: **Influence of wavelength on the light response of excised wheat roots.** *Physiol Plantarum* 1963, **16**:132–141.
62. Feldman LJ, Briggs WR: **Light-regulated gravitropism in seedling roots of maize.** *Plant Physiol* 1987, **83**:241.
63. Walter A, Silk WK, Schurr U: **Effect of soil pH on growth and cation deposition in the root tip of Zea mays L.** *J Plant Growth Regul* 2000, **19**:65–76.
64. Ruts T, Matsubara S, Wiese-Klinkenberg A, Walter A: **Diel patterns of leaf and root growth: endogenous rhythmicity or environmental response?** *J Exp Bot* 2012, **63**:3339–3351.
65. Hund A, Fracheboud Y, Soldati A, Frascaroli E, Salvi S, Stamp P: **QTL controlling root and shoot traits of maize seedlings under cold stress.** *Theor Appl Genet* 2004, **109**:618–629.
66. Trachsel S, Messmer R, Stamp P, Hund A: **Mapping of QTLs for lateral and axile root growth of tropical maize.** *Theor Appl Genet* 2009, **119**:1413–1424.
67. Ruta N, Liedgens M, Fracheboud Y, Stamp P, Hund A: **QTLs for the elongation of axile and lateral roots of maize in response to low water potential.** *Theor Appl Genet* 2010, **120**:621–631.
68. Sorgonà A, Abenavoli M, Gringeri P, Lupini A, Cacco G: **Root architecture plasticity of citrus rootstocks in response to nitrate availability.** *J Plant Nutr* 2007, **30**:1921–1932.
69. Sorgonà A, Abenavoli MR, Cacco G: **A comparative study between two citrus rootstocks: effect of nitrate on the root morpho-topology and net nitrate uptake.** *Plant Soil* 2005, **270**:257–267.
70. Liao M, Fillery IRP, Palta JA: **Early vigorous growth is a major factor influencing nitrogen uptake in wheat.** *Funct Plant Biol* 2004, **31**:121–129.
71. Trachsel S, Stamp P, Hund A: **Growth of axile and lateral roots of maize: response to desiccation stress induced by polyethylene glycol 8000.** *Maydica* 2010, **55**:101–109.
72. Freund H: **Sterilisieren der fertigen Ampullen.** In *Die Ampullenfabrikation.* Berlin Heidelberg: Springer; 1916:61–72.
73. Bohn M, Novais J, Fonseca R, Tuberosa R, Grift T: **Genetic evaluation of root complexity in maize.** *Acta Agronomica Hungarica* 2006, **54**:291–303.
74. Butler D: **asreml: asreml() fits the linear mixed model. R package version 3.0.** 2009. http://www.vsni.co.uk.

12

An ontology approach to comparative phenomics in plants

Anika Oellrich[1†], Ramona L Walls[2†], Ethalinda KS Cannon[3], Steven B Cannon[4,5], Laurel Cooper[6], Jack Gardiner[7], Georgios V Gkoutos[8], Lisa Harper[4], Mingze He[7], Robert Hoehndorf[9], Pankaj Jaiswal[6], Scott R Kalberer[4], John P Lloyd[10], David Meinke[11], Naama Menda[12], Laura Moore[6], Rex T Nelson[4], Anuradha Pujar[12], Carolyn J Lawrence[5,7*] and Eva Huala[13*]

Abstract

Background: Plant phenotype datasets include many different types of data, formats, and terms from specialized vocabularies. Because these datasets were designed for different audiences, they frequently contain language and details tailored to investigators with different research objectives and backgrounds. Although phenotype comparisons across datasets have long been possible on a small scale, comprehensive queries and analyses that span a broad set of reference species, research disciplines, and knowledge domains continue to be severely limited by the absence of a common semantic framework.

Results: We developed a workflow to curate and standardize existing phenotype datasets for six plant species, encompassing both model species and crop plants with established genetic resources. Our effort focused on mutant phenotypes associated with genes of known sequence in *Arabidopsis thaliana* (L.) Heynh. (Arabidopsis), *Zea mays* L. subsp. *mays* (maize), *Medicago truncatula* Gaertn. (barrel medic or Medicago), *Oryza sativa* L. (rice), *Glycine max* (L.) Merr. (soybean), and *Solanum lycopersicum* L. (tomato). We applied the same ontologies, annotation standards, formats, and best practices across all six species, thereby ensuring that the shared dataset could be used for cross-species querying and semantic similarity analyses. Curated phenotypes were first converted into a common format using taxonomically broad ontologies such as the Plant Ontology, Gene Ontology, and Phenotype and Trait Ontology. We then compared ontology-based phenotypic descriptions with an existing classification system for plant phenotypes and evaluated our semantic similarity dataset for its ability to enhance predictions of gene families, protein functions, and shared metabolic pathways that underlie informative plant phenotypes.

Conclusions: The use of ontologies, annotation standards, shared formats, and best practices for cross-taxon phenotype data analyses represents a novel approach to plant phenomics that enhances the utility of model genetic organisms and can be readily applied to species with fewer genetic resources and less well-characterized genomes. In addition, these tools should enhance future efforts to explore the relationships among phenotypic similarity, gene function, and sequence similarity in plants, and to make genotype-to-phenotype predictions relevant to plant biology, crop improvement, and potentially even human health.

* Correspondence: triffid@iastate.edu; huala@phoenixbioinformatics.org
†Equal contributors
[7]Department of Genetics, Development and Cell Biology, Roy J Carver Co-Laboratory, Iowa State University, Ames, IA 50010, USA
[13]Phoenix Bioinformatics, 643 Bair Island Rd Suite 403, Redwood City, CA 94063, USA
Full list of author information is available at the end of the article

Background

Plant phenotypic variation constitutes the raw material for much of plant biology, including research on gene function in model species, breeding of desirable crop varieties, functional investigations from the cellular to ecosystem scale, and inference about the evolution and ecology of both plants and the species that interact with them. Disentangling the relationships among genotypes, phenotypes, and the environment is one of the grand challenges of contemporary biology [1], yet this endeavor is severely limited by our ability to collect, integrate, and systematically analyze phenotypic data [2]. Researchers generally use free text to describe phenotypes, which allows for rich descriptions, but makes it hard to compare phenotypes across species, integrate data into the existing knowledge landscape, or derive information from combined datasets [3]. In recent years, ontologies have become powerful tools for working with phenotypic data, particularly in biomedicine, because standardizing terminology across species and sub-disciplines enables inference based on logical relationships [4-6]. Here we present a new approach to studying plant phenotypes modeled on recent advances in the use of ontologies in biomedical research on animal model systems.

Throughout this paper, we use the words phenotype, phene, and phenome with precise meanings. A 'phenotype' is the composite set of one or more observable characteristics associated with a given organism or cell, that results from the interaction of the genotype and the environment [7,8]. The separate characteristics that make up a phenotype are termed 'phenes' [9,10]. For example, in maize, a *dwarf* phenotype can be defined as a composite of the phenes 'reduced internode length' and 'compact, broad leaves'. Phenes relate to 'phenomes' in the way that genes relate to genomes: an organism's or species' phenome is composed of the complete set of its phenes. Phenomics, therefore, is the study of all phenotypes associated with an organism or species (i.e. its phenotype space). In correspondence with Genome Wide Association Studies (GWAS), Phenome Wide Association Studies (PheWAS) associate a gene with a variety of phenes or phenotypes, which is particularly relevant for genes that have a pleiotropic effect [11].

Biomedical scientists have developed and utilized phenotype ontologies and ontological reasoning to support comparative and predictive phenomics [12,13]. Phenotype ontologies are controlled, hierarchically-related phenotypic descriptions that enable large-scale computation among individuals, populations, and even multiple species [14]. A number of vocabularies and pre-composed phenotype ontologies (in which terms are pre-defined) have been developed for specific taxa or applications [15-18], but comparison across datasets or among different species requires an extensive alignment process whenever different vocabularies/ontologies are used to represent the data. An alternative to phenotype ontology alignment is the use of post-composed phenotypes, in which all the elements of a phenotype are explicitly logically defined or "composed" from existing terms from species-independent ontologies [16]. One method of post-composing a phenotype description is to first break it down into its component phenes, and then define an affected Entity (E) and a describing Quality (Q) for each phene [19,20]. In this method of post-composing phenotypes, Entity-Quality (EQ) statements are composed for all phenes under considerations, and the entire set of phenes is reasoned over simultaneously. Finally, to derive novel insights from curated genotype and phenotype data, semantic similarity measures are applied, based on a consistent ontological representation [21-23].

This approach has been applied successfully to mammalian phenotypes to predict gene function across species, as well as disease, drug, or pathway involvement of genes [5,12,13,24,25]. Two major limitations to adopting a similar approach in plants are the lack of phenotype data curated with species-neutral ontology terms, and the need for standards for creating EQ statements to describe plant phenotypes. Nonetheless, two important existing resources are available to support post-composed ontology analysis of plant phenotype data: 1) well-developed ontologies for plant science [26], particularly the Plant Ontology (PO) [27] and Gene Ontology (GO) [28,29]; 2) curated sets of mutant phenotype descriptions for multiple plant species in model-organism and crop databases such as MaizeGDB [30,31], Oryzabase [32], Gramene [33,34], and the Sol Genomics Network (SGN) [15,35] as well as in the literature (e.g., [36]). In addition, an intellectual framework for logically defining plant traits has been developed in the Plant Trait Ontology (TO) [33].

To push the field of plant phenomics forward, it is clear that there is a need for additional high-quality phenotype descriptions generated by research, as well as for high-confidence predictions of phenotypic associations among equivalent phenotypes, both across species and between phenotypes and their causative genotypic variants and environments. Here we describe how our work to translate existing high-quality phenotypic descriptions across six plant species enabled the prediction of phenotypic associations. Furthermore, we demonstrate that additional curation of such data into ontological representations can expand the phenotypic predictive capacity of plant sciences.

This paper includes methodology, as well as an initial dataset that was used to test and refine the methodology. In brief, we compiled EQ statements for 1,742 phenes from 2,747 genes and gene models in six plant species [*Arabidopsis thaliana* (Arabidopsis), *Zea mays* ssp. *mays* (maize), *Medicago truncatula* (barrel medic or Medicago),

Oryza sativa (rice), *Glycine max* (soybean), and *Solanum lycopersicum* (tomato)] and applied consistency checks to ensure a high-quality phenotype annotation set. The annotated phenotype data set was subjected to an automated semantic similarity analysis, based on PhenomeNET [37,38], and the results are available in a separate plant instance [39]. The semantic similarity dataset was evaluated for its ability to enhance predictions of gene families, gene functions, and shared metabolic pathways across the six species and compared to an existing classification of plant phenotypes [36].

Results and discussion

A method for describing phenotypes with a common semantic representation across six plant species

We include in the Results a brief description of our method, because this is the first report outlining this type of analysis of phenotypes across multiple reference species in plants. For this analysis we limited our species set to the model/crop species Arabidopsis, maize, Medicago, rice, soybean, and tomato, to take advantage of the existing data for these species. However, the method could be applied to any plant for which there are characterized mutant phenotypes associated with sequenced genes. To maximize the ability to compare both phenotypes and genotypes across species, we used only genotypes for which the sequence was known and made efforts to limit our datasets to phenotypes resulting from mutations to a single gene. For genes where phenotype information was available for different alleles, we counted each allele as a separate genotype (Additional file 1). For each species, the authors with the most relevant expertise selected free text phenotype descriptions for inclusion using methods specific to that species (see Methods). The number of genotypes analyzed varied widely among species (maximum 2,393 in Arabidopsis, minimum 30 in soybean), reflecting the availability of phenotypic descriptions for each species.

We first decomposed each free text phenotype description into a set of simple atomized statements corresponding to each component, or "phene", of the phenotype. We then translated each of these components into an EQ (Entity-Quality) statement (Figure 1). As with EQ statements previously developed for mammalian species, we distinguish between structural phenotypes, such as "short plant", and process phenotypes, such as "late flowering" [40]. In a structural phenotype, the Entity is an affected part of the plant, represented with a term from the Plant Ontology (PO) [27] or Gene Ontology (GO) *cellular component* branch [29]. In a process phenotype, the Entity is an altered process represented with a term from the GO *biological process* branch. In both cases, the manner in which the entity is affected was described using Quality terms from the Phenotype and Trait Ontology (PATO) [16]. For example, the atomized statement "short leaves" can be expressed as: Entity = *vascular leaf* from the Plant Ontology (PO:0009025)[a] and Quality = *decreased length* from the Phenotype and Trait Ontology (PATO:0000574).

We found that many of the phenes required more complex EQ statements and terms from additional ontologies to fully represent their meaning. For example, the phene "lack of anthocyanins in aleurone" could be expressed as: *aleurone layer* (PO:0005360) *lacks parts or has fewer parts of type* (PATO:0001999) *anthocyanins* (CHEBI:38697), with the form:

primary E1 – Q – secondary E1

where Q is a relational quality. In addition, we found that the Entities themselves may be complex. For example, an EQ statement for the free text "silks are green" would be phrased: *style* (PO:0009074) *part_of* (BFO:0000050)[b] *ear inflorescence* (PO:0020136) *green* (PATO:0000320), with the form:

[primary E1 – R – primary E2] – Q

where R is a relation from the Relation Ontology (RO) [41,42]. All EQ statements in the dataset could be parsed using the generalized formula:

[(primary E1) – R – (primary E2)] – [Q – QL]
– [(secondary E1) – R –(secondary E2)]

where QL is a qualifier to the quality Q. At a minimum, we required that a primary E1 and Q be present, and any of the other elements were optional. We found that all phenotypes in our dataset could be described with the ontologies listed in Table 1, although we recognize that our dataset does not encompass the entire breadth of possible plant phenotypes, and additional ontologies and development of existing ontologies will be needed to annotate more diverse phenotypes.

Because a phenotype consists of one or more phenes, one or more EQ statements were used to describe it. For example, the phenotype "corngrass" in maize is described as "narrow leaves, extreme tillering, highly reduced ears and tassel; grasslike growth habit, often vegetative leaves in the ear and tassel". This phenotype was broken down into 5 phenes, which were annotated with EQ statements (Figure 1). Likewise, an individual EQ statement can be used to describe more than one phenotype. In the corngrass example, the EQ statement that describes the "narrow leaf" phene is also used in several other phenotypes, such as "narrowleaf" (Additional file 1). Our approach considers each "phenotype" as the sum of its individual EQ statements.

This method allows for highly detailed and species-independent descriptions of phenotypes, but still has several limitations. Creating accurate EQ statements requires knowledge of both the species-specific phenotypes as well

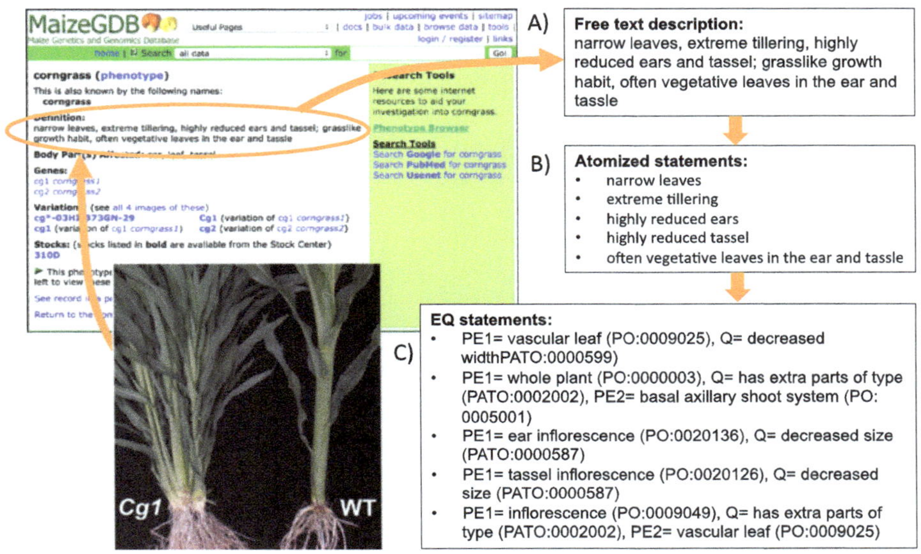

Figure 1 The method applied to annotate mutant phenotypes from textual descriptions. Textual descriptions from the literature or databases **(A)**, based on observations of mutant plants, are first broken down into atomized statements corresponding to phenes **(B)** that are then represented with EQ statements **(C)**.

as the ontologies used to describe them. Brief phenotype descriptions may be available in databases (e.g., MaizeGDB, SGN), but much more complete and accurate descriptions of mutant phenotypes are spread over many publications spanning several years. Collecting and extracting phenotype information is very labor intensive. In addition, EQ statements are created by curators and thus still reflect a certain amount of subjectivity. It is often possible to build more than one EQ statement for the same textual description, and it is not always clear when to use a process entity versus a structural entity. In this project, we established a strict set of rules and protocols and held regular meetings to help ensure consistent construction of comparable EQ statements across species (see Methods). In the future, we would like to evaluate the importance of consistently structured EQ statements for analyzing semantic similarity, and determine whether some variation can be tolerated.

An ontology-based dataset of mutant phenotypes for six reference plant species

The complete list of genes, genotypes, phenotypes, atomized statements, and EQ statements can be found in Additional file 1, which is also included as part of the complete dataset housed in the iPlant Data Commons [44]. The largest set of annotations came from Arabidopsis, followed by maize, rice, and tomato (Table 2). The low numbers of annotations for Medicago and soybean reflect

Table 1 Description of applied ontologies

Ontology	Content	Link
Plant Ontology (PO) [27]	Plant anatomy and morphology and development stages	http://www.plantontology.org/
Gene Ontology (GO) [29]	Biological processes, cellular components and molecular functions	http://geneontology.org/
Chemical Entities of Biological Interest ontology (ChEBI) [43]	Molecular entities focused on 'small' chemical compounds.	http://www.ebi.ac.uk/chebi/
Phenotypic Qualities Ontology (PATO) [16,19]	Phenotypic qualities	http://obofoundry.org/wiki/index.php/PATO:Main_Page
Plant Experimental Conditions Ontology (EO)	Treatments, growing conditions, and/or study types	http://planteome.org/amigo/cgi-bin/crop_amigo/term_details?term=EO:0007359
NCBI taxonomy (NCBITAXON)	A curated classification and nomenclature for all of the organisms in the public sequence databases.	http://www.ncbi.nlm.nih.gov/taxonomy
Relation Ontology (RO) [41]	Core upper-level relations and biology-specific relations	https://code.google.com/p/obo-relations/

Species-independent ontologies used to form EQ statements. All ontologies were downloaded on 15 March 2014.

Table 2 The number of EQ statements, genes, genotypes, and phenotypes they were associated with, for six plant species

Species	#EQs (phenes)	#unique EQs - all genotypes	#genes	#genotypes	#phenotypes
Arabidopsis thaliana	5172	1260	2393	2393*	1385
Zea mays ssp *mays*	373	180	114	169	117
Oryza sativa L.	340	271	92	95	86
Solanum lycopersicum	269	174	72	128	90
Medicago truncatula	149	99	40	45	40
Glycine max	61	39	30	30*	24
Total	6364	2023	2741	2866	1742

The number of EQ statements, genes, genotypes, and phenotypes they were associated with, for each species.
*#Genotypes equals # genes because no information on alleles was available for these species.

the relatively small number of studies on these species and the scarcity of curated phenotypes. There was little overlap of unique phenotypes among species (Additional file 2). In the following sections, we describe some of types of computational analyses that can be done with this dataset.

Quantitative analysis of pairwise semantic phenotype similarity of genotypes across the entire dataset

To determine pairwise semantic phenotype similarity scores, we used the method described for mammalian genotypes [37] where phenotypes are represented by EQ statements that are then integrated using species-independent ontologies and a semantic similarity measure. That is, every phenotype (which is composed of one of more EQ statements) is compared to every other phenotype, and their similarity within the ontological graph is evaluated. For a pair of phenotypes to receive a score of 1, each phenotype would have to contain the same number of identical (or nearly identical) EQ statements. A score of 0 would mean that none of the EQ statements for either phenotype were similar. For the 8,213,956 possible pairs from the 2,866 genotypes, 548,888 (7%) of the genotype pairs yielded phenotype semantic similarity scores greater than zero. Score distributions, overall and on a per-species basis, are provided in the following sub-sections.

Distribution of similarity scores

We calculated semantic similarity scores for 548,888 genotype pairs in the range of >0 – 1. A similarity score of 0 indicates no semantic overlap with respect to the phenotype, while a similarity score of 1 indicates an identical semantic phenotype description (and therefore equivalent sets of EQs). Figure 2A illustrates the distribution of semantic similarity scores for intra- as well as inter-species genotype pairs. For 13% (71,290) of the genotype pairs possessing a semantic similarity score, the score fell into the range 0.9 – 1 (not including the similarity of a genotype to itself, which is always 1). While 13% seems high, some of the nearly identical scores occur because of the limited availability of phenotype information for many genotypes. For example, if two genotypes are annotated with the same single EQ statement, the result is a semantic similarity score of one, even if in reality those mutant genotypes may have many more phenes that were not recorded. Only known phenes that were already curated from the scientific literature were assigned to genotypes, and our method cannot compensate for gaps in the literature (e.g., due to limitations in biological experiments). As the dataset grows, a better separation of genotypes with respect to their semantic phenotype similarity will be possible.

Almost half (241,042 = 44%) of the non-zero semantic similarity scores are below 0.1, indicating that many of the phenotypes show only a small overlap in their description. For example, the rice mutant DWARF4 (Os03g0227700 [45], allele osdwarf4-1) shows a similarity of 0.08 with the rice mutant MADS18 (Os07g0605200 [45]). This results from both genes being annotated with "dwarf"-related phenotypes but also possessing diverging annotations, such as *increased panicle number* for DWARF4 and *early flowering* for MADS18. It is not surprising that most of the phenotypes show at least some marginal overlap, as this is intrinsic to the aim of the study: making phenotypes comparable. This highlights the potential of the method, but, at the same time, raises the need for consistent, coherent, and complete phenotype annotations in order to computationally replicate the underlying biology and derive accurate predictions.

Although there are more complex scoring mechanisms that take frequency of EQ statements into consideration [23], we applied a Jaccard index that determines the overlap of phenes used in the phenotype descriptions. In an earlier study, it was shown that different types of semantic similarity measures do not differ much as long as the results are interpreted carefully [21]. In future work, we intend to investigate the applicability of alternative scoring methods, in combination with the development of benchmark sets for evaluation purposes.

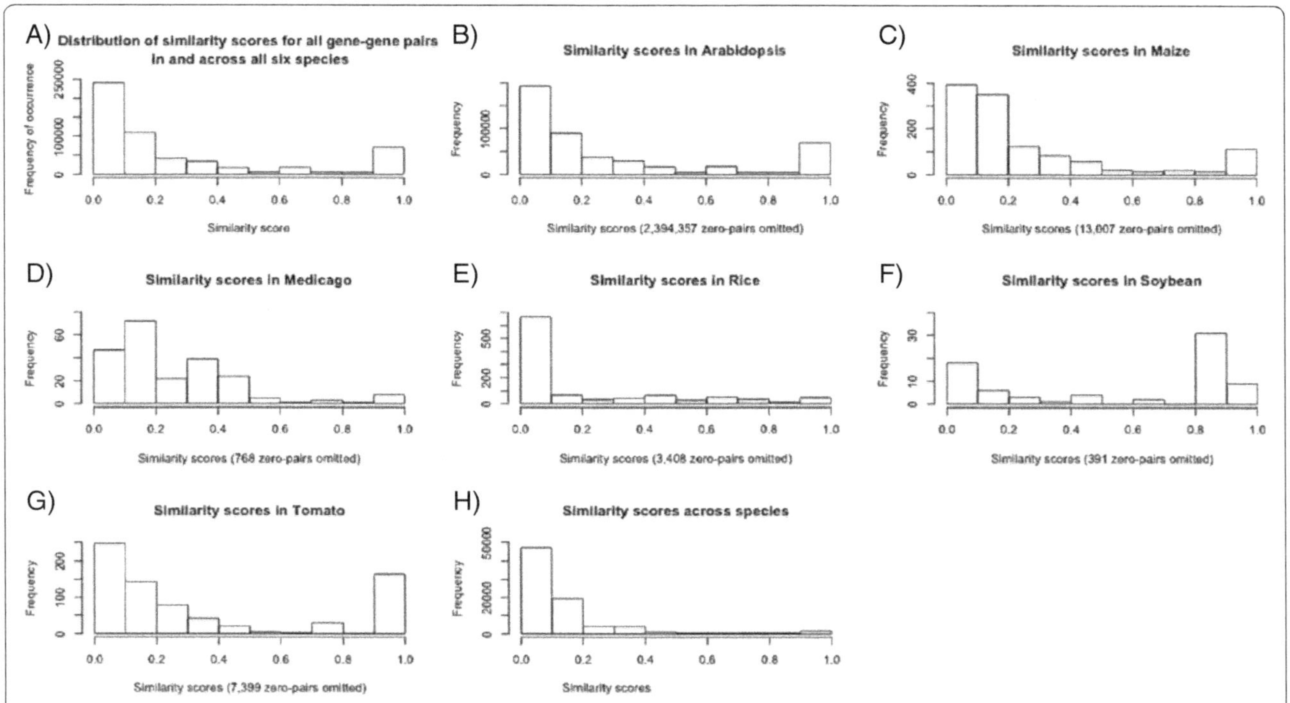

Figure 2 Semantic similarity score distributions for inter- and intraspecific pairwise phenotype similarity. When binning all semantic similarity scores across all species, 44% of semantic similarity scores indicate a relatively low phenotypic overlap between genes (semantic similarity range 0–0.1) while 13% show highly similar phenotypes (similarity score range 0.9-1) **(A)**. Distributions of intraspecific scores (pairwise scores where both genotypes belong to the same species) were similar to the overall distribution of scores **(B-H)**.

Species-specific distribution of scores

To obtain further insights into the distribution of similarity scores, we split similarity scores according to species. If both genotypes that were used to calculate the pairwise similarity score belong to the same species, we recorded the resulting similarity score only for this species. If both genotypes leading to a particular similarity score belong to different species, we recorded it as a cross-species score. The resulting seven similarity score distributions are illustrated in Figure 2B-H. Species-specific score distributions are mostly consistent with the overall score distribution (Figure 2A). There are some differences for soybean and Medicago, but this is likely due to the small sizes of the phenotype annotation sets in these species.

Differences between the semantically-generated phenotype network and a manually derived phenotype grouping

A previous analysis of Arabidopsis used the same set of phenotypes and laid much of the groundwork for this present study [36]. Mutant phenotypes were categorized in a simplified, three-level hierarchy consisting of 4 groups divided into 11 classes and 42 subsets. Each gene was assigned to one of 11 phenotypic classes based on the developmental stage when the phenotype was first observed and what methods and conditions were used to detect it (see more details in Methods). Genes were also assigned to one or more of the 42 subsets, based on the nature of the phenotype (e.g., gametophyte defective, flowering time).

To assess whether our results recapitulate those of [36], we calculated the average similarity scores for each of their classes (higher level grouping) and subsets (lower level grouping). Semantic similarity by class was greater than 0.3 for all classes except Vegetative, and ranged from 0.13 for Vegetative to 0.87 for Chemical and Biological (Additional file 3 and Figure 3). Average semantic similarity scores were lower and more variable across subsets, ranging from 0.10 for GEM (gametophyte, embryo defective) to 0.92 for OBI (other biological interactors), with 25 of 42 subsets having average scores less than 0.3 (Figure 3). Although there were indeed several classes and subsets that had good concordance with the semantic similarity scores, in general, semantic similarity scores within both classes and subsets were low (less than 0.5).

There could be several reasons for low semantic similarity scores within classes or subsets, such as annotations that are not ideally defined, a poor choice of semantic scoring mechanism, or classes/subsets that are too broadly

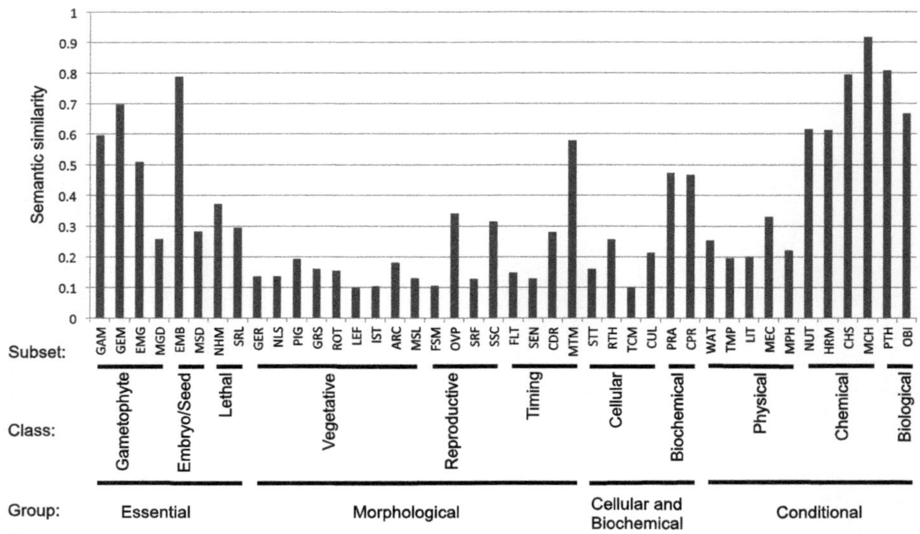

Figure 3 Average semanitic similarity scores for previously derived groupings of Arabidopsis genotypes. The average pairwise semantic similarity for subsets previously identified by [36] ranged from ~0.1 to ~0.9. Subsets are shown grouped by the classes and groups to which they belong.

defined and therefore contain a large variety of phenotypes. In general, we expected pairs of genes within the same subset to have lower semantic similarity scores than pairs of genes within the same class, because genes can belong to multiple subsets, but only a single class. If the phenotype of a gene has multiple phenes, that gene should to belong to multiple subsets, and unless two genes share all of the same phenes (and therefore belong to all of the same subsets), they would have a relatively low similarity score within each subset. Genes in the same class may have somewhat higher similarity scores, because classes aggregate several phenotypic subsets (Figure 3). However, they do not aggregate subsets from other classes (as semantic similarity does) and thus are unlikely to completely mirror semantic similarity scores.

Examination of two contrasting subsets, Flowering time (FLT) and Pathogens/Herbivores (PTH), can help to explain some of the agreement or disagreement between membership in a class or subset and degree of semantic similarity (Figure 3). The PTH subset seems to be more coherent with respect to phenotype annotations than the other groups, which suggests that PTH genes are not documented as having pleiotropic effects. In contrast, pairs of genes in the FLT subset have low average semantic similarity, suggesting that these genes are highly pleiotropic. Consistent with this, the PTH subset genes have on average 1.68 phenes whereas genes in the FLT subset have on average 3.99 phenes.

The categorical system devised by [36] has the distinct advantages of being more intuitive and not requiring an understanding of ontologies to make annotations or carry out an analysis of the data. However, the disadvantages are that category boundaries are sometimes somewhat arbitrary, very disparate phenotypes may be included in a single category (e.g., miscellaneous categories), and each phenotype may be forced into a single class. Although the class/subset classification can capture pleiotropic phenotypes, it does not provide a way to compare pleiotropic phenotypes of multiple genes the way semantic similarity scores based on collections of EQ statements does. In contrast, the ontology approach allows the grouping of phenotypes at any level of the ontology that may be appropriate for a particular analysis, while still allowing each observation (phene) to be separately annotated.

Semantic similarity predicts participation in shared metabolic and regulatory pathways

It is a premise of this work that through computational analysis of EQ statements representing phenotypes, biological processes can be recapitulated, modeled, and even discovered. Were this to be true, one would expect, for example, that gene products in the same metabolic pathways would be annotated with EQ statements that are highly similar. To test this hypothesis, we used the PlantCyc project databases AraCyc (v 11.5) [46], Oryzacyc (v 1.0), SoyCyc (v 4.0) and CornCyc (v 4.0) as well as LycoCyc from SGN (v 3.3 *Solanum lycopersicum*) [47], and MedicCyc from the Noble Foundation [48]. One metabolic pathway that is well populated among those databases and for which our phenotype datasets have representation is the phenylpropanoid biosynthesis initial reactions of flavonoid biosynthesis. 3-hydroxy flavonoids, also called anthocyanins, are pigments. They serve to, e.g., attract pollinators and protect plants from UV-B damage [49,50].

For the gene products involved in the phenylpropanoid biosynthesis pathway – more specifically the initial reactions of flavonoid biosynthesis – we queried Plant PhenomeNET. The most informative query result came from maize, which had only the *c2* gene (*colorless2* converts 4-coumaryl-Coa to 2′, 4, 4′, 6′-tetrahydroxychalcone) curated into the phenylpropanoid biosynthesis initial reactions. When Plant PhenomeNet was queried with GRMZM2G422750 (the gene model identifier for *c2*) a number of maize genes associated with phenotypes were returned:

c2 GRMZM2G422750 similarity score 1 (identity: this is the query)
c1 GRMZM2G005066 similarity score 1
r1 GRMZM5G822829 similarity score 0.6666666667
b1 GRM similarity score 0.5

All three of the identified gene models are involved in the anthocyanin pathway of maize, which controls flavonoid synthesis (reviewed in [51]). More specifically, the gene products of the *c1, r1,* and *b1* loci activate genes in the anthocyanin pathway. This result: (1) indicates that reasoning across curated phenotypes in plants is capable of creating result sets that recapitulate well-characterized biological phenomena, (2) hints that for plant species that are not genetically well-characterized, the ontological reasoning approach to predicting phenotypic associations could assist in forward genetics approaches, and (3) highlights the potential use of reasoning across phenotypic ontological associations to prioritize high-quality data curation where data are missing from or complementary to repositories like the PlantCyc database.

Focusing on (2) – that the suggested approach can help with characterizing understudied species – the reasoning is as follows. Consider a poorly studied species with a number of mutant phenotypes that include an altered seed color phene. The phenotypes of this species would be described and codified using ontological representations. These phenotypic descriptions then could be used as queries to return genes from a well-characterized species (e.g., maize) with phenotypes that have high similarity to the phenotype in the poorly studied species. This result set could indicate to a researcher who is not an expert in pigment biology that the flavonoid and anthocyanin biosynthetic pathways and their regulators were of interest for determining which genes were responsible for the phenotype.

Evaluation of phenotypic similarity across orthologs and gene families

Manual assessment of gene families

We were able to place 2,741 EQ-annotated genes (2,393 Arabidopsis, 30 soybean, 40 Medicago, 92 rice, 72 tomato, 114 maize) into 1,895 gene families, of which 460 families contain two or more genes annotated with EQ statements. The gene families, based on the Phytozome 10 Angiosperm-level families [52], generally contain both dicot and monocot representatives from the species in this study. Forty-two of the families contain between five and 12 genes with EQ statements, allowing us to assess how often homologous genes have similar functions. Further, there are 147 families that contain EQ statements from two or more species. These are of interest because it allows us to assess how often functions are conserved between orthologs.

For most families with multiple EQ-annotated genes, gene function is conserved or similar. For example, in the terpene synthase family (family 54585183, Additional files 4, 5 and 6), with 12 EQ-annotated genes from Arabidopsis, rice, and maize, all genes included aspects of "dwarf" phenotypes (quality "decreased height", PATO:0000569). However, salient phenotypes in maize also include floral hermaphrodism, in contrast to the typical male and female floral separation in wild type domesticated maize. In the Flowering Locus T family (family 54614050, Additional files 4, 5 and 6), there are 12 EQ-annotated genes from five of our study species. All of the characterized mutant phenotypes involve floral development or photoperiod control.

We also observed gene families in which annotated phenotypes are quite different across orthologs. For example, in the family (54614050, Additional files 4, 5 and 6), a leucine-rich repeat, serine-threonine kinase family, the SUNN mutant in Medicago display extra root nodules, while the CLV1 mutant in Arabidopsis displays abnormal leaf phyllotaxy and altered floral morphology [53,54].

Plant phenomeNET: a web interface for searching the plant dataset

We adapted PhenomeNET [37] to provide the results of the computational analysis of the plant data sets to the broader research community in an online form. Plant PhenomeNET is available via [39] and provides access to the genotypes of all six species that possess at least one EQ statement. For each genotype, a detailed genotype page provides information about similarity scores to any of the other genotypes as well as a link to an additional page providing the phenotype assigned by the curator and those inferred via the ontologies. We note here that similarity scores of 0 for genotype pairs are not reported in Plant PhenomeNET.

Using plant phenomeNET – Searching for *tasselseed1*

To illustrate the usage of Plant PhenomeNET, we provide an example search for *tasselseed1* (*ts1*) maize gene. The tassel of maize normally bears only male flowers, but in the *ts1* mutant, female flowers also develop in the spikelets born on the tassel. By entering "ts1" into the search box and submitting the form, we obtain a list of genes that all match the string "ts1" (for search query and results see: panel A and B of Figure 4). For the navigation from the search list, there are two options provided (see last two columns in panel B of Figure 4): one can either show the phenotype or explore phenotypically similar mutants.

By following the first link "show phenotypes", the user obtains the assigned (top list, panel C, Figure 4) as well as the inferred EQ statements (bottom list, panel C, Figure 4) for the *ts1* gene. For example, the curator assigned an EQ statement for the "abnormality of the tassel" as part of the phenotype. One of the EQ statements that was inferred based on the ontology structure is "extra floral organs in spikelet".

Following the second link to "explore" similar mutants, the user obtains an ordered list of phenotypically similar mutants with the most similar at the top and the least similar at the bottom (see panel D, Figure 4). Each of the mutants provided in the list can then be explored further with the links provided for navigation. One interesting case is presented in our list with the rice mutant FOR1, with the atomized statement "extra floral organs in the spikelet". This example illustrates how important correctly assigned EQ statements are, and how, using inferred EQ statements, connections can made between mutants from different species.

Conclusions

After defining a common method for semantic representation of plant phenotypes, we annotated 2,741 genotypes with 2,023 unique EQ statements. This represents the first cross-species plant data set that can readily be integrated with other data via shared ontologies. This use of ontologies to support phenotypic reasoning enables integration beyond plants and would enable generalized analyses to discover phenomena conserved across all domains of life as described in [55]. An example of such cross-domain inference is their finding that the pathways that underlie gravitropism sensing in Arabidopsis root tips are concordant with an inner ear defect in human developmental biology known as Waardberg syndrome [55].

In our initial computational analyses using semantic phenotype similarity scoring, we were able to identify gene sets that are functionally related, i.e. belonging to the same gene family or involved in the same pathway. The method described here can be readily applied to other plant species to suggest genes for analysis in under-studied species or crop wild relatives, or expanded to describe and compare phenotypes across diverse plant species for evolutionary analysis, as has been done for fish [17,56]. Because the ontologies used for comparisons are taxon-neutral, there are no restrictions to expanding this method to non-flowering plant models such as *Physcomitrella patens*, *Selaginella moellendorffi*, *or Ceratopteris* [57-59], and comparisons across widely divergent species (e.g., maize versus moss) may even reveal surprising conservation or co-option of gene function. Even for the species used in this initial study, there is much to learn about the relationships between genotype and phenotype, and we urge the continued curation and EQ-based annotation of plant phenotypes, to expand this data set and increase its utility. Because species- and clade-specific databases will need to curate and hold these statements, a standardized method for storing this data, preferably using a common database schema such as [60], should be implemented.

Methods

Generation of the lists of phenotypes for each plant species

Each of the model plant species represented in this analysis is supported by a database of genomic and other data. These databases are highly individualized, based on the differing needs of their community members. As such, the generation of a list of phenotypes associated with a mutant allele of a known gene was slightly different for each species.

Selection of Arabidopsis phenotypes

The Arabidopsis mutant phenotype dataset, first compiled by [36], includes ~2,400 genes with recessive mutant phenotypes for which the disrupted gene is known. Information for this dataset was previously gathered from: 1) a sequence-based map of genes with mutant phenotypes [61]; 2) the SeedGenes database of essential genes [62], as updated by [63]; 3) a list of genes associated with mutant phenotypes obtained from TAIR [64]; and 4) several thousand publications describing Arabidopsis mutant phenotypes retrieved from the Pubmed Database [65] using appropriate keywords (Arabidopsis, mutant(s), mutation(s), knockout, and null). Short, free-text phenotype descriptions found in column I of Supplemental Table S2 of [36] were used as the input for the Arabidopsis EQ statements. Genes with only a dominant, gain-of-function mutant phenotype [66] were generally excluded. Based on past work, the Arabidopsis phenotypes analyzed here are associated with sequenced genes but not with specific mutant alleles.

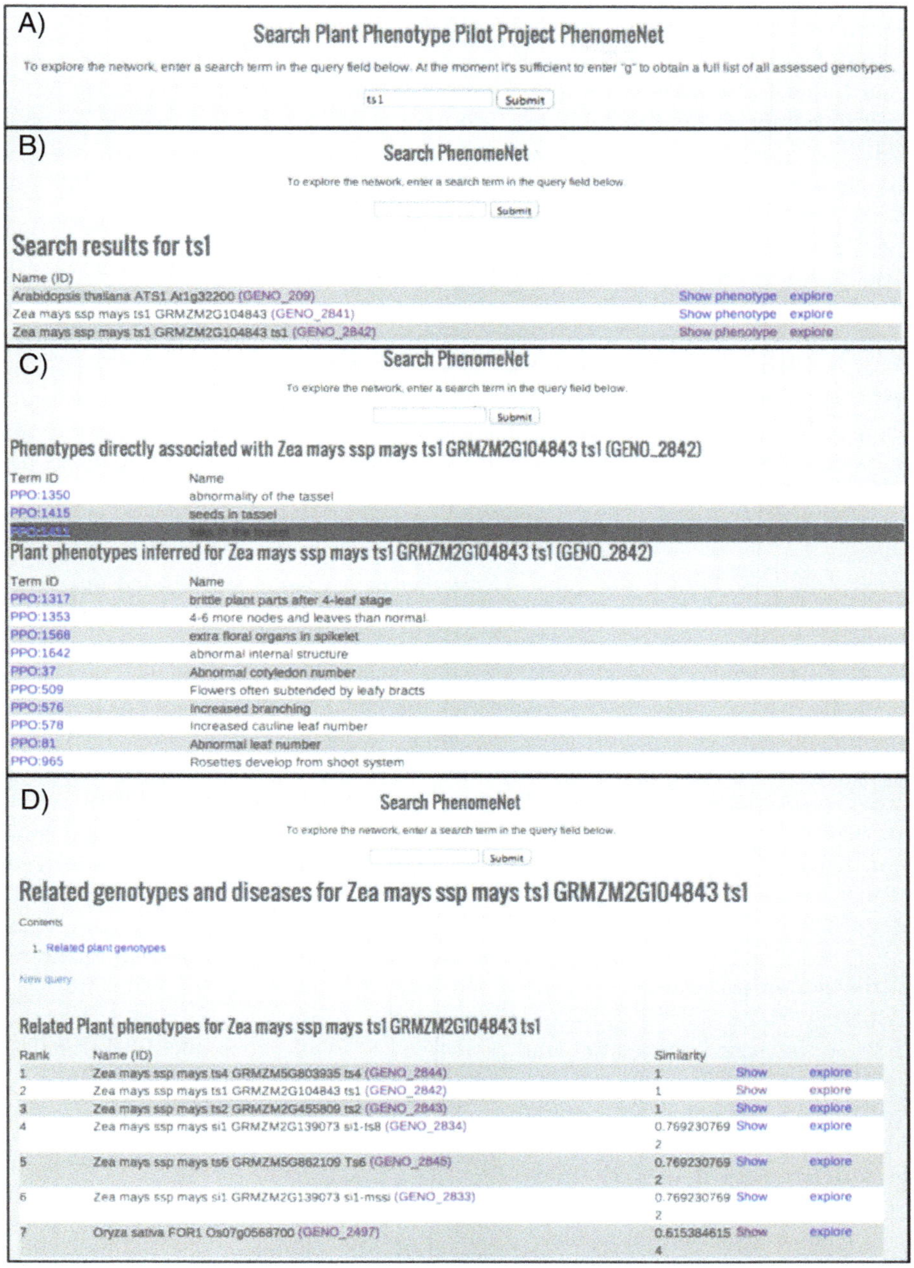

Figure 4 This figures illustrates the usage of Plant PhenomeNET for the maize gene *mac1*. After searching for the gene **(A)**, search results are returned **(B)** and assigned and inferred phenes are shown **(C)**, as well as semantically similar phenotypes from other genes **(D)**. See text for more details.

Selection of maize phenotypes

In the MaizeGDB database [31], maize phenotypes are associated with mutant alleles (variations) of genes based on a maize-specific controlled phenotype vocabulary, consisting of 1,088 phenotypes. Of the 1,088 phenotypes associated with mutant-defined loci, we removed continuous trait phenotypes (e.g., phenotypes that are attributable to quantitative trait loci or QTL), and several other types of phenotypes not likely to be relevant for this analysis, such as gel mobility of a protein on a starch gel. Of the remaining phenotypes, we selected only those associated with gene models (DNA sequences).

Selection of rice phenotypes

In order to create a list of rice mutants that were associated with known genes, data was combined from Gramene [34] and Oryzabase [32,67]. The Oryzabase file was quite large (about 4,800 traits/phenotypes listed, with about 1,600 of those associated with a known locus), while the Gramene list was smaller, with about 160 loci. The information from the two sets was combined and all the mutants with identifiers from both databases were cross-referenced to ensure there was no overlap or duplications. Many of the described mutants had to be eliminated from the master list as they were only described morphologically (i.e. not associated with a known locus or gene). For the remaining mutants, we combined all available phenotypic descriptions from the two sources.

Selection of soybean and Medicago phenotypes

Curated lists of phenotypes for these species are not available in public databases. Thus, in order to create lists of mutant genes in soybean and Medicago for this study, the primary literature was searched for phenotypes and their descriptions.

Selection of tomato phenotypes

Tomato loci with a known phenotype were selected from the Sol Genomics Network database (SGN) [15,35]. Phenotypes are associated with alleles, with some loci having multiple alleles with different phenotypes. The loci were curated manually based on previously described mutants [68] and literature curation of published tomato cloned genes with an associated phenotype. We included only loci with morphological or metabolic phenotypes, excluding isozyme alleles and loci that have a described phenotype but no associated gene sequence.

Quality assurance across the entire data set

In order to provide consistency across species and allow for computational analysis of the entire phenotype data set, we developed a set of rules to define how the EQ statements should be constructed, and employed manual and automated quality checks to verify compliance with the rules. Manual checks determined if the EQ statements were made in a consistent manner across species. We did find consistency in most cases; however, minor inconsistencies have a relatively small effect, as the power of using hierarchical ontologies to describe phenotypes allows similar but not identical EQ statements to have high similarity scores.

Automated quality checks computationally verified the validity of the assigned EQ statements based on our predefined set of rules. An example of such a rule is a requirement that entities be represented with either PO (for structural) or GO IDs (for process phenotypes) and that the type of Quality chosen from PATO must match the Entity (i.e. a structure quality for a structure entity and a process quality for a process entity). Furthermore, the automated checks ensured that valid identifiers were used for each ontology term and that each term label matched its ID, which was useful for correcting typographical errors. More details on the rules we employed are provided in Additional file 7. The automated quality assurance was an iterative process in which the errors were removed continuously as the data set expanded. The data set comprising EQ statements from all six species (Additional file 1) successfully passed the automated checking procedure.

Building a phenotype network using semantic similarities of gene pairs based on assigned EQ statements

The computational analysis relied on the representation of phenotypes as EQ statements. Each phenotype was represented as an affected entity that is further described with a quality. The application of EQ statements has been proven useful for cross-species gene function prediction, as well as pathway involvement and the identification of disease gene candidates [53,69]. As described by [19], species-specific phenes were decomposed into an affected Entity and Quality, and represented using species-independent ontologies. All the ontologies used here for the description of the phenotypes in any of the six species were downloaded on 15 March 2014 and converted to OWL EL. In addition to the ontologies, a set of logical definitions to connect plant structures with biological processes has been downloaded on 29 April 2013 and was also integrated with the ontologies (see Additional file 8). For further details on the applied ontologies see Table 1.

Once the ontologies were transferred into an OWL EL profile, they were combined into one ontology. We applied the method implemented in PhenomeNET [37], to represent the statements in OWL with:

has–part some (E and has–quality some Q)

where Entities and Qualities were used as defined by the curators. Following this approach generates one integrated ontology that then can be used to infer additional phenes using reasoning over the ontology. An inferred phene is an EQ statement that is an ancestor term of the assigned EQ statement. For example, the maize *mac1* (multiple archesporial cells1) gene was curated with an EQ statement named "Male and female infertility" and from the complete list of curator-assigned statements, and one additional EQ statement named "Complete sterility" was inferred.

To determine the semantic phenotype similarity of two genotypes (genotype A and B), a Jaccard index based on the binary vectors is calculated:

$$\text{simphen} = \left(P_{\text{geno}}A \cap P_{\text{geno}}B\right) / \left(P_{\text{geno}}A \cup P_{\text{geno}}B\right)$$

where P_geno_A represents the phenes of genotype A and P_geno_B represents the phenes of genotype B.

Applying this scoring method, phenotype semantic similarity scores fall into the range [0, 1], with 0 indicating no overlap between phenotypes and 1 indicating identical phenotypes. Calculating the semantic similarity score for each possible combination of genotypes results in a 2,866 × 2,866 data matrix. Similarity scores > 0 are provided as Additional file 9. We note here that 10 EQ statements of Arabidopsis genotypes (<0.2% of total EQ statements) were excluded from the computational analysis, because they either needed further discussion among the curators due to the relations used to build the entity or include a term that was removed from the ontology in the period between curation and the computational analysis.

This matrix constituting a genotype network based on phenotype similarities was 1) compared to an existing, manually created phenotype-specific grouping of genes [36], and 2) used to assess gene function (see following sections and Results and Discussion). We note here that this scoring is highly dependent on the assigned EQ statements and that the annotations assigned to date are as complete as can be derived from existing findings. This means that for phenes that have not been tested yet, we assume that this phene is *absent*. With the growth of the data set, more detail will be added to the genotypes, which in consequence will improve the accuracy of semantic phenotype similarity scores and the representation of biological processes.

Employed data and software

We downloaded all the ontologies from the OBO Foundry [70,71] or their respective download site (see Table 1), and used El Vira (version 0.2) [72] to transform ontologies from an OWL DL profile into an OWL EL profile. The application of OWL EL files facilitates faster reasoning over the combined ontologies and is consistent with the description of the method described for mammal data [37]. To integrate the individual annotation files along with the respective ontologies used in annotation into a single ontology, the Brain library version 1.5.2 was used to easily modify OWL EL ontologies [73]. All scripts required for the data analysis were implemented in Groovy (version 2.0.4) [74]. A copy of PhenomeNET was set up to hold the results of the computational analysis, which were uploaded using the PhenomeNET database scheme. Plant PhenomeNET is accessible from [39].

Comparison of semantic similarity and an existing classification of plant phenotypes

For intraspecific comparison of Arabidopsis phenotypes, we used Table S2 from [36]. In this previous work, genes were sorted into a three-tiered hierarchy of phenotypes of groups, class, and subsets. Their classification system was designed for the specific purpose of defining the set of essential genes for an organism, and for this purpose it was not necessary to differentiate among phenotypes of different mutant alleles of the same gene. Genes were placed into a single group and class, prioritized by developmental stage when phenotypes are first observed and what methods and conditions are used to detect them. The lowest rank included phenotypes where detection required a biochemical assay or microscopic examination. When the phenotype of a weak allele was more informative or better characterized than the phenotype of a null allele, the assignment was made on the basis of the better-known phenotype (e.g., *fy* - null is *emb lethal* but known as *flowering time gene*). Genes were also assigned to one or more of 42 phenotypic subsets, such as shoot architecture, flowering time, miscellaneous seed defects, and temperature.

To carry out a comparison of the previous results to the present work, we rearranged the dataset from [36] so that each unique gene/subset combination was on a single row. Because genes could belong to multiple subsets, there were multiple rows per gene. We removed data for 82 genes that were in [36] but not included in the present study. We calculated average semantic similarity of the classes and subsets as the average of all pairs of genes where both genes were in the same class or subset.

Pathway assessment based on phenotype network

The BioCyc databases for Arabidopsis (AraCyc version 11.5), maize (CornCyc version 4.0), rice (OryzaCyc version 1.0), and soybean (SoyCyc version 4.0) were downloaded from Plant Metabolic Network [45,75]. The database for tomato (LycoCyc version 3.3) was downloaded from the Sol Genomics Network [15,35], and the database for Medicago (MedicCyc version 2.0) was requested from and provided by The Samuel Roberts Noble Foundation [47,76].

To identify well-populated pathways across all six species, we divided the number of pathway steps catalyzed by a gene product for which a phenotype was included in our dataset by the average number of reactions in the pathway across the species examined (e.g., number of steps with a curated phenotype divided by number of total steps in the pathway). For instances where more than one gene encoded the enzyme responsible for a single step, that step was counted only once (i.e. the presence or absence of a gene encoding the enzyme was counted, not the number of genes encoding that step in a particular plant genome).

Assessment of gene families using the phenotype network

Gene families are based on the Angiosperm-level families from the Phytozome10 release [50], accessed on August 13, 2014, as multiple-sequence alignments for each family. These gene family alignments included peptide sequences from 43 species, and comprised 29,803 gene families. From these alignments, we calculated HMM-based alignment models using hmmbuild (HMMer package version 3.1 r4562, Eddy, 2011 [77]). We then searched the peptide sequences from each of the seven species discussed in this paper, along with peptide sequences from *Amborella trichopoda* (to serve as an outgroup in phylogenies), against the gene family HMMs, using hmmscan (maximum E-value 1e-4), and then placed each sequence into the family of the top HMM match, giving a multi-fasta file for each gene family. The resulting family files were realigned to the respective HMM using hmmalign. Prior to generating phylogenetic trees, the resulting alignments were trimmed of non-aligning residues (as lower case characters in the output of hmmalign, indicating non-match-state residues in the HMM alignments). Phylogenetic trees were calculated using RAxML (raxmlHPC-PTHREADS-AVX, v. 8.0.26 [78]), using model PROTGAMMAAUTO. Analyses of EQ statements relative to gene families were conducted by generating combined EQ statements for each gene (concatenating multiple EQ statements into a single string separated by ";;") and then joining these combined statements with genes. The resulting analyses are in Additional file 4. Alignments and phylogenetic trees are in Additional files 5 and 6, respectively.

Genome assembly and annotation versions used in these gene families were: *Glycine max* assembly and annotation version Wm82.a2.v1; *Medicago truncatula* assembly v 4.0v1; *Arabidopsis thaliana* v TAIR10; *Oryza sativa* Japonica (Nipponbare) assembly IRGSP-1.0, with the IRGSP-1.0 gene model names; *Zea mays* spp mays B75 RefGen v3, assembly annotation v 6a; *Lycopersicon esculentum* v iTAG2.3; *Amborella trichopoda* v 1.0.

Endnotes

[a]Ontology term identifiers of the form PO:0000925 are shorthand for identifiers of the form http://purl.obolibrary.org/obo/PO_0009025.

[b]Some relations in the Relation Ontology fall within the BFO namespace, because they are imported from the Basic Formal Ontology.

Additional files

Additional file 1: All EQ statements curated for the six species. All EQ statements in tabular form, with explanations of how to fill in each column.

Additional file 2: Overlap among unique phenotypes for sets of species. Unique phenotypes means that if one EQ statement is shared between two species, it is counted as overlap only once, no matter how often it occurs.

Additional file 3: Average similarity scores for previously derived Arabidopsis genes grouped by class. Classes follow [36].

Additional file 4: Gene families with EQ statements. An Excel file with gene family membership by species, concatenated EQ statements for genes with EQ annotations, and gene family descriptions.

Additional file 5: Gene family alignments. An archived, compressed directory of the multi-fasta alignments (text files) for the 1,985 gene families with EQ statements from this study. Access using "tar -xzf alignments_w_EQs.tar.gz" and then with an alignment viewer or standard text editor.

Additional file 6: Gene families with EQ statements. An archived, compressed directory of the phylogenetic reconstructions ("trees"), calculated from the alignments in Additional file 5. Tree files are in Phylip/Newick format. Access using "tar -xzf trees_w_EQs.tar.gz" and then with a phylogenetic tree viewer.

Additional file 7: List of error checks for the EQ statements.

Additional file 8: Logical definitions for biological processes in plants. A subset of logical definitions built to connect biological processes with plant structures were used as part of the computational analysis. While the logical definitions are now part of GO, we used an earlier independent version provided here as an OWL file, which can be opened in a text editor or OWL editor.

Additional file 9: Similarity scores of genotype pairs with similarity >0.

Abbreviations

EQ: Entity-Quality; GO: Gene Ontology; MaizeGDB: Maize Genetics and Genomics Database; OWL: Web Ontology Language; PATO: Phenotype and Trait Ontology; PO: Plant Ontology; QTL: Quantitative trait locus (or Loci); RO: Relation Ontology; SGN: Sol Genomics Network; TAIR: The arabidopsis information resource.

Competing interests

The authors declare that they have no competing interests.

Authors' contributions

EH conceived the study. LH, RLW, LC, DM, SK, JL, RN, NM, AP, EH and LM curated phenotypes for the six plant species from databases and the published scientific literature. AO, GVG, and RH conducted the computational analysis and provided the results in Plant PhenomeNET. SBC calculated gene family alignments and phylogenies, and PJ and SBC manually verified the correspondence of semantic similar genotypes and gene families. AO, RLW, and DM compared the semantic similarity results to Lloyd and Meinke data set [36]. CJL and MH conducted the pathway analyses. EH and CJL coordinated group activities. LC managed requests for and questions about Plant Ontology (PO) and Plant Experimental Conditions Ontology (EO) terms. NM and EC developed methods for storing EQ statements in species and clade databases. All authors engaged in many useful discussions over the course of carrying out the research, wrote sections of the manuscript, and approved the final version.

Acknowledgements

The NSF Phenotype RCN supported this work by allowing plant and computational biologists to come together and develop a common vocabulary and supported the preliminary analysis of generated data. The Samuel Roberts Noble Foundation provided the Cyc database for Medicago (MedicCyc version 2.0). RLW was supported in part by the National Science Foundation under Grant Numbers #DBI-0735191 and #DBI-1265383. PJ and LC contributions are supported by the National Science Foundation Awards #IOS-0822201, #IOS-1127112 and #IOS-1340112.

Author details

[1]Wellcome Trust Sanger Institute, Wellcome Trust Genome Campus, Hinxton CB10 1SA, UK. [2]iPlant Collaborative, University of Arizona, 1657 E. Helen St., Tucson, Arizona 85721, USA. [3]Department of Electrical and Computer Engineering Iowa State University, 1018 Crop Informatics Lab, Ames, Iowa 50011, USA. [4]USDA-ARS Corn Insects and Crop Genetics Research Unit, Iowa State University, Crop Genome Informatics Lab, Iowa State University, Ames, IA 50011, USA. [5]Department of Agronomy, Agronomy Hall, Iowa State University, Ames, IA 50010, USA. [6]Department of Botany and Plant Pathology, 2082 Cordley Hall, Oregon State University, Corvallis, OR 97331, USA. [7]Department of Genetics, Development and Cell Biology, Roy J Carver Co-Laboratory, Iowa State University, Ames, IA 50010, USA. [8]Department of Computer Science, Aberystwyth University, Llandinam Building, Aberystwyth

SY23 3DB, UK. [9]Computer, Electrical and Mathematical Sciences & Engineering Division and Computational Bioscience Research Center, King Abdullah University of Science and Technology, 4700 King Abdullah University of Science and Technology, P.O. Box 2882, Thuwal 23955-6900, Kingdom of Saudi Arabia. [10]Department of Plant Biology, Michigan State University, 220 Trowbridge Rd, East Lansing, MI 48824, USA. [11]Department of Botany, Oklahoma State University, 301 Physical Sciences, Stillwater, OK 74078, USA. [12]Boyce Thompson Institute for Plant Research, 533 Tower Road, Ithaca, NY 14853, USA. [13]Phoenix Bioinformatics, 643 Bair Island Rd Suite 403, Redwood City, CA 94063, USA.

References

1. National Research Council (US) Committee on Research at the Intersection of the Physical and Life Sciences. Research at the Intersection of the Physical and Life Sciences. Washington, D.C: National Academies Press; 2010.
2. Benfey PN, Mitchell-Olds T. From genotype to phenotype: systems biology meets natural variation. Science. 2008;320:495–7.
3. Groth P, Weiss B, Pohlenz H-D, Leser U. Mining phenotypes for gene function prediction. BMC Bioinformatics. 2008;9:136.
4. Gkoutos GV, Schofield PN, Hoehndorf R. Computational tools for comparative phenomics: the role and promise of ontologies. Mamm Genome Off J Int Mamm Genome Soc. 2012;23:669–79.
5. Doelken SC, Köhler S, Mungall CJ, Gkoutos GV, Ruef BJ, Smith C, et al. Phenotypic overlap in the contribution of individual genes to CNV pathogenicity revealed by cross-species computational analysis of single-gene mutations in humans, mice and zebrafish. Dis Model Mech. 2013;6:358–72.
6. Deans AR, Lewis SE, Huala E, Anzaldo SS, Ashburner M, Balhoff JP, et al. Finding our way through phenotypes. PLoS Biol. 2015;13:e1002033.
7. Hartl DL, Jones EW. Essential genetics: s genomics perspective. 4th ed. Sudbury, Massachusettes, USA: Jones and Bartlett Publishers; 2006.
8. Wikipedia phenotype article [http://en.wikipedia.org/wiki/Phenotype]
9. York LM, Nord E, Lynch J. Integration of root phenes for soil resource acquisition. Funct Plant Ecol. 2013;4:355.
10. Serebrovsky AS. "Somatic segregation" in domestic fowl. J Genet. 1925;16:33–42.
11. Denny JC, Bastarache L, Ritchie MD, Carroll RJ, Zink R, Mosley JD, et al. Systematic comparison of phenome-wide association study of electronic medical record data and genome-wide association study data. Nat Biotechnol. 2013;31:1102–11.
12. Zemojtel T, Köhler S, Mackenroth L, Jäger M, Hecht J, Krawitz P, et al. Effective diagnosis of genetic disease by computational phenotype analysis of the disease-associated genome. Sci Transl Med. 2014;6:252ra123.
13. Hoehndorf R, Schofield PN, Gkoutos GV. An integrative, translational approach to understanding rare and orphan genetically based diseases. Interface Focus. 2013;3:20120055.
14. Hoehndorf R, Dumontier M, Gkoutos GV. Evaluation of research in biomedical ontologies. Brief Bioinform. 2013;14:696–712.
15. Bombarely A, Menda N, Tecle IY, Buels RM, Strickler S, Fischer-York T, et al. The Sol Genomics Network (solgenomics.net): growing tomatoes using Perl. Nucleic Acids Res. 2010;39:D1149–55.
16. Gkoutos GV, Green ECJ, Mallon AM, Hancock JM, Davidson D. Using ontologies to describe mouse phenotypes. Genome Biol. 2005;6:R8.
17. Dahdul WM, Balhoff JP, Engeman J, Grande T, Hilton EJ, Kothari C, et al. Evolutionary characters, phenotypes and ontologies: curating data from the systematic biology literature. PLoS One. 2010;5:e10708.
18. Köhler S, Doelken SC, Mungall CJ, Bauer S, Firth HV, Bailleul-Forestier I, et al. The Human Phenotype Ontology project: linking molecular biology and disease through phenotype data. Nucleic Acids Res. 2013;42:D966–74.
19. Mungall C, Gkoutos GV, Smith C, Haendel M, Lewis S, Ashburner M. Integrating phenotype ontologies across multiple species. Genome Biol. 2010;11:R2.
20. Knowlton MN, Li T, Ren Y, Bill BR, Ellis LB, Ekker SC. A PATO-compliant zebrafish screening database (MODB): management of morpholino knockdown screen information. BMC Bioinformatics. 2008;9:7.
21. Pesquita C, Faria D, Falcão AO, Lord P, Couto FM. Semantic similarity in biomedical ontologies. PLoS Comput Biol. 2009;5:e1000443.
22. Mistry M, Pavlidis P. Gene ontology term overlap as a measure of gene functional similarity. BMC Bioinformatics. 2008;9:327.
23. Gan M, Dou X, Jiang R. From ontology to semantic similarity: calculation of ontology-based semantic similarity. ScientificWorldJournal. 2013;2013:793091.
24. Hoehndorf R, Hancock JM, Hardy NW, Mallon A-M, Schofield PN, Gkoutos GV. Analyzing gene expression data in mice with the Neuro Behavior Ontology. Mamm Genome Off J Int Mamm Genome Soc. 2014;25:32–40.
25. Hoehndorf R, Hiebert T, Hardy NW, Schofield PN, Gkoutos GV, Dumontier M. Mouse model phenotypes provide information about human drug targets. Bioinforma Oxf Engl. 2014;30:719–25.
26. Walls RL, Athreya B, Cooper L, Elser J, Gandolfo MA, Jaiswal P, et al. Ontologies as integrative tools for plant science. Am J Bot. 2012;99:1263–75.
27. Cooper L, Walls RL, Elser J, Gandolfo MA, Stevenson DW, Smith B, et al. The plant ontology as a tool for comparative plant anatomy and genomic analyses. Plant Cell Physiol. 2013;54:e1–1.
28. Ashburner M, Ball CA, Blake JA, Botstein D, Butler H, Cherry JM, et al. Gene ontology: tool for the unification of biology. The gene ontology consortium. Nat Genet. 2000;25:25–9.
29. The Gene Ontology Consortium. The gene ontology: enhancements for 2011. Nucleic Acids Res. 2012;40:D559–64.
30. Schaeffer ML, Harper LC, Gardiner JM, Andorf CM, Campbell DA, Cannon EKS, et al. MaizeGDB: curation and outreach go hand-in-hand. Database J Biol Databases Curation. 2011;2011:bar022.
31. Maize Genetics and Genomics Database [http://maizegdb.org/]
32. Yamazaki Y, Sakaniwa S, Tsuchiya R, Nonomura K-I, Kurata N. Oryzabase: an integrated information resource for rice science. Breed Sci. 2010;60:544–8.
33. Jaiswal P. Gramene database: a hub for comparative plant genomics. In: Pereira A, editor. Plant reverse genetics. Volume 678. Totowa, NJ: Humana Press; 2011. p. 247–75.
34. Gramene [http://www.gramene.org/]
35. Sol genomics network [http://solgenomics.net/]
36. Lloyd J, Meinke D. A comprehensive dataset of genes with a loss-of-function mutant phenotype in Arabidopsis. Plant Physiol. 2012;158:1115–29.
37. Hoehndorf R, Schofield PN, Gkoutos GV. PhenomeNET: a whole-phenome approach to disease gene discovery. Nucleic Acids Res. 2011;39:e119–9.
38. PhenomeNET [http://phenomebrowser.net]
39. Plant PhenomeNET [http://phenomebrowser.net/plant/]
40. Hoehndorf R, Oellrich A, Rebholz-Schuhmann D. Interoperability between phenotype and anatomy ontologies. Bioinforma Oxf Engl. 2010;26:3112–8.
41. Smith B, Ceusters W, Klagges B, Kohler J, Kumar A, Lomax J, et al. Relations in biomedical ontologies. Genome Biol. 2005;6:R46.
42. The OBO relations ontology [https://code.google.com/p/obo-relations/]
43. Hastings J, de Matos P, Dekker A, Ennis M, Harsha B, Kale N, et al. The ChEBI reference database and ontology for biologically relevant chemistry: enhancements for 2013. Nucleic Acids Res. 2013;41(Database issue):D456–63.
44. Link to complete dataset on iPlant Data Store. [http://ezid.cdlib.org/id/ark:/87946/p3159c]
45. The Rice Annotation Project Database [http://rapdb.dna.affrc.go.jp/index.html]
46. Zhang P, Foerster H, Tissier CP, Mueller L, Paley S, Karp PD, et al. MetaCyc and AraCyc. Metabolic pathway databases for plant research. Plant Physiol. 2005;138:27–37.
47. SolCyc biochemical pathways [http://solgenomics.net/tools/solcyc/index.pl]
48. Urbanczyk-Wochniak E, Sumner LW. MedicCyc: a biochemical pathway database for *Medicago truncatula*. Bioinforma Oxf Engl. 2007;23:1418–23.
49. Koes RE, Quattrocchio F, Mol JNM. The flavonoid biosynthetic pathway in plants—function and evolution. Bioessays. 1994;16:123–32.
50. Stapleton AE, Walbot V. Flavonoids can protect maize DNA from the induction of ultraviolet radiation damage. Plant Physiol. 1994;105:881–9.
51. Bruce W, Folkerts O, Garnaat C, Crasta O, Roth B, Bowen B. Expression profiling of the maize flavonoid pathway genes controlled by estradiol-inducible transcription factors CRC and P. Plant Cell. 2000;12:65–80.
52. Goodstein DM, Shu S, Howson R, Neupane R, Hayes RD, Fazo J, et al. Phytozome: a comparative platform for green plant genomics. Nucleic Acids Res. 2012;40(Database issue):D1178–86.
53. Schnabel E, Karve A, Kassaw T, Mukherjee A, Zhou X, Hall T, et al. The *M. truncatula* SUNN gene is expressed in vascular tissue, similarly to RDN1, consistent with the role of these nodulation regulation genes in long distance signaling. Plant Signal Behav. 2012;7:4–6.
54. Schnabel E, Journet E-P, de Carvalho-Niebel F, Duc G, Frugoli J. The *Medicago truncatula* SUNN gene encodes a CLV1-like leucine-rich repeat receptor kinase that regulates nodule number and root length. Plant Mol Biol. 2005;58:809–22.

55. Washington NL, Haendel MA, Mungall CJ, Ashburner M, Westerfield M, Lewis SE. Linking human diseases to animal models using ontology-based phenotype annotation. PLoS Biol. 2009;7:e1000247.
56. Mabee BP, Balhoff JP, Dahdul WM, Lapp H, Midford PE, Vision TJ, et al. 500,000 fish phenotypes: The new informatics landscape for evolutionary and developmental biology of the vertebrate skeleton. Z Angew Ichthyol J Appl Ichthyol. 2012;28:300–5.
57. Banks JA, Nishiyama T, Hasebe M, Bowman JL, Gribskov M, de Pamphilis C, et al. The Selaginella genome identifies genetic changes associated with the evolution of vascular plants. Science. 2011;332:960–3.
58. Rensing SA, Lang D, Zimmer AD, Terry A, Salamov A, Shapiro H, et al. The Physcomitrella genome reveals evolutionary insights into the conquest of land by plants. Science. 2008;319:64–9.
59. Sessa EB, Banks JA, Barker MS, Der JP, Duffy AM, Graham SW, et al. Between two fern genomes. GigaScience. 2014;3:15.
60. Mungall CJ, Emmert DB, FlyBase Consortium. A Chado case study: an ontology-based modular schema for representing genome-associated biological information. Bioinforma Oxf Engl. 2007;23:i337–46.
61. Meinke DW, Meinke LK, Showalter TC, Schissel AM, Mueller LA, Tzafrir I. A sequence-based map of Arabidopsis genes with mutant phenotypes. Plant Physiol. 2003;131:409–18.
62. SeedGenes Project [http://www.seedgenes.org/]
63. Muralla R, Lloyd J, Meinke D. Molecular foundations of reproductive lethality in *Arabidopsis thaliana*. PLoS One. 2011;6:e28398.
64. The Arabidopsis Information Resource [http://www.arabidopsis.org/]
65. PubMed database [www.ncbi.nlm.nih.gov/pubmed]
66. Meinke DW. A survey of dominant mutations in *Arabidopsis thaliana*. Trends Plant Sci. 2013;18:84–91.
67. Oryzabase [http://www.shigen.nig.ac.jp/rice/oryzabase/]
68. Tomato Genetics Resource Center [http://tgrc.ucdavis.edu/]
69. Smedley D, Oellrich A, Köhler S, Ruef B, Sanger Mouse Genetics P, Westerfield M, et al. PhenoDigm: analyzing curated annotations to associate animal models with human diseases. Database J Biol Databases Curation. 2013;2013:bat025.
70. The Open Biological and Biomedical Ontologies foundry [http://www.obofoundry.org/]
71. Smith B, Ashburner M, Rosse C, Bard J, Bug W, Ceusters W, et al. The OBO foundry: coordinated evolution of ontologies to support biomedical data integration. Nat Biotechnol. 2007;25:1251–5.
72. Hoehndorf R, Dumontier M, Oellrich A, Wimalaratne S, Rebholz-Schuhmann D, Schofield P, et al. A common layer of interoperability for biomedical ontologies based on OWL EL. Bioinforma Oxf Engl. 2011;27:1001–8.
73. Croset S, Overington JP, Rebholz-Schuhmann D. Brain: biomedical knowledge manipulation. Bioinforma Oxf Engl. 2013;29:1238–9.
74. Groovy [http://groovy-lang.org]
75. PlantCyc [http://www.plantcyc.org]
76. The Samuel Roberts Noble Foundation- *Medicago truncatula* Pathway Database [http://mediccyc.noble.org/]
77. Eddy SR. Accelerated profile HMM searches. PLoS Comput Biol. 2011;7:e1002195.
78. Stamatakis A. RAxML version 8: a tool for phylogenetic analysis and post-analysis of large phylogenies. Bioinforma Oxf Engl. 2014;30:1312–3.

13

Reverse transcriptase droplet digital PCR shows high resilience to PCR inhibitors from plant, soil and water samples

Nejc Rački*, Tanja Dreo, Ion Gutierrez-Aguirre, Andrej Blejec and Maja Ravnikar

Abstract

Background: Detection and quantification of plant pathogens in the presence of inhibitory substances can be a challenge especially with plant and environmental samples. Real-time quantitative PCR has enabled high-throughput detection and quantification of pathogens; however, its quantitative use is linked to standardized reference materials, and its sensitivity to inhibitors can lead to lower quantification accuracy. Droplet digital PCR has been proposed as a method to overcome these drawbacks. Its absolute quantification does not rely on standards and its tolerance to inhibitors has been demonstrated mostly in clinical samples. Such features would be of great use in agricultural and environmental fields, therefore our study compared the performance of droplet digital PCR method when challenged with inhibitors common to plant and environmental samples and compared it with quantitative PCR.

Results: Transfer of an existing Pepper mild mottle virus assay from reverse transcription real-time quantitative PCR to reverse transcription droplet digital PCR was straight forward. When challenged with complex matrices (seeds, plants, soil, wastewater) and selected purified inhibitors droplet digital PCR showed higher resilience to inhibition for the quantification of an RNA virus (Pepper mild mottle virus), compared to reverse transcription real-time quantitative PCR.

Conclusions: This study confirms the improved detection and quantification of the PMMoV RT-ddPCR in the presence of inhibitors that are commonly found in samples of seeds, plant material, soil, and wastewater. Together with absolute quantification, independent of standard reference materials, this makes droplet digital PCR a valuable tool for detection and quantification of pathogens in inhibition prone samples.

Keywords: PCR amplification, Inhibition, qPCR, Droplet digital PCR, Environmental samples

Background

At present, real-time quantitative PCR (qPCR) is the method of choice for detection and quantification of many pathogens and other DNA and RNA targets [1-3]. However, absolute quantification using qPCR depends on the use of standards; i.e., reference materials with known target concentrations, which are rarely commercially available. Even when reference materials are available, quantification accuracy is highly dependent on the amplification efficiencies between the samples and the reference material [4].

* Correspondence: Nejc.Racki@nib.si
Department of Biotechnology and Systems Biology, National Institute of Biology, Večna pot 111, SI-1000, Ljubljana, Slovenia

One of the factors that can significantly influence amplification efficiency is the presence of inhibitors in samples for PCR amplification. Such inhibitors are either co-extracted with nucleic acids from the samples, or can arise from the extraction procedure itself [5-7]. A wide array of inhibitors (and facilitators) of PCR amplification has been identified over the years, although the mechanisms of inhibition are not always known. Interference with cell lysis, sequestration or degradation of nucleic acids, and hindrance of polymerase activity are all common mechanisms of PCR inhibition [8].

Strategies to lower the influence of inhibitors in qPCR include: (i) extensive sample processing and purification; (ii) decrease in the amount of sample matrix, thereby removing or diluting matrix-derived inhibitors [9]; and (iii) application of corrections during the data analysis. In plant material, and particularly when testing for microorganisms in necrotic plant tissue, sample-specific inhibition is common, and the choice of the optimal DNA extraction and purification procedures can be a lengthy and cumbersome process [10].

Droplet digital PCR (ddPCR) has emerged as a promising technology to overcome these limitations of qPCR. Indeed, ddPCR is an endpoint and absolute measurement approach that enables determination of a target concentration without the need for a standard. This envisages cost-effective precise detection that is accessible to a variety of different scientific fields [11-13]. Moreover, assays developed for qPCR can be readily transferred to the ddPCR platform, with little or no additional optimization [14,15].

Despite relying on the same basic process as PCR and qPCR, i.e., PCR amplification of nucleic acids, it has been proposed that ddPCR can provide higher tolerance to inhibitors present in certain samples [11,13,16,17]. A lower susceptibility of ddPCR to inhibition has been reported for quantification of DNA targets in clinical samples [18], and food and feed samples, using certified reference materials [12]. Dingle et al. [18] reported lower susceptibilities of ddPCR to some inhibitors commonly found in serum, plasma and whole-blood specimens, while Morisset at al. [12] reported that ddPCR is tolerant to inhibitors in maize seed powder.

Compared to the clinical field, the fields of plant analysis and environmental monitoring are faced with a much broader range of target organisms and complex matrices, and the demand for absolute quantification is increasing. Accurate, sensitive and highly specific quantification methods are needed to determine pathogen survival and transmission. In environmental samples in particular, quantification of pathogens is gaining importance, as regulating bodies are shifting toward quantitative microbial risk assessment [6], where the allowed limits are based on the quantities of detected pathogens, and not just on their presence. Absolute quantification in such samples might be hindered by a number of different types and a variable content of inhibitors. Plant metabolites such as neutral and acidic polysaccharides and polyphenolic and humic substances are well-known inhibitors of PCR reactions and are abundant in plant materials, soil and polluted water [5-7,19]. We have previously reported greater tolerance of reverse transcriptase (RT)-ddPCR to inhibitors from wastewater, in comparison to RT-qPCR [15]; however, there are limited data available on the performance of RT-ddPCR in the presence of inhibitors common to diverse plant and environmental samples.

Pepper mild mottle virus (PMMoV) is a plant RNA virus that mainly infects pepper (*Capsicum annum*; Wetter et al., [20]), but also occurs naturally in plants, soil and water. PMMoV can survive the passage through the human digestive tract [21], and has been shown to be commonly present at high concentrations in human feces [22]. PMMoV is also known to persist in the environment for long periods of time [23], and consequently it has been proposed to be a good indicator for fecal contamination of water sources [24]. For these reasons, we chose here a PMMoV assay as a model system to study the effects of inhibitors of PCR in plant and environment matrices.

Using this model system, we compared the resilience of RT-ddPCR and RT-qPCR to selected inhibitors. The RT-ddPCR assay was adapted from an already existing PMMoV qPCR assay [25], and it was evaluated for the detection and quantification of PMMoV in complex matrices found in plants and the environment. In addition, the effects on RT-ddPCR of inhibitors typically present in such matrices were assessed.

Results

Transfer of the assay to the ddPCR platform

The transfer of the PMMoV assay to RT-ddPCR was successful. RT-qPCR and RT-ddPCR showed similar analytical sensitivities in parallel analysis of PMMoV RNA serial dilutions, (Figure 1 and Additional file 1: Table S1). RT-ddPCR typically showed better precision (smaller coefficients of variation) than RT-qPCR, especially for the dilutions with lower target concentrations (Figure 1B). Both assays showed good linearity of amplification, which was seen as high correlation coefficients (R^2) obtained by linear regression of the same dilutions (RT-qPCR, 0.998; RT-ddPCR, 0.989) (Figure 1A). Moreover, comparing the slopes of the regression lines showed no significant difference ($p > 0.05$), an additional argument suggesting that performance of both methods is similar. The dynamic range of quantification for the RT-ddPCR spanned three orders of magnitude (10^4- 10 target copies/10 µl reaction), which was narrower than that for the RT-qPCR, but with a lower limit of quantification (Figure 1).

Influence of inhibitors on the different ddPCR parameters

The inhibitors influenced the signals generated in the RT-ddPCR to variable extents. The inhibitors studied here did not considerably interfere with the formation of the droplets. The number of accepted droplets was always above the predefined minimum number (8,000), with the exception of one repetition with added humic acid (7,886 droplets), that was excluded from the subsequent analysis. A mean number of accepted droplets over all of the samples was 12,665 (coefficient of variation, 11%; Additional file 1: Figure S1, top panel). However, some of the inhibitors influenced the RT-ddPCR signals. An increase in the fluorescence of the negative droplets was observed with the seed extract sample, and to a lesser extent in the presence of the higher concentrations of tannic acid (Additional file 1: Figure S2, Figure 2). The observed

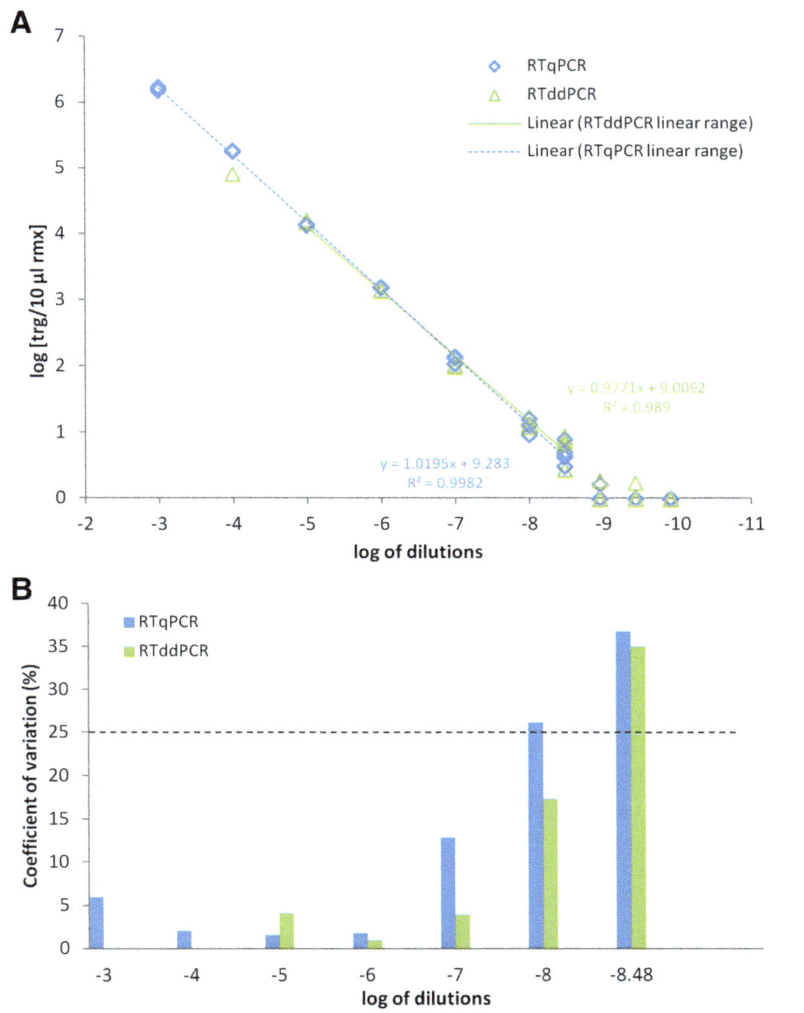

Figure 1 RT-qPCR and RT-ddPCR parallel analysis of PMMoV RNA serial dilutions. (A) Correlation between the log serial dilutions and determined concentrations (log target/10 µl reaction mixture) for the RT-qPCR and RT-ddPCR. For each method, the linear regression is plotted, with the equation for the correlation coefficient (R^2) given. **(B)** Coefficients of variation for the measurement of each of the serial dilutions within the linear range. The dotted horizontal line indicates the maximal acceptable coefficient of variation for the determination of the limit of quantification.

increases in the fluorescence of negative droplets were most likely due to the inherent fluorescence of seed extract and tannic acid in the channel of the fluorescent dye FAM. The mean signal of positive droplets showed a slightly higher variability among the samples; however, this effect did not always correlate with the concentration of the inhibitors (see, e.g., soil extract in Figure 2; Additional file 1: Figure S3). The signals of the positive and negative droplets were well separated in most cases, as assessed by the mean signals of the negative and positive droplets and their three-times standard deviations (Figure 2). A poorer cluster quality, seen only with the highest concentration of pectin and with the higher concentrations of tannic acid (Additional file 1: Figure S2 B), led to failure in the droplet classification by the Quanta Soft automatic analysis algorithm. The 'rain' was low in all cases (maximum 0.01% of accepted droplets; Additional file 1: Figure S1, lower panel). The inhibitors affected the RT-ddPCR in different ways: the seed extract generated an increase in the signal and number of negative droplets and a reduction in the number, but not the signal, of positive droplets; tannic acid generated only a minimal increase in the negative droplet signal, but a marked reduction in the signal and number of positive droplets, which led to their disappearance at the highest inhibitor concentrations; pectin did not have any effects on the signal of negative droplets, nor on the number of positive droplets, but it reduced the signal of the positive droplets, and affected their clustering. These different effects suggest that the different inhibitors can affect the RT-ddPCR process through different mechanisms and at different levels.

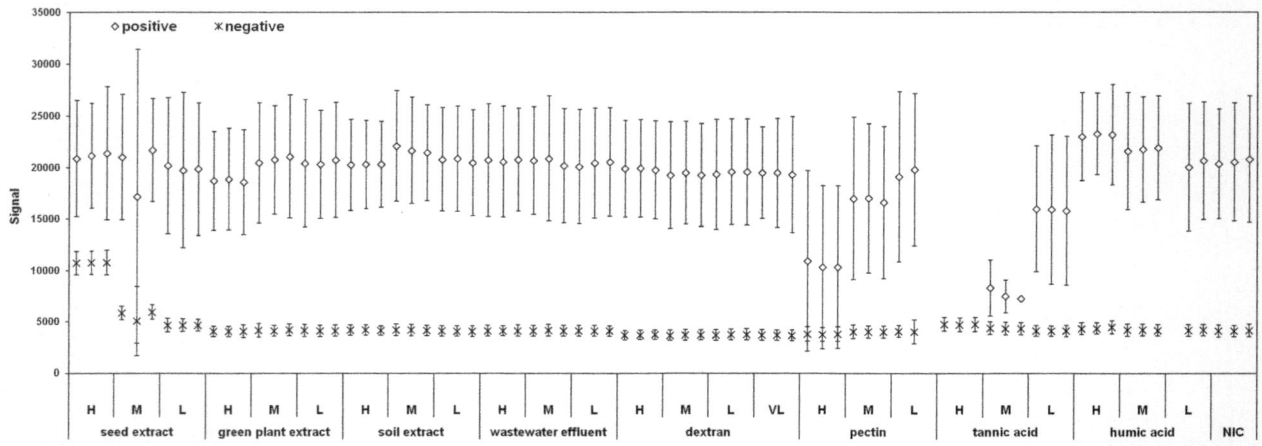

Figure 2 Mean signals of the positive and negative droplets from the RT-ddPCR for detection of PMMoV in samples spiked with serial dilutions of the inhibitor extracts from the selected matrices and from the chemical inhibitors. NIC, target RNA with no inhibitors; H, M, L, (VL), high, medium, low (and very low) concentrations of the inhibitor extracts, respectively. Three technical repeats were analyzed for each sample and inhibitor concentration. Error bars represent standard deviation of each signal.

The mean target copies per partition (droplet) (lambda; [26]) for the RT-ddPCR varied little within replicates for a given inhibitor concentration (average coefficient of variation 7%). However, most of the inhibitors tested led to an underestimation of lambda (Figure 3) in comparison to the no-inhibition control. The seed extract and tannic acid were the inhibitors that most markedly reduced the lambda, compared to the no-inhibition control. This reduction was most likely related to the strong effects exerted by these two inhibitors on the number of positive droplets (Additional file 1: Figure S2), and which resulted in a marked underestimation of the target concentration calculated in these samples (see below, Figures 4A and 5C).

Influence of inhibitors on the concentrations determined by the RT-qPCR and RT-ddPCR

The target concentrations were calculated for all of the samples with both qPCR and ddPCR. The comparisons between the target concentrations obtained in the absence and presence of different quantities of inhibitors were used to estimate the overall inhibition percentages (Figures 4 and 5), and to compare the inhibitor sensitivities inherent to each of these two methods.

Inhibitors from complex matrices

Four matrices were tested for their influence on PCR amplification: seeds, green plant material, soil, and wastewater effluent. The RT-qPCR and RT-ddPCR showed similar susceptibilities to all of the tested concentrations of the seed extract, from total inhibition at the highest tested concentration to partial, but still significant ($p < 0.05$), inhibition at lower concentrations (Figure 4A). In contrast, green plant extract only affected the RT-qPCR (Figure 4B), as the RT-ddPCR was not significantly inhibited. High concentrations of wastewater effluent partially inhibited both the RT-qPCR and RT-ddPCR. While this inhibition of the RT-qPCR was constant across all of the tested dilutions, the RT-ddPCR was no longer significantly inhibited at the medium and low wastewater effluent concentrations (Figure 4D). The soil extract showed concentration-dependent inhibitory effects on the RT-qPCR.

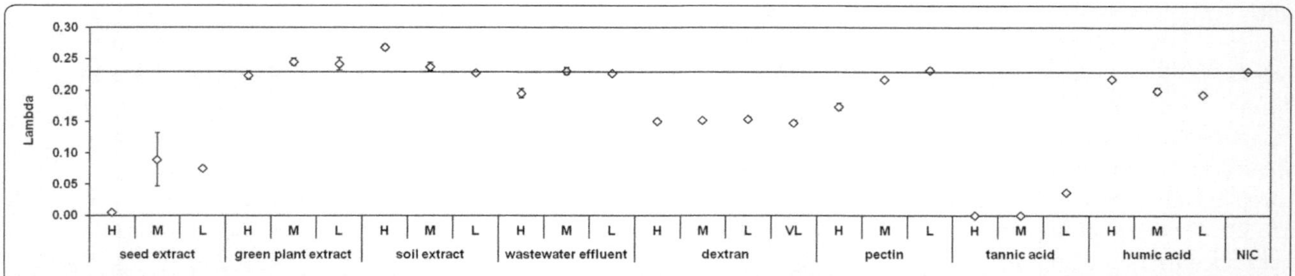

Figure 3 Mean copies per partition (lambda) determined in the control sample (NIC) with no inhibitor added and in the samples spiked with serial dilutions of the inhibitor extracts from the selected matrices and from the chemical inhibitors. H, M, L, (VL), high, medium, low (and very low) concentrations of inhibitors, respectively. Black line denotes the mean value of lambda for NIC.

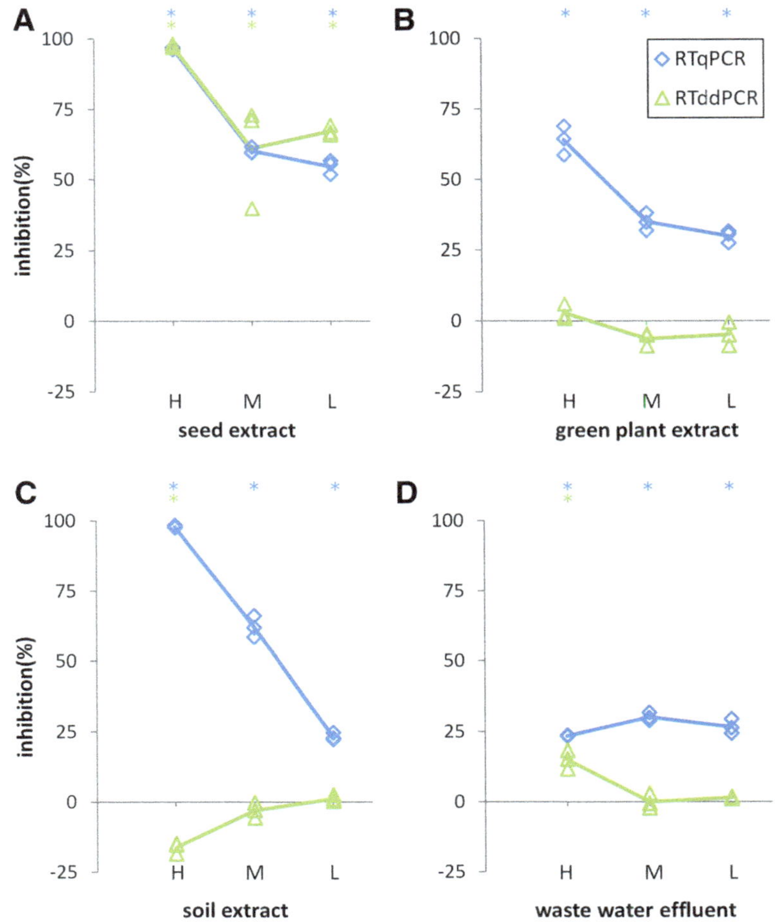

Figure 4 Influence of the samples spiked with serial dilutions of the inhibitor extracts on the PMMoV quantification by RT-qPCR (blue traces - diamonds) and RT-ddPCR (green traces - triangles). H, M, L, high, medium, low concentrations of inhibitors, respectively. The PMMoV concentrations are shown relative to those in the no-inhibition control, where 100% inhibition represents total inhibition (no positive signal obtained), and 0% inhibition represents no inhibition (target concentration the same as the no-inhibition control). **(A)** seed extract. **(B)** green plant extract. **(C)** soil extract. **(D)** wastewater effluent. The measurements were carried out in triplicate. Average values of the measurements for each method are connected by traces for better clarity. Asterisks (*) above each chart denote significantly ($p < 0.05$) different measurements, compared to 0% inhibition (no inhibition control) ± 10%; blue * for RT-qPCR and green * for RT-ddPCR.

In comparison, the highest concentration of the soil extract significantly ($p < 0.05$) enhanced the signal of the RT-ddPCR, which led to an overestimation of the target. Medium and low concentrations of this soil extract did not show this effect (Figure 4C).

Chemical inhibitors

Pectin and dextran sulfate were selected as representatives of neutral and acidic polysaccharides, respectively [5,19]. Tannic acid was selected as a representative of a polyphenolic substance, and humic acid as a major component of the humic substances found in environmental samples that are rich in plant organic matter [7].

The RT-qPCR and RT-ddPCR showed different sensitivities to these selected chemical inhibitors. At concentrations of dextran sulfate and humic acid where the RT-qPCR was totally inhibited, the RT-ddPCR showed significant tolerance to inhibition (Figure 5). Both the RT-qPCR and RT-ddPCR showed high susceptibilities to inhibition by tannic acid (Figure 5C), whereas pectin only significantly ($p < 0.05$) affected quantification of the RT-qPCR and RT-ddPCR reactions at the highest tested concentration (Figure 5A).

Discussion

An RT-qPCR assay that has been used for the detection and quantification of PMMoV, a plant pathogen and a proposed indicator of fecal contamination [27], was successfully transferred to an RT-ddPCR platform. The RT-qPCR and RT-ddPCR both detected low target concentrations; however, the precision of the absolute quantification was greater for the RT-ddPCR. This is in conclusion with our

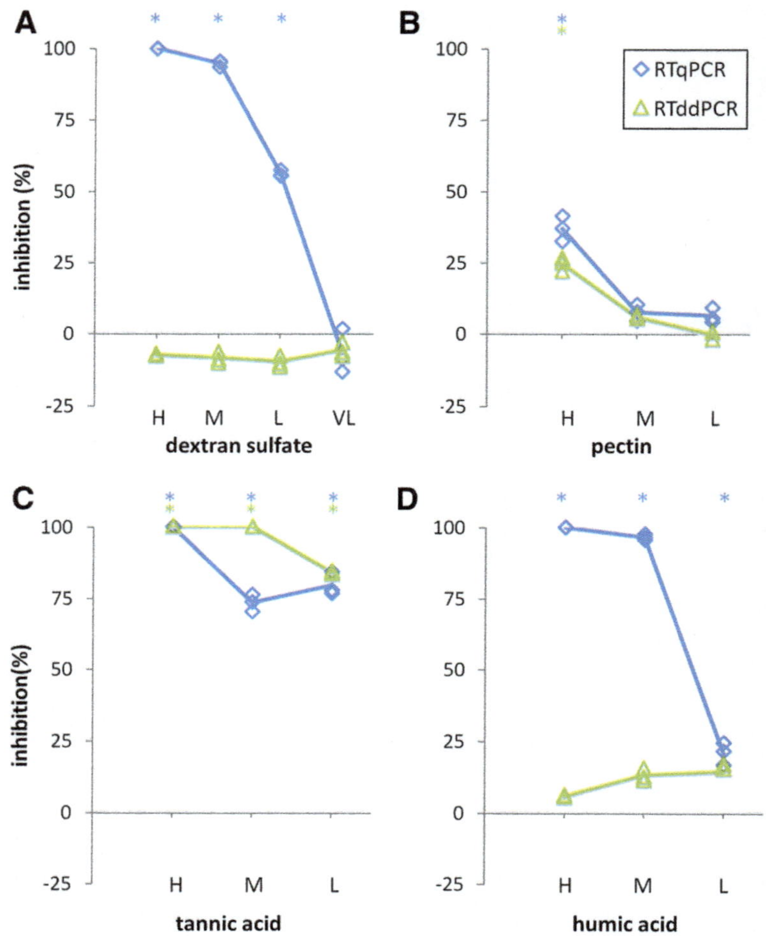

Figure 5 Influence of the samples spiked with serial dilutions of the chemical inhibitors on the PMMoV quantification by RT-qPCR (blue traces - diamonds) and RT-ddPCR (green traces - triangles). H, M, L, (VL), high, medium, low (and very low) concentrations of inhibitors, respectively. The PMMoV concentrations are shown relative to those in the no-inhibition control, where 100% inhibition represents total inhibition (no positive signal obtained), and 0% inhibition represents no inhibition (target concentration the same as the no-inhibition control). **(A)** dextran sulfate. **(B)** pectin. **(C)** tannic acid. **(D)** humic acid. The measurements were carried out in triplicate. Average values of the measurements for each method are connected by traces for better clarity. Asterisks (*) above each chart denote significantly ($p < 0.05$) different measurements, compared to 0% inhibition (no inhibition control) ± 10%; blue *for RT-qPCR and green *for RT-ddPCR.

previous experience with assay transfer from RTqPCR to RTddPCR platform [15]. In addition, the absolute quantification in the RT-ddPCR did not rely on reference materials, which is an advantage in many fields for which such materials are difficult to obtain.

Focusing on the RT-ddPCR, these inhibitors affected the monitored parameters differently. In highly inhibited samples, there were no positive drops, which indicated that the amplification was blocked, thus giving a negative result. At lower inhibition levels, the RT-ddPCR signals of positive and/or negative droplets were affected in terms of both the mean values and their distribution (i.e., the standard deviations of the signals). This can lead to failure of the analysis algorithm and lower accuracy of quantification. It has been previously reported that partial inhibition of PCR can lead to increased amounts of rain droplets; i.e., droplets with intermediate fluorescence. In the present study, the assay used for the RT-ddPCR enabled optimal separation of the signals with low levels of rain, whether the inhibitors were present or not. The seed extract and tannic acid affected both the signal levels and the distribution and number of positive droplets, which was reflected in the concentrations determined. Pectin affected the signal and distribution of the positive droplets, but not their number, and therefore the final concentrations were not affected. At higher inhibitor concentrations, humic acid, and more markedly, the soil extract, resulted in higher calculated target concentrations. All of these effects suggest different mechanisms of interference between these inhibitors and the processes in the RT-ddPCR reaction.

Overall, the RT-ddPCR showed higher resilience to inhibitors from the complex samples of the seed extract, plant extract, soil extract, and wastewater effluent than the RT-qPCR, as the inhibition of the RT-ddPCR measurements was typically lower at the same concentrations of inhibitor when compared to RT-qPCR. The RT-ddPCR also showed higher resilience to these selected chemical inhibitors, which represented those that can be commonly found in the complex matrices considered here. This is in agreement with our findings from previous experiments on rotavirus detection in wastewaters [15]. Together with the data of Dingle et al. [18] and Morisset et al. [12], our findings further support the hypothesis that a higher resilience to inhibitors is an inherent characteristic of ddPCR, when compared to qPCR, for both DNA and RNA quantification.

This higher resilience of the RT-ddPCR to such inhibitors can be largely explained by the different methods of signal acquisition between qPCR and ddPCR. qPCR quantification is solely based on the time point within the PCR reaction when the intensity of the fluorescence from the probe degradation increases above background (the cycle of quantification, or Cq; [1]). Due to logarithmic amplification, small differences between reactions are accumulated through each cycle, and these can significantly shift the Cq. Therefore, in qPCR, the efficiency of the primers and the probe annealing, and/or the amplification efficiency, can have large influences on the final results. On the contrary, ddPCR quantifies the target concentration at the end of the amplification, via the Poisson calculation of the ratio between the positive and negative partitions. The time (cycle) when the fluorescent signal in an individual partition increases above background is not relevant for the calculation of the target concentration. Therefore, this partial inhibition can be tolerated [18].

The inhibition observed with the seed extract and soil extract complex samples and with the pectin and tannic acid chemical compounds led to underestimation or overestimation of the target concentrations. These effects were, however, more pronounced for the RT-qPCR. As reported previously by Dingle et al. [18], such inhibitors can influence the mean signals and dispersion of both the positive and negative droplets. However, if the number of positive droplets is not affected and the separation between the positive and negative clusters is still optimal, data analysis can help to mask the inhibition. For example, although the seed extract spiked in the RT-ddPCR reaction increased the background fluorescence, the influence of the middle and lower concentrations of the seed extract on the quantification of the RT-ddPCR can be alleviated by adjustment of the threshold. In a similar way, a decrease in the threshold level allowed correct quantification at higher pectin concentrations.

Conclusions

Our data confirm a higher resilience of the PMMoV RT-ddPCR to inhibitors that are commonly found in samples of seeds, plant material, soil, and wastewater, in contrast to RT-qPCR. We have also confirmed that different inhibitors affect the ddPCR parameters in a different way, which suggests that there are different mechanisms of interference. Unlike to qPCR, ddPCR data analysis (i.e., the threshold level) can help achieving a correct quantification even in the presence of low to moderate inhibition. In combination with a higher absolute quantification accuracy and the lack of need for standard curve, this makes RT-ddPCR a suitable method for quantification of microbes, being especially beneficial for cases where highly accurate quantification of pathogens in complex matrixes is pursued, i.e., for quantitative microbial risk assessment.

Materials and methods

RNA isolation

PMMoV was provided by M. Botermansa and J. (Ko) Th. J. Verhoeven from the National Plant Protection Organization of The Netherlands, and was propagated in pepper (*Capsicum annuum* L.) plants under standard greenhouse conditions. RNA from PMMoV-infected pepper plants was isolated using RNeasy Plant Mini kits (Qiagen, CA, USA). Initially, 0.2 g plant material was homogenized in 900 µl RNeasy Plant Mini kit buffer using a FastPrep high-speed benchtop homogenizer (MP Biomedicals, CA, USA) and FastPrep™ Lysing Matrix A (MP Biomedicals) at 4.5 m/s for 40 s, and then incubated for 3 min at 56°C. After centrifugation at 13,000x *g* for 1 min, 550 µl lysate was transferred to a QIAshredder spin column placed in a 2 ml collection tube, and the RNA was isolated following the manufacturer protocol. The purified RNA was eluted in 100 µl molecular grade H_2O and stored at −20°C until analysis. The RNA was always denatured at 95°C for 5 min and kept on ice until its addition to the reaction.

Reverse transcription real-time quantitative PCR

The RNA of PMMoV was amplified using the primers and probe described by Haramoto et al. [25]. The reverse transcription and qPCR were combined (i.e., RT-qPCR) into a single step using AgPath-ID™ One-Step RT-PCR kits (Life Technologies, CA, USA). The final reaction volume of 10 µl contained 900 nM primers, 200 nM probe, and 2 µl sample (isolated RNA, or molecular grade RNAse-free water for the no-template controls). Plates were analyzed using a 7900HT Fast Real-Time PCR system (Applied Biosystems, CA, USA). The thermal cycling conditions were as recommended in the AgPath kit manual. The data were acquired and analyzed using the SDS 2.4 software (Applied Biosystems, CA, USA). The threshold was set manually at 0.2 (a level that was above the baseline and sufficiently low to be within the exponential increase region of the amplification curve), and the baseline was set automatically.

Reverse transcription droplet digital PCR

For the RT-ddPCR reaction, One-Step RT-ddPCR kits for probes (Bio-Rad, CA, USA) were used. Each sample was analyzed in at least triplicate. The final reaction volume of 20 μl contained 900 nM primers, 200 nM probe, and 4 μl sample (isolated RNA, or molecular grade RNAse free water for no-template controls). As the volume of RT-ddPCR reaction was twice the volume of RT-qPCR reaction, to ensure comparable concentrations, all the added reagents and samples were twice the volumes of those added to the RT-qPCR. The further procedures were as described by Rački et al. [15]. The positive droplets that contained amplification products were discriminated from the negative droplets by applying the Quanta Soft automatic analysis of individual wells or by manually selecting a fluorescence threshold (Bio-Rad, CA, USA). The threshold was set to 7000 fluorescence units, except where it was necessary to adjust it to include positive droplets that had shifted due to partial inhibition (Additional file 1: Figure S2). The data generated by the QX 100 droplet reader were rejected from subsequent analysis if a low number of total droplets was accepted (measured) per 20 μL PCR (<8,000), or if all of the droplets were positive (saturation of the reaction).

Comparison of RT-qPCR and RT-ddPCR performances

A dynamic range of quantification of the PMMoV assays [25] was compared on the qPCR and ddPCR platforms using 10-fold dilutions, and at lower concentrations (3-fold dilutions) of PMMoV RNA in molecular grade water (Sigma, MO, USA). The same RNA dilutions were used for the preparation of reactions for both of these methods. For the comparison of the results, the Cq values for the RT-qPCR and RT-ddPCR concentrations were expressed as copies of target per 10 μl reaction mix (trg/10 μl rmx). For the conversion of Cq to trg/10 μl rmx, the RT-qPCR PMMoV assay had to be calibrated. Due to a lack of availability of a prequantified PMMoV standard, the concentration of PMMoV target copies in the starting preparation of PMMoV RNA (used to prepare the serial dilutions) was calculated from the RT-ddPCR measurements. The most accurately calculated target concentration (lowest coefficient of variation between repetitions, corresponding to the 10^6 dilution; Figure 1) was chosen and used to calculate back the target concentrations in the rest of the dilutions. Then linear regression (target concentration vs Cq) was used to derive the number of detected targets from the Cq of the samples [28]. The limit of quantification for both of the methods was determined as the lowest dilution of PMMoV RNA that could be quantified with a coefficient of variation between technical repetitions ≤25% [29].

PCR inhibitors

Two groups of inhibitor sources were selected to evaluate their influence in both RT-qPCR and RT-ddPCR. The first type involved complex matrices that are prone to cause inhibition of PCR; i.e., a seed extract, green plant material, soil and wastewater. The second group involved chemical compounds that are known to be present in plant and/or environmental samples: pectin (from citrus fruit; P9135; Sigma-Aldrich; Germany), dextran sulfate (sodium salt, from *Leuconostoc ssp.*; 31403; Sigma-Aldrich; Germany), tannic acid (*acidum tanicum*; 16201; Riedel de Haën; Germany) and humic acid (53680; Sigma-Aldrich; Germany). The extracts were prepared as follows: (i) seed extract: 150 *C. annuum* seeds were soaked overnight in 10 ml 0.1 M phosphate buffer at 4°C; (ii) soil extract: 1 g dry planting soil (Archut Fruchstofer Erde; Germany) was soaked overnight in 10 ml 0.1 M phosphate buffer at 4°C; and (iii) green plant extract: 2 g frozen healthy green leaves of *C. annuum* were added to 10 ml 0.1 M phosphate buffer without incubation. Each of the three samples was homogenized with a FastPrep high-speed benchtop homogenizer (MP Biomedicals) at 4.5 m/s for 40 s, using FastPrep™ Lysing Matrix A (MP Biomedicals), and centrifuged at 10,000x *g* for 10 min at 4°C. The supernatants were collected and stored at −20°C until used. Wastewater effluent was obtained from a local wastewater treatment plant (Central Wastewater Treatment Plant Domžale-Kamnik; Ihan, Slovenia) and filtered (Whatman grade 5 H/N paper; Sartorius) and stored as for the extracts. Then, 5-fold serial dilutions of the wastewater effluent and the extracts were prepared to challenge the RT-qPCR and RT-ddPCR reactions. The chemical compound inhibitors were diluted in ultrapure milliQ water (EMD Millipore; MA, USA), as 1% (w/v) solutions of pectin, dextran sulfate, tannic acid and humic acid. Again, 5-fold serial dilutions were added to the RT-qPCR and RT-ddPCR reactions. Where possible, the starting concentration of the inhibitor was selected to be high enough to cause total inhibition of the PMMoV RT-qPCR assay. At least three dilutions were tested for each inhibitor.

Evaluation of inhibition

The influences of the inhibitors on both the RT-qPCR and RT-ddPCR reactions were assessed relative to the mean measured signals in each sample with no added inhibitors (no-inhibition control, as milliQ water plus PMMoV RNA). The effects of the inhibition on the determined concentrations are given as % inhibition, where 100% inhibition represented total inhibition (no positive signal), and 0% inhibition represented no inhibition (obtained target concentration the same as the no-inhibition control). The same amount of PMMoV RNA was added to each reaction.

Signals obtained from no inhibition control and signals obtained from reactions, challenged with different inhibitor concentrations (Figures 4 and 5) were tested using the general linear hypothesis method (R package multcomp, [30]).

However, because of high repeatability of the results, all the differences were statistically significant ($p < 0.05$), even the ones that in practical term were the same. Therefore to give statistical analysis some practical relevance, tolerance of 0% inhibition ±10% was introduced i.e. a difference of measured signal was considered relevant, if it was significantly ($p < 0.05$) different from the signal obtained from no inhibition control ±10%.

To define the influence of the inhibitors on the different parameters for the RT-ddPCR, the key parameters likely to be affected by the presence of the inhibitors were monitored: (i) signal levels and distribution; and (ii) determined concentrations. The percentage of the 'rain' droplets out of all of the accepted droplets was derived from the number of droplets with signals in the range between the mean amplitude of the negative droplets plus three standard deviations, and the mean amplitude of the positive droplets minus three standard deviations; i.e., the droplets outside the majority of the normally distributed signals.

Additional files

Additional file 1: Table S1. Performance of PMMoV assay by RT-qPCR and RT-ddPCR. Measurements of dilutions of target PMMoV RNA by both platforms are presented. **Figure S1.** Number of accepted droplets in each reaction (top), percent of droplets with intermediate fluorescence ('rain'; bottom), as influenced by the spiked serial dilutions of the inhibitor extracts from the selected matrices and from the chemical inhibitors in the RT-ddPCR quantification of PMMoV. Control, target RNA with no inhibitors; bias, systematic distortion; H, M, L (L*), high, medium, low (and lowest) concentrations of inhibitor, respectively. Black line denotes the minimal number of accepted droplets (8000) that had to be obtained in order for measurement to be considered valid. **Figure S2.** Signals for the ranges of the positive droplets, as influenced by the spiked serial dilutions of the inhibitor extracts from the selected matrices and from the chemical inhibitors in the RT-ddPCR quantification of PMMoV. Darker colors correspond to higher inhibitor concentrations. On the bottom no-template control (NTC) and no inhibition control (NIC) are shown. **Figure S3.** Droplet plots of the representative RT-ddPCR reactions spiked with serial dilutions of the inhibitor extracts (see legend) from the selected matrices (A) and from the chemical inhibitors (B). For each dilution of the inhibitors, only one of three repetitions is shown. Grey droplets, negative for PMMoV; blue droplets, positive for PMMoV; pink line, position of the threshold for each sample. For easier comparison of the influence of each inhibitor on the amplitude plots, the corresponding no inhibition control (NIC) and no-template control (NTC) are shown for every plot.

Competing interests
The authors declare that they have no competing interests.

Authors' contributions
Conceived and designed the experiments: NR, TD, IGA, MR. Performed the experiments: NR. Analyzed the data: NR, TD, AB. Wrote the paper: NR, TD, IGA, MR. All authors read and approved the final manuscript.

Acknowledgments
The authors are grateful to M. Botermansa and J. (Ko) Th. J. Verhoeven from the National Plant Protection Organization of the Netherlands for providing the initial PMMoV material. We also thank H. Prosen from the Faculty of Chemistry and Chemical Technology, University of Ljubljana, for providing the humic acid, S. Leskovec from the Biotechnical Faculty, University of Ljubljana, for providing the tannic acid, and Barbara Brajer Humar and Marjeta Stražar from the Domžale-Kamnik central wastewater treatment plant for providing the wastewater effluent sample. This study was financially supported by the Slovenian Research Agency (contract no. L2-4314 and 34504). The ddPCR equipment used in this study was financed by the Metrology Institute of the Republic of Slovenia, with financial support from the European Regional Development Fund.

References
1. Bustin SA: *A to Z of Quantitative PCR*. International University Line; 2004. [*IUL Biotechnology Series*]
2. Girones R, Ferrús MA, Alonso JL, Rodriguez-Manzano J, Calgua B, Corrêa ADA, Hundesa A, Carratala A, Bofill-Mas S: **Molecular detection of pathogens in water–the pros and cons of molecular techniques.** *Water Res* 2010, **44**:4325–4339.
3. Boonham N, Kreuze J, Winter S, van der Vlugt R, Bergervoet J, Tomlinson J, Mumford R: **Methods in virus diagnostics: from ELISA to next generation sequencing.** *Virus Res* 2014, **186**:20–31.
4. Cankar K, Stebih D, Dreo T, Zel J, Gruden K: **Critical points of DNA quantification by real-time PCR–effects of DNA extraction method and sample matrix on quantification of genetically modified organisms.** *BMC Biotechnol* 2006, **6**:37.
5. Demeke T, Jenkins GR: **Influence of DNA extraction methods, PCR inhibitors and quantification methods on real-time PCR assay of biotechnology-derived traits.** *Anal Bioanal Chem* 2010, **396**:1977–1990.
6. Gibson KE, Schwab KJ, Spencer SK, Borchardt MA: **Measuring and mitigating inhibition during quantitative real time PCR analysis of viral nucleic acid extracts from large-volume environmental water samples.** *Water Res* 2012, **46**:4281–4291.
7. Opel KL, Chung D, Mccord BR: **A study of PCR inhibition mechanisms using real time PCR.** *J Forensic Sci* 2010, **55**:25–33.
8. Wilson IG: **Inhibition and facilitation of nucleic acid amplification.** *Appl Environ Microbiol* 1997, **63**:3741–3751.
9. Trombley Hall A, McKay Zovanyi A, Christensen DR, Koehler JW, Devins Minogue T: **Evaluation of inhibitor-resistant real-time PCR methods for diagnostics in clinical and environmental samples.** *PLoS One* 2013, **8**:e73845.
10. Pirc M, Ravnikar M, Tomlinson J, Dreo T: **Improved fireblight diagnostics using quantitative real-time PCR detection of Erwinia amylovora chromosomal DNA.** *Plant Pathol* 2009, **58**:872–881.
11. Baker M: **Digital PCR hits its stride.** *Nat Methods* 2012, **9**:541–544.
12. Morisset D, Štebih D, Milavec M, Gruden K, Žel J: **Quantitative analysis of food and feed samples with droplet digital PCR.** *PLoS One* 2013, **8**:e62583.
13. Sedlak RH, Jerome KR: **Viral diagnostics in the era of digital polymerase chain reaction.** *Diagn Microbiol Infect Dis* 2013, **75**:1–4.
14. Hayden RT, Gu Z, Ingersoll J, Abdul-Ali D, Shi L, Pounds S, Caliendo aM: **Comparison of droplet digital PCR to real-time PCR for quantitative detection of cytomegalovirus.** *J Clin Microbiol* 2013, **51**:540–546.
15. Rački N, Morisset D, Gutierrez-Aguirre I, Ravnikar M: **One-step RT-droplet digital PCR: a breakthrough in the quantification of waterborne RNA viruses.** *Anal Bioanal Chem* 2013.
16. Hindson CM, Chevillet JR, Briggs H, Gallichotte EN, Ruf IK, Hindson BJ, Vessella RL, Tewari M: **Absolute quantification by droplet digital PCR versus analog real-time PCR.** *Nat Methods* 2013, **10**:1003–1005.
17. Nixon G, Garson JA, Grant P, Nastouli E, Foy CA, Huggett JF: **Comparative study of sensitivity, linearity, and resistance to inhibition of digital and nondigital polymerase chain reaction and loop mediated isothermal**

amplification assays for quantification of human cytomegalovirus. *Anal Chem* 2014, **86:**4387–4394.
18. Dingle TC, Sedlak RH, Cook L, Jerome KR: **Tolerance of droplet-digital PCR vs real-time quantitative PCR to inhibitory substances.** *Clin Chem* 2013, **59:**1670–1672.
19. Peist R, Honsel D, Twieling G, Löffert D: **PCR inhibitors in plant DNA preparations.** *QIAGEN news* 2001, **3:**7–9.
20. Wetter C, Conti M, Altschuh D, Tabillion R v RM: **Pepper mild mottle virus, a tobamovirus infecting pepper cultivars in sicily.** *Phytopathology* 1984, **74:**405.
21. Colson P, Richet H, Desnues C, Balique F, Moal V, Grob J-J, Berbis P, Lecoq H, Harlé J-R, Berland Y, Raoult D: **Pepper mild mottle virus, a plant virus associated with specific immune responses, Fever, abdominal pains, and pruritus in humans.** *PLoS One* 2010, **5:**e10041.
22. Zhang T, Breitbart M, Lee WH, Run J, Wei CL, Wee S, Soh L, Hibberd ML, Liu ET, Rohwer F, Ruan Y: **RNA viral community in human feces: prevalence of plant pathogenic viruses.** 2006, **4.**
23. Mehle N, Ravnikar M: **Plant viruses in aqueous environment - survival, water mediated transmission and detection.** *Water Res* 2012, **46:**4902–4917.
24. Hamza IA, Jurzik L, Uberla K, Wilhelm M: **Evaluation of pepper mild mottle virus, human picobirnavirus and Torque teno virus as indicators of fecal contamination in river water.** *Water Res* 2011, **45:**1358–1368.
25. Haramoto E, Kitajima M, Kishida N, Konno Y, Katayama H, Asami M, Akiba M: **Occurrence of pepper mild mottle virus in drinking water sources in Japan.** *Appl Environ Microbiol* 2013, **79:**7413–7418.
26. Huggett JF, Foy CA, Benes V, Emslie K, Garson JA, Haynes R, Hellemans J, Kubista M, Mueller RD, Nolan T, Pfaffl MW, Shipley GL, Vandesompele J, Wittwer CT, Bustin SA: **Guidelines for minimum information for publication of quantitative digital PCR experiments.** 2013, **000**
27. Rosario K, Symonds EM, Sinigalliano C, Stewart J, Breitbart M: **Pepper mild mottle virus as an indicator of fecal pollution.** *Appl Environ Microbiol* 2009, **75:**7261–7267.
28. Bustin S: **Absolute quantification of mRNA using real-time reverse transcription polymerase chain reaction assays.** *J Mol Endocrinol* 2000, **25:**169–193.
29. European Network of GMO Laboratories : **Definition of minimum performance requirements for analytical methods of GMO testing european network of GMO laboratories.**2008, 1–8.
30. Hothorn T, Bretz F, Westfall P: **Simultaneous inference in general parametric models.** *Biom J* 2008, **50:**346–363.

14

Mapping mutations in plant genomes with the user-friendly web application CandiSNP

Graham J Etherington[1,2†], Jacqueline Monaghan[1†], Cyril Zipfel[1] and Dan MacLean[1*]

Abstract

Background: Analysis of mutants isolated from forward-genetic screens has revealed key components of several plant signalling pathways. Mapping mutations by position, either using classical methods or whole genome high-throughput sequencing (HTS), largely relies on the analysis of genome-wide polymorphisms in F_2 recombinant populations. Combining bulk segregant analysis with HTS has accelerated the identification of causative mutations and has been widely adopted in many research programmes. A major advantage of HTS is the ability to perform bulk segregant analysis after back-crossing to the parental line rather than out-crossing to a polymorphic ecotype, which reduces genetic complexity and avoids issues with phenotype penetrance in different ecotypes. Plotting the positions of homozygous polymorphisms in a mutant genome identifies areas of low recombination and is an effective way to detect molecular linkage to a phenotype of interest.

Results: We describe the use of single nucleotide polymorphism (SNP) density plots as a mapping strategy to identify and refine chromosomal positions of causative mutations from screened plant populations. We developed a web application called CandiSNP that generates density plots from user-provided SNP data obtained from HTS. Candidate causative mutations, defined as SNPs causing non-synonymous changes in annotated coding regions are highlighted on the plots and listed in a table. We use data generated from a recent mutant screen in the model plant *Arabidopsis thaliana* as proof-of-concept for the validity of our tool.

Conclusions: CandiSNP is a user-friendly application that will aid in novel discoveries from forward-genetic mutant screens. It is particularly useful for analysing HTS data from bulked back-crossed mutants, which contain fewer polymorphisms than data generated from out-crosses. The web-application is freely available online at http://candisnp.tsl.ac.uk.

Keywords: High-throughput sequencing, Single nucleotide polymorphisms, Forward-genetics, Mapping, Web application

Background

Carefully designed forward-genetic screens have been an integral part of research programs for decades and remain an important tool for resolving biological pathways. Many proteins contributing to plant immune signalling have been discovered through such screens. As one example, the receptor kinase FLAGELLIN SENSING 2 (FLS2) was identified from a mutagenized *Arabidopsis thaliana* (hereafter, Arabidopsis) population as the receptor for bacterial flagellin [1]. The discovery of FLS2 and other surface-localized immune receptors that detect conserved molecular features of microbes (known as pathogen-associated molecular patterns; PAMPs) revolutionized our understanding of plant immunity [2], and reinforces the importance of genetic screens in modern research.

Genetic screens in all systems are based on similar principles. Individuals containing a phenotype of interest are first isolated from a mutagenized or naturally polymorphic population. Marker-assisted linkage analysis is then performed to identify the genomic region containing the underlying mutation(s). Finally, mutations are identified by sequence analysis and the causative mutation is usually confirmed by complementation with a non-mutated (wild-type) copy of the gene.

The most commonly used mutagenesis strategies in Arabidopsis include the induction of guanine-to-adenine

* Correspondence: dan.maclean@tsl.ac.uk
†Equal contributors
[1]The Sainsbury Laboratory, Norwich Research Park, Norwich NR4 7UH, UK
Full list of author information is available at the end of the article

substitutions using ethylmethane sulfonate (EMS) or the insertion of transfer-DNA using *Agrobacterium tumefaciens*-mediated transformation [3]. The number of mutations identified in a mapped region depends primarily on the mutagenesis. Increasing the strength of the mutagen will likely result in the recovery of more mutants containing the phenotype of interest, however this also results in more mutations in each mutant genome and can complicate correct gene identification. Mapping mutations by position classically involves out-crossing to a polymorphic ecotype and linking the phenotype of recombinant F_2 individuals to molecular markers with known genomic positions, such as insertion/deletions (indels) or single nucleotide polymorphisms (SNPs). Rather than genotyping individual recombinants exhibiting the scored phenotype, linkage analysis can be performed on bulked recombinants. This process, referred to as bulk segregant analysis (BSA) [4], eases genetic analysis and is particularly effective when a large number of molecular markers are available. Although robust, these classical methods are time-consuming and labour-intensive, commonly taking more than a year (in the case of Arabidopsis) to correctly identify the causative mutation. Positional cloning can be particularly tedious in suppressor or modifier screens, where multiple loci are segregating in the mapping population. Correct identification of causative mutations depends greatly on the strength of the mutagenesis, the complexity of the cross, the penetrance of the phenotype, and the availability of molecular markers.

Recent advances in high-throughput sequencing (HTS) technologies have allowed for rapid identification of causative mutations and have been widely adopted in many fields [5]. HTS approaches have many advantages over classical mapping strategies, including a potential reduction in time and personnel needed to identify a causative mutation. Combining HTS with BSA has proven to be particularly useful [5]. Several expert operator methods utilizing HTS of bulked segregants have been described to assist plant genetic screens, including SHOREmap [6,7], Next-Generation Mapping (NGM) [8], MutMap [9-11], and others [5]. However, each of these has restrictions in their functioning that make them limited to certain types of crosses or are difficult to use for non-experts including bench-trained biologists.

The two major tools used by plant researchers start with a classical mapping approach, requiring data generated from out-crosses. SHOREmap [6,7] uses a statistic that explicitly calculates the relative abundance of alleles identified in bulked out-crossed F_2 populations and therefore relies on *a priori* knowledge of polymorphic allele positions in both the parental and out-crossed ecotypes. As a result, this powerful tool cannot be used when such crosses are not performed (for example, in back-crossed populations) or when genetic marker resources are not yet available. Comparatively, the marker-independent, web-based method NGM [8], does not rely on previous knowledge of ecotype-specific polymorphisms, but rather uses the ratio of the expected allele frequency of the causative mutation in bulked out-cross F_2 segregants relative to the background allele frequency of unrelated mutations. This two-step process relies first on identifying a coarse mapping interval of relative SNP paucity in which the mutation should lie; this region is usually in the order of megabases in length [8]. To further reduce the width of the coarse interval and ease identification of the causative mutation, the SNP frequency in different bands of SNP allele frequencies (overlapping groupings of similar allele frequencies; *e.g.*, 0.5-0.6, 0.51-0.61, etc.) are compared. The point that maximises the ratio between bands representing homozygous alleles and heterozygous alleles at 50% frequency is expected to be at or around the causative mutation. The MutMap [10,11] and CloudMap [12] systems avoid the need for coarse mapping as an initial step but do not exist as user-friendly tools such that application and optimisation of parameters requires extensive expertise in a command-line computing environment.

We recently conducted a forward-genetic modifier screen in the immune-deficient *bak1-5* background to identify novel components involved in plant immune signalling [13]. BRASSINOSTEROID INSENSITIVE 1-ASSOCIATED KINASE 1 (BAK1) is a multi-functional co-receptor that interacts with and phosphorylates several surface-localized immune receptors including FLS2 [14-19]. Accordingly, loss-of-function *bak1* alleles are strongly impaired in signalling triggered by several PAMPs [15,17-19]. We mutagenized *bak1-5* seeds (in the Columbia-0 (Col-0) ecotype) with EMS and screened the M_2 generation for *modifier of bak1-5* (*mob*) mutants that restored immune signalling. To uncover the causal *mob* mutation(s), we back-crossed *bak1-5 mob* mutants to the parental line (*bak1-5*) and Illumina sequenced bulked F_2 *mob* segregants. Importantly, as the parent was itself generated through EMS-mutagenesis of a transgenic Arabidopsis line [15,20,21], we additionally sequenced *bak1-5*, which had been back-crossed for three generations prior to mutagenesis, as a reference.

We chose this approach over out-crossing to ease phenotyping and segregation analysis. First, selection of the *mob* mutant phenotype required scoring a quantitative response dependent on immune receptor activity, which varies in different ecotypes. For example, *FLS2* in the Wassilewskija-0 (Ws-0) ecotype contains a deletion mutation resulting in a truncated and non-functional FLS2 receptor [22], while *FLS2* in the Landsberg *erecta*-0 (L*er*-0) ecotype contains polymorphisms that cause FLS2 to bind flagellin about three times stronger than FLS2 in Col-0 [23]. Second, selection of the *mob* mutant phenotype was dependent on *bak1-5*, which would not be present in a polymorphic ecotype such as L*er*-0 and would thus need to be genotyped prior to phenotype scoring. Similar considerations would likely arise in any screen involving second-site modifier or suppressor mutations.

While a back-cross simplifies genetic analysis, bulk segregant sequence analysis is complicated by far fewer segregating SNPs (1 SNP every 65,000 bp) compared to out-crosses (1 SNP every 900 bp) (Additional file 1). Although few in number, we found that simply plotting the position of SNPs with close-to-homozygous alternate allele frequencies along the chromosomes was a convenient and easy way of performing bulk segregant linkage analysis from a back-crossed population. We developed this method into a user-friendly web-based application, called CandiSNP, which generates density plots from SNP data obtained from HTS. We demonstrate the utility of CandiSNP by analysing sequence data generated from two allelic *mob* mutants, *bak1-5 mob1* and *bak1-5 mob2*, which are caused by mutations in the gene encoding the calcium-dependent protein kinase CPK28 [13].

Implementation

CandiSNP is part of a straightforward and flexible workflow

To provide a publicly available easy-to-use tool for mapping mutations, we developed the CandiSNP web application (Figure 1). Prior to using CandiSNP, users must identify SNP positions. Typically this would be done in a workflow that maps quality-controlled (QC) reads to the appropriate reference genome and provides SNP position data as input (Figure 2A). The data must be provided in a simple comma-delimited format and must have the following column headers: 'Chr', 'Pos', 'Ref', 'Alt' and

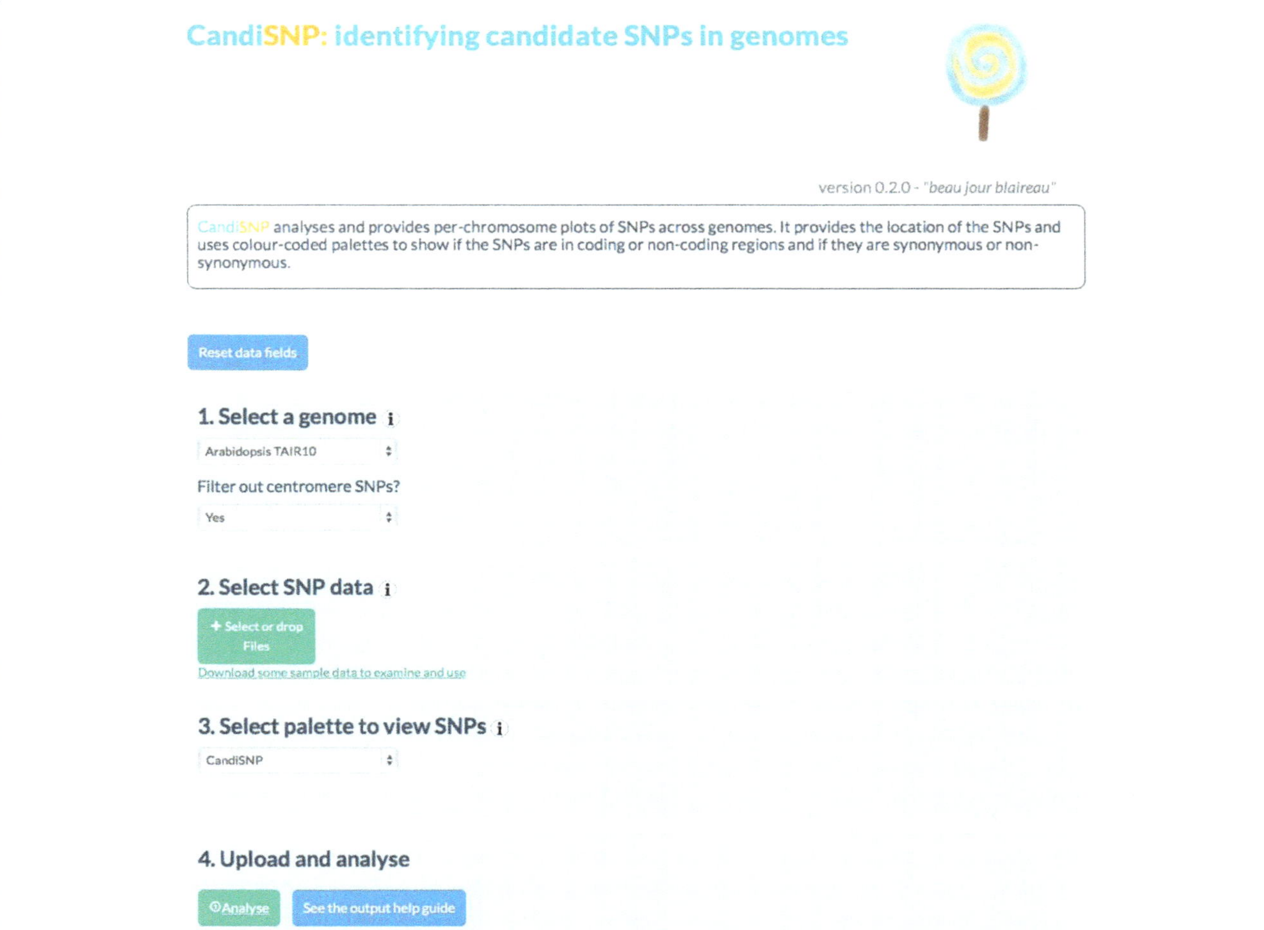

Figure 1 Screen-shot of the CandiSNP web application. CandiSNP is openly accessible online at http://candisnp.tsl.ac.uk. The application is laid out so users can make their way through the application in numbered steps. Users choose which genome they would like to use for comparison (the program currently supports Arabidopsis, rice, tomato, grape, maize and soybean genomes). The option of filtering SNPs concentrated around centromeres is also provided. Users then upload their SNP data file, indicate their preferred allele frequency cut-off, and choose from a number of different palettes for SNP visualization.

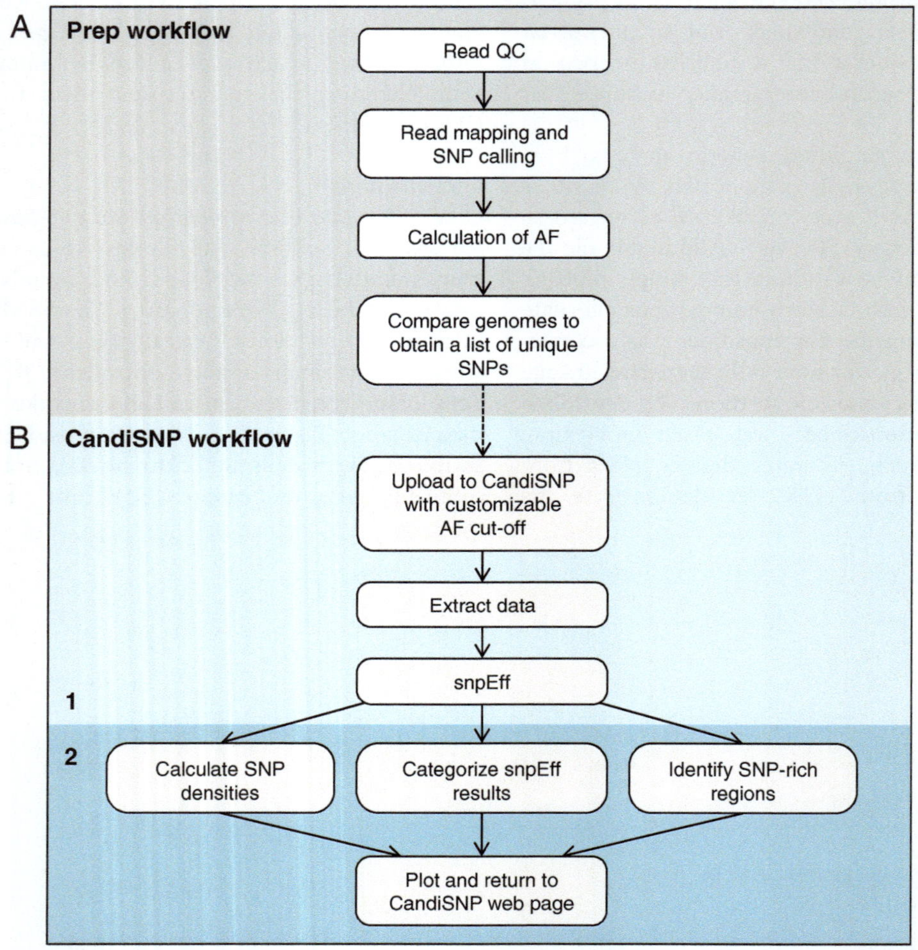

Figure 2 Bioinformatics pipeline for sequence analysis. Pipeline indicating the preparatory steps required by the user **(A)** prior to running the CandiSNP web application **(B)**.

'Allele_Freq' (meaning: Chromosome, Position, Reference base, Alternate base, and Allele Frequency, respectively). To create CandiSNP input files, we suggest using the pileups_to_snps.rb Ruby script (Additional file 2 and https://github.com/danmaclean/candisnp/blob/master/pileup_to_snps.rb). By using this generic and flexible input format, our system allows the user to take advantage of existing pipelines and data, including files generated by external service providers or even datasets originating from other technologies.

CandiSNP analysis is a two-step process (Figure 2B). CandiSNP first uses snpEff [24] to categorise SNPs according to their position in genomic features. The SNP predictions are categorised as: (1) Causing a change in an intergenic (non-annotated) region, (2) Causing a synonymous change in an annotated protein-coding region, or (3) Causing a non-synonymous change in an annotated protein-coding region. CandiSNP then creates a chromosome map visualizing the position of all SNPs meeting a user-selected (and customizable) alternate allele frequency (AF) threshold, and renders this information by colouring SNPs according to category. SNPs in category (3) are represented in a colour that highlights their priority as putative causative SNPs. The density and distribution of SNPs is also visualized as a line graph below each chromosome. The user is provided with a downloadable list of genomic effects for all inputted SNPs and is provided with a Scalable Vector Graphic figure of publication quality that can be easily exported.

Currently, CandiSNP supports several plant genomes including *Arabidopsis thaliana* Col-0 TAIR9 and TAIR10 [25], *Oryzae sativa* v7 [26], *Solanum lycopersicum* v2.40 [27], *Glycine max* 1.09v8 [28], *Vitus vinefera* v1 [29] and *Zea mays* B73 v5b [30].

Design and availability

CandiSNP is available in multiple formats for users with diverse security and confidentiality needs and differing access to computational infrastructure. Primarily, CandiSNP is provided as a web application, available at http://candisnp.tsl.ac.uk. The web application takes text files as input. Instructions are provided on-screen to assist users new to the tool. The web application requires no registration and does not collect user information. For laboratories with bioinformatics support wishing to use an internal and private version of the web application, we provide a package and source code in Perl/HTML/Javascript for free download and use under the GNU GPL3 Licence, from the dedicated code hosting website GitHub at https://github.com/danmaclean/candisnp. For those wishing to run the CandiSNP process on a command line as part of bioinformatics pipelines, a Perl module is also available as part of the source code.

Results and discussion

Case study

Bulk segregant analysis of two mob mutants using Illumina sequencing identifies thousands of polymorphisms

As a case study for CandiSNP we examined HTS data obtained for two allelic recessive mutants, *bak1-5 mob1* and *bak1-5 mob2*, that were isolated from the *modifier of bak1-5* (*mob*) screen [13]. Both *bak1-5 mob1* and *bak1-5 mob2* were back-crossed with *bak1-5* and the F_2 populations were screened for the *mob* phenotype (Figure 3A). F_2 segregants that displayed the *mob* phenotype were bulked and genomic DNA was isolated. The *bak1-5* parental genome was prepared by harvesting individuals from a homozygous back-crossed line. DNA samples were sent to the Beijing Genomics Institute (BGI, Hong Kong) for library construction and 90 bp paired-end sequencing on the Illumina HiSeq platform.

With the aid of FASTQC v 0.10.1 [31] in the Galaxy platform [32-34], all reads were quality controlled (QC) so that reads that contained undefined nucleotides, were not 90 bp long, or were full-length homopolymer runs were removed. Reads containing nucleotides with a Sanger-scaled PHRED quality score of less than 10 at the 3′ end were trimmed to this minimum using Sickle version 1.21.0 [35]. The QC pipeline is available as a Galaxy workflow at http://dx.doi.org/10.6084/m9.figshare.1248898.

QC reads were then mapped to the TAIR10 genome [25] using the BWA v 0.6.1 aligner [36]. For the *bak1-5* genome, we aligned 47.9 million 90 bp paired-end reads, with a mean insert size of 467 bp, to the TAIR10 Arabidopsis reference sequence (98.5% of reads aligned). We similarly aligned 47.2 million 90 bp paired-end reads, with a mean insert size of 452 bp, to TAIR10 for the *bak1-5 mob2* genome (99.6% of reads aligned). Average alignment depth over the nuclear chromosomes was 36 for *bak1-5* and 59.3 for *bak1-5 mob2*. Details regarding the *bak1-5 mob1* genome sequence have been previously described [13].

After alignment, SNPs were identified and allele frequencies were calculated using SAMtools v 0.1.8 [37]. Reads with mapping quality scores less than 20 and individual bases with sequence quality less than 20 were discarded. Genome positions where the reference base was unknown were excluded. Positions were considered SNPs if they had a minimum read coverage of 6 and a maximum of 250. The alignment and SNP calling workflow is available at http://dx.doi.org/10.6084/m9.figshare.1171109. In total, *bak1-5* contained 2,639 SNPs compared to Col-0, while *bak1-5 mob1* and *bak1-5 mob2* contained 4,188 and 3,581 SNPs, respectively (Table 1).

Filtering of non-unique SNPs in the mutants reduces complexity

To reduce the complexity of the *bak1-5 mob1* and *bak1-5 mob2* datasets, we compared SNP calls from the different genome sequences and removed SNPs that either *mob* had in common with each-other or *bak1-5*. Comparing the *bak1-5 mob* genomes to the parental *bak1-5* genome identified 2,111 and 2,132 SNPs shared between *bak1-5* and either *bak1-5 mob1* or *bak1-5 mob2*, respectively (Figure 3B). There were an additional 444 SNPs that were shared between the *bak1-5 mob1* and *bak1-5 mob2* genomes that were not identified in *bak1-5*. We reasoned that these shared SNPs were contributed by *bak1-5* but were not identified due to low sequence coverage in those areas. These analyses identified 2,746 polymorphisms that were shared between at least two of the genomes. Discarding these allowed us to identify over 1,000 SNPs that were uniquely present in *bak1-5 mob1* and *bak1-5 mob2* (Figure 3B). The general value of comparing multiple mutant sequences to remove shared SNPs has been previously demonstrated [12] and the general case discussed in Additional file 3. However, deleting all common SNPs in this way precludes identification of identical causative SNPs in different mutants, which, while extremely rare, is something to consider prior to performing such analysis.

CandiSNP enables easy visual assessment of SNP positions and finds genomic regions with low recombination linked to the phenotype of interest

Positional cloning is based on linking phenotypes to molecular markers with known genomic positions. If recessive, F_2 recombinants that contain the phenotype of interest are homozygous for the unknown mutation. Identifying molecular markers that are invariably homozygous for the mutant type are therefore linked to the scored phenotype and thus the mutation. As we conducted a back-cross rather than an out-cross, the only molecular markers we could use were those resulting from the comparative analysis just described between the parental and mutant genomes.

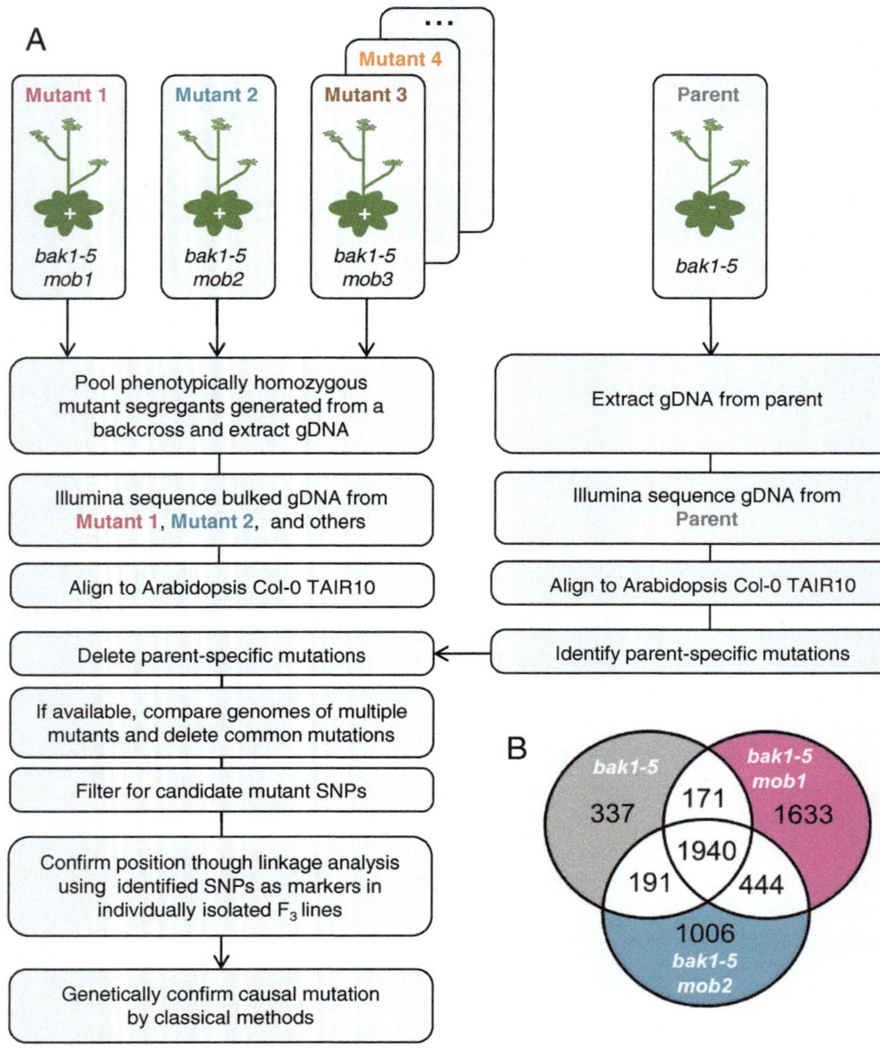

Figure 3 Pipeline for bulking segregants and identification of unique SNPs. (A) The recessive *bak1-5 mob1* and *bak1-5 mob2* mutants were back-crossed to the parent *bak1-5*, allowed to self-fertilize in the F_1, and phenotypically scored in the F_2 for the *mob* phenotype. Positive segregants were bulk harvested and genomic DNA was prepared and sequenced using the Illumina HiSeq platform. For comparison, the *bak1-5* genome was also sequenced. A similar genetics pipeline could be employed for dominant mutants, but material would need to be bulked from segregants that were phenotypically verified as homozygous in the F_3 generation. **(B)** A three-way comparison between the *bak1-5*, *bak1-5 mob1*, and *bak1-5 mob2* genomes identified the total number of unique SNPs in each genome.

To identify which of the over 1,000 unique SNPs in the mutant genomes are linked to the scored phenotype, it is necessary to determine which SNPs are homozygous or close-to-homozygous. Although bulking mutants increases the likelihood of identifying homozygous SNPs (in theory, with an allele frequency of 100%), some margin of error must be allowed to account for sequencing and phenotyping errors. CandiSNP facilitates the easy discovery of a useful frequency cut-off by allowing the user to iteratively refine the allele frequency and view a new plot concurrent with previous ones for comparison. As a further refinement of the CandiSNP web application, we included the option of removing SNPs concentrated around centromeres (in organisms where a centromere is defined in the genome assembly), as these are areas of low recombination frequency and tend to skew density analysis. After selecting an allele frequency threshold, CandiSNP plots the positions of retained SNPs as dots across the chromosomes and highlights SNPs of different classes according to a selected palette. The per-chromosome density and distribution of SNPs is rendered in a second plot to aid in cases with high numbers of SNPs. In our case study we chose 75% as an acceptable frequency cut-off, and used CandiSNP to identify 88 and 143 unique

Table 1 Identification of unique and candidate SNPs in the parental and mutant genomes

	bak1-5	*bak1-5 mob1*	*bak1-5 mob2*
Total SNPs compared to Col-0 TAIR10	2639	4188	3581
Unique SNPs compared to the parent	2639	1633	1006
Unique SNPs, AF >75%	785	88	143
Unique SNPs, AF >75%, annotated coding	240	16	41
Unique SNPs, AF >75%, annotated coding, non-synon	168	9[a]	31[b]

To identify SNPs unique to each genome, the parental and mutant genomes were compared and filtered. Unique SNPs in *bak1-5* refer to those that are not found in the Col-0 TAIR10 genome. For the *bak1-5 mob1* and *bak1-5 mob2* datasets, SNPs shared between any of the three genomes (*bak1-5*, *bak1-5 mob1* and *bak1-5 mob2*; Figure 3B) were removed, resulting in SNPs uniquely found in each of those genomes. Filtering for SNPs with an allele frequency higher than 75% that cause non-synonymous ('non-synon') changes in annotated coding regions resulted in a list of candidate causative mutations.
[a]Candidate causative SNPs for *bak1-5 mob1* are listed in Table 2.
[b]Candidate causative SNPs for *bak1-5 mob2* are listed in Table 3.

SNPs meeting this requirement in the *bak1-5 mob1* (Additional file 4) and *bak1-5 mob2* (Additional file 5) genomes, respectively (Table 1). CandiSNP further identified 9 and 31 candidate causative SNPs (those which cause non-synonymous changes in protein-coding regions) for *bak1-5 mob1* (Table 2; Additional file 4) and *bak1-5 mob2* (Table 3; Additional file 5). By choosing a palette to highlight candidate SNPs (shown in our case study as red dots) we observe putative map positions for both mutants at the bottom of chromosome 5 (Figure 4).

Therefore, CandiSNP visualizes the location of SNPs linked to a phenotype of interest. Moreover, CandiSNP provides annotations describing the genomic feature in which each SNP is located. This function provides useful information for biologists who can make conceptual links between biological knowledge and molecular function of the genomic feature and refine candidate lists further. On its own this annotation function provides a fast and easy way of finding the effect of any mutation in the supported genomes, enhancing the usefulness of CandiSNP beyond that of mutant mapping.

Fine mapping and confirmation of causative SNPs

We confirmed the presence of candidate causative SNPs by Sanger sequencing polymerase chain reaction (PCR)-generated amplicons containing the predicted mutations prepared from individual back-crossed F_3 *bak1-5 mob1* and *bak1-5 mob2* plants compared to the *bak1-5* parent (Tables 2 and 3). Using these SNPs as molecular markers allowed us to further map the mutations by position and narrowed the list of candidate causative SNPs down to 3 in *bak1-5 mob1* and 6 in *bak1-5 mob2*. Primers used for this analysis are available in Additional file 6. Alternative methods for such analysis could include other allele-specific genotyping methods such as designing cleaved amplified polymorphic sequence (CAPS) markers [38] or conducting high-resolution melting (HRM) analysis [39] on PCR amplicons. We previously reported genetic confirmation that the polymorphic *CPK28* alleles contained within these lists of candidate SNPs were causative of the *bak1-5 mob1* and *bak1-5 mob2* mutant phenotypes [13]. Our analysis was simplified by knowledge of allelism between the two mutants, which clearly indicated *CPK28* as the causative locus. In the absence of such knowledge, marker-assisted genotyping of additional homozygous F_3 lines could further reduce the number of candidate mutations and ease genetic confirmation.

CandiSNP performance

As CandiSNP is a predictive classification method for determining whether a given SNP is a causative mutation, it is important that we estimate the accuracy of the

Table 2 Candidate causative SNPs in *bak1-5 mob1*

Chr	Position	Ref/Alt	AF (%)	AGI	Gene ID	AA change	Sanger F_3
1	11892068	C/T	100	At1g32830	Transposable element	n.a.	Absent
1	11892070	G/T	100	At1g32830	Transposable element	n.a.	Hom[a]
1	11892252	T/G	100	At1g32830	Transposable element	n.a.	Hom[a]
1	16516501	T/C	83.3	At1g43745	Transposable element	n.a.	Absent
1	16525522	T/C	77.8	At1g43755	Transposable element	n.a.	Hom[a]
1	24243231	G/A	80.9	At1g65270	Unknown protein	G > S	Hom
5	26457834	G/A	85.0	At5g66210	CPK28	A > V	Hom[b]
5	26458077	G/A	78.5	At5g66210	CPK28	S > L	Hom[b]
5	26474069	G/A	76.5	At5g66270	Zn-finger family protein	P > L	Hom

Unique SNPs in annotated coding regions with allele frequencies (AF) over 75% identified by CandiSNP for *bak1-5 mob1*, listing the Chromosome number (Chr), position, reference base (Ref), sequenced alternate base (Alt), locus number (AGI), gene identification (Gene ID), amino acid change (AA change; 'n.a.' is not applicable). All SNPs were confirmed in at least three independent back-crossed lines (F_3 generation) by Sanger sequencing compared to *bak1-5*. SNPs that were homozygous (Hom) or not present (Absent) are listed.
[a]These SNPs were also identified in *bak1-5* by Sanger sequencing (however, not by Illumina sequencing) and are therefore not unique to *bak1-5 mob1*.
[b]These SNPs are the causative mutations for *bak1-5 mob1* [13].

Table 3 Candidate causative SNPs in *bak1-5 mob2*

Chr	Position	Ref/ Alt	AF (%)	AGI	Gene ID	AA change	Sanger F_3
1	7446564	C/T	76.2	At1g21270	WAK2	P > L	Seg
1	11892984	C/T	100	At1g32830	Transposable element	n.a.	Hom[a,b]
1	16513961	T/G	77.1	At1g43740	Transposable element	n.a.	Hom[a,b]
1	17757465	C/T	76.0	At1g48090	Calcium-dependent lipid-binding protein	D > N	Seg
1	18192647	C/T	76.0	At1g49190	ARR19	R > W	Seg
1	22178447	C/T	82.6	At1g60140	TPS10	D > N	Hom
2	2568811	C/T	83.3	At2g06470	Transposable element	n.a.	Hom[a]
2	5277241	T/G	100	At2g12850	Transposable element	n.a.	Not tested
4	2362567	C/A	100	At4g04655	Transposable element	n.a.	Seg[a]
5	5569896	G/A	94.1	At5g16930	AAA-type ATPase family protein	W > stop	Seg
5	14285189	G/A	81.3	At5g36260	Eukaryotic aspartyl protease family protein	S > L	Absent
5	14579245	G/A	75.0	At5g36935	Transposable element	n.a.	Not tested
5	15751875	G/A	77.2	At5g39350	Tetratricopeptide repeat-like superfamily protein	R > K	Seg
5	17503318	G/A	82.7	At5g43560	TRAF-like superfamily protein	E > K	Seg
5	17597830	G/A	80.0	At5g43800	Transposable element	n.a.	Not tested
5	17820568	G/A	93.3	At5g44240	ALA2	A > T	Seg
5	18251689	G/A	83.3	At5g45140	Nuclear RNAP2	P > S	Seg
5	18261108	G/A	75.0	At5g45150	RTL3	D > N	Seg
5	18399206	G/A	80.9	At5g45400	RPA70C	V > M	Seg
5	21859555	G/A	90.0	At5g53840	F-box/RNI-like/FBD-like domains-containing protein	S > F	Seg
5	21939106	G/A	80.9	At5g54062	Unknown protein	E > K	Seg
5	22002355	G/A	88.9	At5g54203	Transposable element	n.a.	Absent
5	22066915	G/A	79.3	At5g54340	C2H2 and C2HC zinc-finger superfamily protein	V > I	Seg
5	22430866	G/A	90.0	At5g55310	TOP1β	A > V	Seg
5	22565056	G/A	76.5	At5g55750	Hydroxyproline-rich glycoprotein family protein	P > S	Seg
5	26458017	C/T	93.1	At5g66210	CPK28	W > stop	Hom[c]
5	26560691	C/T	95.8	At5g66550	Maf-like protein	G > R	Hom
5	26626055	C/T	78.9	At5g66690	UGT72E2	P > S	Hom
5	26710709	C/T	75.0	At5g66880	SNRK2.3	P > S	Hom
5	26716839	C/T	80.9	At5g66900	CC-NB-LRR family protein	D > N	Hom
5	26935248	C/T	75.0	At5g67500	VDAC2	T > I	Hom

Unique SNPs in annotated coding regions with allele frequencies (AF) over 75% identified by CandiSNP for *bak1-5 mob2*, listing the Chromosome number (Chr), position, reference base (Ref), sequenced alternate base (Alt), locus number (AGI), gene identification (Gene ID), amino acid change (AA change; 'n.a.' is not applicable). All SNPs were confirmed in at least three independent back-crossed lines (F_3 generation) by Sanger sequencing compared to *bak1-5*. SNPs that were homozygous (Hom), segregating (Seg), not identified (Absent), or not tested are listed.

[a]These SNPs were also identified in *bak1-5* by Sanger sequencing (however, not by Illumina sequencing) and are therefore not unique to *bak1-5 mob2*.

[b]These are single base deletion mutations.

[c]This SNP is the causative mutation for *bak1-5 mob2* [13].

classifications made. One useful approach for assessing the overall accuracy of the analysis, rather than each individual prediction, *post hoc*, is to construct a receiver operating characteristic (ROC) curve [40]. In such an analysis, a set of independently verified 'true positive' results are compiled and the ability of the classifier to recall these at different parameters is plotted. To assess CandiSNP we used the *bak1-5 mob1* and *bak1-5 mob2* verified causative SNPs and varied the allele frequency parameter to carry out a standard ROC analysis [40] (Additional file 7). True positives were defined as verified causative SNPs and false positives were defined as any non-causative SNP identified by CandiSNP regardless of location and category. False negatives were defined as causative SNPs not included in that threshold and true negatives as any position in the genome where a SNP was not identified (*i.e.*, genome size – false positives). Sensitivity was calculated as the number of true positives divided

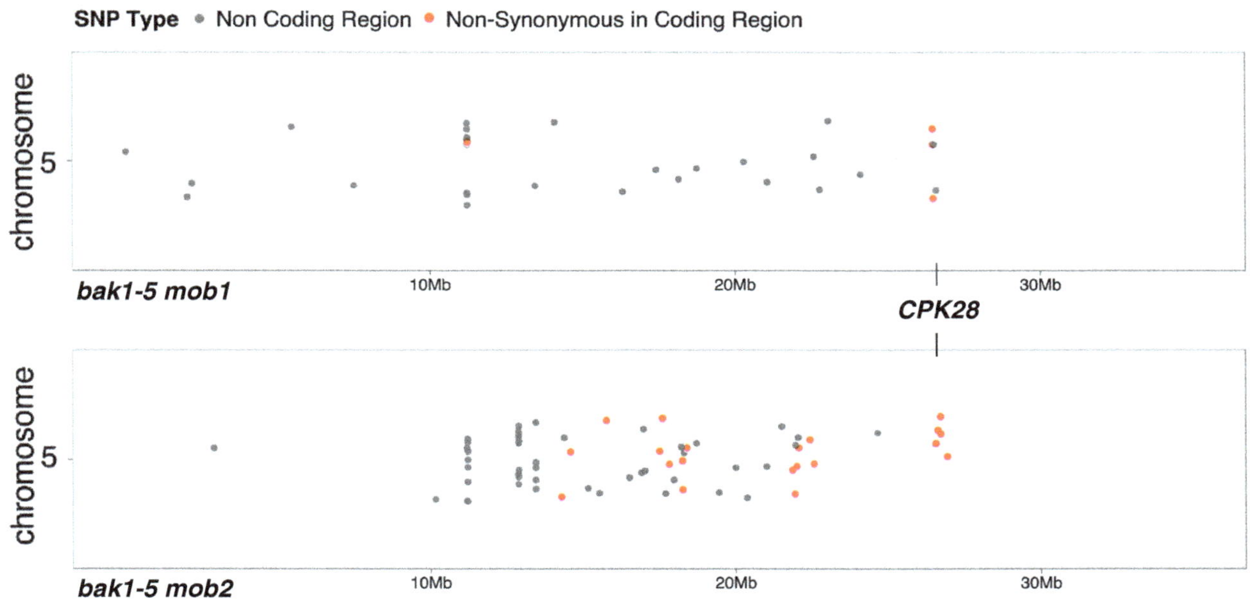

Figure 4 Chromosome 5 SNP density plots for *bak1-5 mob1* and *bak1-5 mob2*. All SNPs with allele frequencies >75% are plotted in grey, while candidate causative SNPs (defined as those causing non-synonymous changes in gene-coding regions) are plotted in red. The position of *CPK28/At5g66210* is indicated.

by the total of true positives and false negatives. Specificity was calculated as the number of true negatives divided by the number of false positives and true negatives. Sensitivity assesses the ability of CandiSNP to recall the verified SNPs whilst specificity assesses the ability of CandiSNP to exclude non-causative SNPs.

For our test case, the sensitivity of CandiSNP drops completely for allele frequencies over 75% (Additional file 7), indicating that accounting for phenotype penetrance and sequencing errors is an important factor in the pipeline. Further, rather than reducing the number of errors, setting an overly stringent allele frequency causes the pipeline to fail by screening out real candidates. Specificity remains high across all possible allele frequencies, mostly due to the masking effect of a very high true negative count. The closely related false positive score shows a decrease to less than 25% of the candidate SNP list after an allele frequency of 62%. While the absolute optimum for our data is at 75% (*i.e.*, the maximisation of sensitivity and minimisation of false positives), taken as a whole the ROC analysis indicates that allele frequencies of 60% to 75% represent a likely 'best trade-off' window for CandiSNP analysis.

Conclusions

Genetic screens have revealed important regulators of signal transduction pathways and remain an important tool in modern research. Although greatly accelerated with the advent of HTS technologies, correct identification of causative mutation(s) remains a bottleneck in forward-genetics. To increase the repertoire of programs available to plant geneticists, we developed the CandiSNP web application, which is particularly useful for datasets containing few SNPs. In our test case, CandiSNP successfully identified causative SNPs in two recessive mutants after bulking phenotypically homozygous F_2 segregants generated from a back-cross. We propose that CandiSNP could additionally be used to identify causative SNPs in dominant mutants as long as they are verified to be phenotypically homozygous in the F_3 prior to bulk segregant analysis. By plotting homozygous and close-to-homozygous SNPs identified from HTS along the chromosome arms, the program visualizes areas of linkage and easily narrows down candidate mutation positions. CandiSNP is both fast and accurate, producing high-quality editable graphics in a matter of minutes. CandiSNP is a user-friendly web application that will facilitate gene discovery in plant genetic screens.

Availability requirements

Project name: CandiSNP.

Project home page: http://candisnp.tsl.ac.uk

(Source code is available under the GPLv3 open-source license at https://github.com/danmaclean/candisnp).

Operating system(s): All systems capable of running a modern web-browser.

Other requirements: Internet connection.

License: GPL3 (http://www.gnu.org/licenses/licenses.html).

Any restrictions: None.

Additional files

Additional file 1: Calculation of expected SNP frequencies in out- and back-crosses. SNP frequencies are calculated from different types of crosses using available data.

Additional file 2: SNP calling script. To create our CandiSNP input files we used the pileups_to_snps.rb Ruby script which relies on the bio-samtools and bio-gngm Ruby Gems. The input data is the FASTA reference sequence (TAIR10 genome in this case) and the SAM/BAM alignment file for each dataset. The output is a comma-delimited file of SNPs formatted for input to the CandiSNP application. The source code for this script is available at https://github.com/danmaclean/candisnp/blob/master/pileup_to_snps.rb.

Additional file 3: Simplifying SNP analysis by comparing multiple mutant genomes. Comparing multiple mutant genomes identifies unique SNPs in each mutant.

Additional file 4: Genome-wide SNP density plots for *bak1-5 mob1*. Using the 'CandiSNP palette', all SNPs with allele frequencies >75% are plotted in grey, while candidate causative SNPs (defined as those causing non-synonymous changes in gene-coding regions) are plotted in red. These plots are also available at http://dx.doi.org/10.6084/m9.figshare.1250028.

Additional file 5: Genome-wide SNP density plots for *bak1-5 mob2*. Using the 'CandiSNP palette', all SNPs with allele frequencies >75% are plotted in grey, while candidate causative SNPs (defined as those causing non-synonymous changes in gene-coding regions) are plotted in red. These plots are also available at http://dx.doi.org/10.6084/m9.figshare.1250027.

Additional file 6: Primers used in this study. Genomic regions containing candidate SNPs in *bak1-5 mob1* and *bak1-5 mob2* were amplified by PCR using these primers and Sanger-sequenced alongside *bak1-5* as a control.

Additional file 7: Receiver operating characteristic curve plot. Receiver operating characteristic curve plot, demonstrating the optimal parameters for use in CandiSNP on *bak1-5 mob1*. The proportion of false positives (non-causative SNPs identified by CandiSNP regardless of location and category) is also plotted as a comparison to the standard sensitivity and specificity [40]. The overlaid dashed line represents the optimal point that maximises the ability of CandiSNP to find real causative SNPs (sensitivity) while minimising the inclusion of false positives.

Abbreviations

AF: Allele frequency; Alt: Alternative sequence; BAK1: BRASSINOSTEROID INSENSITIVE 1-ASSOCIATED KINASE 1; Bp: base pair; BWA: Burrows-Wheeler aligner; CAPS: Cleaved amplified polymorphic sequence; CandiSNP: Candidate SNP; Chr: Chromosome; CPK28: CALCIUM-DEPENDENT PROTEIN KINASE 28; Col-0: Columbia-0; EMS: Ethylmethane sulfonate; F: Filial generation; FLS2: FLAGELLIN SENSING 2; HRM: High-resolution melting; Indel: Insertion/deletion; L*er*-0: Landsberg *erecta*-0; M: Mutant generation; *Mob*: *modifier of bak1-5*; NGM: Next-generation mapping; HTS: High-throughput sequencing; PCR: Polymerase chain reaction; Pos: Position; SAM: Sequence alignment/map; SNP: Single nucleotide polymorphism; TAIR: The Arabidopsis Information Resource; QC: Quality control; Ws-0: Wassilewskija-0.

Competing interests

The authors declare that they have no competing interests.

Authors' contributions

JM performed the mutant screen and bulk segregant analysis and worked with GE and DM for data analysis and program development under the supervision of CZ. GE and DM performed bioinformatics and created the CandiSNP web application. All authors contributed to writing the paper. All authors read and approved the final manuscript.

Acknowledgements

We thank Dr. Ghanasyam Rallapalli and Dr. Frederikke Gro Malinovsky for careful reading of the manuscript and all members of the MacLean and Zipfel laboratories for helpful suggestions. Martin Page is acknowledged for technical assistance. This research was funded by grants from The Gatsby Foundation (DM, CZ) and The European Research Council (CZ). JM was the recipient of a Long-Term Fellowship from the European Molecular Biology Organization.

Author details

[1]The Sainsbury Laboratory, Norwich Research Park, Norwich NR4 7UH, UK. [2]The Genome Analysis Centre, Norwich Research Park, Norwich NR4 7UH, UK.

References

1. Gomez-Gomez L, Boller T: **FLS2: an LRR receptor-like kinase involved in the perception of the bacterial elicitor flagellin in Arabidopsis.** *Mol Cell* 2000, **5**(6):1003–1011.
2. Boller T, Felix G: **A renaissance of elicitors: perception of microbe-associated molecular patterns and danger signals by pattern-recognition receptors.** *Annu Rev Plant Biol* 2009, **60**:379–406.
3. Page DR, Grossniklaus U: **The art and design of genetic screens: Arabidopsis thaliana.** *Nat Rev Genet* 2002, **3**(2):124–136.
4. Michelmore RW, Paran I, Kesseli RV: **Identification of markers linked to disease-resistance genes by bulked segregant analysis: a rapid method to detect markers in specific genomic regions by using segregating populations.** *Proc Natl Acad Sci U S A* 1991, **88**(21):9828–9832.
5. Schneeberger K: **Using next-generation sequencing to isolate mutant genes from forward genetic screens.** *Nat Rev Genet* 2014, **15**(10):662–676.
6. Schneeberger K, Ossowski S, Lanz C, Juul T, Petersen AH, Nielsen KL, Jorgensen JE, Weigel D, Andersen SU: **SHOREmap: simultaneous mapping and mutation identification by deep sequencing.** *Nat Methods* 2009, **6**(8):550–551.
7. Hartwig B, James GV, Konrad K, Schneeberger K, Turck F: **Fast isogenic mapping-by-sequencing of ethyl methanesulfonate-induced mutant bulks.** *Plant Physiol* 2012, **160**(2):591–600.
8. Austin RS, Vidaurre D, Stamatiou G, Breit R, Provart NJ, Bonetta D, Zhang J, Fung P, Gong Y, Wang PW, McCourt P, Guttman DS: **Next-generation mapping of Arabidopsis genes.** *Plant J* 2011, **67**(4):715–725.
9. Takagi H, Uemura A, Yaegashi H, Tamiru M, Abe A, Mitsuoka C, Utsushi H, Natsume S, Kanzaki H, Matsumura H, Saitoh H, Yoshida K, Cano LM, Kamoun S, Terauchi R: **MutMap-Gap: whole-genome resequencing of mutant F2 progeny bulk combined with de novo assembly of gap regions identifies the rice blast resistance gene Pii.** *New Phytol* 2013, **200**(1):276–283.
10. Fekih R, Takagi H, Tamiru M, Abe A, Natsume S, Yaegashi H, Sharma S, Kanzaki H, Matsumura H, Saitoh H, Mitsuoka C, Utsushi H, Uemura A, Kanzaki E, Kosugi S, Yoshida K, Cano L, Kamoun S, Terauchi R: **MutMap+: genetic mapping and mutant identification without crossing in rice.** *PLoS One* 2013, **8**(7):e68529.
11. Abe A, Kosugi S, Yoshida K, Natsume S, Takagi H, Kanzaki H, Matsumura H, Mitsuoka C, Tamiru M, Innan H, Cano L, Kamoun S, Terauchi R: **Genome sequencing reveals agronomically important loci in rice using MutMap.** *Nat Biotechnol* 2012, **30**(2):174–178.
12. Minevich G, Park DS, Blankenberg D, Poole RJ, Hobert O: **CloudMap: a cloud-based pipeline for analysis of mutant genome sequences.** *Genetics* 2012, **192**(4):1249–1269.
13. Monaghan J, Matschi S, Shorinola O, Rovenich H, Matei A, Segonzac C, Gro-Malinovsky F, Rathjen J, MacLean D, Romeis T, Zipfel C: **The calcium dependent protein kinase CPK28 buffers plant immunity and regulates BIK1 turnover.** *Cell Host Microbe* 2014, **16**:605–615.
14. Sun Y, Li L, Macho AP, Han Z, Hu Z, Zipfel C, Zhou JM, Chai J: **Structural basis for flg22-induced activation of the Arabidopsis FLS2-BAK1 immune complex.** *Science* 2013, **342**(6158):624–628.
15. Schwessinger B, Roux M, Kadota Y, Ntoukakis V, Sklenar J, Jones A, Zipfel C: **Phosphorylation-dependent differential regulation of plant growth, cell death, and innate immunity by the regulatory receptor-like kinase BAK1.** *PLoS Genet* 2011, **7**(4):e1002046.
16. Schulze B, Mentzel T, Jehle AK, Mueller K, Beeler S, Boller T, Felix G, Chinchilla D: **Rapid heteromerization and phosphorylation of ligand-activated plant transmembrane receptors and their associated kinase BAK1.** *J Biol Chem* 2010, **285**(13):9444–9451.
17. Roux M, Schwessinger B, Albrecht C, Chinchilla D, Jones A, Holton N, Malinovsky FG, Tor M, de Vries S, Zipfel C: **The arabidopsis leucine-rich repeat receptor-like kinases BAK1/SERK3 and BKK1/SERK4 are required**

for innate immunity to Hemibiotrophic and Biotrophic pathogens. *Plant Cell* 2011, **23**(6):2440–2455.
18. Heese A, Hann DR, Gimenez-Ibanez S, Jones AM, He K, Li J, Schroeder JI, Peck SC, Rathjen JP: **The receptor-like kinase SERK3/BAK1 is a central regulator of innate immunity in plants.** *Proc Natl Acad Sci U S A* 2007, **104**(29):12217–12222.
19. Chinchilla D, Zipfel C, Robatzek S, Kemmerling B, Nurnberger T, Jones JD, Felix G, Boller T: **A flagellin-induced complex of the receptor FLS2 and BAK1 initiates plant defence.** *Nature* 2007, **448**(7152):497–500.
20. Nekrasov V, Li J, Batoux M, Roux M, Chu ZH, Lacombe S, Rougon A, Bittel P, Kiss-Papp M, Chinchilla D, van Esse HP, Jorde L, Schwessinger B, Nicaise V, Thomma BP, Molina A, Jones JD, Zipfel C: **Control of the pattern-recognition receptor EFR by an ER protein complex in plant immunity.** *EMBO J* 2009, **28**(21):3428–3438.
21. Tedman-Jones JD, Lei R, Jay F, Fabro G, Li X, Reiter WD, Brearley C, Jones JD: **Characterization of Arabidopsis mur3 mutations that result in constitutive activation of defence in petioles, but not leaves.** *Plant J* 2008, **56**(5):691–703.
22. Zipfel C, Robatzek S, Navarro L, Oakeley EJ, Jones JD, Felix G, Boller T: **Bacterial disease resistance in Arabidopsis through flagellin perception.** *Nature* 2004, **428**(6984):764–767.
23. Vetter MM, Kronholm I, He F, Haweker H, Reymond M, Bergelson J, Robatzek S, de Meaux J: **Flagellin perception varies quantitatively in Arabidopsis thaliana and its relatives.** *Mol Biol Evol* 2012, **29**(6):1655–1667.
24. Cingolani P, Platts A, le Wang L, Coon M, Nguyen T, Wang L, Land SJ, Lu X, Ruden DM: **A program for annotating and predicting the effects of single nucleotide polymorphisms, SnpEff: SNPs in the genome of Drosophila melanogaster strain w1118; iso-2; iso-3.** *Fly (Austin)* 2012, **6**(2):80–92.
25. Huala E, Dickerman AW, Garcia-Hernandez M, Weems D, Reiser L, LaFond F, Hanley D, Kiphart D, Zhuang M, Huang W, Mueller LA, Bhattacharyya D, Bhaya D, Sobral BW, Beavis W, Meinke DW, Town CD, Somerville C, Rhee SY: **The Arabidopsis Information Resource (TAIR): a comprehensive database and web-based information retrieval, analysis, and visualization system for a model plant.** *Nucleic Acids Res* 2001, **29**(1):102–105.
26. Project IRGS: **The map-based sequence of the rice genome.** *Nature* 2005, **436**:793–800.
27. Consortium TTG: **The tomato genome sequence provides insights into fleshy fruit evolution.** *Nature* 2012, **485**:635–641.
28. Schmutz J: **Genome sequence of the palaeopolyploid soybean.** *Nature* 2010, **463**:178–183.
29. Velasco R, Zharkikh A, Troggio M, Cartwright DA, Cestaro A, Pruss D, Pindo M, FitzGerald LM, Vezzuli S, Reid J, Malacarne G, Iliev D, Coppola G, Wardell B, Micheletti D, Macalma T, Facci M, Mitchell JT, Perassolli M, Eldredge G, Gatto P, Oyzerski R, Moretto M, Gutin N, Stefanini M, Chen Y, Segala C, Davenport C, Dematte L, Mraz A, Battilana J, Stormo K, Costa F, Tao Q, Si-Ammour A, Harkins T, Lackey A, Perbost C, Taillon B, Stella A, Solovyev V, Fawcett JA, Sterck L, Vandepoele SK, Fontana P, Gutin A, Ven De Peer Y, Salamini F, Viola R: **A high-quality draft concensus sequence of the genome of a heterozygous grapevine variety.** *PLoS One* 2007, **2**(12):e1326.
30. Schnable PS: **The B73 maize genome: complexity, diversity, and dynamics.** *Science* 2009, **326**(5956):1112–1115.
31. Andrews S: *FastQC: A Quality Control Tool for High Throughput Sequence Data*; 2010. http://www.bioinformatics.babraham.ac.uk/projects/fastqc.
32. Goecks J, Nekrutenko A, Taylor J: **Galaxy: a comprehensive approach for supporting accessible, reproducible, and transparent computational research in the life sciences.** *Genome Biol* 2010, **11**(8):R86.
33. Giardine B, Riemer C, Hardison RC, Burhans R, Elnitski L, Shah P, Zhang Y, Blankenberg D, Albert I, Taylor J, Miller W, Kent WJ, Nekrutenko A: **Galaxy: a platform for interactive large-scale genome analysis.** *Genome Res* 2005, **15**(10):1451–1455.
34. Blankenberg D, Von Kuster G, Coraor N, Ananda G, Lazarus R, Mangan M, Nekrutenko A, Taylor J: **Galaxy: a web-based genome analysis tool for experimentalists.** *Curr Protoc Mol Biol* 2010, **Chapter 19**:Unit 19 10 11–21.
35. Joshi NA, Fass JN: **Sickle: A Sliding-Window, Adaptive, Quality-Based Trimming Tool for FastQ Files (Version 1.21)** *[Software]*; 2011. Available at https://github.com/najoshi/sickle.
36. Li H, Durbin R: **Fast and accurate long-read alignment with Burrows-Wheeler transform.** *Bioinformatics* 2010, **26**(5):589–595.
37. Li H, Handsaker B, Wysoker A, Fennell T, Ruan J, Homer N, Marth G, Abecasis G, Durbin R: **The Sequence Alignment/Map format and SAMtools.** *Bioinformatics* 2009, **25**(16):2078–2079.
38. Glazebrook J, Drenkard E, Preuss D, Ausubel FM: **Use of cleaved amplified polymorphic sequences (CAPS) as genetic markers in Arabidopsis thaliana.** *Methods Mol Biol* 1998, **82**:173–182.
39. Tucker EJ, Huynh BL: **Genotyping by high-resolution melting analysis.** *Methods Mol Biol* 2014, **1145**:59–66.
40. Fawcelt T: **An introduction to ROC analysis.** *Pattern Recogn Lett* 2006, **27**(8):861–874.

15

The leaf angle distribution of natural plant populations: assessing the canopy with a novel software tool

Mark Müller-Linow*, Francisco Pinto-Espinosa, Hanno Scharr and Uwe Rascher

Abstract

Background: Three-dimensional canopies form complex architectures with temporally and spatially changing leaf orientations. Variations in canopy structure are linked to canopy function and they occur within the scope of genetic variability as well as a reaction to environmental factors like light, water and nutrient supply, and stress. An important key measure to characterize these structural properties is the leaf angle distribution, which in turn requires knowledge on the 3-dimensional single leaf surface. Despite a large number of 3-d sensors and methods only a few systems are applicable for fast and routine measurements in plants and natural canopies. A suitable approach is stereo imaging, which combines depth and color information that allows for easy segmentation of green leaf material and the extraction of plant traits, such as leaf angle distribution.

Results: We developed a software package, which provides tools for the quantification of leaf surface properties within natural canopies via 3-d reconstruction from stereo images. Our approach includes a semi-automatic selection process of single leaves and different modes of surface characterization via polygon smoothing or surface model fitting. Based on the resulting surface meshes leaf angle statistics are computed on the whole-leaf level or from local derivations. We include a case study to demonstrate the functionality of our software. 48 images of small sugar beet populations (4 varieties) have been analyzed on the base of their leaf angle distribution in order to investigate seasonal, genotypic and fertilization effects on leaf angle distributions. We could show that leaf angle distributions change during the course of the season with all varieties having a comparable development. Additionally, different varieties had different leaf angle orientation that could be separated in principle component analysis. In contrast nitrogen treatment had no effect on leaf angles.

Conclusions: We show that a stereo imaging setup together with the appropriate image processing tools is capable of retrieving the geometric leaf surface properties of plants and canopies. Our software package provides whole-leaf statistics but also a local estimation of leaf angles, which may have great potential to better understand and quantify structural canopy traits for guided breeding and optimized crop management.

Keywords: Stereo imaging, 3-d reconstruction, Leaf segmentation, Leaf modeling, Canopy, Structural canopy straits, Leaf angle, Leaf angle distribution

Introduction

Canopies of plant populations are featured by functional designs with a complex arrangement of leaves and stems, which are subject to temporal and spatial fluctuations on various scales. The potential of leaf display is defined by the genetic framework, which sets up the range for development and environmental responses. Different temporal scales are involved: (i) leaves and canopy element are passively moved by wind causing a highly fluctuating light environment within the canopy [1]; (ii) on the diurnal time scale several leaf movement and sun tracking strategies were described aiming for either optimal light interception or avoidance of high light conditions [2-5]; (iii) growth and developmental processes change and alter canopy structure during the seasonal cycle and

*Correspondence: m.mueller-linow@fz-juelich.de
Institute of Bio and Geosciences, IBG-2: Plant Sciences, Forschungszentrum Jülich GmbH, Wilhelm-Johnen-Str, 52425 Jülich, Germany

as a reaction to environmental stresses [6,7]. During their seasonal development most plants display strong morphological changes, which depend on the availability of resources and on the fluctuation of abiotic factors. Structural properties which are altered on the diurnal and seasonal basis may affect the efficiency of light interception within the canopy and thus may influence canopy light use efficiency [5,8,9]. Regarding the spatial scales structural shaping and adaptive reactions do not happen uniformly but strongly depend on the vertical and horizontal distribution of stems and leaves. Light availability becomes increasingly limited and fluctuating in the lower canopy layers and plants may adapt to this by layer-specific distributions of leaf orientations. These structure-function relations have recently being put in the focus of breeding strategies as potential yield improvements using biochemical optimization of photosynthesis have reached an optimum, while structural optimization may still pose some potential to improve canopy light use efficiency [10].

One of the geometric key measures, which has been used most commonly, is the leaf area index (LAI), which simply relates plant surface to soil surface. While this parameter has been intensively studied (potential and limitations of LAI measurements are discussed in [11-13]), the LAI does not provide any information on leaf architecture or the distribution of leaf orientations. Leaf orientation can be greatly affected by environmental influences like drought, which makes this an interesting trait for breeders when comparing drought stress tolerance between cultivars. Most of the available studies are dealing with single plants [14], are implemented under lab conditions [15,16] or use virtual plant models [17-19]. Others use labor-intensive manual methods and thus are based on a limited sample size [20]. Especially when dealing with lab-based applications and single plants, reconstructions of the entire canopy may be accomplished. Only a few studies focus on a quantitative assessment of leaf orientations of the outer canopy layers of plants grown under field conditions [21-23]. Despite the importance of measuring structural traits such as the leaf angle distribution, there is currently no method available, that (i) can easily be used under field conditions, (ii) works on changing canopies that are often moved by wind or (iii) delivers a high sample size of a representative number of leaves within a canopy with high spatial resolution. All these requirements are essential for knowledge-guided crop breeding [24,25], where the rating by visual judgment is still the most common method. Despite the variety of sensor-supported methods, which entered this field of phenotyping in the recent years with a high potential to assist the labor-intensive work and to replace some of the subjective ratings with automatic registration routines, only a minor method and knowledge transfer has taken place [26]. For the non-manual estimation of leaf angles several methodical approaches are available which include structured light approaches [14], stereo imaging [7,27] and laser scanning techniques [28-30]. A suitable 3-d system for rapid plant phenotyping should facilitate easy segmentation of leaves and be applicable under field conditions, thus should yield stable results also under windy conditions.

With this communication we present a fundamental step forward in using stereo camera approaches to quantify the outer canopy layer of different experimental plant systems. Our method was first developed by Biskup et al. [27] and then further refined and applied in Rascher et al. [9]. This approach uses a set of two consumer cameras that are mounted a few meters top-of-canopy. Stereo images are taken without major constraints making it an easy to use field set-up. A pipeline of computer routines later allows the calculation of depth maps, allows for the segmentation of leaf sections and the estimation of leaf orientations. However, this first scientific version required advanced computer knowledge to handle the processing routines and the system only fitted a planar leaf model and did not allow for natural leaf geometry. Herewith, we introduce a fundamentally revised and in many aspects extend version of the system of Biskup et al. [27]. Software routines were re-programmed and equipped with graphical user interfaces within the MatLab environment now being substantially more robust. A user guided semi-automated leaf segmentation routine was added and the 3-d images are now fitted with realistic and flexible leaf models allowing for a local derivation of leaf orientations. To cope with the susceptibility of field-acquired data to noise we included several filters and alternative processing paths which may be selected according to the particular problem. Different levels of user interaction are implemented reaching from default paths up to precise adjustment of parameters.

Implementation

Setup of a stereo camera system

Our software package has been developed to derive 3-d surface models and leaf surface parameters from stereo images of plants and plant canopies taken in a nadir perspective using ordinary cameras and lenses with a fixed focal length *f*. Such a setup is easy to handle and needs only a few additional components like a calibration pattern (for target calibration) and optionally an inclinometer and a compass to correct the direction and misalignment of the stereo rig. Figure 1 shows our example of the stereo rig. The two cameras need to be aligned in a fixed geometry (two clamps at the rigid metal profile in Figure 1). This setup is calibrated and must not be changed during image acquisition. This means also that all camera internal automation, especially auto-focus, need to be turned off. If applied on non-rigid objects

Figure 1 Stereo rig. Two cameras are fixed with a profile system and slightly turned to each other to capture the same scene. A remote-control release allows for synchronous triggering. In this setup the baseline was adjusted to $b \approx 200\ mm$.

like a moving canopy simultaneous exposure is of crucial importance and therefore cameras should be triggered via a remote-control release. To find the best depth resolution and depth of focus we provide a tool which helps to estimate the distances between cameras (which is the baseline b) and the distance between stereo rig and plants.

Software

Our software package has been developed with Matlab R2012b on a Windows-based platform. Three external toolboxes for image calibration [31-33] and a tool for unstructured 2-d triangular surface meshing [34] are included in the package. Image, calibration and subsequent computed data are organized within a project structure, which builds on a fixed stereo setup. The software consists of 3 essential modules, which control the 3-d reconstruction (i), the leaf segmentation (ii) and the surface modeling (iii), all of them featuring individual graphical user interfaces (GUIs). An outline of each module is given in the block diagram in Figure 2. All modules are interlinked, i.e. some work only with the particular input data (B-C top: green-framed boxes), which come from the first two modules (4 green boxes in Figure 2). Alternative processing options are indicated as dotted boxes, which may be applied. A more vivid view on the overall process is depicted in Figure 3 which uses an example of our case study. The outcome is a 3-d polygon mesh on the base of fitted (planar, quadratic and cubic surface function) or smoothed (Laplacian or curvature flow) leaf surface models, which then can be used for further surface statistics, e.g. estimation of the leaf angle distribution and leaf area index. In addition to surface reconstruction this tool also provides linear, quadratic and cubic modeling of leaf axes and calculation of the respective leaf axes angles. We tested this to be useful for modeling grass-like species, but do not go into details here. We included four additional tools each equipped with a GUI, which will also be outlined here only briefly. The first one (depicted in Figure 4) uses a manual segmentation approach to separate plant pixels from the background. This segmentation, which helps to improve the result in module (i) and (ii), is performed in the HSV color space [35]. The second supplementary tool helps to select the right settings for individual cameras and stereo rig. The the third tool is a visualization tool, which displays the highlights of each processed part together with a summary on settings and estimated parameters, like average leaf inclination and leaf area index. We are also providing an additional tool (with a GUI), which allows for an easy manual post-editing of prior segmentations. The main output of the complete processing pipeline is a surface mesh data file in the well-established ply-format and the leaf angle statistics as an excel-file.

Module 1: 3-d reconstruction

Background Figure 2A and Figure 5 display the outline and the GUI of the 3-d reconstruction process. Our method of a 3-d reconstruction via the stereo imaging is based on the detection of analogous image information around corresponding points in left (reference) and right (matching) stereo images. We implemented a target-calibrated and uncalibrated (also referred to as auto-calibrated) process both aiming at transformations of the stereo images, which convert all epipolar lines into horizontals (image rectification [36,37]). The target calibration estimates camera geometries (intrinsics) and the geometry of the stereo rig (extrinsics) using a regular calibration pattern [31,38,39]. The auto-calibration mode works solely image-based [32]. Provided that enough and fairly distributed points are detected in both images this method obtains similar results as the target calibration. In the resulting rectified images the correspondence problem [40,41] is reduced to a 1-dimensional search for correspondences. The relative positions of corresponding points, where point coordinates are given with respect to the principal points of left and right camera, denote the depth information (disparity D) of the respective pixels. We implemented a block matching method, which makes use of the properties of the surrounding pixels and which uses a couple of statistical measures [40,41], e.g. the correlation C of grey values, where corresponding pixels are given by the maximum correlation C_{max} of blocks [42]. The outcome of the correspondence analysis is raw pixel disparities (depth map), which have to be filtered and finally converted to a 3-d metric point cloud in coordinates (x, y, z).

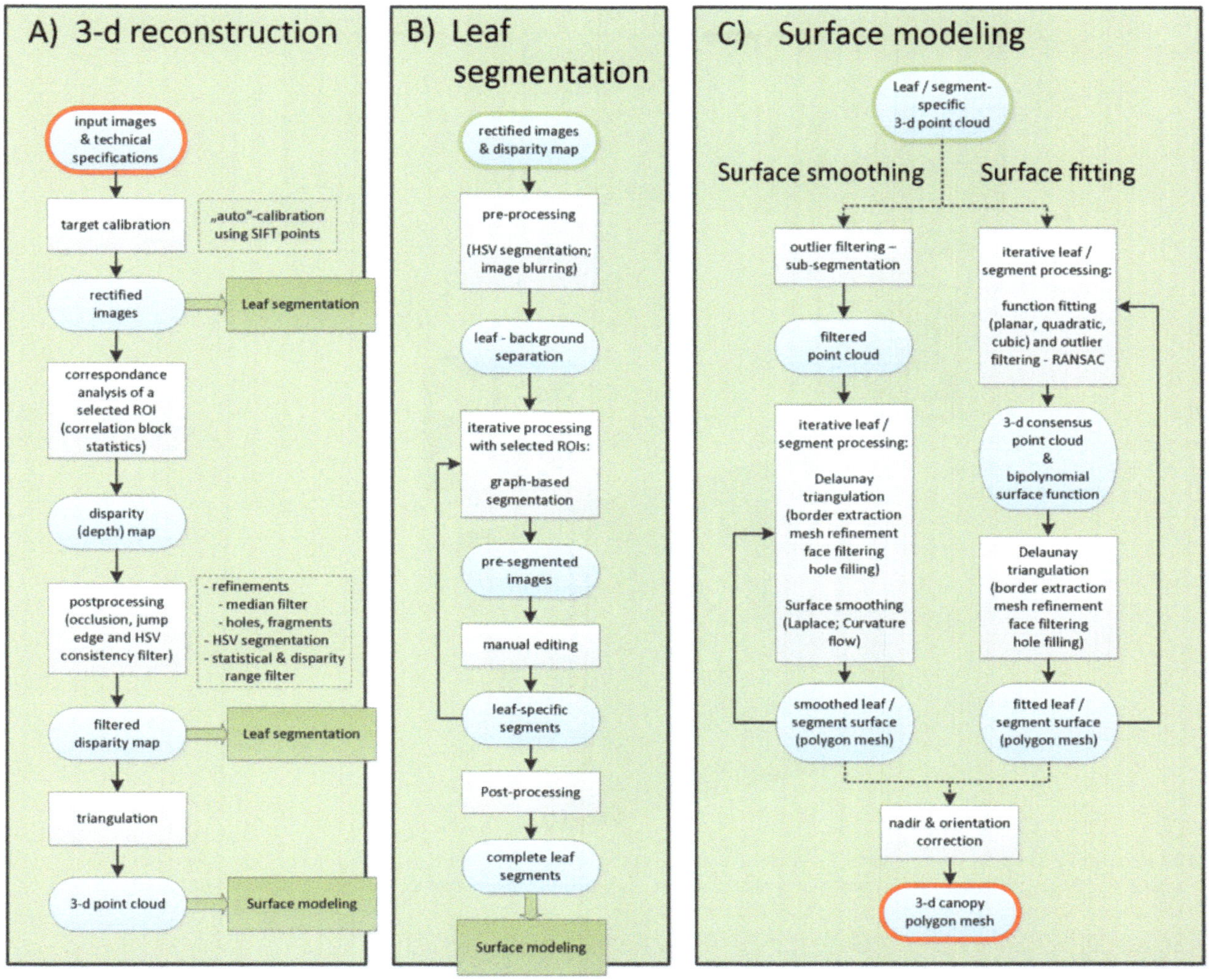

Figure 2 Box diagram of module 1-3. General outline of the leaf angle processing toolbox: The structure of this toolbox comprises 3 basic modules **(A-C)**, all of them controlled by individual graphical user interfaces (these GUIs are depicted in Figures 3, 4, and 5). Optional GUI-supported tools for HSV segmentation, stereo rig settings and result summary are not illustrated here. Module 3 comprises two alternative processing paths (left: via surface smoothing; right: via surface fitting). Alternative and optional modes in module 1 are indicated on the right sides as dotted boxes. Round boxes indicate the input and outcome of a process, rectangular boxes the processes themselves. Green boxes point to the subsequent processes in the other modules. The 3-d reconstruction **(A)** starts with the input of stereo images, calibration images and technical specifications. Outputs are rectified images and disparity maps on the one hand, which serve as the input data for the subsequent segmentation process. On the other hand, the 3-d point cloud data is transferred to the surface modeling process. With the data provided by the 3-d reconstruction the full or partial recognition of leaves is the intention of the image segmentation **(B)**. On the base of segment-specific points clouds leaf surface structures are modeled in the third module **(C)** either using smoothing operations or bipolynomial surface functions. The resulting polygon mesh of the canopy provides a basis for further statistical analysis of particular plant traits like leaf angles or leaf area.

Calibration & rectification - default: We developed a robust auto-tracking algorithm to correctly identify and sort the target pattern features, which are then transferred to the calibration toolbox of Bouguet [31]. The toolbox interfaces have been modified to allow smooth integration to our software. Intrinsics and extrinsics of the camera rig are finally used to rectify the stereo images.

Auto-calibration - optional: Auto-calibration requires additional information on the cameras' focal length f and the baseline b as well as a certain number of pixel coordinates of corresponding points in the left and right input object image. We combined a scale invariant feature tracking (SIFT) method [33,43] for the detection of corresponding pixels and the auto-calibration toolbox of Fusiello [32]. Rectification is also

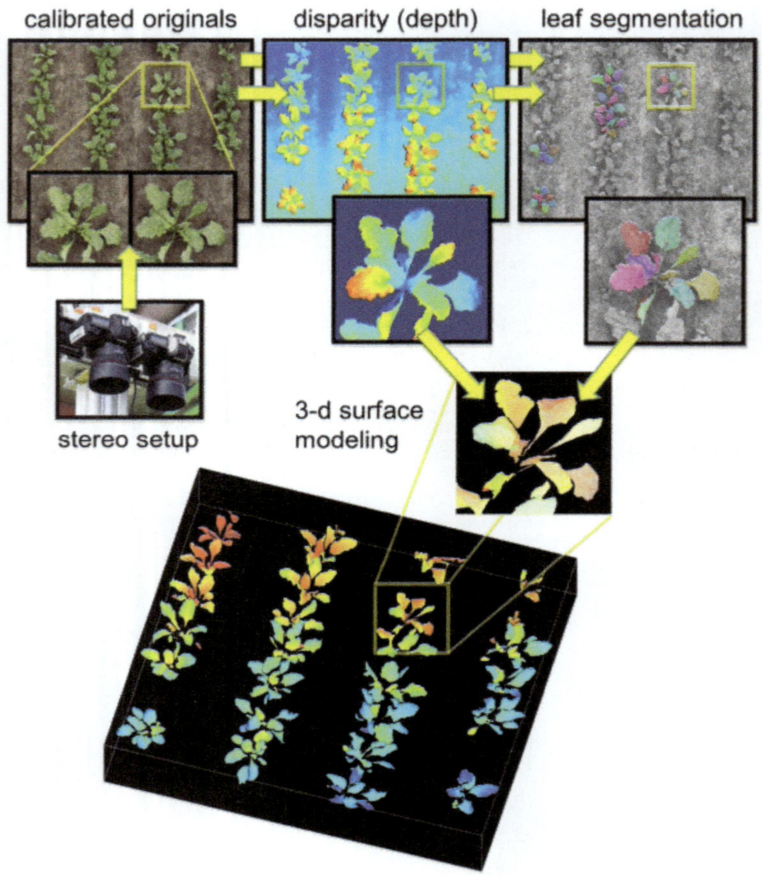

Figure 3 Example of a complete canopy surface reconstruction. Illustration of a complete leaf surface modeling process using the example of a young sugar beet population: The modeling starts with the computation of calibrated stereo images taken with an ordinary stereo setup (top left). These images are used in depth estimation (disparity map - top middle) as well as in leaf segmentation (top right). Jet coloration in the disparity map encodes the distance from the camera with blue values being the farthest. The slight off-nadir position of the stereo rig is reflected in the smooth transition of ground values. The enlarged detail displays the image after post-processing, i.e. filtering of the background, occlusions and outliers. Identification of leaf-specific pixels is illustrated in the segmentation image (top right - color-indexed leaves). Results are used for leaf-specific surface modeling (bottom - in this example surfaces have been reconstructed with curvature flow smoothing). After correcting the off-nadir position all surfaces yield the complete 3-d canopy surface reconstruction.

performed according to the Fusiello method. The estimation process has to be applied on each image pair separately. The Fusiello algorithm shows the tendency of slight image rotations from time to time. We automatically correct these rotations in the module 3: surface reconstruction.

Correspondence analysis: Depending on the image resolution and the plant's surface properties the computation of dense depth maps from stereo images using block method statistics can be computationally demanding. We decided to implement a cross-correlation measure based on the Pearson correlation coefficient. We also tested other block statistics like the sum of squared differences (SSD) and the normalized sum of squared differences (NSSD) for comparison and found only slight deviations from the correlation measure. The computation time and the results of the algorithm mainly depend on 3 parameters, which can be tuned to a certain degree without loosing too much depth information. This parameter set includes the region of interest (ROI), the size of the block B used to compare left and right image pixels and the range R of the horizontal line which is scanned to find the maximum overlap (e.g. highest correlation C_{max}) of both blocks. In the default mode B and R are calculated from the camera calibration information together with user-provided plant parameters. So far we examined the two plant species sugar beet and barley, for which standard parameter sets are included in the program. The ROI is defined by the user in the left reference image.

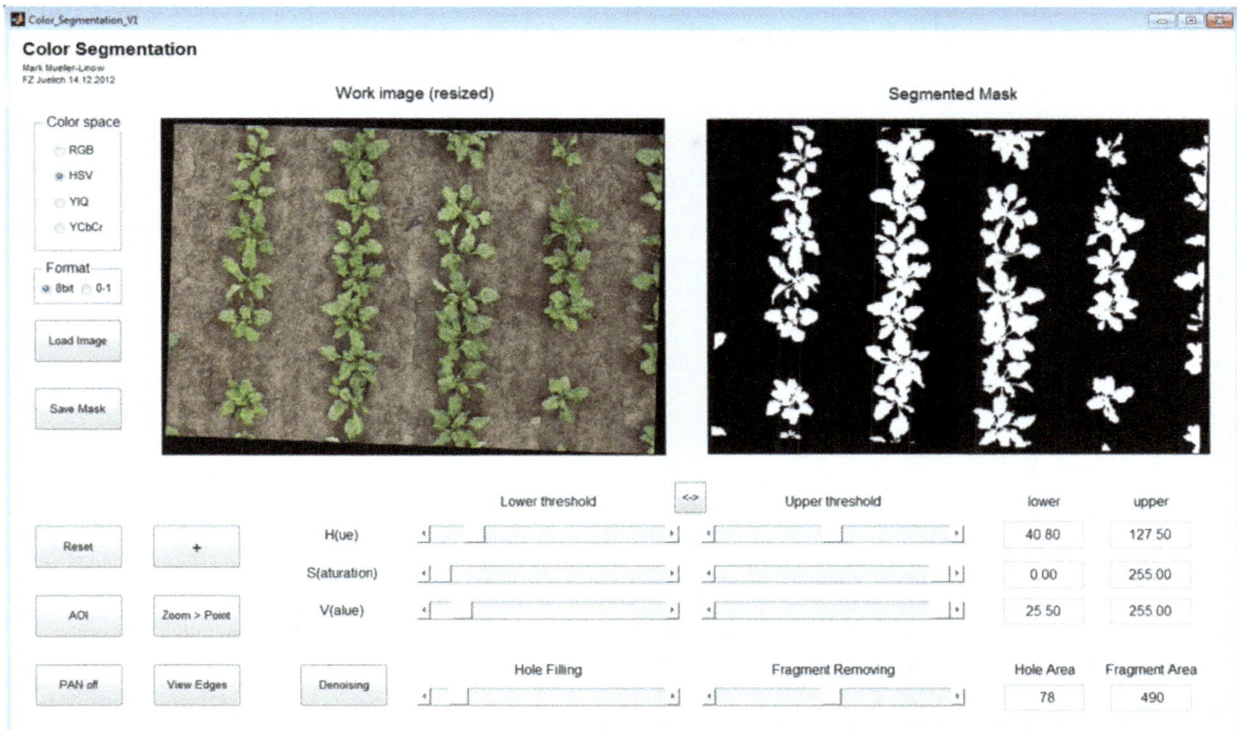

Figure 4 GUI for background segmentation. Graphical user interface for background segmentation: RGB image (left) and resulting binary images (right) are applied in 3-d reconstruction (module 1) and leaf segmentation (module 2). Background segmentation is performed in the HSV color space and includes filters for undersized fragments and for the completion of imperfectly filled segments.

The position and range of the corresponding block in the matching image is computed using SIFT-detected [33,43] plant-specific pixels pairs and their relative positions. Another default feature is the increase of depth estimation accuracy beyond the discrete pixel size using sub-pixel fitting [44]. To achieve this the statistical data around C_{max} is analyzed more deeply using a parabola fit on the neighboring values and recomputing C_{max} according to the maximum of the fitted curve. The selected region in the rectified reference image is now the basis for all follow up computations like the depth map estimation and the image segmentation. Additionally the correspondence analysis is performed as a two-step process producing two disparity maps. In the first step the left image serves as the reference for the detection of corresponding points in the right one, while in the second step this process is performed the other way around. This additional map is later used when excluding wrong disparity estimations due to occlusions.

Post-processing & triangulation: Four default and six optional filters have been included for post-processing: The first two default filter deal with wrong disparity estimations close to leaf edges due to occlusions or strong disparity discontinuities, so-called jump edges. The occlusion filter [45,46] detects inconsistencies between the disparity maps of the correspondence pair analysis and removes the respective pixels. The jump edge filter [47,48] considers the metric distance between a pixel and the 8 neighboring pixels (in the disparity map) for the detection and exclusion of jump edges. The third default filter compares the HSV data of corresponding pixels and removes those with a high discrepancy. We observed this filter to be efficient in suppressing fattening of edges. The fourth default filter is a median filter with a box size of 3×3 pixels [49]. The first optional filter analyzes the distribution of disparity values and removes disparities out of range, the second optional filter removes pixel with low statistical significance (i.e. correlation values below a certain threshold). The third optional filter performs a HSV segmentation to separate the plants from the background either automatically or with a pre-defined binary image mask (computed using the GUI in Figure 4). The other three optional filters correct for (i) outliers with a weighted median filter [50], (ii) missing pixels via filling of small gaps, (iii) undersized disparity fragments by removing them. The depth map is finally converted into a 3-d point cloud via triangulation [42].

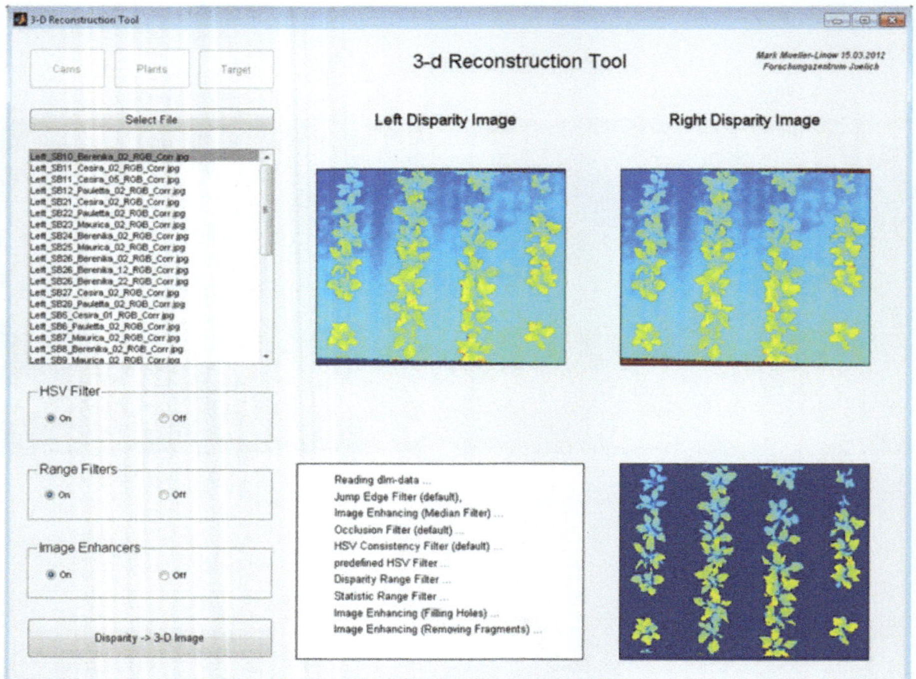

Figure 5 GUI for 3-d reconstruction. Graphical user interface for 3-d reconstruction: Several processes are managed with this graphical user interface (module 1), starting with the calibration of the stereo setup, the (auto)calibrated rectification, the 3-d reconstruction of stereo images and ending with several post-processing steps. The current state in the image processing pipeline displays the results of two 3-d reconstructions (top), executed separately for each of the stereo images and the post-processed disparity image (bottom right) after application of all filter options (as listed on the bottom left).

Module 2: Leaf segmentation

Background Figure 2B and Figure 6 display the outline and the GUI of the leaf segmentation process. To calculate leaf angle distribution each pixel has to be associated with a single leaf and then pixels have to be fitted by a realistic 3-d leaf model. For the planar leaves of soy bean, leaf segmentation was implemented as a graph partitioning method [27]. This method, also referred to as the Felzenszwalb-Huttenlocher (FH) algorithm [51], applies a graph structure on any pre-processed (usually blurred) image information considering pixels as nodes and differences in pixel properties as weighted edges. In the HSV color space, the pixel properties are hue (H), saturation (S) and value (V). We optimized this approach for a better identification of single leaves in various plant species and canopies, as described below.

Pre-processing: For blurring we included a Gauss kernel ($\sigma_G = 0.75$) of 5×5 pixel as the default setting, which works well in most situations. For further user-driven optimization optional filters are available (such as Wiener or averaging filters) as well as pre-defined background segmentation (see HSV segmentation in Figure 4). The removal of non-plant image pixels facilitates the manual post-editing as well as the correct separation for dark image regions.

FH-Algorithm: As any image information may be used for segment computation we combined the three HSV channels and the disparity values in this GUI, each image property equipped with its own graph partitioning threshold (see Figure 6 left). Single channels or any combination of channels may be used in the FH computation, as from every channel an individual segmentation is calculated, assigning a single label to each segment. The resulting segmentations are then combined to a new segmentation, such that a new segment contains a single label from each channel, only.

Manual Editing: Particularly with regard to a more complex leaf surface structure (e.g. sugar beet leaves) the raw segmentation results of each ROI are often not satisfying. We included different manual editing modes, which can be used to achieve a correct segmentation from the pre-segmented image. Over-segmented regions (a leaf consists of several segments) can easily be joined, under-segmented regions (two or more leaves share one segment) can be manually partitioned. In most cases only

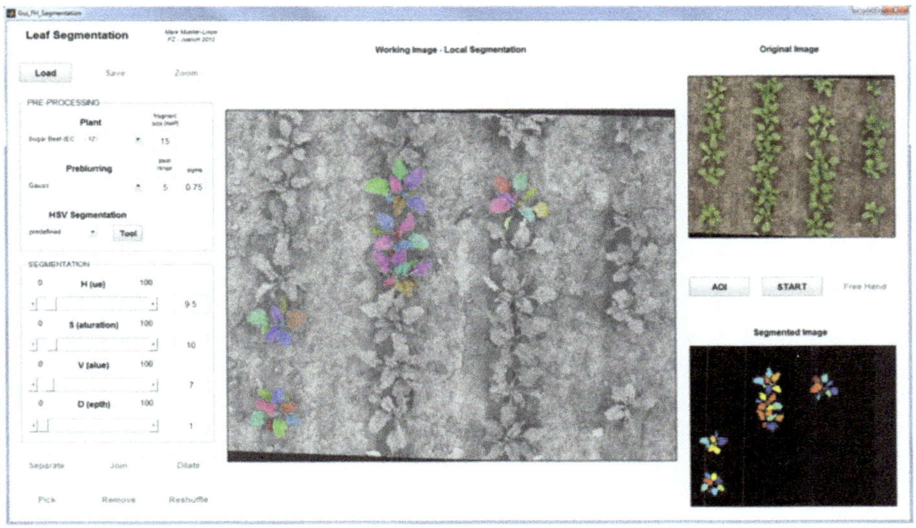

Figure 6 GUI for leaf segmentation. Graphical user interface for leaf segmentation: This GUI from module 2 provides a half-automated graph-based method (FH-algorithm) to segment leaves or leaf sections in RGB images. Selection of small interest regions within the input RGB image (top right) allows for a fast FH segmentation with a subsequent detailed editing (with the tools on the bottom left). Segments of acceptable quality may then be transferred to the final output (bottom right). FH segmentation parameters are regulated in the HSV color space together with a threshold for depth separation (left). Pre-processing specifications (e.g. smoothing and background segmentation) are inserted on the top left side. The current state shows an intermediate result of a segmentation for complete leaves.

small junctions have to be cut in order to separate two segments. In the case of strong fragmentations segments may be joined filling the interspaces at the same time. Disconnected segments, which occur due to other overlapping leaves, may also be joined to improve the accuracy of the subsequent surface modeling. An additional free-hand mode may be used for segment completion or dissection. Correct segments are exported to a live-view control image, which is finally post-processed enhancing the segmentation quality (e.g. by filling holes).

Module 3: Surface modeling

Background We implemented a surface smoothing and a surface fitting approach to model leaf angle distribution. A schematic view on both alternatives is depicted in Figure 2C. Surface smoothing (left) applies curvature flow smoothing [52] or Laplacian smoothing [53,54] to a meshed (via e.g. Delaunay triangulation [55,56] or unstructured triangular surface meshing [34]) set of points to approximate the segment surface. Surface fitting (right) employs different leaf models (3-d surface functions $z(x, y)$) for the point cloud fitting process via singular value decomposition (SVD) [57] inside a RANSAC loop [58].

Surface smoothing: For pre-processing we included a filter, which removes disjoined point clusters up to a particular size and distance to the major clusters. The resulting point clouds are firstly meshed via 2-d Delaunay triangulation on the base of x and y-coordinates to ensure the formation of only such faces reflecting the leaf surface. Faces with edge lengths or areas above predefined thresholds are removed in order to prevent cross-linking in non-plant sections (Figure 7 left - mesh parameters). The final 3-d triangulation, which is performed via unstructured triangular surface meshing, follows on the identification of the leaf border and holes. Holes within each segment, which appear due to fragmentary disparity maps and missing segmentation information, are identified and automatically removed, if they are below a predefined size. Resulting segment meshes are then individually smoothed with the curvature flow or the Laplace method depending on the user's choice. Curvature flow requires 3 parameters, which are preset by the user and which regulate smoothing iteration, smoothing strength and neighborhood impact (Figure 7 left - point cloud).

Surface fitting: Surface fitting aims at adapting coefficients c of a user-selected leaf surface model $z_c(x, y)$ to a point cloud segment, in order to compute leaf angle and curvature. It joins two processes: Finding a consensus set of points in the given point cloud segment via RANSAC thereby deleting outliers and the estimation of surface model coefficients c via singular value decomposition (SVD). The amount of points in the consensus set strongly depends on the number of iterations used to determine the surface coefficients in randomly composed subsets. This number is proportional to the ratio of

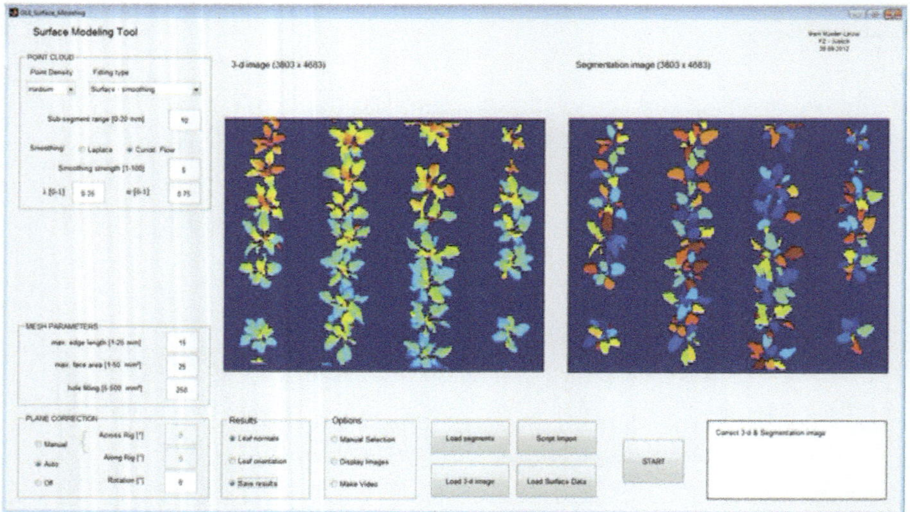

Figure 7 GUI for surface modeling. Graphical user interface for surface modeling: Disparity data (left) and segmentation data (right) are combined in module 3 to approximate surfaces to 3-d point clouds according to different modeling options. Surfaces are either fitted according to planar, quadratic or cubic functions or smoothed using curvature flow or Laplace smoothing.

segment size to subset size. A distance criterion is used to determine whether a subset contributes to the consensus set or not. The surface coefficients, which are then computed from the consensus set via SVD, determine the curvature of the segment. Meshing steps run analogous to the path described in the last section.

Post-processing: For further computations (see case study below) it is important to adjust the final outcome with respect to the inclination and orientation of the stereo rig. The latter one can be achieved by rotating the surface points around the *z*-axis, while the first one needs a transformation of the *xy*-plane (Figure 7 left - plane correction). We included two options to correct the data: (i) The user may record and provide the inclination and direction of the stereo rig manually (with the help of an inclinometer and a compass) or (ii) he may use the automatic mode. Assuming that plants have been grown homogeneously in a near-flat soil, a plane will be fitted through the plant pixel coordinates and used for the computation of the plot inclination.

Leaf surface traits

Leaf area and leaf area index: The software provides a leaf-specific area estimation and the computation of the leaf area index (LAI) by relating the estimated total leaf area to the area covered by the selected image region. In sparse canopies, where leaf occlusion is neglectable these values can be taken as face values. In denser canopies with occlusion and leaf clumping the visible proportion of leaves has to be extrapolated to total leaf area, which requires assumptions on the non-visual part of the canopy. Various approaches are available in the literature and the output of our program can be used for these forward calculations [59,60]. Because of the large number of possible approaches that are developed for the various canopies, we did not implement any method in our software.

Leaf angle distribution: This software tools allow for a statistical assessment of leaf orientations of single plants and small plant populations with a representative number of individuals. A mathematical equivalent for leaf orientation is the leaf angle or leaf angle distribution. Our methods and algorithms are designed for the local derivation of leaf angles on the base of individual leaf models. The leaf angle distribution summarizes the properties of individual leaf surface models. The orientation of each face within such a Delaunay-triangulated surface can be expressed by 2 angles in a spherical coordinate system, the azimuth θ and zenith φ. Typically the zenith angle varies from -90° $< \varphi <$90°. In our convention we assume that each face normal and the normal of the soil span an angle of less than 90°, meaning face normals point 'up'. Therefore the zenith angle will only range from $0 < \varphi <$90°, with a value of $\varphi = 0°$ reflecting a horizontal face. Azimuth angles range from $0° < \theta < 360°$ starting at the right-hand side in the image and then turning counterclockwise.

Error estimation

We tested the accuracy of leaf angle estimation in two experiments. The first one uses an artificial plant with 8 green-colored flat leaves made form plywood which can be adjusted to any zenith angle. The second one employs a sugar beet leaf fixed on a flat surface, which could be oriented arbitrarily. Targets were imaged from nadir position (3.5 m distance) with two Canon EOS 5D Mark II ($f = 50$ mm; $b \approx 200$ mm). We set the leaves of the artificial plant target to different zenith angles, such that most parts of each leaf were in camera view. Individual leaf angles were manually measured using a high resolution dual-axis digital inclinometer (Level Developments LD-2M). Inclination of the sugar beet leaf was manually changed and measured between each imaging step. Here we applied 7 different orientations. Images were processed using the target calibration pipeline. Leaves were segmented and fitted with a planar surface model. Figure 8 summarizes results for both tests. The deviation from the identity line was quantified for the accessible zenith angle interval [0°, 70°]. Steeper leaf parts are not well visible and thus do not give reliable angle estimates [27]. The normalized root mean square error (NRMSE) is approx. 2.5% for the artificial plant and approx. 4.6% for the fixed leaf. Moreover, we computed the sugar beet leaf area of all orientations and estimated the error using the normalized variation coefficient, which is approx. 2.8%. We do not observe a bias towards fronto-parallel surfaces well known for other stereo reconstruction approaches [61].

Case study

Our stereo imaging approach has been tested with different plants demonstrating the functionality of our software across species and applications (Figure 9). Our test cases ranged from the small rosette plant *Arabidopsis thaliana* (Figure 9A) to single trees in apple orchards (Figure 9B) to the agricultural crops sugar beet and barley, which are the main focus species of the Crop.Sense.net network (Figure 9C, D). Currently further studies with *Arabidopsis* are on the way to better understand gene-phenotype interactions and with apple trees to assess fruit traits by 3-d stereo imaging (results will be published elsewhere). In this manuscript we focus on a detailed investigation of four different sugar beet varieties that were subjected to different nitrogen availability. We performed a detailed case study demonstrating the potential of our stereo approach to distinguish subtle seasonal, variety and treatment specific differences in leaf display.

Plants The cultivars *Cesira, Pauletta, Maurica* and *Berenika* have been grown in 6 × 20 meter plots with 4 repetitions in the central experiment of CROP.SENSE.net in the agricultural study site of Campus Klein-Altendorf (NRW, Germany). Two nitrogen treatments (80 kg/ha and 150 kg/ha) have been applied (in the following denoted

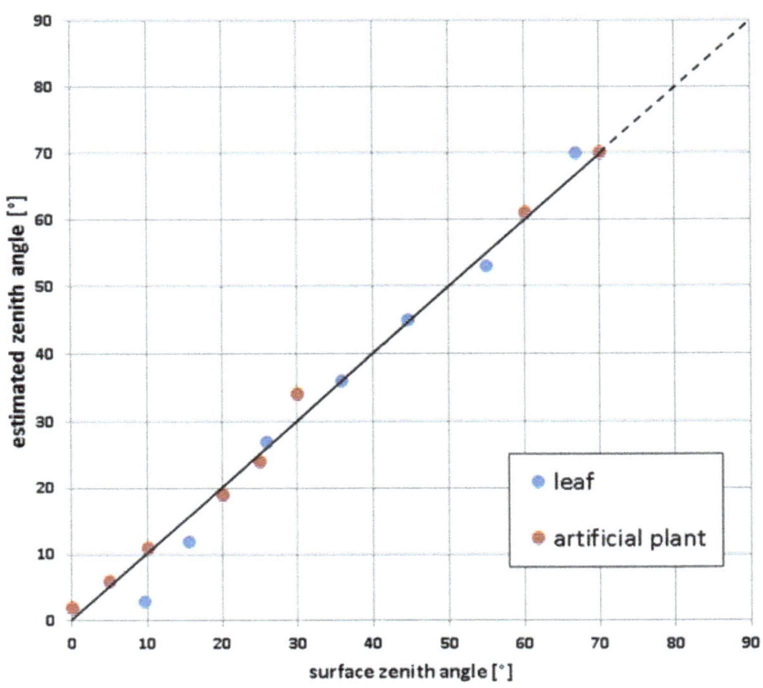

Figure 8 Error estimation. Surface zenith angles of artificial setups were measured with an inclinometer and compared to respective estimated leaf angles from stereo reconstruction. Estimation errors were quantified by the normalized root mean square error (NRMSE): artificial plant leaves (red) displayed a NRMSE of 2.5%; planarly-fixed sugar beet leaf in various orientations (blue) displayed a NRMSE of 4.6%. The dashed line indicates where angle estimates are known to be unreliable [27].

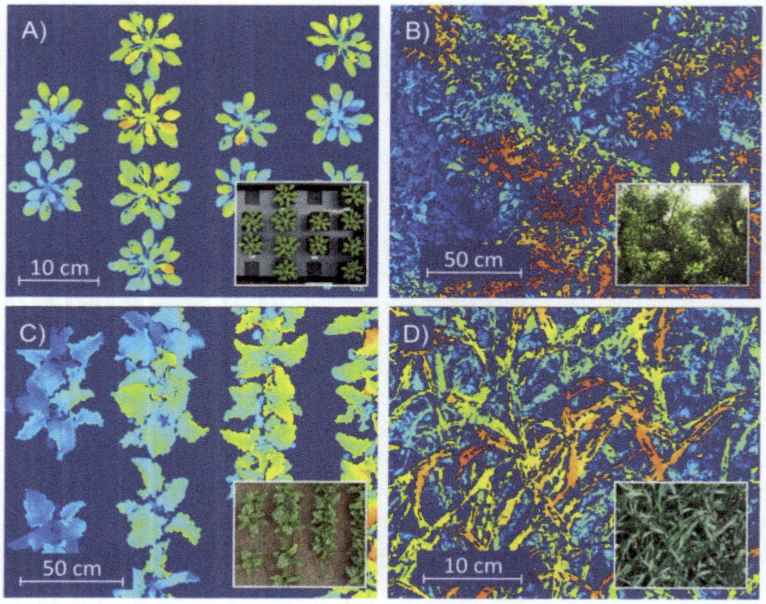

Figure 9 Field of applications. Depth maps and respective RGBs (inlay) of different experimental plant systems: Pixel disparities in the depth map are color-coded ranging from red (closer to cameras) to blue (further away); distance ranges are given in brackets: **(A)** trays of *Arabidopsis thaliana* were monitored in studies of diurnal leaf movement ($\approx$ 20 *mm*); **(B)** single trees (apple orchards) were analyzed with respect to the leaf and fruit stratification; Klein-Altendorf, 2013 ($\approx$ 700 *mm*); **(C)** small plot of sugar beet; case study from CROP.SENSE.net central experiment (Campus Klein-Altendorf); 2012, June 14 ($\approx$ 500 *mm*); **(D)** small barley populations; Crop Garden experiment at FZ Juelich; 2011, July 5 ($\approx$ 400 *mm*).

as N- and $N+$). The cultivar and treatment combinations each with 4 repetitions were planted in plots of 6 $\times$ 20 m.

Measurements We used two Canon EOS 5D Mark II with $f = 50$ mm lenses mounted on a sliding bar with a baseline of $b \approx 200$ mm (Figure 1). The stereo rig was mounted on a bucket truck for imaging at nadir position 3.5 m above the canopy. The remote-control release for simultaneous triggering consisted of two pre-trigger remote cables fixed to one trigger button. Camera calibrations were performed as usual with regular 2-d pattern. For our purposes we designed a rectangular frame with a square-like alignment of 9 $\times$ 6 black circles in front of a white background. The pattern was attached to a flat surface. Calibration quality depends on the number of different target positions, which can be achieved by rotating the calibration pattern. We took at least 20 images covering the whole plant volume. The target imaging distance was similar to the imaging distance to the canopy. The three measurement days in 2012 were May 30 (s), June 14 (m) and September 5 (l). 2 repetitions of each cultivar & treatment combination were imaged at a time resulting in a total of 48 stereo images. These dates correspond to different soil coverage ranging from minor coverage (as depicted in Figure 4) up to complete soil coverage. Images were taken with the same platform orientation around noon ($\pm 2h$). Distance to the canopy was fixed with a laser range finder (Bosch PLR 25) and inclination of the platform was monitored with a high resolution dual-axis digital inclinometer (Level Developments LD-2M). As the cherrypicker platform was always facing east, no further viewing direction data for stereo rig correction has been recorded.

Image processing setup

Images were processed starting with the default calibration pipeline. Figure 3 displays exemplarily the overall surface reconstruction of a sugar beet plant population in an early state (*Berenika* - May 30). The reconstruction worked well for all images, however problems occurred at strongly reflecting leaf surfaces. These parts tend to display hot spots, which move with the viewpoint (non-Lambertian surface effect) leading to wrong disparity estimations especially along the midribs that are orientated parallel to the stereo rig axis. These sections are mostly excluded by the filtering processes. In the following, we tested both surface modeling modes (smoothing and fitting) with regard to the plausibility of results. For surface smoothing we used segmented point clouds of entire leaves. Leaf segments were pre-filtered separately eliminating point clusters with spatial separation of $>$ 5 mm and resulting meshes were smoothed using curvature flow. Surface fitting was performed on leaf segments cut at the midrib. This additional step is necessary as sugar beet leaves tend to be folded along the longitudinal leaf axis depending on variety and stage. The resulting segmented point cloud was then fitted with a quadratic surface function inside a RANSAC loop using subsets of 50 points and a distance criterion of 1.75 mm. In both approaches the orientation of the surface meshes was finally corrected with the recorded inclination data of the stereo rig position. We also checked the automatic correction mode, which produced similar results. Both methods were in principle capable of capturing essential leaf properties which reflect seasonal and genetic differences, however the smoothed leaf model results displayed a higher visual plausibility, which is why we will focus on these in the following.

Estimation of the leaf angle distribution

The orientation of each face (as given by the face normal) within the reconstructed meshes can easily be converted into azimuth (θ) and zenith (φ) angles and - if related to the individual face area - summed up to the distribution of leaf angles. Distributions are calculated as normalized histograms with 1° bin width, if not stated differently. The interpretation may be carried out on the basis of φ or θ angle distributions separately or as joint distributions, assuming that there is a dependency between both distributions. (i) Figure 10A shows the distribution of φ for *Berenika* (May 30/$N+$). Further statistics measures like the mean or median of the φ distribution may help to characterize drought stress, diseases or diurnal leaf movements. The shape of the φ distribution displayed characteristic differences between the cultivars, which were distinctly increased in the older developmental states, while for the θ angle distributions we recorded seasonal differences particularly reflected in the location of the angle averages (data not shown). (ii) On a higher level θ and φ distributions are combined in a two-dimensional histogram with the angular axis denoting θ and the radial axis denoting φ (Figure 10B). The θ distribution is not uniform and there are clusters of increased frequency, in this representation in the southwestern hemisphere with a maximum around $\theta = 220°$. This in turn means that a comparatively large amount of leaf parts are facing

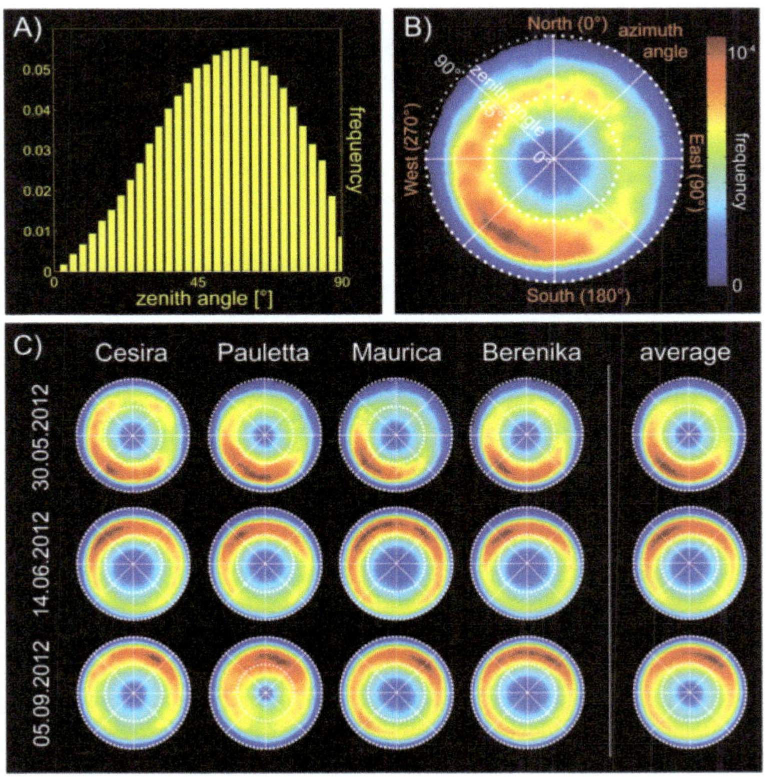

Figure 10 Leaf angle distribution of a sugar beet sample. Leaf angle distributions of sugar beet populations: **(A)** frequency of zenith angles; **(B)** combined zenith and azimuth angles in a 2-d histogram plot for a single image of *Berenika* (May 30/$N+$), the example given in Figure 3. Curvature flow smoothing has been applied on the 3-d data. In this representation the azimuth angles $\theta = 0°, 90°, 180°$ and 270° correspond to the cardinal points North, East, South and West. **(C)** compares state effects (date of measurement) and cultivar effects, each diagram representing the average of 2 repetitions and 2 different nitrogen treatments: As indicated by the overall averages (right), the growth state differences are most obviously reflected in the location and center of the azimuth distribution. (All images have been generated with an extra visualization tool).

Southwest. In the following we checked the consistency of results with respect to genetic and environmental conditions. The setup of the Central Experiment of Campus Klein-Altendorf displays four major factors with potential influence on the phenotype, namely state, genotypic, site, and fertilization effects. Most obvious are the strong morphological changes of all sugar beet cultivars during seasonal development (state effects), while variety-specific differences become more apparent the older the leaves are (genotypic effects). We expected less impact from site (repetitions) and from fertilization (nitrogen treatment) effects, which is why we averaged over these two factors in Figure 10C, displaying the distributions of all state-cultivar combinations. Looking at the state effects one observes a cultivar-independent preferred direction in θ (also illustrated by the averages on the right) which changes distinctly over the season especially from the youngest to the older states. The center of the azimuth distribution, which is the median $\tilde{\theta}$ of the distribution, is shifted from Southwest ($\tilde{\theta}_s = 217°$) to North ($\tilde{\theta}_m = 354°$ and $\tilde{\theta}_l = 4°$). We do not know yet the exact cause for this effect, but there are several reasons, that have to be taken into account: (i) Strong west winds at the location of our study site at Campus Klein-Altendorf may affect the leaf orientation and plants may react to windy conditions differently according to their susceptibility (leaf size or closeness of the canopy due to the sawing pattern); (ii) sugar beets are sown in rows from east to west with a row distance of 50 cm and a plant distance of 20 cm within the rows. Plants in the youngest state s do not have any contact to their neighbors. Plants in the intermediate state m have contact to the plants within their rows, plants in the oldest state l also across the rows. Therefore, young plants with small leaves and no neighborhood competition may orientate their leaves most efficiently to the south, while in older plants this deviation from the optimal distribution may be caused by a combination of several factors. Concerning the distribution of φ the results are rather homogeneous, however differences between states and cultivars are present, particularly in the shape of the distributions. The width of the φ distribution for instance differs strongly between *Pauletta* and *Berenika* in all states being much broader in the case of *Pauletta* and also the youngest states display lower average φ angles with medians of $\tilde{\varphi}_s = 49 \pm 1°$ compared to the older states with $\tilde{\varphi}_m = 56 \pm 2°$ and $\tilde{\varphi}_l = 53 \pm 5°$. Again, this finding may result from a combination of different effects: (i) leaf stages differ distinctly between plants at younger or older states and (ii) leaf erection as a result of neighborhood interaction. Young plants have enough space to adjust their leaves optimally for light interception, while older plants have to compete for light on the one hand and are featured by very large and heavy leaves with planophile leaf sections on the other.

As these results displayed only slight differences between the cultivars, we analyzed the leaf angle distributions of the 48 combinations (measurement date, cultivar, nitrogen treatment and repetition) more deeply by performing a principle component analysis (PCA). To this end we interpret each angle distribution with its N bins as a point in an N-dimensional space, i.e. we populate this space with 48 points. PCA then delivers directions of main variations in this N-dimensional space. We investigated clustering effects when using θ- and φ-distributions separately or jointly. As effects were more pronounced for the latter case, we focus on this analysis in the following. The first two components of the PCA have been depicted in Figure 11A. Most apparently, the three plant states are well clustered (as indicated by the three colors) and also separated in the case of the youngest state s. There is no systematics within the distribution of nitrogen treatments, but clustering of cultivars is present within each sub-group. For a detailed analysis we repeated the PCA separately for each measurement date (Figure 11B-D). As indicated before, all states are featured by a fairly good separation of cultivars, especially for the last state l, while nitrogen treatment effects seem to be negligible. This study was also carried out with a planar leaf model and the quadratic surface function model. The results were comparable but less pronounced than with the model-free surface smoothing option.

Discussion

The reconstructibility of plant canopies from stereo images captured in the lab or under field conditions strongly depends on hardware setup, imaging conditions, plant and planting type. (i) The spatial relation between cameras and between stereo rig and plants affect the field of view, the spatial accuracy (depth and lateral resolution), the degree of occlusions and the reconstructable depth range due to the depth of focus. In the following, we name a few parameters, which should be handled in a balanced manner (our additional stereo rig setup tool is intended for this purpose): Increasing baseline b results in a better depth resolution but produces more occlusions. Using lenses with a higher focal length f also increases the depth resolution, but decreases the field of view, which is also decreased by smaller distances between stereo rig and canopy, while the depth resolution is increased. A higher F-number (aperture) gives a better depth of focus, but also increases the exposure time, which can cause problems in a shifting canopy. (ii) Alternating light conditions and heterogeneous illumination across the canopy are sources of erroneous depth estimations and may cause problems in automatized segmentation processes. Strong incident solar radiation together with specular properties of the leaf surface often produce hot spots and a decrease of visible texture and should therefore be avoided, e.g. by

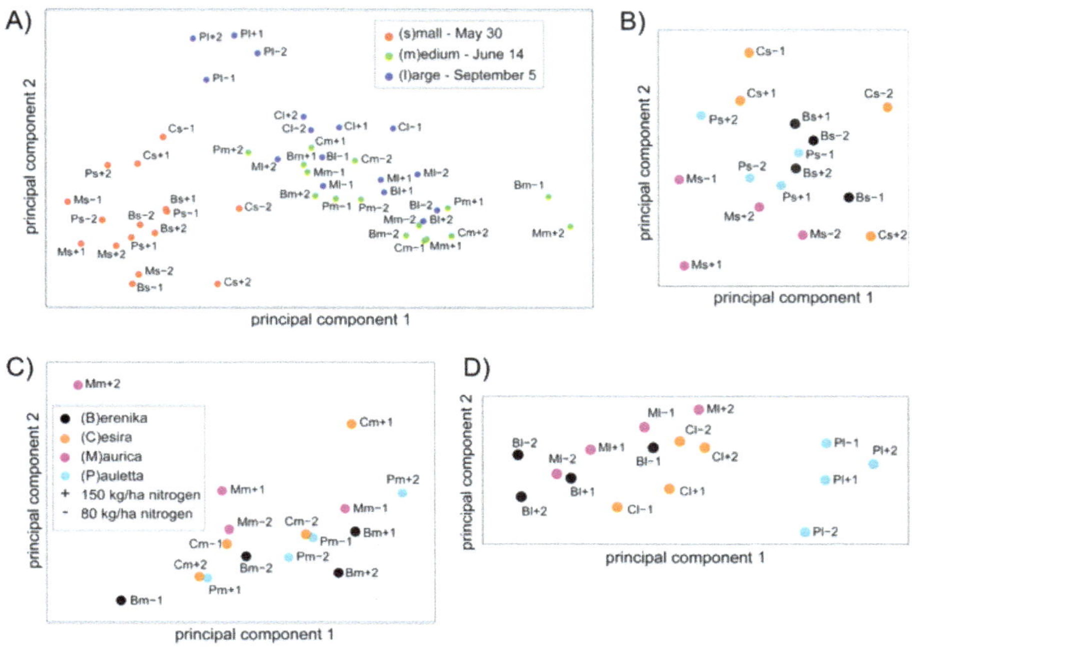

Figure 11 Principal component analysis (PCA) of plot-specific leaf angle distributions. PCA reveals distinct clustering of states and cultivars, while nitrogen treatment effects are lacking. **(A)** Complete set of 48 analyzed images representing 2 repetitions (1,2) of 24 parameter constellations - 4 cultivars (B,C,M,P), 3 states (s, m, l) and 2 nitrogen treatments (+, -): The states (color-indexed measurement days) display strong clustering with the strongest separation of the youngest plant state. **(B-D)** Analysis of each state indicates variety-specific leaf angle distributions resulting in a pronounced clustering of all 4 cultivars (color-indexed) regardless of the measurement day.

measuring under diffuse light conditions (clouded sky). Windy conditions during image acquisition strongly affect the quality of reconstruction and should also be avoided. Short exposure times are required in such cases. (iii) The type of plant, plant age and planting density strongly affect the results. Taking the camera perspective into account the degree of self and of mutual occlusion (leaf clumping) defines the proportion of reconstructable plant material. The amount of image pixels covering the narrow side of the leaf blade should be large enough for a correct surface reconstruction (e.g. barley leaves should be imaged from a shorter distance or with a higher f). Plant leaves with homogeneous surface properties (e.g. barley) may cause wrong disparity values. This effect is particularly strong for specular reflections of sun light, where the hot spots move with the camera perspective, leading to wrong assignments of corresponding image regions. iv) Calibration issues may play a role when the camera system is not used in a fixed setup, which needs to be calibrated only once. If used more flexible, the setup needs to be re-calibrated after every change. If deficient rectifications occur, images can still be analyzed using the auto-calibration mode. v) For the surface fitting process the right choice between smoothing or surface functions strongly depends on the leaf surface properties and on the amount of noise in the data and should therefore be taken accordingly. Approximating the leaf surface from smoothed data might be appropriate for complex leaf geometries, for low-noise data sets and for studies, where a leaf angle resolution on the pixel-level is required. For the estimation of a single leaf-specific angle or for noisy data surface model fitting should be preferred.

Conclusions

With the development of this software package we want to provide a comprehensive tool for the analysis of leaf surface properties within the outer canopy layers using off-the-shelf hardware, which can easily be assembled to a stereo camera rig. With little methodic effort, the generation of dense depth maps, the identification of single leaves and the modeling of the leaf surface structure is feasible. The automatic segmentation of single leaves is possible, however for dense canopies (e.g. older sugar beet plants in the stock) leaf segmentation requires some manual editing despite the information available on the vertical leaf distribution. This is a step, where user interaction is required, and we think that it can be reduced to a minimum with plants that have a more homogeneous shape and curvature. Surface modeling is the delicate step in the processing pipeline and the method of choice strongly depends on the quality of data, the complexity of leaf structure and the scientific questions. Despite the methodic restriction of the reconstructibility to the outer canopy layers, these observations can be of great use as these layers are of utmost importance for photosynthetic activity. Therefore shifting states in that part may serve as a proxy for the physiological and the health status, respectively. Employing the properties of the outer canopy typical seasonal and genotypic differences between our plant populations were clearly demonstrated and we think that this method is applicable to other fields up to the level of agricultural and horticultural plant systems.

Availability and requirements

Project name: Leaf Angle Distribution Toolbox
Operating system: Windows
Programming language: Matlab
Other requirements: Matlab Compiler Runtime (MCR)
License: proprietary - the software is restricted to academic use only. The software is available from the authors upon request. Research projects, which benefit from the Leaf Angle Distribution Toolbox, are obliged to cite this paper.

Abbreviations

b: baseline, distance between the camera centers; *B*: size of the block used in the block matching method of the correspondence analysis; *R*: search range for corresponding points in the correspondence analysis; *f* or f_p: focal length [mm / pixel]; D_{ij}: disparity of pixel in column *i* and row *j*; x_{ij}, y_{ij}, z_{ij}: 3-d coordinates of a pixel in column *i* and row *j*; σ_G: Gaussian kernel segmentation; *N-* / *N+*: nitrogen treatment (80 kg/ha or 150 kg/ha); θ: azimuth; φ: zenith; *s*, *m*, *l*: plant states: small (30.05.2012), medium (14.06.2012), large (05.09.2012).

Competing interests

The authors declare that they have no competing interests.

Authors' contributions

MML carried out the development of the software package and the case study and wrote the manuscript. FP carried out the measurements of the case study at Campus Klein-Altendorf. HS gave advice in all technical aspects. UR conceived the study and participated in the coordination of the study. HS and UR helped to draft the manuscript. All authors read and approved the final manuscript.

Acknowledgements

We acknowledge Kai Schmidt, Manfred Trimborn and Agim Ballvora for the design of the central experiment of CropSense.net at Campus Klein-Altendorf and the team of Campus Klein-Altendorf for cultivation and technical support during the measurements. We also acknowledge Shizue Matsubara for providing the *Arabidopsis* plants. This work has been conducted in frame of a research program funded by BMBF with project number 315309/CROP.SENSe. The authors also thank the German Plant Phenotyping Network (DPPN) and PhenoCrops, funding code 005-1105-0035, which provided additional funding.

References

1. Rascher U, Nedbal L. Dynamics of plant photosynthesis under fluctuating natural conditions. Curr Opin Plant Biol. 2006;9:671–8.
2. Rosa LM, Forseth IN. Diurnal patterns of soybean leaf inclination angles and azimuthal orientation under different levels of UVb radiation. Agric For Meteorol. 1995;78:107–19.
3. Wofford TJ, Allen FL. Variation in leaflet orientation among soybean cultivars. Crop Sci. 1982;22:999–1004.
4. Bawhey CI, Grant RH, Gao W. Digital measurement of heliotropic leaf response in soybean cultivars and leaf exposure to solar uvb radiation. Agric For Meteorol. 2003;120:161–76.
5. Long SP, Zhu XG, Naidu SL, Ort DR. Can improvement in photosynthesis increase crop yields? Plant Cell Environ. 2006;29:315–30.
6. Schurr U, Walter A, Rascher U. Functional dynamics of plant growth and photosynthesis ? from steady-state to dynamics ? from homogeneity to heterogeneity. Plant Cell Environ. 2006;29:340–52.
7. Biskup B, Kuesters R, Scharr H, Walter A, Rascher U. Quantification of plant surface structures from small baseline stereo images to measure the three-dimensional surface from the leaf to the canopy scale. Nova Acta Leopoldina. 2009;96(357):31–47.
8. Loomis RS, Williams WA, Duncan WG. Community architecture and the productivity of terrestrial plant communities. New York, USA: New York Academic; 1967, pp. 291–308.
9. Rascher U, Biskup B, Leakey ADB, McGrath JM, Ainsworth EA. Altered physiological function, not structure, drives increased radiation-use efficiency of soybean grown at elevated CO2. Photosynth Res. 2010;105(1):15–25.
10. Zhu X-GZ, Long SP, Ort DR. Improving photosynthetic efficiency for greater yield. Annu Rev Plant Biol. 2010;61:235–61.
11. Hosoi F, Omasa K. Factors contributing to accuracy in the estimation of the woody canopy leaf area densitiy profile using 3d portable LIDAR imaging. J Exp Bot. 2007;58:3463–73.
12. Hosoi F, Nakai Y, Omasa K. Estimation and error analysis of woody canopy leaf area density profiles using 3-d airborne and ground-based scanning lidar remote-sensing techniques. IEEE Trans Geosci Remote Sens. 2010;48(5):2215–23.
13. Hosoi F, Omasa K. Estimation of vertical plant area density profiles in a rice canopy at different growth stages by high-resolution portable scanning LIDAR with a lightweight mirror. ISPRS J Photogram Remote Sens. 2012;74:11–9.
14. Bellasio C, Olejnickova J, Tesar R, Sebela D, Nedbal L. Computer reconstruction of plant growth and chlorophyll fluorescence emission in three spatial dimensions. Sensors. 2012;12:1052–71.
15. Mullen JL, Weinig C, Hangarter RP. Shade avoidance and the regulation of leaf inclination in Arabidopsis. Plant Cell Environ. 2006;29:1099–106.
16. Dornbusch T, Lorrain S, Kuznetsov D, Fortier A, Liechti R, Xenarios I, et al. Measuring the diurnal pattern of leaf hyponasty and growth in Arabidopsis - a novel phenotyping approach using laser scanning. Func Plant Biol. 2012;39:860–9.
17. Watanabe T, Hanan JS, Room PM, Hasegawa T, Nakagawa H, Takahashi W. Rice morphogenesis and plant architecture: Measurement, specification and the reconstruction of structural development by 3d architectural modelling. Ann Bot. 2005;95:1131–43.
18. Vos J, Evers JB, Buck-Sorlin GH, Andrieu B, Chelle M, de Visser PHB. Functional-structural plant modelling: a new versatile tool in crop science. J Exp Bot. 2010;61(8):2101–15.
19. Sarlikioti V, de Visser PHB, Marcelis LFM. Exploring the spatial distribution of light interception and photosynthesis of canopies by means of a functional-structural plant model. Ann Bot. 2011;107:875–83.
20. Deckmyn G, Nijs I, Ceulemans R. A simple method to determine leaf angles of grass species. J Exp Bot. 2000;51(349):1467–70.
21. Ivanov N, Boissard P, Chapron M, Valery P. Estimation of the height and angles of orientation of the upper leaves in the maize canopy using stereovision. Agronomie. 1994;2:183–94.
22. Hosoi F, Omasa K. Voxel-based 3-d modeling of individual trees for estimating leaf area density using high-resolution portable scanning LIDAR. IEEE Trans Geosci Remote Sensing. 2006;44:3610–8.
23. Pisek J, Ryu Y, Alikas K. Estimating leaf inclination and g-function from leveled digital camera photography in broadleaf canopies. Trees. 2011;25: 919–24.
24. Duvick DN, Smith JSC, Cooper M. Long-Term Selection in a Commercial Hybrid Maize Breeding Program. Hoboken, New Jersey, USA: John Wiley & Sons, Inc.; 2010, pp. 109–51.
25. Ford ED, Cocke A, Horton L, Fellner E, Van Volkenburgh M. Estimation, variation and importance of leaf curvature in zea mays hybrids. Agric Forest Meteorol. 2008;148(10):1598–610.
26. Paproki A, Sirault X, Berry S, Furbank R, Fripp J. A novel mesh processing based technique for 3d plant analysis. BMC Plant Biol. 2012;12(63):13.

27. Biskup B, Scharr H, Schurr U, Rascher U. A stereo imaging system for measuring structural parameters of plant canopies. Plant Cell Environ. 2007;30(10):1299–308.
28. Ehlert D, Heisig M, Adamek R. Assessment of a Laser Scanner on Agricultural Machinery. In: Engineering for Rural Development - 9th International Scientific Conference. University of Agriculture, Jelgava, Latvia; 2010. p. 32–36.
29. Sanz-Cortiella R, Llorens-Calveras J, Escolà A, Arnó-Satorra J, Ribes-Dasi M, Masip-Vilalta J, et al. Innovative LIDAR 3d dynamic measurement system to estimate fruit-tree leaf area. Sensors. 2011;11(6):5769–91.
30. Lumme J, Karjalainen M, Kaartinen H, Kukko A, Hyyppä J, Hyyppä H, et al. Terrestrial laser scanning of agricultural crops. In: Proceedings of the XXI. IS-PRS Congress: Silk Road for Information from Imagery. Bejing, China; 2008. p. 563–66.
31. Bouguet J-Y. Calibration toolbox for Matlab. 2005. http://vision.caltech.edu/bouguetj/calib_doc/index.html.
32. Fusiello A, Trucco E, Verri A. A compact algorithm for rectification of stereo pairs. Mach Vis Appl. 2000;12(1):16–22.
33. Vedaldi A, Fulkerson B. VLFeat: An Open and Portable Library of Computer Vision Algorithms. 2008. http://www.vlfeat.org/.
34. Engwirda D. MESH2D - Automatic Mesh Generation. 2009. http://www.mathworks.de/matlabcentral/fileexchange/25555-mesh2d-automatic-mesh-generation.
35. Agoston M. Computer Graphics and Geometric Modeling: Implementation and Algorithms. London, UK: Springer; 2005, pp. 300–306.
36. Shapiro L, Stockman G. Computer Vision. Upper Saddle River, New Jersey, USA: Prentice Hall; 2001.
37. Hartley R, Zisserman A. Multiple View Geometry in Computer Vision. Cambridge, United Kingdom: Cambridge University Press; 2003.
38. Zhang Z. Flexible camera calibration by viewing a plane from unknown orientations. In: ICCV. Kerkyra, Corfu, Greece; 1999. p. 666–73.
39. Zhang Z. A flexible new technique for camera calibration. IEEE Trans Pattern Anal Mach Intell. 2000;22:1330–4.
40. Brown MZ, Burschka D, Hager GD. Advances in computational stereo. IEEE Trans Pattern Anal Mach Intell. 2003;25(8):993–1008.
41. Scharstein D, Szeliski R. A taxonomy and evaluation of dense two-frame stereo correspondence algorithms. Int J Comput Vis. 2002;47(1-3):7–42.
42. Trucco E, Verri A. Introductory Techniques for 3D Computer Vision. New Jersey, USA: Prentice Hall; 1998.
43. Lowe DG. Distinctive image features from scale-invariant keypoints. Int J Comput Vis. 2004;60(2):91–110.
44. Fusiello A, Roberto V, Trucco E. Efficient Stereo with Multiple Windowing. In: IEEE Computer Society Conference on Computer Vision and Pattern Recognition. San Juan, Puerto Rico; 1997. p. 858–863.
45. Fua P. Combining Stereo and Monocular Information to Compute Dense Depth Maps that Preserve Depth Discontinuities. In: 12th International Joint Conferences on Artificial Intelligence; 1991. p. 1292–1298.
46. Fua P. A parallel stereo algorithm that produces dense depth maps and preserves image features. Mach Vis Appl. 1993;6:35–49.
47. Khongsab P. Signal processing and performance evaluation of a PMD camera for space docking. Kiruna: Master's thesis, Lulea University of Technology; 2009.
48. Lejeune A, Piérard S, Van Droogenbroeck M, Verly J. A new jump edge detection method for 3D cameras. In: International Conference on 3D Imaging (IC3D). Liège, Belgium; 2011. http://hdl.handle.net/2268/103459.
49. Huang T, Yang G, Tang G. A fast two-dimensional median filtering algorithm. IEEE Trans Acoust Speech Signal Process. 1979;27(1):13–8.
50. Ma Z, He K, Wei Y, Sun J, Wu E. Constant Time Weighted Median Filtering for Stereo Matching and Beyond. In: IEEE International Conference on Computer Vision. Sydney, NSW, Australia; 2013. p. 49–56.
51. Felzenszwalb PF, Huttenlocher DP. Efficiently computing a good segmentation. In: Proceedings DARPA Image Understanding Workshop. Monterey, California, USA; 1998. p. 251–8.
52. Desbrun M, Meyer M, Schröder P, Barr AH. Implicit fairing of irregular meshes using diffusion and curvature flow. In: Proceedings of the 26th Annual Conference on Computer Graphics and Interactive Techniques. New York, NY, USA: ACM Press/Addison-Wesley Publishing Co.; 1999. p. 317–324.
53. Zhang Y, Hamza AB. Vertex-based diffusion for 3-d mesh denoising. IEEE Trans Image Process. 2007;16(4):1036–45.
54. Hansen GA, Douglass RW, Zardecki A. Mesh enhancement: selected elliptic methods, foundations and applications. London, UK: London Imperial College Press; 2005.
55. Delaunay BN. Sur la sphere vide. Bull Acad Sci USSR. 1934;7:793–800.
56. de Berg M, Cheong O, van Kreveld M, Overmars M. Computational Geometry: Algorithms and Applications. London, UK: Springer; 2008.
57. Golub G, van Loan C. Matrix Computations. Baltimore, Maryland, USA: Johns Hopkins University Press; 1996.
58. Fischler M, Bolles R. Random sample consensus: A paradigm for model fitting with applications to image analysis and automated cartography. Comm ACM. 1981;24(6):381–95.
59. Welles JM. Some indirect methods of estimating canopy structure. Remote Sensing Rev. 1990;5:31–43.
60. Breda NJJ. Ground-based measurement of leaf area index: a review of methods, instruments and current controversies. J Exp Bot. 2003;54(392): 2403–17.
61. Li G, Zucker SW. Differential Geometric Consistency Extends Stereo to Curved Surfaces. In: ECCV (3). London, UK: Springer; 2006. p. 44–57.

Phenotypic analysis of the Arabidopsis heat stress response during germination and early seedling development

Joana Silva-Correia[1,2†], Sara Freitas[1†], Rui M Tavares[1], Teresa Lino-Neto[1*] and Herlânder Azevedo[1,3]

Abstract

Background: Phenotypic characterization of *Arabidopsis thaliana* gain- and loss-of-function mutants is a delicate and meticulous task that often involves the analysis of multiple parameters. Arabidopsis heat tolerance has been evaluated based on direct assessments that include seed germination, seedling survival, hypocotyl and root elongation, or indirect measurements such as chlorophyll content or ion leakage.

Results: In an attempt to simplify the detection of heat stress-associated phenotypes, a collection of protocols for analysis of seed germination and seedling survival to heat treatment is proposed. Temperatures and lengths of heat treatments were combined into several heat tolerance assays, to be used as a primary approach for the search and characterization of basal and acquired heat tolerance-associated phenotypes at early developmental stages. The usefulness of this methodology was illustrated through the characterization of heat-related phenotypes in different Arabidopsis ecotypes as well as in gain- and loss-of-function mutants.

Conclusions: The use of standardized experimental protocols designed to detect temperature-related phenotypes is proposed. The suggested plate-based assays provide an appropriate framework of experimental conditions for detection of variability amongst natural accessions or mutants lines. Functional studies could be facilitated by using this inexpensive and undemanding approach.

Keywords: Arabidopsis, Heat stress, Mutants, Phenotype characterization, Thermotolerance

Background

With the onset of post-genomic capabilities in *Arabidopsis thaliana*, and particularly the availability of large collections of genome-indexed insertion mutants, the plant biology community has been thoroughly employing reverse genetic strategies [1]. The most critical step in this approach is the identification of a phenotype that is distinct from the wild-type, thus providing new insights on gene function [2,3]. Searching for a noticeable phenotype can be a strenuous process, often hindered by gene redundancy, natural variability, or subtle imperceptible changes in growth or development [4]. Previously, Boyes et al. [5] developed a sensitive and dynamic method to detect and interpret phenotypic differences over the entire span of Arabidopsis development, based on a time-dependent map of well-defined morphological traits. However, detection of differential phenotypes is often dependent on the application of particular experimental conditions, specific environmental challenges or a combination of several inputs [4]. Thus, selection of the most suitable experimental conditions is critical for success in the identification of a phenotype.

In the specific case of heat tolerance, a standardized experimental protocol designed to detect temperature-related phenotypes has not been reported to the best of our knowledge. Distinct assays have been used by different laboratories to distinguish responses to temperature, including the analysis of seedling survival [6-14], adult plant survival/fresh weight [9,15], seed germination and cotyledon greening [9,10,13], chlorophyll accumulation [16,17], shoot elongation [15], hypocotyl elongation [6,7,9,18-20], root growth [9,13,17], ion leakage [6],

* Correspondence: tlneto@bio.uminho.pt
†Equal contributors
[1]Center for Biodiversity, Functional & Integrative Genomics (BioFIG), Plant Functional Biology Center, University of Minho, Campus de Gualtar, 4710-057 Braga, Portugal
Full list of author information is available at the end of the article

TBARS accumulation [9], and reporter enzyme activity [18]. In addition to the way the phenotype is scored, other variables lead to differentiated thermotolerance assays, including the plant species, the plant's developmental stage and the type of heat stress regime (reviewed by [21]). The importance of different time and temperature treatments was precisely depicted in pioneer studies in soybean, leading to important advances in the study of heat stress resistance (e.g. [22]). Unfortunately, the use of such different conditions promotes high variability in the results, with implications in the capacity to detect differential phenotypes.

Plants can rapidly acquire heat tolerance to otherwise lethal temperatures, if they are pre-exposed to a moderate high-temperature or exposed to a gradual increase in temperature. This phenomenon, known as acquired heat tolerance, is clearly distinct from basal heat tolerance, which refers to the innate ability of plants to survive exposure to temperatures above those optimal for growth [23]. As differences have also been detected when using different recovery periods between the acclimation treatment and the high temperature challenge, the term 'thermotolerance diversity' was proposed to describe the multiple mechanisms that plants use in response to changes in the environment [21]. The type of heat tolerance involved (basal or acquired) and the type of shifting temperature (instant or gradual) must then be considered for the experimental design, in order to enclose most of the potential conditions that could generate a visible phenotype. From the diverse methods used for evaluating thermotolerance phenotypes, the basal thermotolerance and short-term acquired thermotolerance (usually less than 2 h) are the most common forms of phenotyping strategies [21]. Using in vitro assays, the acclimation treatment is usually performed at 37-38°C for 60–90 min [6-9,13,14,17-20]. Temperature and duration of the stress imposition, application of a pre-conditioning treatment (which can differ in temperature and/or in time length) and duration of the recovery period are only some of the conditions that could be arranged in multiple ways to establish a phenotypic assay. An often underestimated aspect is the time point for determining plant traits. Even for similar morphological assessments, some researchers obtain their results a few days after heat treatment, while others wait for several days. Therefore, it is significantly important to perform a time-course analysis of the phenotypic traits in question, which can reveal a phenotype per se, to facilitate the discovery of masked alterations between mutant and wild-type plants. The growth stage of the treated plant also interferes with the recognition of phenotypic alterations. While the analysis performed in seeds is normally restricted to the germination stage, different assessments could be achieved both in early (2.5-10 day-old) or late (10–25 day-old) seedlings.

The diversity of existing protocols compels the creation of comprehensive surveys for heat-associated phenotypes; however these can be time-consuming and unproductive. In order to create a simplified framework for the phenotypic detection of heat-associated mutants, it is necessary to first assess the heat tolerance response of Arabidopsis wild-type plants. Motivated by the same purpose, Burke et al. [16] characterized the acquired heat tolerance of Col seedlings by chlorophyll accumulation assays. However, to our knowledge there is no study that characterizes the heat response of Arabidopsis plants, based on the assessment of important and highly practical germination and seedling survival traits. The present report provides a comprehensive primary approach to identify and interpret phenotypic differences of heat-associated mutants at early developmental stages. In seedlings, basal tolerance was characterized by temperature and exposure time variations, and acquired heat tolerance by characterizing survival after different acclimation conditions. We introduced a seed-based assay that was thoroughly surveyed in terms of temperature and exposure time variations, and subsequently validated by characterizing several natural occurring ecotypes and heat stress-associated functional mutants.

Results

Evaluation of basal heat tolerance by seedling survival assays

The heat tolerance in photosynthetically active seedlings can be evaluated as a function of seedling survival after imposition of heat stress. Seedlings that remain green and actively growing are considered to be viable. Initial experiments were conducted in the wild-type ecotype Columbia (Col), the reference genotype for Arabidopsis since the sequencing of its genome [24]. Basal heat tolerance in 7-day-old Col seedlings was evaluated in terms of heat shock (HS) duration (10–45 min, at 45°C) and temperature (38-50°C, for 20 min), estimated as the percentage of surviving seedlings after a recovery period of 6 days at 23°C (Figure 1). In order to avoid the introduction of another stress factor, all the assays were performed under a dim light. Results show that, when subjected to HS, seedling viability was lost in a very short time frame (Figure 1A,B): loss was noticeable within 10 min and by min 25 plants became unviable. A similar sigmoidal-like inhibition profile was obtained using 20 min shocks at different HS temperatures (Figure 1C,D). Specifically, seedling viability was lost within the 42-46°C range.

Evaluation of acquired heat tolerance by seedling survival assays

To characterize acquired heat tolerance in Arabidopsis, 7-day-old Col seedlings were subjected to a range of acclimation temperatures (23-40°C) for 60 min, followed

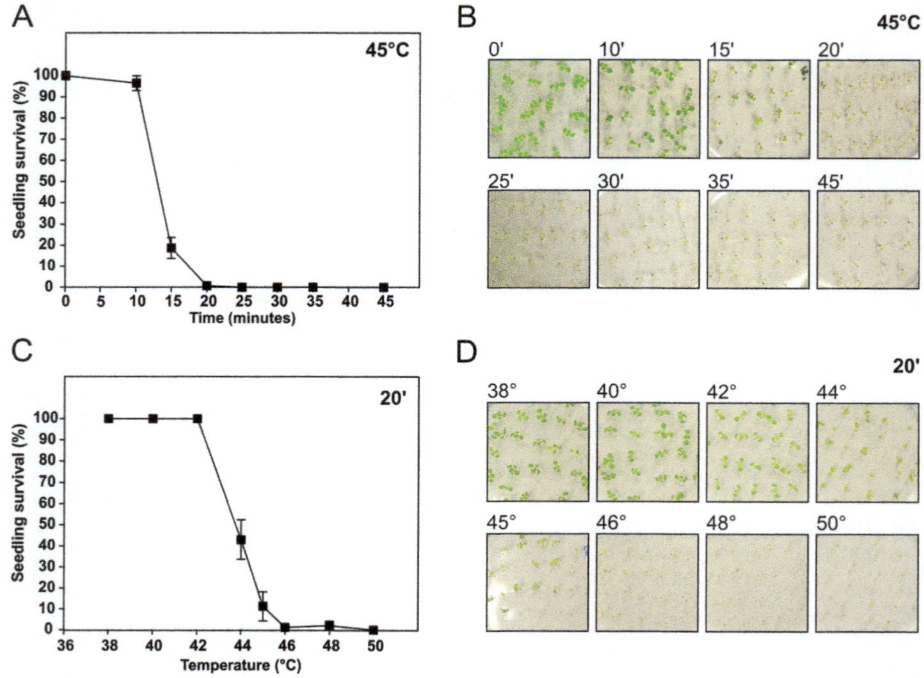

Figure 1 Assessment of basal heat tolerance in 7-day-old Col seedlings. (A-B) Seedlings were subjected to a 45°C heat shock for different time periods (0–45 min). **(C-D)** Seedlings were subjected to a 20 min heat shock in a range of different temperatures (38-50°C). Seedling survival was scored after a 6-day recovery period, and represented as a percentage of initial viable seedlings. Error bars represent SEM (n = 5 plates with ~25 seedlings each).

by recovery for 120 min at 23°C, prior to a 45°C and 20 min HS, which was previously established as an eminently lethal shock. Results indicate that acclimation efficiency positively correlated with acclimation temperatures up to 37°C, when survival was ~97% (Figure 2A,B). Subsequently, the acclimation capacity was quickly lost. We also performed a time-dependent analysis of a 37°C acclimation shock (Figure 2C,D). Results demonstrate that 30 min of acclimation resulted in viability of most seedlings, but a more extended period guaranteed complete survival of seedling to a subsequent heat shock. They also evidenced the sub-lethal nature of the 37°C treatment, since heat shock treatments for up to 75 min did not result in viability loss.

Evaluation of basal heat tolerance through seed germination

We just described how basal and acquired heat stress responses could be analysed by seedling survival, modulating the intensity of the stimulus using ranges of temperatures and exposure times (Figures 1 and 2). However, this experimental design can be laborious when facing for instance different temperatures and an adequate number of replicas. To simplify the search for heat resistance and sensitivity phenotypes, we resorted to seed germination-based assays. As seeds normally exhibit higher heat tolerance than seedlings [19], the temperature and duration of treatments were altered accordingly.

A thorough map of the Arabidopsis seed germination response to heat shock imposition was once more performed in the wild-type ecotype Col. Seed heat tolerance assays were performed either with increasing periods (0–300 min) of heat shock treatment at 50°C (Figure 3A-D), or with increasing temperatures (40-57°C) of a 60 min treatment (Figure 3E-H). Results were highly consistent between radicle emergence and green cotyledon indicators, with an approximate 2-day delay between parameters. As expected, germination was compromised in both a time-dependant and temperature-dependent manner. By profiling germination on a daily-basis for up to 12 days, we observed that seeds tended to recover from HS with time rather than quickly display their full germination capacity (Figure 3A,B,E,F). We chose to calculate radicle emergence after 12 days, as a function of exposure time and temperature variables (Figure 3C,G). Results show that HS generated a sigmoidal-like inhibition curve for both time and temperature variables. HS duration immediately impacted on the germination capacity, generating a gentle time-dependent inhibition slope (Figure 3C). Meanwhile HS temperature generated a more dramatic inhibition, at ~50°C, with the germination rate remaining impervious to HS up until

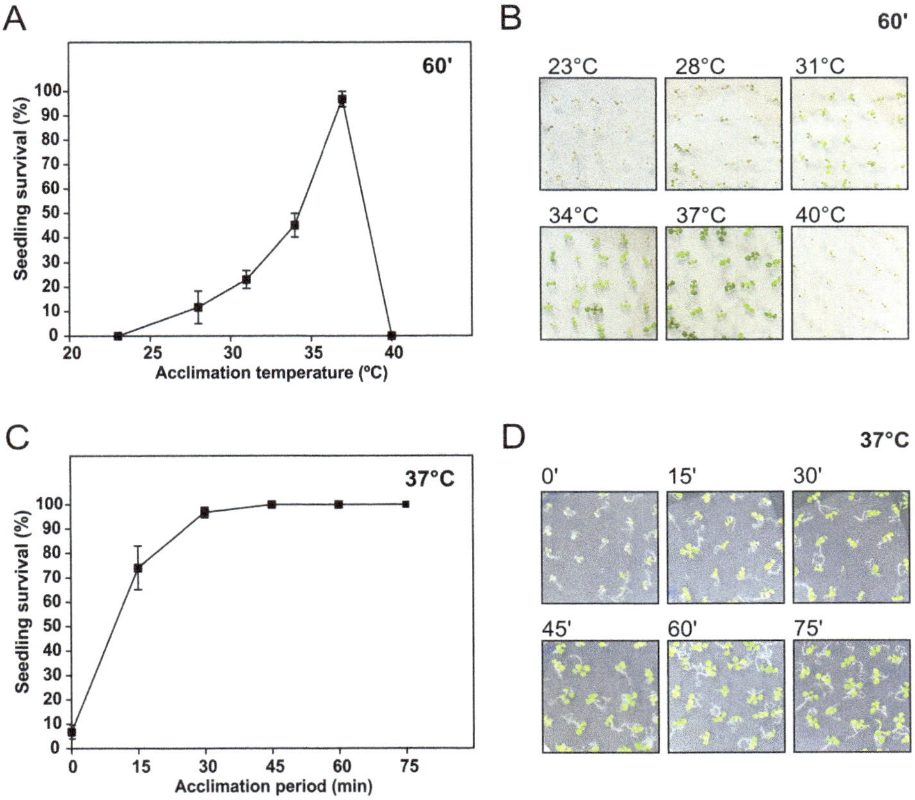

Figure 2 Assessment of acquired heat tolerance in 7-day-old Col seedlings. (A-B) Seedlings were acclimated for 60 min in a range of different temperatures (23-40°C), recovered at 23°C for 120 min and heat-treated at 45°C for 20 min. **(C-D)** Seedlings were acclimated at 37°C for different time periods (0–75 min), recovered at 23°C for 120 min and heat-treated at 45°C for 20 min. Seedling survival was scored after a 6-day recovery period, and represented as a percentage of initial viable seedlings. Error bars represent SEM (n = 5 plates with ~25 seedlings each).

48°C, yet diminishing almost 80% within a 2°C interval (between 50-52°C).

Impact of natural variability and functional mutations on seed heat stress resistance

To validate the current approach, we tested the tolerance of seeds to HS (50°C and 60 min) in various different genetic backgrounds. First, five different ecotypes were analysed to determine the assay's suitability to natural variability studies (Figure 4). As determined by radicle emergence and formation of green cotyledons after 12 days, ecotypes WS and C24 were more resistant to HS, with ~90% germination, immediately followed by Col (~85%). Ecotypes Cvi and particularly Ler were significantly more sensitive to heat stress. Meanwhile, analysis on a daily basis allowed for the establishment of informative profiles (Figure 4A,B). For instance, C24 displayed early germination in comparison to WS, suggesting it is the most heat resistant ecotype. However, statistical analysis of germination after 12 days placed Col, WS and C24 as identically tolerant ecotypes (Figure 4C). Conversely, the initial profile of Cvi and Ler suggested equal responses to HS, which were only resolved later in the assay and therefore deemed statistically different. As before, green cotyledon estimation was as informative as radicle emergence (Figure 4B). Interestingly, germination profiles using the green cotyledon indicator seemed to stabilize at the end of the experiment (days 10-12), indicating that a 12-day analysis is suitable to resolve germination phenotypes in what concerns the heat stress response.

In parallel, we implemented the same strategy on functional mutants, consisting mostly of null alleles of known heat stress-related Arabidopsis genes. As a negative control, the mutant *tso2*, defective on the ribonucleotide reductase small subunit gene, was assayed in parallel. Mutant seeds subjected to a 50°C and 60 min HS were differentially affected by heat stress (Figure 5), and we were able to categorize mutants according to their sensitivity. Mutants were termed as highly sensitive (*hot1-3*), virtually losing seed germination capacity; moderately sensitive (*rbohD*; *NahG*; *siz1-2*), displaying 30-70% germination capacity; or insensitive to heat (*hsa3-2*; *hsfa2*), when germination rates within 12 days were statistically similar to the wild-type. As expected, the negative control mutant

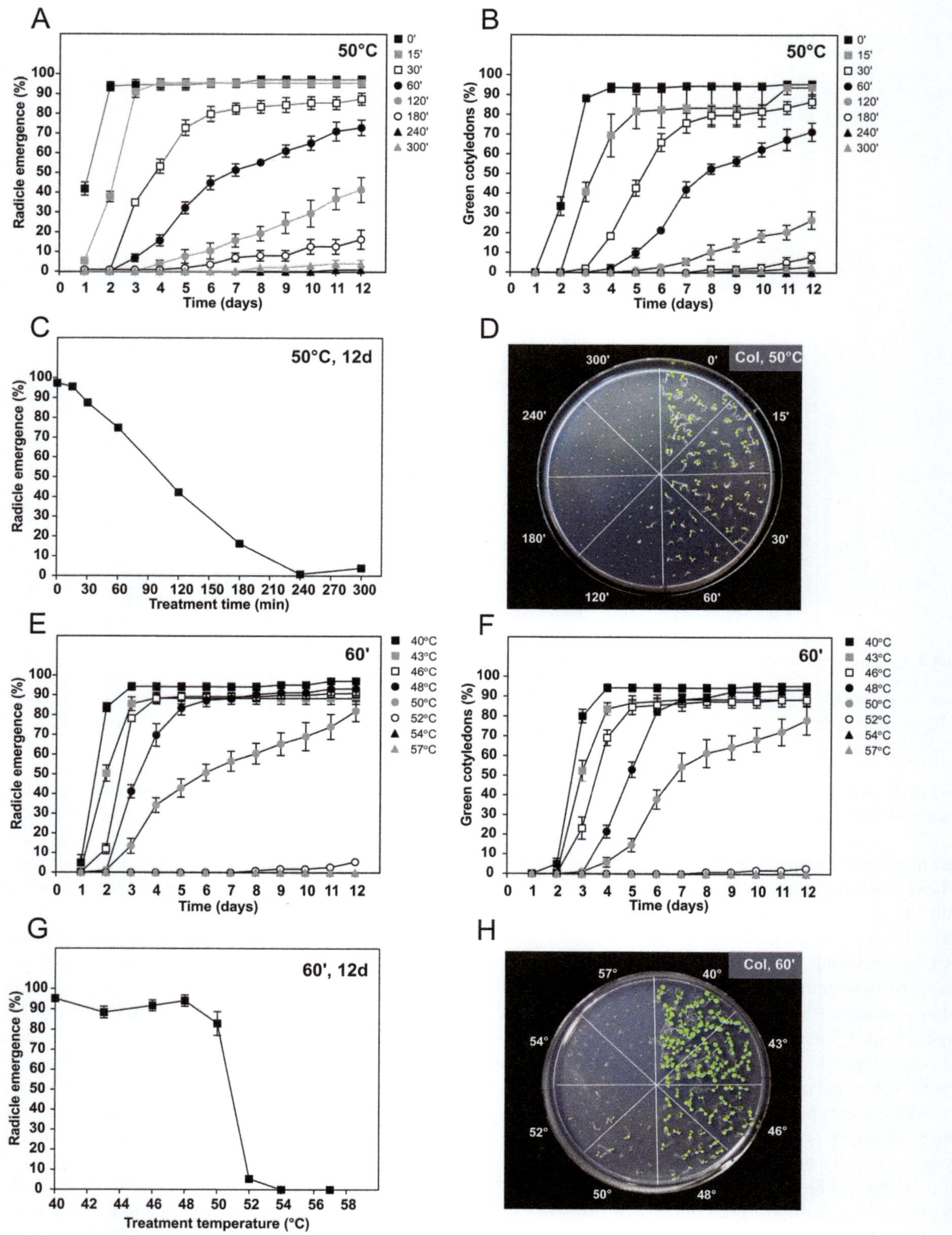

Figure 3 (See legend on next page.)

(See figure on previous page.)
Figure 3 Assessment of basal heat tolerance in Col seeds. (A-D) Seeds were subjected to 50°C heat shock for different time periods (0–300 min). **(A)** Radicle emergence and **(B)** green cotyledon emergence during the 12 day post-heat shock period. **(C)** Germination levels (radicle emergence parameter) after 12 days, plotted as a function of treatment period. **(D)** Morphology of 12-day-old seedlings. **(E-H)** Seeds were subjected to a 60 min heat shock in a range of different temperatures (40-57°C). **(E)** Radicle emergence and **(F)** green cotyledon emergence during the 12 day post-heat shock period. **(G)** Germination levels (radicle emergence parameter) after 12 days, plotted as a function of heat-shock temperature. **(H)** Morphology of 12-day-old seedlings. Percentages were determined in relation to the total number of sown seeds. Error bars represent SEM (n = 5 plates with ~20 seedlings each).

tso2 displayed the same heat resistance level as the wild-type. As with previous results, germination rates stabilized within 12 days, validating this approach. However, daily scoring of the germination profile once again proved to be informative. For instance, the insensitive mutant *tso2* displayed a significant delay in germination though final germination rates were similar to the wild-type; the insensitive mutant *hsfa2* displayed a consistently slight defect in germination that did not overcome statistical analysis at day 12 (Figure 5A,B) but was statistically different when considering the time-course germination profile (paired Student *t*-test; data not shown).

Discussion

Uniformity in plant trait measuring is an important aspect of current plant research [5], being equally valid for the analysis of developmental and stress-based traits. In the current report, several phenotypic assays were carried out to map changes in the heat tolerance capacity of Arabidopsis. The Col ecotype, which has become the ecotype standard within the Arabidopsis community and now congregates most functional studies [2], was characterized in what concerns seedling basal and acquired heat tolerance, as well as seed basal heat tolerance. These responses were analysed systematically, exploiting the two defining variables in a heat shock: temperature and exposure time. Later, seed-based analysis was validated using both Arabidopsis natural occurring ecotypes and functional mutants.

Seedling-based assays for basal and acquired heat tolerance

A basic aspect to heat-related phenotype assays is the range of temperatures and treatment length. Studies in *Arabidopsis thaliana* normally specify 45°C as the standard temperature to detect alterations in plant fitness

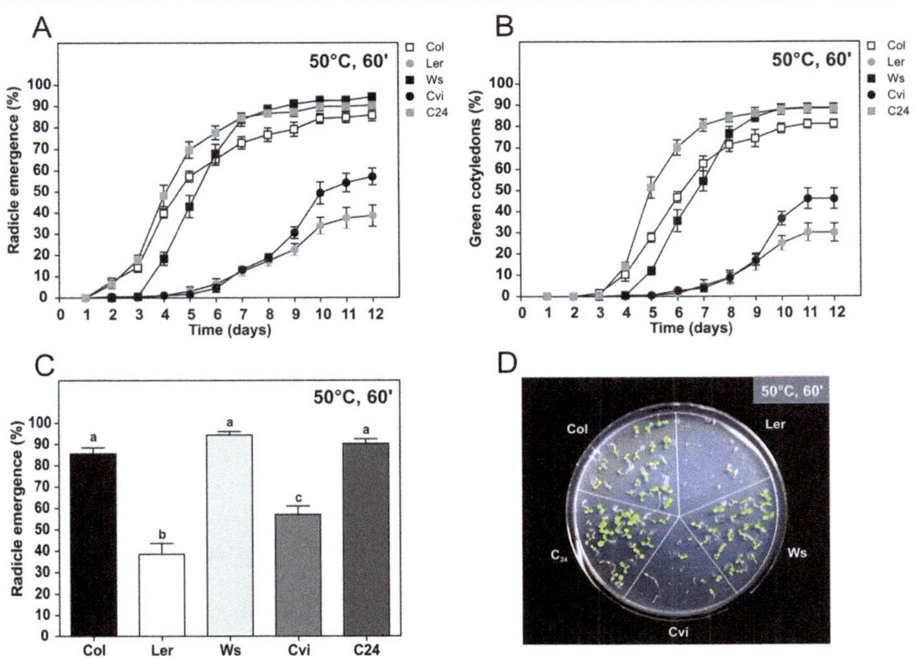

Figure 4 Assessment of seed basal heat tolerance in different *Arabidopsis thaliana* ecotypes. (A-D) Seeds were subjected to a 50°C heat shock for 60 min. **(A)** Radicle emergence and **(B)** green cotyledon emergence during the 12 days post-heat shock period. **(C)** Germination levels (radicle emergence parameter) after 12 days. **(D)** Morphology of 12-day-old seedlings. Percentages were determined in relation to the total number of sown seeds. Error bars represent SEM (n = 7 plates with ~25 seedlings each). In C, letters represent statistically different genotypes (One-way ANOVA with Tukey's multiple comparison test for all pairs of columns; P < 0.05).

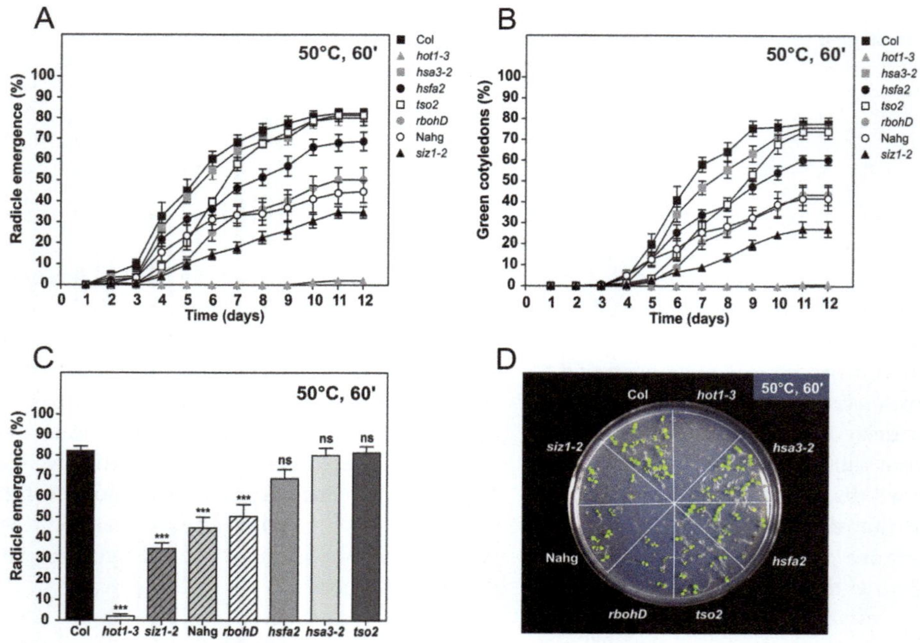

Figure 5 Assessment of seed basal heat tolerance in different *Arabidopsis thaliana* functional mutant genotypes. (A-D) Seeds were subjected to a 50°C heat shock for 60 min. **(A)** Radicle emergence and **(B)** green cotyledon emergence during the 12 day post-heat shock period. **(C)** Germination levels (radicle emergence parameter) after 12 days. **(D)** Morphology of 12-day-old seedlings. Percentages were determined in relation to the total number of sown seeds. Error bars represent SEM (n = 7 plates with ~25 seedlings each). Asterisks represent mutant genotypes with statistically significant differences in relation to the wild-type Col (Student *t*-test; ns, non-significant; ***, P < 0.001).

[6-9,11-13,17-19,25]; however, the duration of the heat treatment is highly variable in these reports, ranging from 15 to 220 min. Moreover, the reason why this temperature is the most commonly applied is not completely understood, since no extensive assessment of heat tolerance in Arabidopsis wild-type plants has been performed to the best of our knowledge. In the present report we showed that the 20 min heat shock period was the minimum necessary input to obtain nearly full lethality in Col. Lethality is likely the result of non-reversible denaturation of proteins or cellular structures [26,27]. A 45°C and 20 min treatment could thus be successfully used as a single shock treatment to resolve resistance in basal heat tolerance that didn't overextend on the stress stimulus. The use of time- and/or temperature-dependant inhibition profiles, as performed in the present study, is even more suited to resolve basal heat resistance or hypersensitivity responses. Even though less exposure time may help resolve hypersensitivity to heat, the sensitivity of the method to slight variations in temperature or time, as depicted by the rapid inhibition curves, could compromise the establishment of a real heat phenotype.

A conditioning pre-treatment with moderate temperatures prior to a lethal temperature treatment has been shown to partially improve survival in several plant species [14,23]. This phenomenon, known as acquired heat tolerance, enables plants to withstand excessively high temperatures that would be damaging or lethal without pre-acclimation [26]. In the current work, variable preconditioning temperatures and exposure times were tested in Arabidopsis ecotype Col, using a subsequent recovery period of 120 min at 23°C and the predetermined experimental heat shock conditions (45°C for 20 min). By varying acclimation temperatures, we observed a very clear peak in seedling survival associated with preconditioning at 37°C. This would suggest that a significant heat stress input, in this case 15°C above the optimum growth temperature of Arabidopsis, is required to fully trigger signalling events leading to pre-acclimation responses. Also, tested pre-acclimation temperatures higher than 37°C (40°C) generated sufficient damage to compromise viability, confirming previous reports for the use of 37-38°C as the standard pre-acclimation temperature [7,14]. In support of the sub-lethal nature of a 37°C treatment, exposure of seedlings to this temperature over extended periods (higher than 45 min) did not induce any decrease in seedling viability.

As previously detailed, we demonstrated the critical necessity for a controlled experimental set up. Technically we suggest testing to be carried out by immersion of sealed plates in a water bath, as opposed to the use of heat chambers. We observed that heat transfer is significantly more effective over water in comparison to air (not shown). This procedure also avoids the loss in temperature that results from opening heat chambers, leading to underestimation of the stress input. In support, from three described heating devices (heating blocks, water baths, and growth chambers/ovens), Yeh and co-workers [21] also conveyed the water bath as the most efficient method to deliver heat stress temperatures. We additionally suggest that piling up plates be avoided, as it prevents adequate heat transfer, generating significant variability within replicates (not shown).

Seed-based assays for basal heat tolerance

Seed-based assays provide an alternative approach to the analysis of heat stress resistance or hypersensitivity. They are less laborious and allow the analysis of a higher number of plants or genotypes. They also overcome most of the previously referred problems in delivering heat shock in a precisely controlled manner, by allowing treatments to be performed in microtubes. We highlight the importance of performing experiments using synchronized seed stocks to avoid variability in seed germination and fitness.

In seeds, heat challenges can affect both the overall germination capacity and germination progression. In the present report, when germinating seeds were daily scored, slower germination profiles were observed as a function of stronger treatments, suggesting a correlation between heat stress intensity and seed dormancy. As the germination process is defined as the set of events occurring after imbibition until a visible protrusion of the radicle occurs [28], a seed was considered germinated when emergence of the radicle was observable. Often however, formation of green cotyledon leaves is considered as the germination parameter [29]. Present results were consistent between both parameters with an observable 2-day delay for green cotyledon formation, suggesting that heat stress affects seed dormancy/germination but not photomorphogenesis of green cotyledons.

Even though plant trait phenotyping requires extremely controlled conditions [5], an often underestimated aspect is the time point at which plant traits are scored. Daily estimation of germination rates can facilitate phenotype discovery, but for a less laborious approach, we suggest data acquisition after 12 days, as the germination capacity became stabilized following that period for a significant number of genotypes and heat stress intensities. Consistent with previous reports [19], seed lethality required a significant increase in stress intensity compared to seedlings. For recognition of basal heat stress phenotypes, we propose evaluation of seed germination by varying shock time (50°C; 15–180 min) rather than shock temperature (47-53°C; 60 min), as it provides a gentler inhibition slope that facilitates phenotype discovery and mathematical establishment of EC_{50}-like parameters. When facing experimental limitations, a 50°C and 60 min shock can be adopted as standard to amply resolve sensitivity to heat while maintaining sufficient resolution to identify heat stress resistance phenotypes.

Addressing natural variability and functional heat stress-related mutant phenotypes

The ability to detect differential phenotypes using seed-based assays was validated using Arabidopsis ecotypes from various geographical locations. Differences can be associated to the physiological variation of individual ecotypes [30] with ecotypes Landsberg erecta (Ler) and Cape Verde Islands (Cvi) clearly manifesting higher sensitivity to heat stress. The observed variability can be attributed to the complex admixture patterns that have been associated to *Arabidopsis thaliana*, as a consequence of the recolonization of Europe following the last glaciation [31]. The enclosed methodologies were also designed to help screen for phenotype differences in mutants that have altered basal or acquired heat tolerance responses. In particular, seed-based phenotypic assays were sufficiently sensitive to characterize various degrees of basal heat tolerance, as mutants were easily resolved between insensitive, moderately sensitive or highly sensitive to heat stress imposition. Specifically, *hot1-3* was significantly affected by high temperature, confirming previous evidence that implicated *HOT1*, a major heat shock protein, in both basal and acquired heat tolerance [20]. Conversely, loss-of-function mutants for *HSFA2* (heat shock transcription factor) and *HSA32* (heat stress associated protein), both functionally associated to acquired but not to basal resistance [6,7], presented basal heat tolerance similar to the wild-type and the negative control mutant *tso2* (ribonucleotide reductase small subunit gene). Moderate sensitivity to heat stress was observed for *NahG*, a mutant ectopically expressing a bacterial salicylate hydrolase that scavenges salicylic acid [32]; the null allele of *SIZ1*, a major SUMO E3 ligase that is involved in several abiotic stress responses [33]; and *RBOHD*, the main NADPH oxidase in Arabidopsis, involved in ROS generation and signalling [34]. Results support previous plant level evidence for *NahG* and *rbohD* sensitivity to basal and acquired heat tolerance and *siz1-2* sensitivity to basal but not acquired heat tolerance ([9], reviewed by [33]). Interestingly, the present analysis was able to resolve a potential basal seed heat tolerance phenotype in *rbohD* which was undetected in a previous analysis by Larkindale et al. [9].

Conclusions

The proper characterization of basal and acquired heat tolerance is crucial to address the molecular aspects underlying heat stress resistance. For instance, several QTLs related to germination have been reported in different crops (e.g. [35]) and in model plants such as Arabidopsis [36] and *Medicago truncatula* [37]. From these studies several candidate genes have been identified that could be associated with germination in high temperatures (e.g. [37] and references therein). These traditionally conform to the seven canonical classes of proteins associated with heat stress responses in eukaryotes [38]. The use of model plant species, such as *Arabidopsis thaliana*, will speed up the identification of more key determinant genes, either underlying QTLs related to germination, or involved in heat resistance mechanisms at plant level. For such studies the use of a simple set of protocols to detect seed and plantlet thermotolerance will be greatly beneficial. Also, the outlined methodologies proved to be effective in facilitating the detection of variability amongst geographically distinct natural accessions. In this context, they can be extended to studies addressing allelic variability in Arabidopsis, which are being potentiated by recent NGS-based efforts such as the *Arabidopsis thaliana* 1001 Genomes Project [31].

The present assays provide a framework of experimental conditions that are also helpful as an inexpensive and undemanding approach for functional studies. They are particularly appropriate for reverse genetic strategies that require sensitive phenotyping in the search for elusive phenotypic traits in the researcher's mutants-of-interest. A comprehensive discussion of the diverse heat stress phenotyping methods and their importance for plant molecular genetic studies has already performed elsewhere [21]. Although the complete understanding of heat stress responses should consider different heat stress regimens and evaluation of different output traits [21], by focusing on the establishment of a set of heat response protocols we hope to provide a simple and standard way of detecting thermotolerance phenotypes.

The utility of plate-based analysis has been highlighted in the mapping of developmental phenotypes [5]. By using plate-based assays, one can detect early phenotypes that are often overlooked in soil-based phenotypic analyses, while simultaneously minimizing natural variability. Furthermore, time consumption and space requirements are maximized, making data collection amenable to high throughput experimental set-ups such as forward genetic screens. The basis for the isolation of novel loss- and gain-of-function mutations that are determinant for heat stress resistance is thus facilitated.

Methods

Plant material

Wild-type seeds of *Arabidopsis thaliana* ecotypes Columbia (Col), Wassilewskija (Ws), Landsberg erecta (Ler), Cape Verde Islands (Cvi) and C24 were ordered from the Nottingham Arabidopsis Stock Centre (NASC) (http://arabidopsis.info/; [39]). Homozygous T-DNA insertion mutant seeds of *hot1-3* [20] and *atrbohD* [40] were kindly provided by E. Vierling (Department of Biochemistry & Molecular Biophysics, University of Arizona, USA) and M. A. Torres (Department of Biology, University of North Carolina, USA), respectively. An RNAi line of *Hsa32* [7] and the *tso2* mutant [16] were kindly provided by Y. Charng (Agricultural Biotechnology Research Center, Academia Sinica, Taiwan) and J. Burke (USDA-ARS Cropping Systems Research Laboratory, USA), respectively. T-DNA insertion line of *HsfA2* (SALK_008978) was obtained from the Arabidopsis Biological Resource Center (ABRC; Ohio State University). The T-DNA insertion mutant *siz1-2* (SALK_065397) was ordered from the NASC European Arabidopsis Stock Centre (arabidopsis.info). The transgenic line *NahG*, that expresses a bacterial SA hydroxylase, was kindly provided by Miguel Botella (University of Malaga, Spain). All genotypes were grown in synchronized fashion and bulk seed stocks were maintained in the dark, at room temperature, in firmly sealed individual containers to prevent rehydration.

Seed sterilization and in vitro germination

Arabidopsis seeds were stratified by immersion in water and incubation at 4°C for 3 days. Surface-sterilization was initiated by soaking seeds in 1 ml of 70% (v/v) ethanol for 5 min. Ethanol was replaced by 1 ml of bleach solution [20% (v/v) commercial bleach with 3.5% (w/v) effective chloride; 0.1% (v/v) Tween-20] and seeds were further incubated for 10 min. After being rinsed five times with sterile water, seeds were resuspended in a sterile 0.08% (w/v) agarose solution. When required, seeds were subjected to heat shock. Sterilized seeds were dispersed onto plates containing MS-based medium [41] composed of 1x basal salt mixture (Duchefa), 1.5% (w/v) sucrose, 0.5 g l^{-1} MES and 0.8% (w/v) agar at pH 5.7. Plates were sealed with parafilm to prevent dehydration and placed horizontally in a growth room, under the following conditions: photoperiod of 16 h light/8 h darkness; light intensity of 100 μmol photon m^{-2} s^{-1} and 23°C.

Seedling survival assays

Seeds were plated onto MS medium and grown horizontally under the same conditions as previously referred. After 7 days of growth, heat stress was imposed by submersion of parafilm-sealed plates into a temperature-controlled water bath. Heat treatment was performed 3 h into the light/day photoperiod, under a dim light. Four kinds of assays were performed: a) treatment at different extents of time (5–45 min) under a constant temperature of 45°C; b) incubation at different temperatures (38-50°C) for 20 min; c) pre-incubation at different temperatures (28-40°C) for 60 min, recovery at 23°C for 120 min, and heat-treatment at 45°C for 20 min; d) pre-incubation at different extents of time (15–75 min) under a constant temperature of 37°C, recovery at 23°C for 120 min, and heat-treatment at 45°C for 20 min. After heat treatments, seedlings were returned to the growth room and allowed to recover for 6 days. The number of viable seedlings was then quantified to determine the survival rate. Seedlings that were still green and generated new leaves were scored as survivors.

Germination assays

Seed heat tolerance assays were performed immediately after stratification and sterilization. Seeds were heat-stressed by submersion of the corresponding microtubes into a temperature-controlled water bath. Three sets of assays were performed: a) treatment of Col at different extents of time (0–300 min) under a constant temperature of 50°C; b) incubation of Col at different temperatures (40-57°C) for 60 min; c) treatment of different ecotypes and functional mutants at 50°C for 60 min. Subsequently, seeds were plated onto MS medium and monitored daily for germination (radicle emergence and formation of green cotyledons) over a 12-day period. The germination frequency was determined in relation to total sown seeds per each condition.

Abbreviations

HS: Heat shock; ROS: Reactive oxygen species; NGS: Next-generation sequencing.

Competing interests

The authors declare that they have no competing interests.

Authors' contributions

JSC and SF carried out all the experiments and drafted the manuscript, RMT participated in the study design and contributed to data analysis, TLN conceived the study, contributed to its design and revised critically the manuscript, HA coordinated the study, contributed to the design, analysis and interpretation of data and revised critically the manuscript. All authors read and approved the final manuscript.

Acknowledgments

Joana Silva-Correia was supported by FCT, grant reference SFRH/BD/16663/2004. Herlânder Azevedo was supported by FCT, grant reference SFRH/BPD/17198/2004, and the project "Genomics and Evolutionary Biology" cofinanced by North Portugal Regional Operational Programme 2007/2013 (ON.2 – O Novo Norte), under the National Strategic Reference Framework (NSRF), through the European Regional Development Fund (ERDF).

Author details

[1]Center for Biodiversity, Functional & Integrative Genomics (BioFIG), Plant Functional Biology Center, University of Minho, Campus de Gualtar, 4710-057 Braga, Portugal. [2]Present address: 3B's Research Group – Biomaterials, Biodegradables and Biomimetics, University of Minho Headquarters of the European Institute of Excellence on Tissue Engineering and Regenerative Medicine, AvePark, 4806-909 Guimarães, Taipas, Portugal. [3]Present address: CIBIO, Centro de Investigação em Biodiversidade e Recursos Genéticos, Universidade do Porto, Campus Agrário de Vairão, 4485-661 Vairão, Portugal.

References

1. Azevedo H, Silva-Correia J, Oliveira J, Laranjeira S, Barbeta C, Amorim-Silva V, Botella MA, Lino-Neto T, Tavares RM: **A strategy for the identification of new abiotic stress determinants in Arabidopsis using web-based data mining and reverse genetics.** *OMICS* 2011, **15:**935–947.
2. Feng CP, Mundy J: **Gene discovery and functional analyses in the model plant *Arabidopsis*.** *J Integr Plant Biol* 2006, **48:**5–14.
3. Kuromori T, Takahashi S, Kondou Y, Shinozaki K, Matsui M: **Phenome analysis in plant species using loss-of-function and gain-of-function mutants.** *Plant Cell Physiol* 2009, **50:**1215–1231.
4. Bouché N, Bouchez D: **Arabidopsis gene knockout: phenotypes wanted.** *Curr Opin Plant Biol* 2001, **4:**111–117.
5. Boyes DC, Zayed AM, Ascenzi R, McCaskill AJ, Hoffman NE, Davis KR, Görlach J: **Growth stage-based phenotypic analysis of Arabidopsis: a model for high throughput functional genomics in plants.** *Plant Cell* 2001, **13:**1499–1510.
6. Charng Y, Liu H, Liu N, Chi W, Wang C, Chang S, Wang T: **A heat-inducible transcription factor, HsfA2, is required for extension of acquired thermotolerance in *Arabidopsis*.** *Plant Physiol* 2007, **143:**251–262.
7. Charng Y, Liu H, Liu N, Hsu F, Ko S: **Arabidopsis Hsa32, a novel heat shock protein, is essential for acquired thermotolerance during long recovery after acclimation.** *Plant Physiol* 2006, **140:**1297–1305.
8. Suzuki N, Bajad S, Shuman J, Shulaev V, Mittler R: **The transcriptional co-activator MBF1c is a key regulator of thermotolerance in *Arabidopsis thaliana*.** *J Biol Chem* 2008, **283:**9269–9275.
9. Larkindale J, Hall JD, Knight MR, Vierling E: **Heat stress phenotypes of Arabidopsis mutants implicate multiple signaling pathways in the acquisition of thermotolerance.** *Plant Physiol* 2005, **138:**882–897.
10. Wu SJ, Locy RD, Shaw JJ, Cherry JH, Singh NK: **Mutation in Arabidopsis *HIT1* locus causing heat and osmotic hypersensitivity.** *J Plant Physiol* 2000, **157:**543–547.
11. Liu HT, Gao F, Li G-L, Han J-L, Liu D-L, Sun D-Y, Zhou R-G: **The calmodulin-binding protein kinase 3 is part of heat-shock signal transduction in *Arabidopsis thaliana*.** *Plant J* 2008, **55:**760–773.
12. Liu H-T, Li G-L, Chang H, Sun D-Y, Zhou R-G, Li B: **Calmodulin-binding protein phosphatase PP7 is involved in thermotolerance in Arabidopsis.** *Plant Cell Environ* 2007, **30:**156–164.
13. Chen J, Burke JJ, Velten J, Xin Z: **FtsH11 protease plays a critical role in Arabidopsis thermotolerance.** *Plant J* 2006, **48:**73–84.
14. H-c L, Y-y C: **Acquired thermotolerance independent of heat shock factor A1 (HsfA1), the master regulator of the heat stress response.** *Plant Signal Behav* 2012, **7:**547–550.
15. Ludwig-Muller J, Krishna P, Forreiter C: **A glucosinolate mutant of Arabidopsis is thermosensitive and defective in cytosolic Hsp90 expression after heat stress.** *Plant Physiol* 2000, **123:**949–958.
16. Burke JJ, O'Mahony PJ, Oliver MJ: **Isolation of Arabidopsis mutants lacking components of acquired thermotolerance.** *Plant Physiol* 2000, **123:**575–587.
17. Su PH, Li HM: **Arabidopsis stromal Hsp70s are essential for plant development and important for thermotolerance of germinating seeds.** *Plant Physiol* 2008, **146:**1231–1241.
18. Hong SW, Lee U, Vierling E: **Arabidopsis *hot* mutants define multiple functions required for acclimation to high temperatures.** *Plant Physiol* 2003, **132:**757–767.
19. Hong SW, Vierling E: **Mutants of *Arabidopsis thaliana* defective in the acquisition of tolerance to high temperature stress.** *Proc Natl Acad Sci U S A* 2000, **97:**4392–4397.
20. Hong SW, Vierling E: **Hsp101 is necessary for heat tolerance but dispensable for development and germination in the absence of stress.** *Plant J* 2001, **27:**25–35.
21. Yeh C-H, Kaplinsky NJ, Hu C, Charng Y-y: **Some like it hot, some like it warm: phenotyping to explore thermotolerance diversity.** *Plant Sci* 2012, **195:**10–23.
22. Lin C-Y, Roberts JK, Key JL: **Acquisition of thermotolerance in soybean seedlings. Synthesis and accumulation of heat shock proteins and their cellular localization.** *Plant Physiol* 1984, **74:**152–160.
23. Larkindale J, Mishkind M, Vierling E: **Plant responses to high temperature.** In *Plant Abiotic Stress.* Edited by Jenks MA, Hasegawa PM. Oxford: Blackwell Publishing; 2005:100–144.
24. The-Arabidopsis-Genome-Initiative: **Analysis of the genome sequence of the flowering plant *Arabidopsis thaliana*.** *Nature* 2000, **408:**796–815.
25. Larkindale J, Vierling E: **Core genome responses involved in acclimation to high temperature.** *Plant Physiol* 2008, **146:**748–761.
26. Vierling E: **The roles of heat shock proteins in plants.** *Annu Rev Plant Physiol Plant Mol Biol* 1991, **42:**579–620.
27. Kampinga HH, Brunsting JF, Stege GJJ, Burgman PWJJ, Konings AWT: **Thermal protein denaturation and protein aggregation in cells made thermotolerant by various chemicals: role of heat shock proteins.** *Exp Cell Res* 1995, **219:**536–546.
28. Nonogaki H, Chen F, Bradford KJ: **Mechanisms and genes involved in germination sensu stricto.** In *Seed Development, Dormancy and*

Germination. Edited by Bradford KJ, Nonogaki H. Oxford: Blackwell Publishing; 2007:264–304.

29. Fujii H, Verslues PE, Zhu J-K: **Identification of two protein kinases required for abscisic acid regulation of seed germination, root growth, and gene expression in Arabidopsis.** *Plant Cell* 2007, **19:**485–494.
30. Swindell WR, Huebner M, Weber AP: **Plastic and adaptive gene expression patterns associated with temperature stress in *Arabidopsis thaliana*.** *Heredity* 2007, **99:**143–150.
31. Weigel D, Mott R: **The 1001 genomes project for *Arabidopsis thaliana*.** *Genome Biol* 2009, **10:**107.
32. Friedrich L, Vernooij B, Gaffney T, Morse A, Ryals J: **Characterization of tobacco plants expressing a bacterial salicylate hydroxylase gene.** *Plant Mol Biol* 1995, **29:**959–968.
33. Castro PH, Tavares RM, Bejarano ER, Azevedo H: **SUMO, a heavyweight player in plant abiotic stress responses.** *Cell Mol Life Sci* 2012, **69:**3269–3283.
34. Marino D, Dunand C, Puppo A, Pauly N: **A burst of plant NADPH oxidases.** *Trends Plant Sci* 2012, **17:**9–15.
35. Argyris J, Truco MJ, Ochoa O, McHale L, Dahal P, Van Deynze A, Michelmore RW, Bradford KJ: **A gene encoding an abscisic acid biosynthetic enzyme (LsNCED4) collocates with the high temperature germination locus Htg6.1 in lettuce (*Lactuca* sp.).** *Theor Appl Genet* 2011, **122:**95–108.
36. Clerkx EJM, El-lithy ME, Vierling E, Ruys GJ, Vries HB-d, Groot SPC, Vreugdenhil D, Koornneef M: **Analysis of natural allelic variation of Arabidopsis seed germination and seed longevity traits between the accessions Landsberg erecta and Shakdara, using a new recombinant inbred line population.** *Plant Physiol* 2004, **135:**432–443.
37. Menna P, Dias B, Du C: **QTL analysis of seed germination and pre-emergence growth at extreme temperatures in *Medicago truncatula*.** *Theor Appl Genet* 2011, **122:**429–444.
38. Richter K, Haslbeck M, Buchner J: **The heat shock response: life on the verge of death.** *Mol Cell* 2010, **40:**253–266.
39. Scholl RL, May ST, Ware DH: **Seed and molecular resources for Arabidopsis.** *Plant Physiol* 2000, **124:**1477–1480.
40. Torres MA, Dangl JL, Jones JDG: ***Arabidopsis* gp91^{phox} homologues *AtrbohD* and *AtrbohF* are required for accumulation of reactive oxygen intermediates in the plant defense response.** *Proc Natl Acad Sci U S A* 2002, **99:**523–528.
41. Murashige T, Skoog F: **A revised medium for rapid growth and bioassay with tobacco issue cultures.** *Physiol Plantarum* 1962, **15:**473–497.

Ion Torrent sequencing as a tool for mutation discovery in the flax (*Linum usitatissimum* L.) genome

Leonardo Galindo-González[1*†], David Pinzón-Latorre[1†], Erik A Bergen[1], Dustin C Jensen[2] and Michael K Deyholos[3]

Abstract

Background: Detection of induced mutations is valuable for inferring gene function and for developing novel germplasm for crop improvement. Many reverse genetics approaches have been developed to identify mutations in genes of interest within a mutagenized population, including some approaches that rely on next-generation sequencing (e.g. exome capture, whole genome resequencing). As an alternative to these genome or exome-scale methods, we sought to develop a scalable and efficient method for detection of induced mutations that could be applied to a small number of target genes, using Ion Torrent technology. We developed this method in flax (*Linum usitatissimum*), to demonstrate its utility in a crop species.

Results: We used an amplicon-based approach in which DNA samples from an ethyl methanesulfonate (EMS)-mutagenized population were pooled and used as template in PCR reactions to amplify a region of each gene of interest. Barcodes were incorporated during PCR, and the pooled amplicons were sequenced using an Ion Torrent PGM. A pilot experiment with known SNPs showed that they could be detected at a frequency > 0.3% within the pools. We then selected eight genes for which we wanted to discover novel mutations, and applied our approach to screen 768 individuals from the EMS population, using either the Ion 314 or Ion 316 chips. Out of 29 potential mutations identified after processing the NGS reads, 16 mutations were confirmed using Sanger sequencing.

Conclusions: The methodology presented here demonstrates the utility of Ion Torrent technology in detecting mutation variants in specific genome regions for large populations of a species such as flax. The methodology could be scaled-up to test >100 genes using the higher capacity chips now available from Ion Torrent.

Keywords: *Linum usitatissimum*, Reverse genetics, Ion Torrent, EMS

Background

Flax (*Linum usitatissimum* L.) is cultivated as a source of either oil or fiber, both of which have distinct properties that make flax a valuable crop. The oil of flax seeds (i.e. linseed) is rich in polyunsaturated fatty acids including alpha-linolenic acid, which has purported health benefits and is also useful as a drying oil in manufacture of resins, finishes, and flooring. The stem phloem fibers (i.e. bast fibers) of flax are remarkably long and strong and are used for textiles and increasingly as substitutes for fiberglass in composite materials. The commercial potential of flax, as well as interesting aspects of its biology (including well-documented phenotypic and genomic plasticity of some accessions [1]), have led to an increase in research activity in this species, highlighted by the release of an assembly of its whole genome sequence [2]. To accelerate the development of novel germplasm and to better exploit the available DNA sequence resources for flax, we sought to develop a mutant population and a reverse genetics platform for this crop.

Mutations can be induced by treating individuals with physical, biological or chemical mutagens [3]. Ethyl methanesulfonate (EMS) is widely used for inducing point mutations in plants [4-12], and results mostly in G/C to A/T transitions [4] that show a nearly random distribution throughout the genome. While one study showed that the frequency of EMS-induced mutations

* Correspondence: galindo@ualberta.ca
†Equal contributors
[1]Department of Biological Sciences, University of Alberta, Edmonton, AB, Canada, T6G 2E9
Full list of author information is available at the end of the article

was estimated at about 1 mutation/300 kb screened [4], the density of mutations can vary for different plants and treatments [12]. Therefore the frequency of Single Nucleotide Variants (SNVs) for sequence length becomes an important factor in the probability of finding a phenotypic effect.

Two main approaches have been developed to relate genotype to phenotype in mutated populations. Forward genetics aims to evaluate the phenotype of hundreds or thousands of individuals to find abnormalities in characteristics like growth or development. Once a phenotypically abnormal individual is identified, map-based cloning or other molecular analyses must be used to identify the DNA sequence that was altered by mutation [13]. In reverse genetics, researchers start with a known DNA sequence of interest, and try to determine the effects of a mutation on the phenotype of the organism [3]. One advantage of reverse genetics is that it overcomes some of the limitations of forward genetics that are caused by functional redundancy [13]. In reverse genetics, mutations in a gene of interest can be obtained even in absence of a clear phenotypic effect, and therefore mutants of related genes can be combined to determine the impact of simultaneous loss-of-function or alteration of two or more genes.

Both forward and reverse genetics require researchers to screen a large number of individuals for the mutation of interest. Several methods have been developed to screen for mutations in a gene of interest within hundreds or thousands of individuals in parallel. TILLING (Targeting Induced Local Lesions in Genomes) was devised as one such methodology. In TILLING, the gene of interest is amplified by PCR of pools of DNA from members of an EMS-mutagenized population. Polymorphisms in the PCR amplicons are detected using denaturing high-performance liquid chromatography (DHPLC), or using a CEL I nuclease preparation, which cleaves the heteroduplexes that form between mutant and non-mutant DNA within the amplicons from the pooled DNA; the activity of the nuclease is then detected by gel electrophoresis [14,15]. TILLING has been used in diverse species including Arabidopsis [4], rice [8], soybean [6], sorghum [9] and tomato [10]. Other alternatives to CEL I-based TILLING have also been described including high resolution DNA melting and conformation sensitive capillary electrophoresis [5,12,16]. However, with the advent of Next Generation Sequencing (NGS) technologies [17-19], the possibilities to improve the efficiency of reverse-genetic screening have increased. NGS provides direct information about the mutated sequence and does not require formation of heteroduplexes. While the cost of sequence is still too high to allow for whole-genome sequencing of every individual in a mutant population of a species such as flax with a genome size of 373 Mb [2], the cost per-reaction may be reduced by incorporation of specific tags or barcodes in the primer sequences, allowing pooling of many samples in a single sequencing run by targeting specific regions of interest. An early approach using NGS to detect EMS mutations on tomato with GS FLX sequencing allowed screening of over 15000 plants [20]; GS FLX has also been used in the evaluation of Tef (*Eragrostis tef*) to examine genes related to lodging resistance [11]. Additional studies have used Illumina technology to perform TILLING by sequencing in rice and wheat [21], and in tobacco [22].

Here we present the first study using an Ion Torrent Personal Genome Machine (PGM) to discover single nucleotide variants (SNVs) or rare variants (these two terms – along with "mutation" - are used interchangeably throughout the text) in an EMS mutant population of an elite linseed variety of flax. The Ion Torrent has one of the lowest instrument and per-run costs of the major NGS platforms [23,24], and its sequencing output is on a scale consistent with the expected requirements of this application. Ion Torrent works with chips bearing millions of microwells with transistor sensors that allow detection of changes in current produced by the release of hydrogen once new nucleotides are incorporated to the clonally amplified DNA strands attached to each one of the beads residing in each well [25]. Massive parallelism can be achieved with this technology and the sequencing capacity limit depends on the number of sensors in the array. During the development of the technology the increases in sequencing throughput have been achieved by growth of the chip size, closer packing of features (e.g. wells) and shrinking of features. In this way for example a 5.2 fold increase in sensor count has been achieved when moving from a 314 chip to a 316 chip [25].

We demonstrate the utility of this approach by identifying SNVs in eight genes of interest, after performing a pilot experiment in three genomic regions with known SNVs to validate the methodology. Our methodology allows identification of putative variants on the target genes and can be scaled up in the number of genes and individuals to screen large populations.

Results

We conducted three experiments to develop an Ion Torrent-based method for discovery of single nucleotide variants (SNVs) in flax: (i) a pilot experiment with combinations of known SNVs (using an Ion 314 chip); (ii) a proof of concept experiment with a mutagenized population of flax (also using an Ion 314 chip); and (iii) a scale-up experiment using the higher capacity, Ion 316 chip.

Experiment I: Pilot

To evaluate our ability to detect known variants in selected regions of DNA, we used DNA from two non-mutagenized cultivars of linseed flax: CDC Bethune and Macbeth [26]. We designed primers (Additional file 1) encompassing SNVs that had been previously identified in a comparison of CDC Bethune and Macbeth DNA sequences [27] and designated these regions as S20, S411 and S900 using their scaffold of origin (e.g. S20 = scaffold 20 of the published genome assembly [2]). We mixed DNA from CDC Bethune with DNA from Macbeth to simulate a total of 28 pools from either 64 or 96 individuals, in which one individual in the pool was polymorphic (i.e. carried a SNV not present in any other member of the pool). As a negative control, we also constructed simulated pools that consisted of only DNA from CDC Bethune.

We amplified the three target regions using a two-step PCR (Figure 1 and Additional file 2A-B). The two-step PCR was used because it allowed us to incorporate specific barcodes for each pool (Additional file 3). After the second PCR step, we gel-purified the amplification products to eliminate primer dimers, which could otherwise be preferentially amplified during subsequent emulsion PCR. Gel-purified DNA was diluted to 1 ng μL^{-1}, and pooled before diluting all mixed products to 26 pM. This pooled sample was diluted one time to obtain a second pool of half the concentration (13 pM), which was used to perform a second emulsion PCR. We measured the percent of templated Ion Sphere particles (ISPs), as 37.3% for the 26 pM sample and 27.6% for the 13 pM sample. The latter sample was selected for sequencing since the template ISPs fell in the acceptable range of 10 to 30% [28]. The loading of the 26 pM sample was deemed too high for sequencing.

A total of 119.38 Mbp of sequence were obtained which represented 678,532 reads after filtering for polyclonal, dimer and low quality sequences (Table 1). While the modal read length for tested genes was > 200 bp, the mean read size was 176 bp due to a large number of reads in the 50 bp range. These short reads were comprised of incomplete sequence extensions, and sequence artifacts that were filtered out during the subsequent mapping step, leaving 47.35% of all reads to be mapped to the 28 pools (Tables 1 and 2), in each one of the three genomic regions evaluated (Additional file 4). Average read coverage was 4,103, 2,392 and 4,794 for sequences S20, S411 and S900 respectively (Table 2). While coverage did not seem to vary along the sequence, the coverage between pools did vary (Figure 2).

Analysis of mapped reads from the simulated pools identified expected SNVs in two of the three targeted loci. For S20, a SNV was identified in base 65 (A > C) (Figure 2A). For S900 (Figure 2B), a SNV was identified in base 28 (G > A), but the expected SNV at position 120 (C > T) of S411 that was previously reported was not found in any of the pools. However, novel SNVs (i.e. polymorphisms between Macbeth and CDC Bethune that had not been previously reported) were found in two of the targeted loci: we discovered an additional SNV at position 30 (T > C) in S20, and an additional SNV at position 121 (G > A) in S900 (Figure 2). All of these observations were confirmed by Sanger sequencing of targeted loci (Additional file 5).

Discovery of EMS-induced mutations in PME genes

Having demonstrated in the pilot experiment that we could detect known SNVs within simulated pools of DNA, we next attempted to discover novel SNVs within pools of DNA obtained from an actual mutagenized population (proof of concept). We used 10 ng of DNA from each of 768 individuals and pooled the DNA as explained in materials and methods. Because of the way our experiment was designed, each one of the 768 individual DNA samples was present in three pools; when a SNV was found in three intersecting pools we could pinpoint the sample of origin. We targeted four genes of the pectin methylesterase (PME) family for discovery of SNVs (*LuPME10*, *LuPME73*, *LuPME79*, *LuPME105*,

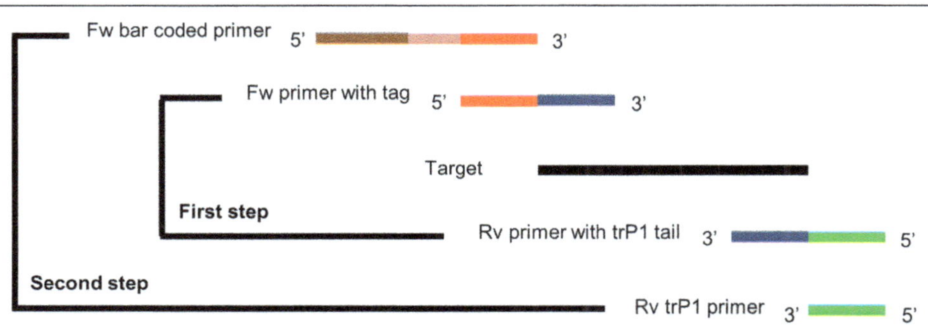

Figure 1 Two-step PCR strategy adopted for high throughput sequencing. On the first-step PCR the specific gene section (target) is amplified with a forward primer (blue) bearing a universal tag (red), and a reverse primer (blue) carrying a trP1 adapter (green). For the second-step PCR the amplicons of the first step are amplified with a reverse primer that matches the trP1 adapter (green) and a forward bar-coded (pink) primer (brown) for each desired pool of individuals and genes.

Table 1 Sequencing statistics of the three experiments performed

Experiment/replicate	Chip type	Percentage of wells with beads in chip	Total number of bases (Mbp)[1]	Total number of reads[1]	Percentage of mapped reads to all genes in experiment	Mean read length (bp)
Pilot	314	74%	119.38	678,532	47.35	176
Proof of concept-1	314	79%	71.80	459,888	60.31	156
Proof of concept-2	314	76%	85.40	543,659	63.02	157
Scale up	316	74%	649.00	3,403,220	92.04	190

[1]After filtering polyclonal wells, test fragments, adapter dimers sequences and low quality reads.

Table 2). These genes were selected because they are relevant to ongoing cell wall research in our laboratory [29]. To minimize the amplification of primer dimers, we tested the PME primers (Additional file 1), under a range of annealing temperatures and found that the optimal temperature range for the touch-up first-step PCR was 56-66°C (this was higher than the annealing temperature range 50-60°C in the pilot experiment), and the optimal second-step PCR annealing temperature was 68°C. This highlighted the importance of empirically testing PCR conditions for any new set of primers. Amplicons were analyzed and purified on agarose electrophoretic gels, eluted, and quantified (as in the pilot experiment) before Ion Torrent sequencing.

For sequencing, we diluted the pooled PME amplicon DNA to 13 pM. This DNA was sequenced in two replicate runs (to test for consistency). The percent of template ISPs for the two replicates was 23.87% and 20.13%, which made both samples suitable for sequencing. A total of 71.8 Mbp and 85.4 Mbp were obtained with an average read length of 156 bp and 157 bp for the two technical replicates (Table 1). The total number of usable reads after performing filtering of polyclonal, low quality reads and primer dimers were 459,888 and 543,659. However, we found that that even after read filtering, there remained a fraction of short reads in the 50 bp range, which presumably represented primer dimers and incomplete products (Additional file 6).

When comparing to the pilot experiment the percentage of mappable reads increased to over 60% for both replicates (Table 1), but the average read coverage per pool in each of the evaluated genes was proportionally lower than in the pilot since reads in this case were distributed among four genes (Table 2). Furthermore, the proportion of reads mapped was not equally distributed among the four genes in any of the pools. When using the mapped reads in all pools to calculate the coefficient of variation (CV) for each gene and replicate, variability

Table 2 Read statistics of the three experiments performed

Experiment/replicate	Scaffold ID/gene ID	Gene name/ annotation[1]	Number of reads mapped in all 28 pools	Percentage of mapped reads in all 28 pools[2]	Average read coverage per pool
Pilot	S20	N/A	115,998	17.09	4,103.69
	S411	N/A	67,971	10.02	2,392.08
	S900	N/A	137,314	20.23	4,794.04
Proof of concept-1	Lus10004720	*LuPME10*	91,755	19.95	2,975.50
	G25305*	*LuPME73*	94,242	20.49	2,942.75
	Lus10031470	*LuPME79*	38,154	8.30	1,059.81
	Lus10043035	*LuPME105*	53,216	11.57	1,621.20
Proof of concept-2	Lus10004720	*LuPME10*	87,889	16.17	2,815.02
	G25305*	*LuPME73*	143,832	26.46	4,404.28
	Lus10031470	*LuPME79*	47,620	8.76	1,255.62
	Lus10043035	*LuPME105*	63,282	11.64	1,941.29
Scale up	Lus10016751	ALS-1	1,443,997	42.43	44,199.34
	Lus10029955	ALS-2	424,425	12.47	13,027.62
	G24175*	CLE	534,255	15.70	16,146.73
	Lus10017825	UGT	729,799	21.44	20,472.63

*Gene Id correspond to first draft assembly of flax (unpublished).
[1]PME = Pectinmethylesterase, ALS = acetolactate synthase, CLE = cyclic peptide, UGT = glucoronosyl/glucosyl transferase.
[2]Percentage from total number of reads in Table 1.

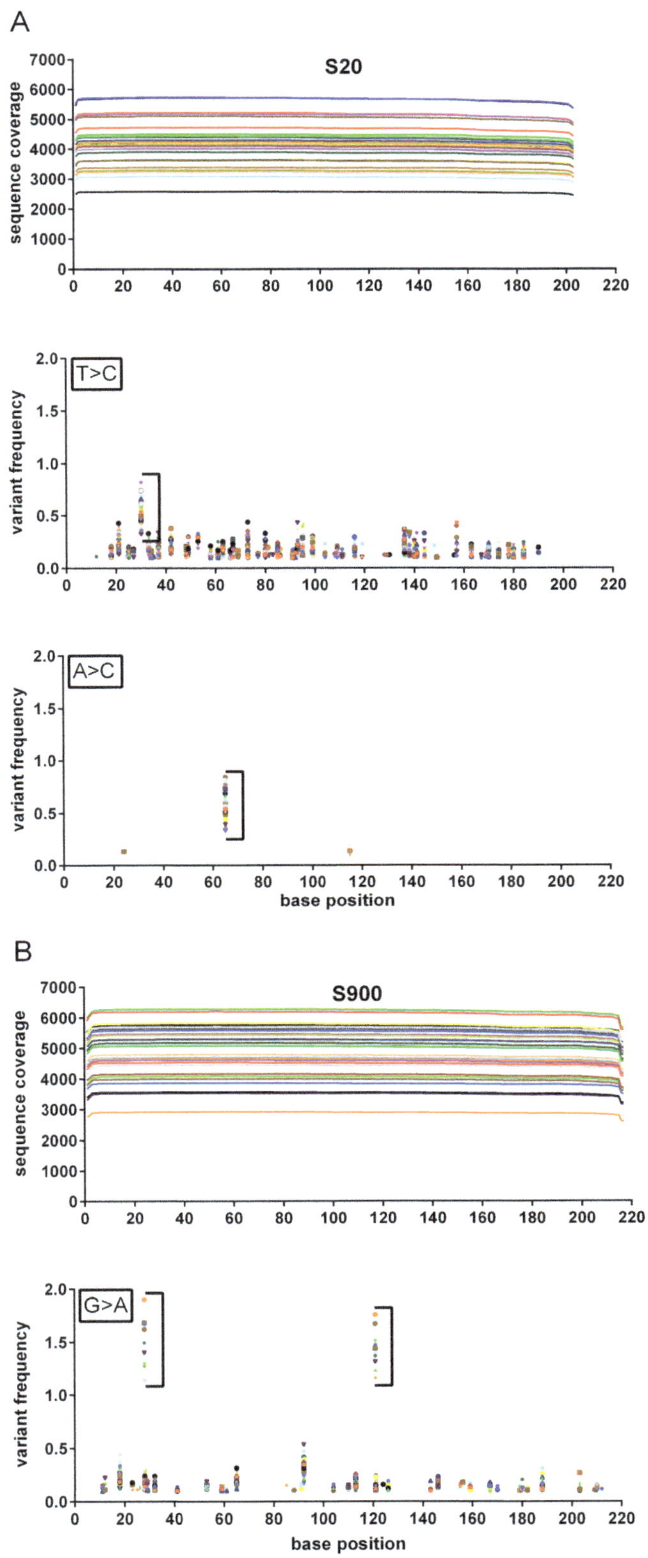

Figure 2 (See legend on next page.)

(See figure on previous page.)
Figure 2 Sequence coverage and frequency of variants in gene sections of the pilot experiment. DNA from the cultivar Macbeth was diluted (1:64 or 1:96) in CDC Bethune DNA in several simulated pools as described in methods. Each line in the sequence coverage graphs represents one of 28 pools. The frequencies of the variants are plotted against the position in the respective reference sequence section. Each symbol in the frequency graphs represents a frequency of at least 0.1% for each of the 28 pools in each position. **A.** S20, **B.** S900; no graphs for S411 are shown since no variants were detected in that sequence.

was evident among genes (Additional file 7), but the variation was constant among the two replicates for each gene.

The coverage per position for each gene was high throughout the sequence, with the exception of *LuPME79*, where a drastic decrease in coverage was observed after position 162 of the reads (Figure 3). Analysis of the sequence with Mfold [30] (not shown), did not predict a secondary structure that would explain this apparent hard stop in sequencing. Additionally, GC content of *LuPME79* (57.14%) was similar to *LuPME73* (57.34%), so a bias in GC content could not explain this difference either.

Based on our experience in the pilot experiment, we selected a minimum coverage per position of 500×, with a frequency of at least 0.5% in three intersecting pools, for defining putative mutations. When only two intersecting pools were found with the expected minimum frequency, all individuals from the intersection where sequenced. There was consistency between replicate runs for most SNVs but some of the SNVs were detected by complementary intersecting pools between both replicates. There was no correlation of false positives with the technically consistent SNVs or the ones found by complementarity.

In our analysis of four targeted PMEs amplified from 768 individuals, we found a total of 13 putative SNVs. Sanger sequencing on the original DNA from the pooled individuals confirmed only five of the 13 putative SNVs (Table 3). When the sequenced sections from the original and mutated individuals were translated, it was found that neither of the two non-synonymous changes found was within the predicted enzyme active sites [31] (Figure 4). Nevertheless, the methodology proved useful for finding mutations in pooled mutated populations when testing several genes at the same time.

Increased read depth for discovery of EMS-induced mutations

Because the previous experiment showed a large variation in mapped reads between genes and read depth among pools in each gene (Table 2 and Additional file 7) and less than half of the predicted SNVs could be

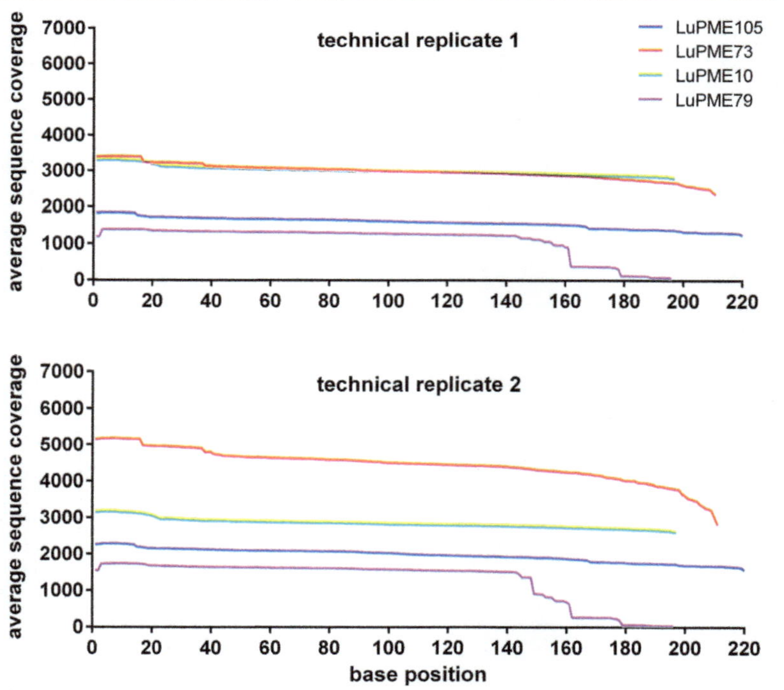

Figure 3 Coverage of four PME genes in two technical replicates. The average sequence coverage from 28 pools in each one of the base positions for the four PME genes is shown.

Table 3 SNVs found in four PME genes

Gene	Base No.	Change	Sanger confirmation	Nucleotide substitution	Amino acid substitution
LuPME79	33	G > A	No	N/A	N/A
LuPME79	96	G > A	Yes	Heterozygous	Non-synonymous
LuPME73	25	G > A	No	N/A	N/A
LuPME73	54	G > A	Yes	Heterozygous	Non-synonymous
LuPME73	81	T > A*	No	N/A	N/A
LuPME73	88	C > T	Yes	Heterozygous	Synonymous
LuPME73	97	A > G*	No	N/A	N/A
LuPME73	139	C > T	Yes	Homozygous	Synonymous
LuPME73	189	C > T	No	N/A	N/A
LuPME10	154	C > T	Yes	Homozygous	Synonymous
LuPME105	34	A > G*	N/A	N/A	N/A
LuPME105	57	A > G*	No	N/A	N/A
LuPME105	115	G > A	No	N/A	N/A

N/A – sequence could not be obtained by Sanger.
*Mutations not expected by EMS, but discovered using the technique.

confirmed by Sanger sequencing, we decided to increase read depth by switching from Ion 314 chips to the higher capacity Ion 316 chips (scale up). We used four genes related to flax metabolism (Additional file 1 and Table 2). These genes are related to characteristics related to bitter taste in flax (cyclic peptides), targeting of group 2 herbicides (acetolactate syntases), or important as major components of cell wall formation (glucoronosyl/glucosyl transferases). We selected regions in these genes based on previous studies showing critical sections and/or amino acids for the function of these proteins [32-35].

We tested again two dilutions at 13 and 26 pM to assess which of these would give a better percentage of template ISPs. We obtained 12.15 and 17.35% of templated ISPs respectively and sequenced only the latter sample, which had the highest percent loading. A total of 649 Mbp were obtained with an average read length of 190 bp (Table 1). The total number of usable reads after filtering was 3,403,220 which was an approximately 5-fold increase from the 314 chips used in the first two experiments. Although the coverage of pools among genes fell slightly towards the end of the sequences (result not shown), the average coverage for the four evaluated genes was 10 times higher than in the previous experiment (Table 2), and therefore the depth was sufficient to assess variants in any position throughout pools and genes.

Using similar parameters as for the PMEs, we found a total of 16 putative SNVs from which 11 were confirmed by Sanger sequencing (Table 4). From these, two were found to be homozygous. One of the heterozygous mutations resulted in the generation of a stop codon.

We tested the heritability of the SNVs discovered in ALS1, ALS2, UGT and CLE by Sanger sequencing of the

```
LuPME79-original  (1)  PMAHQFNAITAQSRTDPNQNTGISIQNCSIKAAKDLAESNGTTRSYLGRP
LuPME79-3H7       (1)  PMAHQFNAITAQSRTDPNQNTGISIQNCSIKTAKDLAESNGTTRSYLGRP

LuPME79-original (51)  WKAYSRTVVMNSYI
LuPME79-3H7      (51)  WKAYSRTVVMNSYI

LuPME73-original  (1)  LTFRNTAGPAKHQAVAVRNSADMSAFFNCSFEGYQDTLYVHSLRQFYRDC
LuPME73-6D11      (1)  LTFRNTAGPAKHQAVAVKNSADMSAFFNCSFEGYQDTLYVHSLRQFYRDC

LuPME73-original (51)  DIYGTIDYIFGNAA
LuPME73-6D11     (51)  DIYGTIDYIFGNAA
```

Figure 4 Alignment of amino acid sections from individuals bearing non-synonymous mutations (Table 3) to the original non-mutated sequences. Gene IDs are followed by an identifier given to the sequenced individuals. Circles below the alignment indicate enzyme active sites. Blue background indicates the amino acid change.

Table 4 SNVs found in four genes of interest

Gene	Base No.	Change	Sanger confirmation	Nucleotide substitution	Amino acid substitution	Sanger confirmation on M_4	Nucleotide substitution in progeny[1]	Script confirmation[2]
ALS-1	119	C > T	Yes	Heterozygous	A/V	Yes	2 homozygous	Yes^{++}
ALS-1	140	C > T	No	N/A	P/L	N/A	N/A	No
CLE	89	G > A	Yes	Heterozygous	G/D	Yes	3 heterozygous	Yes^{+++}
CLE	94	G > A	Yes	Heterozygous	E/K	Yes	1 homozygous, 1 heterozygous	No
CLE	134	G > A	Yes	Heterozygous	R/H	Yes	1 homozygous	Yes^{+++}
ALS-2	26	G > A	No	N/A	G/E	N/A	N/A	No
ALS-2	43	G > A	Yes	Heterozygous	E/K	Yes	2 homozygous	No
ALS-2	100	G > A	No	N/A	E/K	N/A	N/A	No
ALS-2	161	C > T	Yes	Heterozygous	A/V	Yes	2 homozygous	Yes^{+++}
ALS-2	161	C > T	Yes	Homozygous	A/V	Yes	2 homozygous	Yes^{+}
UGT	27	C > T	No	N/A	P/S	N/A	N/A	No
UGT	33	C > T	No	N/A	H/Y	N/A	N/A	No
UGT	81	C > T	Yes	Homozygous	L/F	Yes	3 homozygous	Yes^{+++}
UGT	99	G > A*	Yes	Heterozygous	E/STOP	Yes	3 heterozygous	N/A
UGT	99	G > A*	Yes	Heterozygous	E/STOP	Yes	2 heterozygous	N/A
UGT	184	G > A	Yes	Heterozygous	G/E	Yes	1 homozygous, 1heterozygous	No

*Mutation was found by looking at intersecting pools with frequencies below the set threshold.
[1]Six individuals from progeny examined per mutation.
[2]The frequencies of the variants were used to run a Phyton script which automatically detects the source individual bearing the mutation (Additional file 8). Parameters used in the script were: 2 lower SD cutoff, 10 upper SD cutoff, 1 min. pools. N/A rows were not picked by the script since they were found by a different methodology. The confirmed points by the script had confidence intervals of: $^{+}$75%, $^{++}$85% and $^{+++}$99%.
Mutations on position 161 on ALS-2 and 99 on UGT were found in two different individuals.

progeny of plants in which the mutations were initially identified. The presence of the mutation was confirmed in the progeny of all of the lines (Table 4).

Discussion

Ion Torrent technology in SNV detection

The advent of next-generation sequencing technologies has opened new doors for genomic-scale analyses [18,19]. Among common sequencing platforms IonTorrent offers potential advantages including low instrument cost, low cost per base, and fast output (up to 333 Mbp/h) [23,36]. Ion Torrent has also been reported to be superior for variant calling than Illumina [23], although other studies report similar or slightly higher sensitivity for MiSeq [37]. While Ion Torrent has a high rate of indels caused by homopolymeric runs, and long-range sequence quality can be lower than that of other instruments [23,36,38,39], this is not a problem for calling SNVs with high read depth.

We used Ion 314 and Ion 316 chips and were able to reach reads in the 200 bp range with a total sequence throughput that guaranteed high depth. The size reached by these reads facilitated the evaluation of critical gene regions without the need for post-sequencing assembly. Likewise, we were able to achieve a high-throughput in the number of samples analyzed for several genes. A similar approach used Ion Torrent to map mutations in mice, but examined a large number of regions in a few samples [40].

In the pilot experiment, we detected the expected mutations and additional unreported changes in the tested genomic regions (Figure 2 and Additional file 5). Our experiment with PMEs had a larger variability in read depth among pools and genes (Additional file 7), which may have had an influence on the number of false positives. When we increased our read depth by using the larger Ion 316 chip, the number of false positives decreased significantly.

Several variables were optimized during our experiment. Since sample pooling was used, high-quality DNA in equal amounts was needed to increase the probability of detecting one mutated individual among the population. We standardized a high-throughput CTAB protocol yielding high quality DNA and further quantified the samples by fluorescence to add equal amounts of DNA from each individual in each pool. Additionally, the first step PCR required addition of different PCR additives (ethylene glycol and DMSO), which decreased the formation of secondary structures since preliminary tests showed that a standard PCR resulted in a high proportion of primer dimers (or secondary structures) due to the length of the primers; nevertheless some residual dimers were still unavoidable (Additional file 2). A PCR cleanup did not suffice to get rid of all dimers, which were still carried over to the emulsion PCR, resulting in a preferential amplification of these smaller products in preliminary runs (result not shown). Eluting the specific products from agarose gels improved the detection of the larger specific products but still with some residual carryover. A better primer removal method like a solid-phase reversible immobilization (SPRI) technology was suggested for the process of PCR cleanup for Illumina [41], and could also be used in future experiments.

Differences in read coverage were detected in our different experiments, but this is not uncommon in NGS technologies [21]. As a general trend we found that shorter amplicons resulted in higher average coverage over all positions in all pools for the evaluated gene sections of the three experiments. This can be related especially to emulsion PCR, since this constitutes a step where all genes are mixed and there can be a preferential amplification of shorter amplicons. Factors like shorter denaturation times and faster extension on smaller products may lead to this preferential amplification [42].

However, the exact same relationship was not found for the PMEs (Figure 3). The two PME amplicons corresponding to *LuPME79* and *LuPME105* had a lower average coverage, and while the later does correspond to the larger amplicon of this gene set, the former is the smallest. Therefore different factors may have had an influence on the variability in read count that we encountered. When we calculated GC content it was seen that *LuPME105* had the lowest GC content (43%) of the four PME sections evaluated. Ion Torrent read coverage has been shown to decrease upon high or low GC content or under different levels of genome complexity [23,25,43]. Likewise, gel extraction has been shown also to have a bias for recovery of GC-rich double stranded templates that have higher affinity for kit columns, than AT-rich amplicons which become single stranded upon agarose melting conditions [41].

Neither length, nor GC, seemed correlated with the lower coverage of *LuPME79*, however, the primers from *LuPME79* and *LuPME105* (Additional file 1) had a lower value of Gibbs energy – ΔG - (−11.32 and −10.24 kcal mol^{-1} respectively) compared with the primers of *LuPME10* and *LuPME73* (−5.19 and −8.48 kcal mol^{-1}). Since a lower ΔG favors the formation of secondary structures, this could have had an effect of the PCR resulting in a differential amount of amplicons before pooling. Unforeseen changes like EMS mutations in priming sites can also contribute to differential amplification among samples. We also encountered a drop in coverage after position 162 in *LuPME79*; while we could not detect any evident secondary structure after the hard stop in the sequence reads, 16 out of the 20 previous nucleotides before the read coverage fall are G or C and this could be related to the formation of a secondary structure that impairs the sequencing polymerase from continuing.

Length and GC content of the target locus are not entirely under control of the researcher. However, other factors can be better controlled to achieve near-homogeneous coverage when pooling samples for analysis. For example, an accurate quantification of the PCR products after the first round of PCR by comparison to a standard [44], would decrease biases in amounts of amplicons before pooling. Although previous studies have shown that non-normalized samples are suitable to detect high-frequency variants [44], our study comprised the detection of mutations in pooled samples, where a homozygous mutation could theoretically have a low frequency: aprox. 1% for a 1 in 96 dilution, and of 0.5% for an heterozygous allele. Additionally, very small amounts of DNA (10 ng) were used in the pools. Since differences in the amount of starting DNA can also result in differential amplification [42], it is important to guarantee close to equimolar amounts of starting DNA to avoid losing a variant due to a PCR deficiency.

Overall, the coverage on the 11 different gene regions that were tested allowed for the detection of SNVs in regions of up to 200 bp. Although coverage varied slightly between some pools and genes, coverage along the length of amplicons was generally even but dropped only towards the ends of the sequences (Figure 3), which is common of sequencing-by-synthesis technologies [25]. With further optimization, the Ion 314 chips could easily accommodate the evaluation of 16 amplicons at an average of more than 500× coverage per pool under the 28-pool scheme we utilized. Theoretically, for the Ion 316 chip, 160 amplicons could be evaluated under the same conditions, but adjusting the technical parameters under the current technology to guarantee little variability from pooling to sequencing becomes harder unless a similar system to the Ampliseq, used for human genes [25], can be rapidly implemented for plants.

Another factor that came into consideration was how to decrease the level of false positives in data. Whereas it is has been reported that the level of false positives in Ion Torrent data is larger than Illumina, it has also been shown that Ion Torrent can detect more true positives given enough coverage [23]. From our data it was evident that the increase in depth upon using the Ion 316 chip was concomitant with a decrease in false positives (compare Tables 3 and 4), showing that read depth is key for separating real mutations from noise in SNV studies [21], and in studies where allele frequency needs to be resolved [45]. While our proof of concept and scale experiments differed in the amplicons used, read mapping statistics (Tables 1 and 2) demonstrated that the experiment with the 316 chip had an increase in the number of reads by over 5 fold when compared to the 314 chip.

Since other factors like GC content and small differences in amplicon size are still difficult to control for and do not have a clear correlation always with read depth, we believe that the technical increase on read number by selecting a higher capacity chip is the key factor in obtaining more rare variants. Unfortunately, most second generation NGS technologies still present high error rates (see below). While we tried to control for equimolar amounts of DNA when pooling samples, PCR steps result in uncontrollable dilutions of some samples before sequencing which will result in the loss of some SNVs among noise. Improvements to eliminate such technical variability will increase our ability for SNV detection.

There are additional elements to have into account. For example, it was noticed that the background frequency of base substitution varied between the type of substitutions and among genes. While an A > C change had little or no background over 0.1% for the S20 region (see Figure 2), a T > C change in the same region and a G > A in the S900 region had larger substitution noise. A differential rate of substitution has been linked to the PGM from Ion Torrent upon studying bacteria, with G > A and T > C transitions presenting the higher rates of substitution with the Ion OneTouch 200 template kit [39].

Interestingly while homopolymer errors are the most common error type from this technology [23,36,39], an study with bacteria found that substitutions have the highest variation frequencies, with standard deviations ranging from 26%-56% of the mean substitution rate [39]. This has implications in the detection of rare variants (including false negatives) which may come up at lower frequencies as we detected in our study due to sample dilution in pools. For example, a 0.3% frequency was found for S20 variants in the pilot experiment (Figure 2), and since some random error can reach this frequency, this can lower our detection ability. While there were a few false positives embedded into homopolymeric tracts which constitute the bigger source of error of Ion Torrent technology [39], no specific position or sequence-specific bias in the SNVs that were not confirmed by sanger could be inferred to make any generalization.

Compared to Roche/454 technologies where mutagenized populations are used to discover rare variants [11,20] the Ion Torrent technology offers a higher read depth in short times, which results in a higher probability to find the mutated bases. While one of the main advantages of 454 sequencing over other technologies was their read length (>400 bp), Ion Torrent is quickly catching up to offer similar read lengths of high quality [25]. Our study achieved similar throughput (314 chips) in the number of reads obtained as the aforementioned 454 studies, but the time required for a run on an Ion Torrent PGM is just over 2 hours while the most basic 454 sequencers use at least 10 hours and with prices of equipment and cost per base which are above the ones of Ion Torrent [36,46]. Furthermore, higher throughput was achieved (5-fold) when changing to a 316 chip without a change in runtime while for 454 technology the use of higher-end sequencers may require up to 23 hours for a run [46]. Finally, the rate of false positives seems to be on a similar range for these two technologies.

Our methodology was based on a previous Illumina study [21] and therefore some conclusions can be drawn by comparing the two. It was clear that because of the similar methodology similar results were obtained in several fields. For example low coverage resulted in increased noise which impaired SNV detection in both studies. The throughput of Illumina is generally higher resulting in detection of many more mutations. However, this is achieved in longer runtimes (days) and with more expensive equipment, although cost per base is lower on Illumina [36,37,46]. Because of the higher error rates of the Ion Torrent technology the amount of false positives is usually higher than on Illumina [23].

Conclusion

We have demonstrated that the Ion Torrent can be used in a scalable, amplicon-based approach for efficient discovery of mutations in a small number of genes. The efficiency of the method is limited by the rate of false positives, which may be decreased by higher read-depth and further technical optimization. Ion Torrent technology has been demonstrated to introduce biases at errors at specific steps during sequencing [43], as have other sequencing technologies, especially when PCR step is used in sample preparation [47]. Nevertheless, the Ion Torrent PGM platform detected rare variants with as low as 0.3% frequency per pool according to our results, which is above the substitution error of 0.1% calculated for the technology [25], and we showed that 768 individuals could be easily pooled per run. The Ion Torrent is one of the first technologies that does need optical systems to detect nucleotide incorporation, and does not use modified nucleotides [25,48]; in addition it has a good combination of throughput, cost and time saving compared to other systems [23,36,44,49]. The use of chips with larger capacity [25], will allow increasing both the number of genes and/or pooled samples. Additionally, the availability of 400 bp kits now allows exploring larger regions of interest without the need of using paired ends.

Methods

Plant material

Seeds of *Linum usitatissimum* L. (var. CDC Bethune), an elite linseed cultivar, were obtained from Gordon Rowland (Crop Development Center, Saskatoon, SK). Seeds were soaked in 5 volumes (liquid volume/seed volume) of 0.5%

ethyl methyl sulfonate (EMS) in 25 mM phosphate buffer (pH 7.6) for 4 h at room temperature and then were rinsed with distilled water (three times), and air dried prior to storage. These M_1 seeds were sown at the University of Alberta farm (Edmonton, AB), and their M_2 progeny were harvested as individual families. Approximately four seeds from each M_2 line were sown in rows at Kernen Farm (Saskatoon, SK) in summer 2010. Leaves were harvested and lyophilized for subsequent DNA extraction, and their progeny (i.e. M_3 families) were harvested from individual plants, then threshed and stored at ambient temperature in envelopes until screening.

DNA extraction and pooling

DNA extraction was performed for 96 samples at a time using CTAB [50] with some modifications. Lyophilized leaf samples (10–20 mg) were placed in 8-strip 1.2 mL collection tubes (QIAGEN, Hilden, Germany) containing a sterile 3 mm tungsten carbide beads. Tubes were capped and ground for 2 minutes at 25 Hz using a Retsch MM301 mixer mill (Retsch, Haan, Germany). Samples were centrifuged at 1,450 × g for 1 minute. CTAB was prepared with 2% CTAB (w/v), 2% PVP-40, 100 mM Tris-Cl pH 8.0, 25 mM EDTA pH 8.0, 1 M NaCl and 0.5 g L^{-1} of spermidine. The buffer was pre-warmed (60°C) and supplemented with 10 μg mL^{-1} of RNAse A (Sigma-Aldrich, St. Louis, MO, USA), 100 μg mL^{-1} of proteinase K (Fermentas, Waltham, MA, USA) and 5% mercaptoethanol, before adding 500 μL to each sample. The tubes were re-capped with new caps and mixed by inversion (20 times) and incubated at 60°C for 2 hours, mixing the tubes by inversion every 20 minutes. After incubation, samples were centrifuged for 5 minutes at 5,800 × *g* and the supernatants were transferred to new tubes. Five hundred microliters of chloroform : isoamyl (24:1) were added to each sample and re-capped tubes were mixed by inversion (60 times) before centrifugation for 5 minutes at 5,800 × *g*. The supernatant was transferred to new tubes. Chloroform : isoamyl extraction was repeated once again. Three hundred microliters of ice-cold isopropanol were added to the tubes with the supernatant and the samples were mixed by inversion (20 times) before transferring to −20°C for 2 hours. Incubation was followed by centrifugation at 5,800 × g for 15 minutes. Supernatants were decanted to waste and 500 μL of ice cold 95% ethanol were added to the pellets and samples were gently vortexed prior to centrifugation at 5,800 g for 5 minutes. The previous step was repeated with 500 μL of ice cold 70% ethanol. The ethanol was decanted and the samples were air-dried, and resuspended in 125 μL of TE 10:1 and stored at −20°C.

DNA samples were quantified using Picogreen (Invitrogen, Carlsbad, CA, USA) in a Fluorostar BMG plate reader (BMG labtech, Ortemberg, Germany), using a standard curve of flax DNA. Samples were diluted to 10 ng μL^{-1} and 1 μL or each pooled in groups of 64 or 96. According to the formula used by Tsai et al. [21] using 10 ng of DNA with a flax genome of 2C = 0.764 pg [2] would yield 204.5 copies per allele (for each individual) for the 1 in 64 dilution or 136.3 copies per allele for the 1 in 96 dilution. This is higher than the minimum number of recommended copies (40) to avoid absence or fluctuation of copies among individuals [21].

The pools were created using the following methodology: each sample was in an individual well of eight 96-well plates; all individuals from each plate were pooled creating the first eight pools containing 96 individuals each (designated pools A1 to A8). Then all the individuals from the same column in each plate were pooled creating 12 pools of 64 individuals each (designated pools B1 to B12). And finally all individuals from the same row in each plate were pooled creating the last eight pools of 96 individuals each (designated pools C1 to C8). In this way each individual was part of three different pools and therefore a mutation detected in three intersecting pools would allow us to pinpoint the source individual.

Primer design

The primers were designed using Primer 3 [51] with the following parameters: Two C's or G's in the last five nucleotides towards the 3′ end; up to three nucleotides long homopolymers; a delta G lower than −9 kcal/mole, Tm between 59 and 61°C and size between 19 and 21 nucleotides (when conditions were not met parameters were relaxed). For the forward primer the universal primer tag 5′-CAGTCGGGCGTCATCA-3′ was added (designed by Travis Glenn, Univ. of SC, http://www.gvsu.edu/dna/universal-primer-tag-6.htm) and for the primer Rv the adaptor trP1 was added (5′-CCTCTCTATGGG CAGTCGGTGAT–3′) (Additional file 1).

Pilot experiment

A pilot experiment was performed using known SNVs showing polymorphisms between cultivars CDC Bethune and Macbeth. Since for the pilot experiment we were not trying to discover new mutations we did not have to pool DNA from different individuals as described in the DNA extraction and pooling section; instead Macbeth DNA was diluted 1:64 or 1:96 in Bethune DNA to simulate the presence of an individual with a mutation within the population. Three genomic regions with previously reported SNVs [27], from three flax scaffolds were chosen to test the methodology and were named S20, S411 and S900 (names were derived from the names of the scaffolds that contained them - see Additional file 1).

PCR amplification and barcoding

A two-step PCR strategy was adopted: the first step used a PCR with two sets of cycles at different temperatures (below) to amplify the target gene and the second step incorporated a barcode oligonucleotide to distinguish different pools (Figure 1). First step PCR was performed on each of 28 sample pools and three genes per pool (pilot experiment), with forward primers bearing a universal tag at their 5′ end (Additional file 1), and reverse primers carrying an additional adaptor tail which binds the ionospheres used in the sequencing step.

For the pilot experiment, first-step amplifications were performed on a template that either included DNA with known SNVs S20, S411 and S900 (Macbeth DNA in 1:64 or 1:96 dilution in CDC Bethune DNA) or homogeneous template without the SNVs (i.e. only CDC Bethune DNA). The PCR products were diluted 1:100 and equal amounts (5 μL) of PCR product from each of the three target regions were mixed in each of 28 pools for barcoding; the changes of Macbeth genic regions SNVs where introduced in all pools for the gene of S20, and in 10 pools for genes S411 and S900. For the second step, a total of 28 PCRs were performed with forward barcoded primers (Figure 1 and Additional file 2), and a trP1 primer complementary to the tail from the reverse primer of the first-step PCR. The 28 bar codes allowed us to discriminate between the respective DNA pools after sequencing.

First step PCR was carried under the following conditions: 1X PCR buffer, 2 mM Mg, 0.2 mM dNTPs, 0.2 μM of each forward and reverse primers (Additional file 1), 1 M ethylene glycol, 4% dimethyl sulfoxide (DMSO), 10 ng of DNA and 1.25 units of Platinum Taq DNA polymerase (Invitrogen, Burlington, ON, Canada). The PCR protocol included an initial denaturation step at 94°C for 3 minutes, followed by 5 cycles of 94°C for 20 seconds, 50°C for 30 seconds and 72°C for 1 minute. Finally, an additional 25 cycles were done by changing the annealing temperature to 60°C (touch-up), and a final extension step was performed at 72°C for 10 minutes.

The second-step PCR was carried out as follows: 1X PCR buffer, 2 mM Mg, 0.2 mM dNTPs, 0.2 μM forward bar-coded primer and 0.2 μM reverse trP1 primer (Additional files 1 and 3), 1 μL of the DNA amplicon dilution (three genes combined per pool) and 1.25 units of Platinum Taq DNA polymerase (Invitrogen, Burlington, ON, Canada). The PCR protocol included an initial denaturation step at 94°C followed by 30 cycles of 94°C for 30 seconds, 62°C for 30 seconds and 72°C for 1 minute, and a final extension of 72°C for 10 minutes.

Purification and quantification of PCR products

The products of PCR reactions of 28 pools and 3 genes per pool were run on 1.5% agarose gels. To eliminate primer dimers, a band of the expected size was gel purified using the Wizard SV gel and PCR clean-up system (Promega, Madison, WI, U.S.A). Samples were quantified using a Qubit 2.0 fluorometer using a dsDNA HS assay (Invitrogen, Burlington, ON, Canada). All quantified samples were diluted to 1 ng μL^{-1} and equal amount of the PCR products were combined and were re-measured on the Qubit to confirm that concentration was still 1 ng μL^{-1}. The sample was diluted with low Tris EDTA buffer (TE 10:1) to obtain 15.5×10^6 molecules per microliter (26 pM) which is the recommended concentration for template preparation using Ion Torrent technology.

Sequencing

All procedures for emulsion PCR and next-generation sequencing were performed with Ion Torrent equipment and Ion Torrent kits under the manufacturer specifications (Life Technologies, Carlsbad, CA, U.S.A.): emulsion PCR was performed with the Ion OneTouch 200 template kit in an Ion OneTouch. Enrichment of template positive Ionospheres (ISPs) was performed with an Ion OneTouch ES (Life Technologies, Carlsbad, CA, U.S.A.). Sequencing of enriched templates bound to ionospheres was done using the Ion PGM 200 sequencing kit in an Ion PGM Sequencer with either 314 or 316 chips. FASTQ files of each barcoded group of sequences were recovered from the Ion Torrent server for further analysis. Reads have been deposited in the Sequence Read Archive (SRA) from NCBI under study accession number: SRP052626.

Detection of induced mutations in a population of EMS mutagenized flax

Procedures were similar to the pilot experiment, unless stated otherwise. The experimental design was adapted from Tsai et al., 2011 [21]. A total of 28 pools of DNA from distinct individuals (768 lyophilized leaf samples) were created to facilitate detection of mutations as described in DNA extraction and pooling.

A total of eight primer pairs that amplified pectin methylesterases (PMEs) were designed to target conserved regions presumed to be essential for enzymatic function and tertiary structure stability of these genes [29,31]. Preliminary tests showed that four of the primers pairs (Additional file 1), gave stronger products, and these were used for second-step PCR as described above. Separately, 12 primer pairs from three different metabolism-related genes that constitute important breeding traits (cyclic peptides, acetolactate synthase and UDP - glucuronosyl/glucosyl transferases) were also designed and four primer pairs were selected after testing them by PCR (Additional file 1).

Analysis of Single Nucleotide Variants (SNVs)

Reads obtained from Ion Torrent PGM sequencing were uploaded to the CLC Genomics workbench platform (CLCbio, Aarhus N, Denmark). Reads were mapped to reference sequences previously confirmed by Sanger capillary sequencing of target amplicons (data not shown), using the following parameters: masking mode = no masking, mismatch cost = 2, insertion cost = 3, deletion cost = 3, length fraction = 0.8 and similarity fraction = 0.8, global alignment = no, non-specific match handling = map randomly, output mode = create stand-alone read mapping, create report = yes, collect unmapped reads = yes. Once the reads were mapped to the reference, the mapped reads files were used as input to discover rare variants using quality score with the following parameters: neighborhood radius = 5, maximum gap and mismatch count = 5, minimum neighborhood quality = 15, minimum central quality = 20, ignore non-specific matches = yes, ignore broken pairs = yes, minimum coverage = 100, minimum variant frequency (%) = 0.1 (selected according to a previous study [25]), maximum expected alleles = 4, advanced = no, require presence in both forward and reverse strands = no, filter454/ion homopolymer indels = yes, create track = yes, create annotated table = yes, genetic code = 1 standard. The whole process was automated by creating a CLC workbench workflow. We also generated a Python script to automate mutation identification in the future (Additional file 8).

Frequency tables were created manually using Microsoft Excel per each gene and per each pool after filtering homopolymeric tracts and indel artifacts created by the sequencing technology. Graphs showing the frequency changes by position for the 28 pools in a specific base change (e.g. G to A) were done to detect outliers visually which were indicative of a rare variant. SNV candidates were chosen if the mutation was present in three intersecting pools or in two intersecting pools with high frequency (>0.3% in pilot experiment and >0.5% in remaining experiments). DNAs from individuals (M_3 generation) with potential mutations were re-sequenced using Sanger sequencing to confirm the analysis performed with Ion Torrent.

Additional files

Additional file 1: Primer sequences and adaptors used for pilot and test studies.

Additional file 2: Two-step PCR from pilot experiment. A. Products of first step PCR from the pilot experiment. Amplifications were performed with 5 or 10 ng of pooled DNA in the different dilutions that simulated the inclusion of the mutated individual. Gel was run in 1.5% agarose in TAE 1X at 90 V for 40 minutes. Size of marker bands is given in bp. Negative control refers to a PCR with no DNA template. B. Products of second step PCR from the pilot experiment. Amplifications were performed with 1:100 dilutions of the first-step PCR mixed products of the three genes (S20, S411, S900). Gel was run in 1.5% agarose in TAE 1X at 90 V for 60 minutes. Size of marker bands is given in bp. Negative control refers to a PCR with no DNA template.

Additional file 3: Primer sequences (with barcodes) and adaptors used for second step-pcr.

Additional file 4: Sanger-sequenced fragments of references genes used in the different experiments.

Additional file 5: Alignment of sequenced fragments of the pilot experiment. The number 1 over the alignment indicates position 1 for reference of mutations found.

Additional file 6: Ion Sphere Particles (ISPs) and read identification summary. The data is given for two technical replicates runs (A and B) of PMEs using the Ion Torrent PGM.

Additional file 7: Average read count and dispersion among the mapped reads for two replicate sequencing runs of PME genes. Statistics are given for the 28 pools for all mapped reads or for each individual gene.

Additional file 8: Python script algorithm description. The methodology, results and discussion using this script are presented.

Competing interests

The authors declare that they have no competing interests. We have no affiliation or support from any of the manufacturers of equipment named in this manuscript.

Authors' contributions

LGG designed the experiments and wrote the manuscript. LGG, DPL and EAB carried out the experiments and data analyses. EAB and DCJ wrote the Python script. MKD supervised the experiments and edited the manuscript. All authors read and approved the final manuscript.

Acknowledgements

We thank Sophie Dang from the Molecular Biology facility (MBSU) at the University of Alberta for her training in the use of the Ion Torrent PGM. Funding was provided by Genome Canada (TUFGEN: Total Utilization of Flax Genomics program) and the Natural Sciences and Engineering Research Council (NSERC) of Canada.

Author details

[1]Department of Biological Sciences, University of Alberta, Edmonton, AB, Canada, T6G 2E9. [2]Department of Computing Sciences, Kings University College, Edmonton, AB, Canada, T6B 2H3. [3]IK Barber School of Arts & Sciences, University of British Columbia, Okanagan campus, Kelowna, BC, Canada, V1V 1 V7.

References

1. Cullis CA. Mechanisms and control of rapid genomic changes in flax. Ann Bot. 2005;95:201–6.
2. Wang Z, Hobson N, Galindo L, Zhu S, Shi D, McDill J, et al. The genome of flax (Linum usitatissimum) assembled de novo from short shotgun sequence reads. Plant J. 2012;72:461–73.
3. Alonso JM, Ecker JR. Moving forward in reverse: genetic technologies to enable genome-wide phenomic screens in Arabidopsis. Nat Rev Genet. 2006;7:524–36.
4. Greene EA, Codomo CA, Taylor NE, Henikoff JG, Till BJ, Reynolds SH, et al. Spectrum of chemically induced mutations from a large-scale reverse-genetic screen in Arabidopsis. Genetics. 2003;740:731–40.
5. Botticella E, Sestili F, Hernandez-lopez A, Phillips A, Lafiandra D. High resolution melting analysis for the detection of EMS induced mutations in wheat Sbella genes. BMC Plant Biol. 2011;11:156.
6. Cooper JL, Till BJ, Laport RG, Darlow MC, Kleffner JM, Jamai A, et al. TILLING to detect induced mutations in soybean. BMC Plant Biol. 2008;8:9.
7. Rowland GG. An EMS-induced low-linolenic-acid mutant in McGregor flax (Linum usitatissimum L.). Can J Plant Sci. 1991;71:393–6.

8. Till BJ, Cooper J, Tai TH, Colowit P, Greene EA, Henikoff S, et al. Discovery of chemically induced mutations in rice by TILLING. BMC Plant Biol. 2007;7:19.
9. Xin Z, Wang ML, Barkley NA, Burow G, Franks C, Pederson G, et al. Applying genotyping (TILLING) and phenotyping analyses to elucidate gene function in a chemically induced sorghum mutant population. BMC Plant Biol. 2008;8:103.
10. Okabe Y, Asamizu E, Saito T, Matsukura C, Ariizumi T, Brès C, et al. Tomato TILLING technology: development of a reverse genetics tool for the efficient isolation of mutants from Micro-Tom mutant libraries. Plant Cell Physiol. 2011;52:1994–2005.
11. Zhu Q, Smith SM, Ayele M, Yang L, Jogi A, Chaluvadi SR, et al. High-throughput discovery of mutations in tef semi-dwarfing genes by next-generation sequencing analysis. Genetics. 2012;192:819–29.
12. Kurowska M, Daszkowska-Golec A, Gruszka D, Marzec M, Szurman M, Szarejko I, et al. TILLING: a shortcut in functional genomics. J Appl Genet. 2011;52:371–90.
13. Østergaard L, Yanofsky MF. Establishing gene function by mutagenesis in Arabidopsis thaliana. Plant J. 2004;39:682–96.
14. McCallum CM, Comai L, Greene EA, Henikoff S. Targeting induced local lesions IN genomes (TILLING) for plant functional genomics. Plant Physiol. 2000;123:439–42.
15. Oleykowski CA, Bronson Mullins CR, Godwin AK, Yeung AT. Mutation detection using a novel plant endonuclease. Nucleic Acids Res. 1998;26:4597–602.
16. Gady ALF, Hermans FWK, Van de Wal MHBJ, van Loo EN, Visser RGF, Bachem CWB. Implementation of two high through-put techniques in a novel application: detecting point mutations in large EMS mutated plant populations. Plant Methods. 2009;5:13.
17. Liu L, Li Y, Li S, Hu N, He Y, Pong R, et al. Comparison of next-generation sequencing systems. J Biomed Biotechnol. 2012;2012:1–11.
18. Bräutigam A, Gowik U. What can next generation sequencing do for you? Next generation sequencing as a valuable tool in plant research. Plant Biol. 2010;12:831–41.
19. Deschamps S, Campbell MA. Utilization of next-generation sequencing platforms in plant genomics and genetic variant discovery. Mol Breed. 2010;25:553–70.
20. Rigola D, van Oeveren J, Janssen A, Bonné A, Schneiders H, van der Poel HJA, et al. High-throughput detection of induced mutations and natural variation using KeyPoint technology. PLoS One. 2009;4:e4761.
21. Tsai H, Howell T, Nitcher R, Missirian V, Watson B, Ngo KJ, et al. Discovery of rare mutations in populations: TILLING by sequencing. Plant Physiol. 2011;156:1257–68.
22. Reddy TV, Dwivedi S, Sharma NK. Development of TILLING by sequencing platform towards enhanced leaf yield in tobacco. Ind Crops Prod. 2012;40:324–35.
23. Quail MA, Smith M, Coupland P, Otto TD, Harris SR, Connor TR, et al. A tale of three next generation sequencing platforms: comparison of Ion Torrent. Pacific Biosciences and Illumina MiSeq sequencers. BMC Genomics. 2012;13:341.
24. Hamilton JP, Buell CR. Advances in plant genome sequencing. Plant J. 2012;70:177–90.
25. Merriman B, Rothberg JM. Progress in ion torrent semiconductor chip based sequencing. Electrophoresis. 2012;33:3397–417.
26. Duguid SD, Kenaschuk EO, Rashid KY. Macbeth flax. Can J Plant Sci. 2003;83:803–5.
27. Kumar S, You FM, Cloutier S. Genome wide SNP discovery in flax through next generation sequencing of reduced representation libraries. BMC Genomics. 2012;13:684.
28. Life Technologies Ion OneTouch System User Guide. Life technologies: Carlsbad; 2012.
29. Pinzon-Latorre D, Deyholos MK. Pectinmethylesterases (PME) and pectinmethylesterase inhibitors (PMEI) enriched during phloem fiber development in flax (Linum usitatissimum). PLoS One. 2014;9:e105386.
30. Zuker M. Mfold web server for nucleic acid folding and hybridization prediction. Nucleic Acids Res. 2003;31:3406–15.
31. Di Matteo A, Giovane A, Raiola A, Camardella L, Bonivento D, De Lorenzo G, et al. Structural Basis for the Interaction between Pectin Methylesterase and a Specific Inhibitor Protein. Plant Cell. 2005;17:849–58.
32. Tranel PJ, Wright TR. Resistance of weeds to ALS-inhibiting herbicides : what have we learned ? Weed Sci. 2002;50:700–12.
33. Gui B, Shim YY, Datla RSS, Covello PS, Stone SL, Reaney MJT. Identification and quantification of cyclolinopeptides in five flaxseed cultivars. J Agric Food Chem. 2012;60:8571–9.
34. Bruhl L, Matthaus B, Fehling E, Wiege B, Lehmann B, Luftmann H, et al. Identification of bitter off-taste compounds in the stored cold pressed linseed oil. J Agric Food Chem. 2007;55:7864–8.
35. Walsh DT, Babiker EM, Burke IC, Hulbert SH. Camelina mutants resistant to acetolactate synthase inhibitor herbicides. Mol Breed. 2011;30:1053–63.
36. Loman NJ, Misra RV, Dallman TJ, Constantinidou C, Gharbia SE, Wain J, et al. Performance comparison of benchtop high-throughput sequencing platforms. Nat Biotechnol. 2012;30:434–9.
37. Li X, Buckton AJ, Wilkinson SL, John S, Walsh R, Novotny T, et al. Towards clinical molecular diagnosis of inherited cardiac conditions: a comparison of bench-top genome DNA sequencers. PLoS One. 2013;8:e67744.
38. Koshimizu E, Miyatake S, Okamoto N, Nakashima M, Tsurusaki Y, Miyake N, et al. Performance comparison of bench-top next generation sequencers using microdroplet PCR-based enrichment for targeted sequencing in patients with autism spectrum disorder. PLoS One. 2013;8:e74167.
39. Bragg LM, Stone G, Butler MK, Hugenholtz P, Tyson GW. Shining a light on dark sequencing: characterising errors in Ion Torrent PGM data. PLoS Comput Biol. 2013;9:e1003031.
40. Damerla RR, Chatterjee B, Li Y, Francis RJB, Fatakia SN, Lo CW. Ion Torrent sequencing for conducting genome-wide scans for mutation mapping analysis. Mamm Genome. 2014;25:120–8.
41. Quail MA, Kozarewa I, Smith F, Scally A, Stephens PJ, Durbin R, et al. A large genome center' s improvements to the Illumina sequencing system. Nat Methods. 2008;5:1005–10.
42. Walsh PS, Erlich HA, Higuchi R. Preferential PCR amplification of alleles: mechanisms and solutions. Genome Res. 1992;1:241–50.
43. Ross MG, Russ C, Costello M, Hollinger A, Lennon NJ, Hegarty R, et al. Characterizing and measuring bias in sequence data. Genome Biol. 2013;14:R51.
44. Chan M, Ji SM, Yeo ZX, Gan L, Yap E, Yap YS, et al. Development of a next-generation sequencing method for BRCA mutation screening: a comparison between a high-throughput and a benchtop platform. J Mol Diagn. 2012;14:602–12.
45. Hartwig B, James GV, Konrad K, Schneeberger K, Turck F. Fast isogenic mapping-by-sequencing of ethyl methanesulfonate-induced mutant bulks. Plant Physiol. 2012;160:591–600.
46. Thudi M, Li Y, Jackson SA, May GD, Varshney RK. Current state-of-art of sequencing technologies for plant genomics research. Brief Funct Genomics. 2012;11:3–11.
47. Aird D, Ross MG, Chen W-S, Danielsson M, Fennell T, Russ C, et al. Analyzing and minimizing PCR amplification bias in Illumina sequencing libraries. Genome Biol. 2011;12:R18.
48. Rothberg JM, Hinz W, Rearick TM, Schultz J, Mileski W, Davey M, et al. An integrated semiconductor device enabling non-optical genome sequencing. Nature. 2011;475:348–52.
49. Glenn TC. Field guide to next-generation DNA sequencers. Mol Ecol Resour. 2011;11:759–69.
50. Doyle JJ, Doyle JL. A rapid DNA isolation procedure for small quantities of fresh leaf tissue. Phytochem Bull. 1987;19:11–5.
51. Untergasser A, Cutcutache I, Koressaar T, Ye J, Faircloth BC, Remm M, et al. Primer3–new capabilities and interfaces. Nucleic Acids Res. 2012;40:e115.

18

Phenoplant: a web resource for the exploration of large chlorophyll fluorescence image datasets

Céline Rousseau[1†], Gilles Hunault[2†], Sylvain Gaillard[3], Julie Bourbeillon[4], Gregory Montiel[5], Philippe Simier[5], Claire Campion[6], Marie-Agnès Jacques[1,3], Etienne Belin[1,7†] and Tristan Boureau[1,6*†]

Abstract

Background: Image analysis is increasingly used in plant phenotyping. Among the various imaging techniques that can be used in plant phenotyping, chlorophyll fluorescence imaging allows imaging of the impact of biotic or abiotic stresses on leaves. Numerous chlorophyll fluorescence parameters may be measured or calculated, but only a few can produce a contrast in a given condition. Therefore, automated procedures that help screening chlorophyll fluorescence image datasets are needed, especially in the perspective of high-throughput plant phenotyping.

Results: We developed an automatic procedure aiming at facilitating the identification of chlorophyll fluorescence parameters impacted on leaves by a stress. First, for each chlorophyll fluorescence parameter, the procedure provides an overview of the data by automatically creating contact sheets of images and/or histograms. Such contact sheets enable a fast comparison of the impact on leaves of various treatments, or of the contrast dynamics during the experiments. Second, based on the global intensity of each chlorophyll fluorescence parameter, the procedure automatically produces radial plots and box plots allowing the user to identify chlorophyll fluorescence parameters that discriminate between treatments. Moreover, basic statistical analysis is automatically generated. Third, for each chlorophyll fluorescence parameter the procedure automatically performs a clustering analysis based on the histograms. This analysis clusters images of plants according to their health status. We applied this procedure to monitor the impact of the inoculation of the root parasitic plant *Phelipanche ramosa* on *Arabidopsis thaliana* ecotypes Col-0 and Ler.

Conclusions: Using this automatic procedure, we identified eight chlorophyll fluorescence parameters discriminating between the two ecotypes of *A. thaliana*, and five impacted by the infection of *Arabidopsis thaliana* by *P. ramosa*. More generally, this procedure may help to identify chlorophyll fluorescence parameters impacted by various types of stresses. We implemented this procedure at http://www.phenoplant.org freely accessible to users of the plant phenotyping community.

Background

In plant science, computer vision applied to the monitoring of plants receives increasing interest. Measurements based on automatic image analysis provide a calibrated analysis, thereby eliminating any subjectivity of the raters and ensuring reproducibility [1-4]. Moreover, image analysis is more accurate than manual annotations [5,6]. Actually, an image is an association of pixels that display various intensities and create the colors. To each grayscale image corresponds a unique histogram featuring the number of pixels for each level of gray. The histograms allow a pixel-by-pixel analysis. Image analysis is used for characterizing the architecture of plants as well as roots or the venation of leaves [7-9], their tolerance to heavy metals [10], the cold tolerance [11] or the impact of a pathogen [5,6].

Moreover, the automation of image analysis and eventually of image acquisition allows high-throughput phenotyping [1]. Several software or applications devoted to image analysis have been developed to answer specific questions such as the measuring of the area of leaves, plants and grains [12-14], the impact of pests and abiotic stresses [5,12,13] and the callose deposition [15]. In practice, such methods have to be user-friendly and automated to match the needs of

* Correspondence: tristan.boureau@univ-angers.fr
†Equal contributors
[1]PHENOTIC, SFR 4207 QUASAV, F-49045 Angers, France
[6]Institut de Recherche en Horticulture et Semences, UMR1345, Université d'Angers, SFR 4207 QUASAV, F-49045 Angers, France
Full list of author information is available at the end of the article

the community of biologists, especially when large datasets are treated.

In plant science, imaging can be achieved both by visible imaging (e.g. photographs or scans) and by other types of imaging such as chlorophyll fluorescence, thermographic or hyperspectral imaging. Chlorophyll fluorescence (CF) imaging has been used to study the impact of both biotic and abiotic stresses on photosynthesis and hence on plant physiology [5,11,16-25]. Moreover, CF imaging was reported to allow monitoring phenotypes that are not visible to the human eyes [5,11,22,26,27]. Indeed, CF transients and CF parameters may be measured on plants in dark- and light-adapted states [28-30]. Some CF parameters display a robust contrast between healthy and unhealthy tissues while others seem not to be impacted by stresses [18]. For example, the maximum quantum yield of PSII photochemistry (F_v/F_m), measured from dark-adapted leaves, can be used to quantify the severity of symptoms induced on bean by *Xanthomonas fuscans* subsp. *fuscans* [5] or to monitor the *Arabidopsis* health status in various drought stress conditions [31]. Some CF parameters from light-adapted leaves, such as the effective quantum efficiency of PSII (Φ_{PSII}) and the non-photochemical quenching (NPQ), may also be good discriminating parameters between drought-resistant and drought-sensitive tomatoes [32]. Also, both F_v/F_m from dark-adapted leaves and the fluorescence decrease ratio from illuminated leaves (R_{FD}), that were related to photosystem integrity, can be used to discriminate cold tolerant from cold sensitive accessions of *A. thaliana* [23].

In the perspective of phenotyping the plant response to various stresses, a recurring question is to identify which CF parameters are impacted in response to stresses, thus providing a contrast that can be used to monitor the plant response to stresses. Such contrasted CF parameters may also allow the quantification of the plant resistance to stresses and therefore is of high interest for breeders [5]. Most studies only focused on the few CF parameters with known biological significance. Many other CF parameters can be measured or calculated even though their biological significance may remain obscure ([28,30]; Table 1). However, these could be useful in a phenotyping perspective using image analysis.

Among the numerous pests impacting photosynthesis of the affected plants, the parasitic plants are atypical examples [33-35]. Some of them, including the broomrapes (*Orobanche* spp and *Phelipanche* spp), are harmful parasitic weeds causing devastating yield reductions in many important crops throughout the world. *Phelipanche ramosa* is by far the most widespread broomrape species [36]. *P. ramosa* depends on its host for all its nutrients and water supply, acquired from the host phloem and its life cycle has been well described in regard to its major host plants [37]. After the induction of seed germination by molecules exuded in the rhizosphere by host roots, a radicle emerges from the seed and attaches to the host root surface. The parasitic phase starts with the penetration of the parasite radicle into the host root through a differentiating haustorium, which connects to the host vascular tissues and serves as an attaching organ and as a bridge for water and nutrient transfer from the host. The parasite develops a tubercle, which gives rise to a subterranean shoot and then, after emergence from the soil, a branched flowering spike.

Success in stopping increased infestations needs a reliable strategy of integrated management [38,39], where breeding resistant crops should be one of the key elements. In this context and since host derived resistance remains rare, large scale screenings of genetic resources are of main importance for crop breeders. Classical descriptors for host resistance against broomrapes today mainly consist in the assessment of the number of emerged flowering spikes during field trials [40,41], and in the assessment of the number of *P. ramosa* attachments to host roots as well as of the kinetics of infection in pots or mini-rhizotrons experiments [42,43]. The measure of these classical descriptors remains time consuming and unsuitable for large scale and high-throughput screenings.

Therefore, the development of high-throughput phenotyping system based on images is of major interest as it would be a dramatic improvement in the screening process. Root infection by *P. ramosa* has a systemic impact that could probably be observed on leaves. The identification of such a foliar phenotype may greatly facilitate phenotyping of the response of *A. thaliana* to *P. ramosa*. Thus, in this paper, we aimed at identifying CF parameters impacted during the infection of *A. thaliana* by *P. ramosa*. We measured and calculated various CF parameters (Table 1) on the two ecotypes Columbia-0 (Col-0) and Landsberg (Ler) of *A. thaliana* non-inoculated vs inoculated with *P. ramosa*. For each plant, 55 images and their associated histograms were recorded. We developed a script in the R software aiming at facilitating the identification of CF parameters impacted during the infection. We implemented this R script as a web application at http://www.phenoplant.org. The R script described in the present paper was associated to a script previously described [5] in order to provide easy-to-use web resources for the analysis of CF imaging datasets.

Results

To describe the impact of *P. ramosa* on the photosynthetic performance of *A. thaliana*, CF parameters were compared: inoculated vs. non-inoculated plants of ecotypes Col-0 and Ler. CF parameters were measured or calculated at

Table 1 CF parameters used in this study

Symbol	Name§	Formula	Col-0 control vs. Ler control	Col-0 inoculated vs. Col-0 control	Ler inoculated vs. Ler control
F_O	Minimal chlorophyll fluorescence intensity measured in the dark-adapted state measured in the dark-adapted state (F_O) and during the dark relaxation ($FO_{(85)}$)	Measured			
F'_O	Minimal chlorophyll fluorescence intensity measured in the light-adapted state measured during the light adaptation ($F'_{O(n)}$) and at the steady-state ($F'_{O(74)}$)	Measured			
F_m	Maximal chlorophyll fluorescence intensity measured in the dark-adapted state measured in the dark-adapted state (F_m) and during the dark adaptation ($Fm_{(85)}$)	Measured	$Fm_{(85)}^{*}$		
F'_m	Maximum chlorophyll fluorescence intensity measured in the light adapted state measured during the ligth-adaptation ($F'_{m(n)}$) and at the steady-state ($F'_{m(74)}$)	Measured	$F'_{m(25)}{}^{**}$, $F'_{m(38)}{}^{**}$, $F'_{m(50)}{}^{*}$, $F'_{m(62)}{}^{*}$, $F'_{m(74)}{}^{*}$		
F_P	Peak fluorescence during the initial phase of the Kautsky effect measured at the begining of the light adaptation	Measured			
F_T	Instantanueous fluorescence measured in the dark-adapted state ($F_{T(85)}$), during the light adaptation ($F_{T(n)}$) and at the steady-state ($F_{T(74)}$)	Measured		$F_{T(25)}{}^{*}$, $F_{T(50)}{}^{*}$, $F_{T(62)}{}^{**}$, $F_{T(74)}{}^{*}$	
F_v/F_m	Maximum PSII quantum yield measured at the dark-adaptated state	$(F_m\text{-}F_O)/Fm$	$F_v/F_m{}^{**}$	$F_v/F_m{}^{*}$	
F'_v/F'_m	PSII quantum yield of light adapted sample calculated during the light adaptation ($F'_v/F'_{m(n)}$) and at the steady-state ($F'_v/F'_{m(74)}$)	$(F'_m\text{-}F'_O)/F'_m$	$F'_v/F'_{m(25)}{}^{**}$, $F'_v/F'_{m(38)}{}^{**}$, $F'_v/F'_{m(50)}{}^{**}$, $F'_v/F'_{m(62)}{}^{**}$, $F'_v/F'_{m(74)}{}^{**}$		
NPQ	Non-photochemical quenching calculated in the dark-adapted state ($NPQ_{(85)}$during the light adaptation ($NPQ_{(n)}$) and at the steady-state ($NPQ_{(74)}$)	$(F_m\text{-}F'_m)/F_m$	$NPQ_{(38)}{}^{*}$	$NPQ_{(25)}{}^{*}$	$NPQ_{(85)}{}^{*}$, $NPQ_{(62)}{}^{**}$, $NPQ_{(74)}{}^{*}$
qL	estimator of the fraction of open PSII centers calculated during the light adaptation ($qL_{(n)}$) and at the steady-state ($qL_{(74)}$)	$qP(F'_O/F_T)$	$qL_{(25)}{}^{**}$, $qL_{(38)}{}^{**}$, $qL_{(50)}{}^{**}$, $qL_{(62)}{}^{**}$, $qL_{(74)}{}^{**}$		
qP	Coefficient of photochemical quenching calculated in the dark-adapted state ($qP_{(85)}$), during the light adaptation (qP(n)) and at the steady-state ($qP_{(74)}$)	$(F_m\text{-}F_T)/(F_m\text{-}F_O)$	$qP_{(85)}{}^{**}$, $qP_{(25)}{}^{**}$, $qP_{(38)}{}^{**}$, $qP_{(50)}{}^{**}$,$qP_{(62)}{}^{*}$	$qP_{(25)}{}^{**}$, $qP_{(38)}{}^{**}$, $qP_{(50)}{}^{**}$,$qP_{(62)}{}^{**}$, $qP_{(74)}{}^{**}$	$qP_{(74)}{}^{*}$
Q_y	Instantaneous PSII quantum yield calculated during the light adaptation ($Qy_{(n)}$) and at the steady-state ($Qy_{(74)}$)	$(F_m\text{-}F_t)/F_m$	$Qy_{(85)}{}^{**}$, $Qy_{(25)}{}^{*}$,$Qy_{(38)}{}^{**}$, $Qy_{(50)}{}^{*}$, $Qy_{(62)}{}^{*}$, $Qy_{(74)}{}^{*}$	$Q_{y(25)}{}^{**}$,$Q_{y(38)}{}^{**}$, $Q_{y(50)}{}^{**}$, $Q_{y(62)}{}^{**}$, $Q_{y(74)}{}^{**}$	
R_{FD}	Fluorescence decline ratio calculated during the light adaptation ($R_{FD(n)}$) and at the steady-state ($R_{FD(74)}$)	$(F_p\text{-}F_T)/F_T$	$R_{FD(25)}{}^{**}$, $R_{FD(38)}{}^{**}$, $R_{FD(50)}{}^{**}$, $R_{FD(62)}{}^{**}$, $R_{FD(74)}{}^{*}$		$R_{FD(50)}{}^{**}$, $R_{FD(62)}{}^{**}$, $R_{FD(74)}{}^{**}$

Plants of ecotypes Ler and Col-0 of *A. thaliana* non-inoculated or inoculated with *P. ramosa* were imaged for 13 CF parameters (55 transient measures). This table presents the symbols and the names of the measured and the calculated CF parameters (§: n can be 25, 38, 50 or 62. For example, F_o was measured four times during the light adaptation at 25 s, 38 s, 50 s and 62 s).

Significant differences between the global CF values of non-inoculated and inoculated plants of ecotype Col-0, non-inoculated and inoculated plants of ecotype Ler or non-inoculated plants of ecotype Col-0 and non-inoculated plants of ecotype Ler are indicated respectively in the fourth, fifth and sixth columns (*: Mann–Whitney *U* test, p-value < 0.05, **: Mann–Whitney *U* test, p-value < 0.01).

the dark-adapted state, during the light-adaptation, at the light-adapted steady-state and during the dark-adaptation (Table 1). Finally, every measuring day, 55 images were recorded for each plant. Given our experimental design, our dataset reached 9900 image files and 9900 text files containing the histograms associated to each image. Such a dataset is too big to be manually analyzed. Therefore, we aimed at developing a script that first, automatically sorts out the data to provide an easy overview of the data; second, facilitates the identification of CF parameters significantly impacted during infection of *A. thaliana* by *P. ramosa* (Figure 1); and third, is available as an interface to the plant phenotyping community.

Overview of the dataset by creating contact sheets of images and associated histograms

The aim of such an overview of the data is to make the visual inspection of all the images and associated histograms easier. Therefore, the script sorts out the images and/or the histograms on three types of contact sheets (Figure 1A).

First, to control the quality of images, each image faces its associated histogram. One contact sheet per plant and per

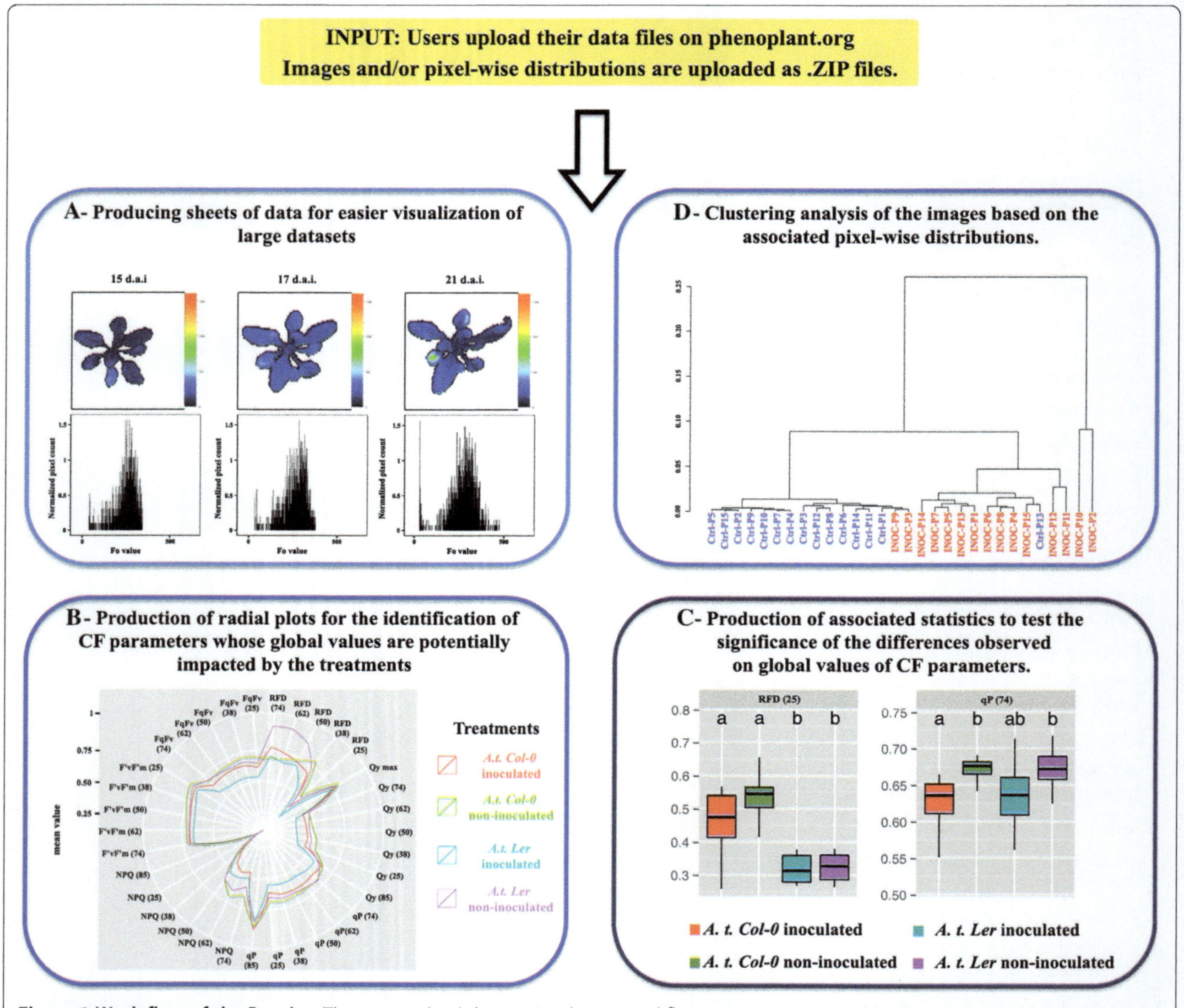

Figure 1 Work flow of the R script. The users upload the previously captured fluorescence images and/or their associated histograms as input. The data are sorted out according to the information indicated by the user on the web interface. To get an overview of the data, the data are represented in the form of contact sheets of images or of histograms **(A)**. Here, *A. thaliana* plants were imaged 15, 17 and 21 days after inoculation (d.a.i.) by *P. ramosa*. Images and histograms are displayed on the contact sheet to visualize a variation during time. The mean of the global CF values for each treatment are graphically represented in a radial plot where each point represents the mean for one treatment and each radius represents a CF parameter **(B)**. In order to detect a significant difference in the global CF values between the treatments, the means of the global CF values for each treatment are represented in the form of a box plot **(C)** in which horizontal lines indicate the 0.25 and 0.75 percentile from bottom to top, the interior line indicates the median and the exterior shapes represent the overall distribution. The results of the Mann–Whitney U tests are indicated in box plots. To compare the various histograms, a dendrogram based on the histograms is produced **(D)**.

CF parameter is produced. Second, contact sheets are produced that enable the comparison of the images (or histograms) of plants having undergone different treatments. Thus, on these contact sheets, all the images (or histograms) from one treatment face those from another treatment. These sheets enable the rapid visual identification of CF parameter(s) potentially impacted by a treatment. For example, the rapid visual inspection of such contact sheets suggests that the coefficient of photochemical quenching during light adaptation (qP) is potentially impacted by the infection by *P. ramosa*. Indeed, *A. thaliana* plants ecotype Col-0 inoculated with *P. ramosa* display values of qP lower than non-inoculated plants (Additional file 1).

Finally, contact sheets are produced that enable the visual inspection of the potential intensity dynamics of CF parameters over time. These contact sheets display the kinetics of images (or histograms) for one transient measure on the same plant over the duration of the monitoring. Such contact sheets allow the identification of CF parameters declining over time due to the inoculation with *P. ramosa*. For example, qP declined for plants of ecotype Col-0 inoculated with *P. ramosa*, contrary to non-inoculated plants (Additional file 2).

Detection of CF parameters impacted by a treatment using global values

Global fluorescence values, i.e. the mean of the values of all the pixels in the image, are well suited for a rapid comparison of the impact of *P. ramosa* on the CF parameters. First, we plotted the global fluorescence values of all the CF transient measures for each treatment on a same radial plot (Figure 1B). Such a plot provides a synthetic view of the comparisons between treatments for all transient measures. The observation of the radial plots highlighted the variation of various transient measures, apparently due to the infection of *A. thaliana* ecotypes Col-0 and Ler by *P. ramosa*. Interestingly, several transient measures may also highlight differences between non-inoculated plants of Col-0 and Ler ecotypes.

In order to test whether these variations are significant, a detailed view of the comparisons between treatments is provided for each transient measure in the form of box plots (Figure 1C). The box plots display more information than the radial plots as minimum and maximum values observed, lower and upper quartiles and percentiles are represented for each transient measure. Moreover, the significance of the differences observed is tested using the Mann–Whitney *U* test [44]. Such a detailed view shows that several CF parameters vary significantly between the various conditions studied.

First, we compared non-inoculated plants from Col-0 and Ler ecotypes and identified CF parameters that could discriminate between these ecotypes at 21 days after inoculation (d.a.i.). Eight CF parameters presented global fluorescence values significantly different among non-inoculated plants of the ecotypes Col-0 and Ler (Mann–Whitney *U* test, p-value < 0.05; Table 1): the maximal chlorophyll fluorescence intensity in light (F'_m) and in dark-adaptation ($F_{m(85)}$), the maximum PSII quantum yield (F_v/F_m), the PSII quantum yield in light (F'_v/F'_m), the instantaneous non-photochemical quenching during the light-adaptation (NPQ), the estimator of the fraction of open PSII centers (qL), the coefficient of photochemical quenching (qP), the instantaneous PSII quantum yield (Q_y) and the instantaneous fluorescence decline ratio in light (R_{FD}).

Second, five CF parameters altered by the infection by *P. ramosa* were identified for the two ecotypes at 21 d.a.i. (Mann–Whitney *U* test, p-value < 0.05; Table 1). For ecotype Ler, NPQ in light and during dark-adaptation, qP in steady-state and R_{FD} in light were significantly lower among plants inoculated with *P. ramosa* and non-inoculated plants. For ecotype Col-0, the instantaneous fluorescence (F_T), NPQ in light, qP and Q_y in light are significantly negatively impacted by the infection.

Discriminating between treatments using a clustering analysis based on the histograms

Results based on the global values provided good candidate CF parameters meant to be impacted by the infection by *P. ramosa*. However, images contain far more information compared to the global values. Therefore, for each CF transient measures, we sorted out all the images by performing a clustering approach on histograms. For this, we calculated distances between images using the Bhattacharyya coefficient [45]. Then, we used six available agglomeration methods to construct dendrograms (Figure 1D): single linkage, complete linkage, UPGMA, WPGMA, WPGMC and UPGMC.

In most cases, the dendrograms obtained with single linkage, WPGMC and UPGMC agglomeration methods did not cluster on a same branch either images of plants having undergone the same treatments, or plants sharing a similar apparent health status (Additional file 3). Alternatively, using the WPGMA, complete linkage and UPGMA agglomeration methods, images clustered on a same branch in the dendrograms corresponded to plants sharing a same treatment or a same apparent health status. Some images however were mis-clustered regarding either the treatments or the health status of plants. Rates of mis-clustered images are reported in Additional file 3.

Among the candidate CF parameters identified for plants of ecotype Col-0 using global values, qP measured during the light-adaptation ($qP_{(25)}$, $qP_{(38)}$, $qP_{(62)}$ and $qP_{(74)}$) and Q_y measured in the beginning of the light-adaptation ($Q_{y(25)}$ and $Q_{y(38)}$) allowed clustering of the images according to the inoculation status of plants. Using the agglomeration WPGMA, complete linkage or

UPGMA agglomeration methods for images obtained with these CF transient measures, we observed an average of 20% of mis-clustering when considering the plant treatment (Additional file 3). However, when considering the apparent health status of plants as defined by visual observation of plants by experts, less than 10% of the images appeared mis-clustered on the dendrograms (Additional file 3). For example, for the CF parameter $qP_{(25)}$, two major clusters can be observed on the dendrogram (Figure 2). Images of all non-inoculated plants except one clustered with images of two inoculated plants to form a first major cluster. The inoculated plants corresponding to the two mis-clustered images in this first cluster appeared in much better health than the other inoculated plants (Figure 2). Infection level was checked for these plants. Macroscopic observations confirmed that inoculation failed, thus explaining their position in the dendrogram. The second major cluster is composed by inoculated plants and one non-inoculated plant. The growth of the mis-clustered non-inoculated plant was strongly altered, suggesting that this plant had undergone a heavy stress (Figure 2). For plants of ecotype Ler, results obtained with the clustering approach are less contrasted for any of the candidate CF parameter.

P. ramosa attachments to host roots does not result in clearly delimited eye-visible symptoms on *A. thaliana* leaves. Therefore, in order to further check the correct sorting out of the CF images by the clustering procedure, we used a set of images of bean leaves inoculated

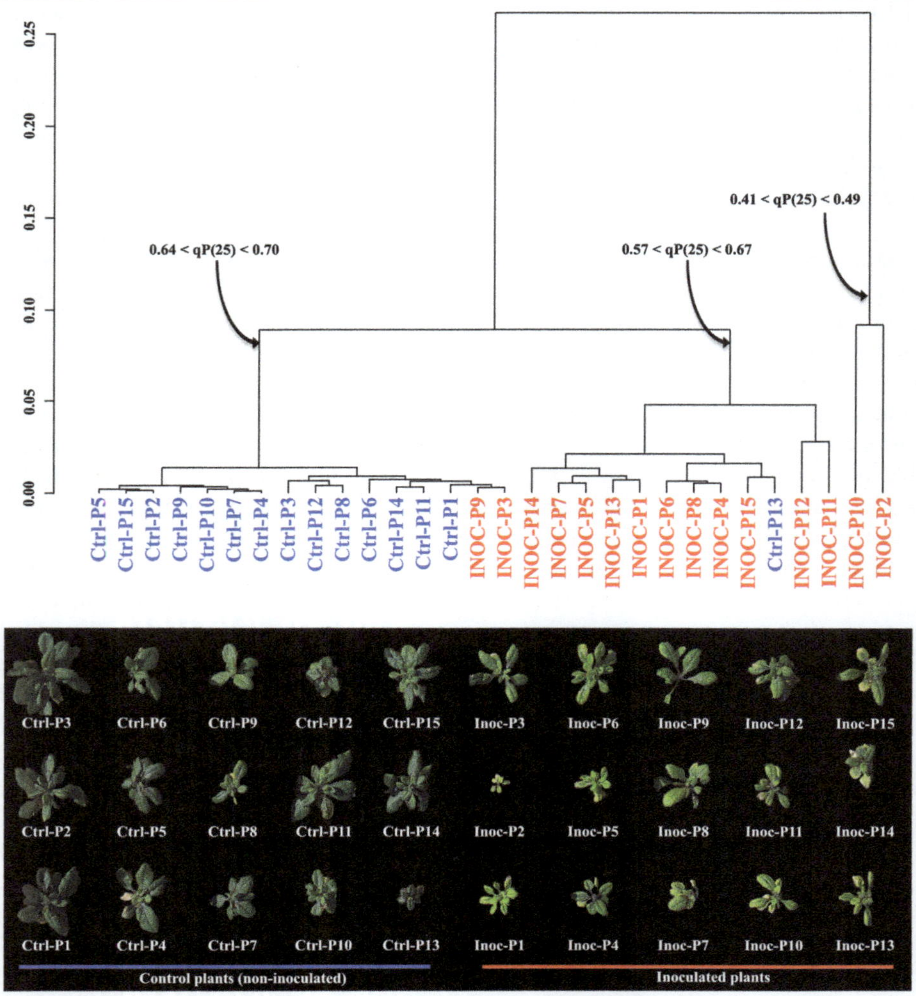

Figure 2 Clustering analysis of plants of ecotype Col-0 non-inoculated or inoculated with *P. ramosa*. Fifteen plants of *A. thaliana* of ecotype Col-0 were inoculated with *P. ramosa* and 15 plants were not inoculated. The plants were imaged for 55 fluorescence parameters. Dendrograms were built based on the histograms of the CF parameter $qP_{(25)}$ using the WPGMA agglomeration method. Inoculated plants (red) and non-inoculated plants (blue) globally belong to different branches of the dendrogram. However, inoculated plants INOC-P9 and INOC-P3 group with non-inoculated plants. These plants are visually healthier than other inoculated plants, as seen on photo at the bottom of the figure. The non-inoculated plant Ctrl-P13 groups with inoculated plants. This plant is visually less healthy than other non-inoculated plants.

with the plant pathogenic bacterium *Xanthomonas fuscans* subsp. *fuscans*, previously described [5]. In contrast to *A. thaliana* infected by *P. ramosa*, leaves in these pictures displayed clearly recognizable symptoms of common bacterial blight. On bean leaves, lesions are clearly delimited and visible to the eye, which enables to control that the CF parameter displays contrasts between impacted and healthy plant tissues. Most healthy control plants clustered together, whereas most inoculated plants formed a separate cluster (Additional file 4). Using this dataset as well, we could explain all the apparent mis-clustering events: inoculated leaves that clustered with non-inoculated ones corresponded to leaves for which inoculation failed. Moreover, non-inoculated leaves that clustered with inoculated ones corresponded to leaves that displayed obvious stresses or growth defects (Additional file 5). The second advantage using a bean image dataset is that we can quantify the amount of symptoms. We previously showed that the CF parameter F_v/F_m could actually be reliable to quantify the amounts of symptoms on bean leaves [5]. Interestingly, in this dataset, the obtained dendrogram clustered images according to their respective amounts of symptoms (Additional file 5).

A web application available at http://www.phenoplant.org

In order to make it available to the plant phenotyping community, the procedure described in the present paper was implemented in a web resource available at http://www.phenoplant.org. The web resource gathers two scripts aiming at making the analysis of fluorescence images easier. We also implemented a second script previously described aiming at the quantitative phenotyping of plant resistance to a biotic and abiotic stresses [5]. The source files are available for download at http://www.phenoplant.org.

Users may upload their datasets as zip files on the web site, and may select their preferences among the proposed options for the analysis. It was designed for images of individual leaves or plants and it accommodates images in TIFF format and histograms in text format. Once the analysis is done, users receive an e-mail indicating where to retrieve the results of the analysis. Results are returned in the portable form of PDF and Excel files.

Discussion

We propose a web resource devoted to the image analysis of large CF image datasets. This resource was designed for CF image datasets but could probably be adapted also to other types of image dataset, if relevant. This resource requires grayscale images in TIFF format and their associated histograms. Datasets are uploaded at http://www.phenoplant.org as zip files. Users just have to follow a specific nomenclature for image files or folders that is explained in the user's guidelines.

This application contains two types of image analysis procedures that answer two different questions. A previously described script aims at the quantitative phenotyping of plant resistance to biotic and abiotic stresses [5]. Quantitative phenotyping is important in plant breeding in order, for example to screen accessions for resistance to pathogens as plant resistance to pathogens may either be qualitative or quantitative. Thus, when a given CF parameter can be used as a descriptor of plant stress, the quantitative procedure implemented in this script enables the evaluation of the severity of the stress.

In the present paper, a new script is described to provide three complementary approaches to get an overview of large image datasets, and to identify CF parameters that can be used as descriptors of plant stresses. Several other algorithms are described in the literature to perform specific image analyses such as the measuring of the area of leaves, plants or grains [12-14], the impact of pests and abiotic stresses [5,12,13] or the callose deposition [15]. Free web-based applications such as PhenoPhyte [12] are available to the plant phenotyping community to analyze plant images obtained by conventional imaging techniques. However, to our knowledge, no software is providing an overview of large image datasets. Given the size expected for high-throughput phenotyping datasets, even simple procedures are problematic if not automated. In our view, a simple algorithm that sorts out the images to provide relevant contact sheets will help the user to get acquainted with his data. Therefore, the first output by the script described in the present paper produces contact sheets of images or histograms meant to visually compare plants having undergone different treatments.

The second output of the script aims at the identification of transient measures that significantly discriminate between two groups of plants (i.e. Col-0 vs. Ler plants, inoculated vs. non-inoculated plants, healthy vs. stressed plants). Most CF parameter measures could significantly discriminate between ecotypes Col-0 and Ler (Table 1). Therefore, CF imaging can thus be used to discriminate between ecotypes of *A. thaliana*. Such a discrimination between plant genotypes was already proposed in a previous study on the use of CF imaging for the discrimination between plant species from the genus *Origanum* [46].

We then compared the global CF values obtained for plants having undergone different treatments. Many studies reported the interest of using CF imaging to monitor the impact of biotic and non biotic stresses on plant tissues [18,21]. In the case of the infection of *A. thaliana* by *P. ramosa*, we identified five CF parameters as potential markers of *A. thaliana* stress by *P. ramosa*. Interestingly, Col-0 and Ler ecotypes seemed to respond differently to *P. ramosa*, as CF parameters impacted by the infection differ from one ecotype to another (Table 1).

To our knowledge, there is no study about such a differential response of *A. thaliana* to a biotic stress. However, this shows that for each plant or each ecotype, an initial analysis aiming at identifying a good CF parameter candidate is necessary prior to quantifying the impact of infection on the plant physiology. Future studies should involve the screening of a large number of ecotypes, in order to select relevant CF parameters to use for screening for resistance against *P. ramosa*. Relevant CF parameters should be impacted on most sensitive ecotypes and such an impact should depict the severity of the disease by *P. ramosa*. Moreover, for each crop of interest (e.g. sunflower or rape), similar screenings should be performed on many varieties to identify what relevant parameters to use for phenotyping. According to the results of the present study, the CF parameter qP of light-adapted leaves may constitute a good candidate. Interestingly in some cases, variations of CF parameters may be recorded prior to the occurrence of eye-visible symptoms [5,26,27], thereby greatly speeding up the process for the selection of resistant crop varieties.

The diversity of CF parameters impacted among various ecotypes of *A. thaliana* during the infection of *A. thaliana* by *P. ramosa* may represent various types of response to the infection. Actually, after the infection by *P. ramosa*, the ecotype Col-0 of *A. thaliana* displayed clearly visible symptoms of growth defect, chlorosis and in some cases marginal necrosis. For the ecotype Ler, no clear visible symptoms were observed in our experiments, despite the fact that attachment to roots and infection were successful (data not shown). Although we do not have comprehensive data on the behavior of ecotype Ler of *A. thaliana*, such observations suggest that this ecotype may be tolerant to the infection by *P. ramosa*. Future studies should help establish whether the different CF parameters impacted in the Col-0 or Ler situations may be used to screen either for sensitivity or tolerance to *P. ramosa*. Nonetheless, it is interesting to note that among the CF parameters impacted by the infection, a significant decrease of R_{FD} occurred in the ecotype Ler that is potentially tolerant, but not in ecotype Col-0 that is sensitive to *P. ramosa*. In response to cold treatment, on the contrary, cold tolerant accessions such as Col-0 displayed a smaller decrease of R_{FD} compared to sensitive or intermediate accessions such as Ler [23]. The identification of CF parameters impacted by the infection may provide interesting hypotheses for studying the mechanisms of interaction between *A. thaliana* and *P. ramosa*. A decrease of qP reveals a decrease in the proportion of open reaction centers, *i.e.* in the proportion of reaction centers that are able to accept further electrons [30]. So with Col-0, the observed decrease of qP following the infection by *P. ramosa* shows that the reoxydation of quinones during photosynthesis is altered. As R_{FD} is directly correlated to the net CO_2 assimilation rate [47], a decrease in R_{FD} may reveal a decrease in CO_2 assimilation in response to the infection by *P. ramosa*.

In a third approach, we intended to better take into account the information contained in the images. For this purpose, the entire histograms of each image were compared. Indeed, shapes of histograms differ on images of stressed plants compared to healthy plants [5,16]. Here, we proposed to discriminate between treatments following the assumption that the distance between the histograms of plants under different treatments is higher than between plants under the same treatment. A similar clustering approach was recently described for the automated sorting of fluorescence microscopy images [48]. Using this procedure, we could retrieve images of plants giving false negative or false positive results, i.e. inoculated plants for which inoculation failed or control plants that were stressed in an uncontrolled manner, respectively. Experiments conducted on the control dataset of bean images suggest that such a clustering approach may allow a grouping of images displaying similar stress intensity. In the present paper, such a classification procedure was based on images obtained from single measures of CF parameter. The use of such a clustering approach on images that would have been reconstructed through a combinatorial imaging approach [11,23,32,49] may reveal a very powerful tool for an automated classification of plants according to their health status.

Conclusions

In this paper, we describe a web resource available at http://www.phenoplant.org, devoted to the image analysis of large CF image datasets. We propose two types of image analysis. A first analysis aims at the quantitative phenotyping of plant resistance to a stress, as described in [5]. A second analysis, described in this paper, aims at providing an overview of large image datasets and at the identification of CF parameters impacted by a stress. On the one hand, CF parameters of interest are first identified based on a comparison of their global values (i.e. the mean value of the parameter over the whole imaged plant). On the other hand, CF parameters of interest may be identified using a clustering approach based on pairwise distances between images. In the latter, each pixel-value is taken into account, in an attempt to exploit maximum information in each image. Such an analysis allows both the identification of CF parameters impacted by a stress and the detection of unhealthy plants or leaves.

Methods

Biological material

P. ramosa (L.) Pomel seeds were collected from mature flowering spikes on boomrape parasitized oilseed rape field (*Brassica napus*) in Saint-Jean-d'Angély (France) in 2012, and stored at 25°C in darkness until use. Seeds were surface-sterilized for 5 min in sodium hypochlorite (12%), and thoroughly rinsed three times for 1 min and three times for 5 min with sterile distilled water. Seeds were then suspended in conditioning medium containing 1.10^{-3} M Na/K phosphate buffer (pH 7.5, adjusted with KOH) and PPM 0,1% (Plant Preservative Mixture, Kalys, Bernin, France), with a ratio of 10 mg seeds.mL^{-1}. Seeds were then placed in the dark at 21°C for 7 days for conditioning. The conditioned seeds were stimulated by adding the synthetic strigolactone GR24 (germination stimulant) at a final concentration of 10^{-9} M in 0.1% v/v acetone. GR24 treatments were carried out at 21°C in the dark for 48 hours and subsequently used for *A. thaliana* root inoculation.

For co-cultivation experiments, *A. thaliana* seeds (Columbia-0 (Col-0) and Landsberg (Ler) ecotypes) were surface sterilized in a closed container with chlorine gas for 3 h (http://www.plantpath.wisc.edu/fac/afb/vapster.html). Surface-sterilized seeds were transferred to 9.4 cm Ø plates containing MS medium supplemented with 0.6% w/v agar. Following stratification for 3 days at 4°C, seeds were incubated at 21°C in a growth chamber (16 h light, 8 h dark) for 15 days. Seedlings were then transferred to 15 cm Ø plates containing MS medium supplemented with 0.6% w/v agar. Plates were incubated vertically at 21°C in a growth chamber (16 h light, 8 h dark) for 7 days. Plantlets (3 per plate) were then transferred onto filter papers each covered or not with 20 mg of GR24 treated *P. ramosa* seeds placed in cut 12 × 12 cm square plates containing a uniform layer of rockwool moisturized with 50 mL of 0.5× Tadano and Tanaka growth medium Tadano and Tanaka, 1980. Five plates per condition (inoculated or not) were used giving a total of 15 non-inoculated plants and 15 plants inoculated with *P. ramosa*. Plates were incubated vertically at 21°C in a growth chamber (16 h light, 8 h dark, 70% humidity) for 15 days and watered every 2 days with 10 mL TT medium.

A second set was used to test the relevance of the method described and was composed by 40 plants of bean (*Phaseolus vulgaris*, cv. Flavert). Twenty plants were mock-inoculated and twenty plants were inoculated with *Xanthomonas fuscans* subsp. *fuscans* strain CFBP4834-R. Technical details are given in [5].

Technical setup and image acquisition

The PSI Open FluorCam FC 800-O (PSI, Brno, Czech Republic) was used to image plants of ecotypes Col-0 and Ler of *A. thaliana*. Both non-inoculated plants and plants inoculated with *P. ramosa* were imaged. The plants were imaged 15, 17 and 21 days after inoculation (d.a.i.). The system sensor is a CCD camera with a pixel resolution of 512 by 512 and a 12-bit dynamic. The system includes four LED panels divided in two pairs. One pair provides actinic light in orange wavelength of around 618 mm, with an intensity that can vary from 200 to 400 μmol/m^2/s. The other pair provides a saturating pulse during 1 s in blue wavelength, typically 455 mm, with an intensity of up to 3000 μmol/m^2/s.

Plants were dark adapted for 45 minutes before taking image series. At the beginning of the light protocol, the minimum fluorescence (F_O) was measured. A saturating pulse allowed the measuring of the maximum fluorescence (F_m) and was followed by a dark relaxation period of 10 s. Then, an actinic light (230 μmol/m^2/s) was applied during 60 s. At the beginning of the application of the actinic light (at 16 s), the fluorescence peak during the initial phase of the Kautsky effect (F_p) was measured. At 25, 38, 50, 62 and 74 s, during the application of the actinic light, the instantaneous fluorescence (F_T) was measured, a saturating pulse was applied to measure the maximum fluorescence at light (F'_m) and a short exposure to far-red irradiance allows the measuring of the minimum fluorescence at light (F'_O) during the light adaptation and at the steady-state (74 s). The actinic light period was followed by a dark relaxation period of 20 s. At 85 s, F_O, F_m and F_T were measured in a dark relaxation state.

Some images come from the equation of measured parameters [50-54]: the PSII quantum yield of light-adapted plants in light or in steady state (F'_v/F'_m), the instantaneous non-photochemical quenching in light during dark relaxation or light adaptation or at the steady-state (NPQ), the coefficient of photochemical quenching of variable fluorescence based on the lake model of PSII in light or in steady-state (qL), the coefficient of photochemical quenching of variable fluorescence based on the puddle model of PSII during dark relaxation, light adaptation or at steady state (qP), the instantaneous PSII quantum yield during dark relaxation, light adaptation and in the steady state (Q_y), the maximum PSII quantum yield (Q_{y_max}), the instantaneous fluorescence decline ratio in light and at the steady-state (R_{FD}).

Thus, for each plant, 55 images and their associated histogram were produced during the protocole, giving 9900 images and 9900 histograms (2 ecotypes of *A. thaliana* × 15 plants × 2 treatments × 3 days × 55 CF transient measures). A figure expliciting the various CF parameters used in this study is available in the FluorCam7 user manual, page 9 (downloadable at http://www.psi.cz/downloads/).

Creation of contact sheets of images and associated histograms

The application draws contact sheets of the images and the histograms. First, the images and histograms are sorted according to the treatment made on the plants. Then, the images are read and displayed using the R package EBImage [55]. The histograms are displayed using the R package ggplot2 [45]. One contact sheet per parameter, per treatment and per day of measurement is returned.

Comparison of the global values of the CF parameters

For each image, the mean of the values of all the pixels is calculated to obtain the global value of the image. Then, for each treatment the mean of the global values is represented on radial plots using the R package ggplot2 [45]. Four radial plots are drawn separately to display parameters whose mean is less than 1, between 1 and 25, between 25 and 300, and superior to 300. Box plots describing the minimum and maximum values observed, the lower and upper quartile and the percentile are drawn using the R package ggplot2 [45]. A Mann–Whitney *U* test [44] is performed using the wilcox.test function implemented in R to statistically compare the global CF values observed for each treatment.

Clustering based on the histograms

The script builds dendrograms in which each leaf represents a histogram. First, the distances between all the pairs of histograms were calculated from the Bhattacharyya coefficient [45] using the R package StatMatch [56]. Second, a clustering step is performed according to six agglomeration methods, *i.e.* single linkage ("single"), complete linkage ("complete"), UPGMA ("average"), WPGMA ("mcquitty"), WPGMC ("median") and UPGMC ("centroid") using the function hclust of the R package stats [57]. Finally, a dendrogram is calculated based on the clustering step, using the R package stats [57].

The relevance of the dendrograms produced through the various agglomeration methods was tested in relation to either the treatment of the plants (i.e. inoculated vs. non-inoculated), or according to their apparent health status. The apparent health had previously been estimated by experts after visual inspection of color photographs for each plant. For both treatment and apparent health, the rate of mis-clustered images was determined for each dendrogram obtained, using the various agglomeration methods.

Technical limitations of the www.phenoplant.org web service are mentioned on the website, and may improve overtime. At the present date, data upload is limited to 2048 Mo (2 Go). The script is implemented on a 16Go RAM server.

The dataset used in this paper gives a zip file of 114 Mo and the complete analysis run during 10 hours.

The source files for the R scripts are published on the web site under CeCILL FREE SOFTWARE LICENSE AGREEMENT, and may be downloaded at the following URL: http://www.phenoplant.org/#tdm8.

Additional files

Additional file 1: Comparison of images of plants of *A. thaliana* ecotype Col-0 non-inoculated or inoculated with *P. ramosa*. Plants of ecotypes Col-0 of *A. thaliana* non-inoculated or inoculated with *P. ramosa* were imaged for $qP_{(25)}$ 21 days after inoculation (d.a.i.). Each color of the image represent a $qP_{(25)}$ value. Non-inoculated plants display $qP_{(25)}$ globally higher than inoculated plants. Indeed, on images of non-inoculated plants, most of the pixel correspond to $qP_{(25)}$ values between 0.7 and 0.8. On images of inoculated plants, most of the pixel correspond to $qP_{(25)}$ values between 0.55 and 0.65.

Additional file 2: Dynamics of $qP_{(25)}$ over time for plants of *A. thaliana* ecotype Col-0 inoculated with *P. ramosa*. Plants of ecotypes Col-0 of *A. thaliana* inoculated with *P. ramosa* were imaged for $qP_{(25)}$ 15, 17 and 21 days after inoculation (d.a.i.). Each color of the image represent a $qP_{(25)}$ value. At 15 and 17 d.a.i., the plant displays $qP_{(25)}$ values superior to 0.65. At 21 d.a.i., plant displays $qP_{(25)}$ values inferior to 0.65.

Additional file 3: Relevance of the dendrograms. Various agglomeration methods (single linkage, WPGMC, WPGMA, complete linkage, UPGMA and UPGMC) were used to construct images based-dendrograms. The relevance was tested in relation to either the treatment of the plants (i.e. inoculated vs. non-inoculated), or according to their apparent health status. This table indicates the percentages of mis-clustered plants for each agglomeration method (nd: not discriminant). CF parameters presenting less than 10% of mis-clustering on the base of the apparent health of the plants are indicated in bold.

Additional file 4: Robustness of the clustering procedure. Bean leaflets of the cultivar Flavert inoculated with *X. fuscans* subsp. *fuscans* CFBP4834-R (1.10^6 $CFU.ml^{-1}$) or mock were imaged for the fluorescence parameter F_v/F_m. A dendrogram (on the left) based on the histograms was built from the UPGMC agglomeration method. Globally, CFBP4834-R-inoculated leaflets (red) and mock-inoculated leaflets (blue) belong to different groups. Two main clusters can be observed. (i) Images of most of the mock-inoculated leaflet and of the CFBP4834-R-inoculated leaflet 4834R-p3 form a first cluster. The histograms (on the right) of these images are similar and present a peak around the F_v/F_m value of 0.8 (indicated by an orange dotted-line). (ii) Images of most of the CFBP4834R-inoculated leaflet and of the mock-inoculated leaflets H_2O-p16, H_2O-p6 and H_2O-p12 form a second cluster. The histograms of these images are right-shifted with regard to those of the first cluster and can present different shapes.

Additional file 5: Images of bean leaves cluster according to the severity of symptoms. Bean leaflets of the cultivar Flavert inoculated with *X. fuscans* subsp. *fuscans* CFBP4834-R (1.10^6 $CFU.ml^{-1}$) or mock were imaged for the fluorescence parameter F_v/F_m. The figure represents an enlargement of the shaded part of the dendrogram found in Additional file 4. Five clusters of histograms can be defined according to the amount of symptoms (indicated by coloured boxes). The image of one leaflet representative of each group is displayed at the bottom. Low intensity pixels are displayed by high intensities of black and correspond to unhealthy tissues (the scale of intensities of F_v/F_m is displayed on the right of the images). The leaflet 4834R-p5 forms a full cluster and displays 68.5% of symptoms. The grey box is composed by the histograms of leaflets displaying between 20.5 and 24.1% of symptoms. The purple box is composed by the histograms of leaflets displaying between 14.2 and 15.6% of symptoms. The leaflet 4834R-p11 forms a full cluster and displays 9.7% of symptoms. The orange box is composed by the histograms of leaflets displaying less than 6.2% of symptoms.

Competing interests

The authors declare that they have no competing interests.

Authors' contributions

CR conceived and designed the experiments, carried out acquisition of chlorophyll fluorescence data, conceived the R script, interpreted the data, wrote and revised the manuscript. GH helped to conceive the R script, conceived the web resource and revised the manuscript. SG and JB helped to conceive the web resource and revised the manuscript. GM, PS and MAJ conceived and designed the experiments, wrote and revised the manuscript.

GM was also in charge of Arabidopsis root inoculation. CC helped setting up the web resource. EB helped setting up the acquisition of chlorophyll fluorescence data, and revised the manuscript. TB conceived and designed the experiments, conceived the R script, interpreted the data, and wrote and revised the manuscript. All authors read and approved the final manuscript.

Acknowledgements
This work was supported by the project PHENOROB from the SFR 4207 QUASAV. Céline Rousseau was funded by the SFR 4207 QUASAV and the University of Angers. We thank Installations Expérimentales Mutualisées (INEM) for plant cultures. We thank the PHENOTIC Platform for allowing image acquisitions. We thank Philippe Simoneau and Jean-Pierre Renou for their support to Phenoplant by the PHENOTIC platform. We also thank C. Collomb-Boureau for editing English.

Author details
[1]PHENOTIC, SFR 4207 QUASAV, F-49045 Angers, France. [2]Université d'Angers, Laboratoire d'Hémodynamique, Interaction Fibrose et Invasivité tumorale hépatique, UPRES 3859, IFR 132, F-49045 Angers, France. [3]Institut de Recherche en Horticulture et Semences, UMR1345, INRA, SFR 4207 QUASAV, F-49071 Beaucouzé, France. [4]Institut de Recherche en Horticulture et Semences, UMR1345, AgroCampus-Ouest, SFR 4207 QUASAV, F-49045 Angers, France. [5]Université de Nantes, Laboratoire de Biologie et de Pathologie Végétales EA 1157, SFR 4207 QUASAV, F-44322 Nantes, France. [6]Institut de Recherche en Horticulture et Semences, UMR1345, Université d'Angers, SFR 4207 QUASAV, F-49045 Angers, France. [7]Laboratoire Angevin de Recherche en Ingénierie des Systèmes (LARIS), Université d'Angers, F-49000 Angers, France.

References

1. Bock CH, Poole GH, Parker PE, Gottwald TR. Plant Disease Severity Estimated Visually, by Digital Photography and Image Analysis, and by Hyperspectral Imaging. CRC Crit Rev Plant Sci. 2010;29:59–107.
2. Poland JA, Nelson RJ. In the eye of the beholder: the effect of rater variability and different rating scales on QTL mapping. Phytopathology. 2011;101:290–8.
3. Bock CH, Parker PE, Cook AZ, Gottwald TR. Visual Rating and the Use of Image Analysis for Assessing Different Symptoms of Citrus Canker on Grapefruit Leaves. Plant Dis. 2008;92:530–41.
4. Sherwood RT, Berg CC, Hoover MR, Zeiders KE. Illusions in visual assessment of Stagonospora leaf spot of Orchardgrass. Phytopathology 1983:173–177.
5. Rousseau C, Belin É, Bove E, Rousseau D, Fabre F, Berruyer R, et al. High throughput quantitative phenotyping of plant resistance using chlorophyll fluorescence image analysis. Plant Methods. 2013;9:17.
6. Xie W, Yu K, Pauls KP, Navabi A. Application of image analysis in studies of quantitative disease resistance, exemplified using common bacterial blight-common bean pathosystem. Phytopathology. 2012;102:434–42.
7. Chéné Y, Rousseau D, Lucidarme P, Bertheloot J, Caffier V, Morel P, et al. On the use of depth camera for 3D phenotyping of entire plants. Comput Electron Agric. 2012;82:122–7.
8. Pace J, Lee N, Naik HS, Ganapathysubramanian B, Lübberstedt T. Analysis of Maize (Zea mays L.) Seedling Roots with the High-Throughput Image Analysis Tool ARIA (Automatic Root Image Analysis). PLoS One. 2014;9:e108255.
9. Green WA, Little SA, Price CA, Wing SL, Smith SY, Kotrc B, et al. Reading the leaves: A comparison of leaf rank and automated areole measurement for quantifying aspects of leaf venation. Appl Plant Sci. 2014;2:1–14.
10. Sighicelli M, Guarneri M. Assessing the poplar photochemical response to high zinc concentrations by image processing and statistical approach. Photosynth Res. 2014;122:315–22.
11. Mishra A, Mishra KB, Höermiller II, Heyer AG, Nedbal L. Chlorophyll fluorescence emission as a reporter on cold tolerance in Arabidopsis thaliana accessions. Plant Signal Behav. 2011;6:301–10.
12. Green JM, Appel H, Rehrig EM, Harnsomburana J, Chang J-F, Balint-Kurti PJ, et al. PhenoPhyte: a flexible affordable method to quantify 2D phenotypes from imagery. Plant Methods. 2012;8:45.
13. De Vylder J, Vandenbussche F, Hu Y, Philips W, Van Der Straeten D. Rosette tracker: an open source image analysis tool for automatic quantification of genotype effects. Plant Physiol. 2012;160:1149–59.
14. Whan AP, Smith AB, Cavanagh CR, Ral J-PF, Shaw LM, Howitt CA, et al. GrainScan: a low cost, fast method for grain size and colour measurements. Plant Methods. 2014;10:23.
15. Zhou J, Spallek T, Faulkner C, Robatzek S. CalloseMeasurer: a novel software solution to measure callose deposition and recognise spreading callose patterns. Plant Methods. 2012;8:49.
16. Bauriegel E, Giebel A, Herppich WB. Hyperspectral and chlorophyll fluorescence imaging to analyse the impact of Fusarium culmorum on the photosynthetic integrity of infected wheat ears. Sensors (Basel). 2011;11:3765–79.
17. Belin É, Rousseau D, Boureau T, Caffier V. Thermography versus chlorophyll fluorescence imaging for detection and quantification of apple scab. Comput Electron Agric. 2013;90:159–63.
18. Berger S, Benediktyová Z, Matouš K, Bonfig KB, Mueller MJ, Nedbal L, et al. Visualization of dynamics of plant-pathogen interaction by novel combination of chlorophyll fluorescence imaging and statistical analysis: differential effects of virulent and avirulent strains of P. syringae and of oxylipins on A. thaliana. J Exp Bot. 2007;58:797–806.
19. Chaerle L, Hagenbeek D, De Bruyne E, Valcke R, Van Der Straeten D. Thermal and chlorophyll-fluorescence imaging distinguish plant-pathogen interactions at an early stage. Plant Cell Physiol. 2004;45:887–96.
20. Pérez-Bueno ML, Ciscato M, VandeVen M, García-Luque I, Valcke R, Barón M. Imaging viral infection: studies on Nicotiana benthamiana plants infected with the pepper mild mottle tobamovirus. Photosynth Res. 2006;90:111–23.
21. Pineda M, Olejníčková J, Cséfalvay L, Barón M. Tracking viral movement in plants by means of chlorophyll fluorescence imaging. J Plant Physiol. 2011;168:2035–40.
22. Woo NS, Badger MR, Pogson BJ. A rapid, non-invasive procedure for quantitative assessment of drought survival using chlorophyll fluorescence. Plant Methods. 2008;4:27.
23. Mishra A, Heyer AG, Mishra KB. Chlorophyll fluorescence emission can screen cold tolerance of cold acclimated Arabidopsis thaliana accessions. Plant Methods. 2014;10:38.
24. Chaerle L, Van Der Straeten D. Seeing is believing: imaging techniques to monitor plant health. Biochim Biophys Acta. 2001;1519:153–66.
25. Sankaran S, Mishra A, Ehsani R, Davis C. A review of advanced techniques for detecting plant diseases. Comput Electron Agric. 2010;72:1–13.
26. Rodríguez-Moreno L, Pineda M, Soukupová J, Macho AP, Beuzón CR, Barón M, et al. Early detection of bean infection by Pseudomonas syringae in asymptomatic leaf areas using chlorophyll fluorescence imaging. Photosynth Res. 2008;96:27–35.
27. Pérez-Bueno ML, Pineda M, Díaz-Casado E, Barón M: Spatial and temporal dynamics of primary and secondary metabolism in Phaseolus vulgaris challenged by Pseudomonas syringae. Physiol Plant 2015.
28. Baker NR. Chlorophyll fluorescence: a probe of photosynthesis in vivo. Annu Rev Plant Biol. 2008;59:89–113.
29. Gorbe E, Calatayud A. Applications of chlorophyll fluorescence imaging technique in horticultural research: A review. Sci Hortic (Amsterdam). 2012;138:24–35.
30. Baker NR, Rosenqvist E. Applications of chlorophyll fluorescence can improve crop production strategies: an examination of future possibilities. J Exp Bot. 2004;55:1607–21.
31. Bresson J, Vasseur F, Dauzat M, Labadie M, Varoquaux F, Touraine B, et al. Interact to survive: Phyllobacterium brassicacearum improves Arabidopsis tolerance to severe water deficit and growth recovery. PLoS One. 2014;9:1–12.
32. Mishra KB, Iannacone R, Petrozza A, Mishra A, Armentano N, La Vecchia G, et al. Engineered drought tolerance in tomato plants is reflected in chlorophyll fluorescence emission. Plant Sci. 2012;182:79–86.
33. Cameron DD, Geniez J-M, Seel WE, Irving LJ. Suppression of Host Photosynthesis by the Parasitic Plant Rhinanthus minor. Ann Bot. 2008;101:573–8.
34. Mauromicale G, Monaco AL, Longo AMG. Effect of Branched Broomrape (Orobanche ramosa) Infection on the Growth and Photosynthesis of Tomato. Weed Sci. 2008;56:574–81.
35. Li L, Xu X, Sun Y, Han W, Tu P. Effects of parasitic plant Cistanche deserticola on chlorophyll a fluorescence and nutrient accumulation of host plant Haloxylon ammodendron in the Taklimakan Desert. J Arid Land. 2012;4:342–8.
36. Parker C. Observations on the current status of Orobanche and Striga problems worldwide. Pest Manag Sci. 2009;65:453–9.

37. Joel D, Hershenhorn J, Eizenberg H, Aly R, Ejeta G, Rich P, et al. Biology and management of weedy root parasites. Hortic Rev (Am Soc Hortic Sci). 2007;33:267–349.
38. Rubiales D, Fernández-aparicio M. Innovations in parasitic weeds management in legume crops . A review. Agron Sustain Dev. 2012;32:433–49.
39. Goldwasser Y, Rodenburg J. Integrated Agronomic Management of Parasitic Weed Seed Banks. In: Joel D, Musselmann M, editors. Parasit Orobanchaceae. Heidelberg New York Dordrecht London: Springer; 2013. p. 393–413.
40. Fernandez-Aparicio M, Sillero J, Pérez-de-Luque A, Rubiales D. Identification of sources of resistance to crenate broomrape (Orobanche crenata) in Spanish lentil (Lens culinaris) germplasm. Weed Res. 2008;48:85–94.
41. Molinero-Ruiz ML, García-Ruiz R, Melero-Vara JM, Domínguez J. Orobanche cumana race F: performance of resistant sunflower hybrids and aggressiveness of populations of the parasitic weed. Weed Res. 2009;49:469–78.
42. Labrousse P. Several Mechanisms are Involved in Resistance of Helianthus to Orobanche cumana Wallr. Ann Bot. 2001;88:859–68.
43. Gauthier M, Véronési C, El-Halmouch Y, Leflon M, Jestin C, Labalette F, et al. Characterisation of resistance to branched broomrape, Phelipanche ramosa, in winter oilseed rape. Crop Prot. 2012;42:56–63.
44. Sokal RR, Rohlf FJ. Biometry. The Principles and Practice of Statistics in Biological Research. San Francisco: W. H. Free; 1969.
45. Bhattacharyya A. On a measure of divergence between two statistical populations defined by their probability distribution. Bull Calcutta Math Soc. 1943;35:99–110.
46. Mishra A, Matous K, Mishra KB, Nedbal L. Towards discrimination of plant species by machine vision: advanced statistical analysis of chlorophyll fluorescence transients. J Fluoresc. 2009;19:905–13.
47. Lichtenthaler HK, Buschmann C, Knapp M. How to correctly determine the different chlorophyll fluorescence parameters and the chlorophyll fluorescence decrease ratio RFd of leaves with the PAM fluorometer. Photosynthetica. 2005;43:379–93.
48. Van Gestel J, Vlamakis H, Kolter R. New Tools for Comparing Microscopy Images: Quantitative Analysis of Cell Types in Bacillus subtilis. J Bacteriol. 2015;197:699–709.
49. Matous K, Benediktyová Z, Berger S, Roitsch T, Nedbal L. Case study of combinatorial imaging: what protocol and what chlorophyll fluorescence image to use when visualizing infection of Arabidopsis thaliana by Pseudomonas syringae? Photosynth Res. 2006;90:243–53.
50. Horton P, Ruban A. Regulation of Photosystem II. Photosynth Res 1992:375–385.
51. Horton P, Ruban aV, Walters RG. Regulation of Light Harvesting in Green Plants. Annu Rev Plant Physiol Plant Mol Biol. 1996;47:655–84.
52. Genty B, Briantais J-M, Baker NR. The relationship between the quantum yield of photosynthetic electron transport and quenching of chlorophyll fluorescence. Biochim Biophys Acta - Gen Subj. 1989;990:87–92.
53. Oxborough K. Imaging of chlorophyll a fluorescence: theoretical and practical aspects of an emerging technique for the monitoring of photosynthetic performance. J Exp Bot. 2004;55:1195–205.
54. Lichtenthaler HK, Miehe J. Fluorescence imaging as a diagnostic tool for plant stress. Trends Plant Sci. 1997;2:6–10.
55. Pau G, Fuchs F, Sklyar O, Boutros M, Huber W. EBImage - an R package for image processing with applications to cellular phenotypes. Bioinformatics. 2010;26:979–81.
56. D'Orazio M. StatMatch: Statistical Matching (aka Data Fusion). 2014.
57. Team RC. R: A Language and Environment for Statistical Computing. 2013.

19

Automated phenotyping of plant shoots using imaging methods for analysis of plant stress responses – a review

Jan F Humplík[1†], Dušan Lazár[2†], Alexandra Husičková[2] and Lukáš Spíchal[1*]

Abstract

Current methods of in-house plant phenotyping are providing a powerful new tool for plant biology studies. The self-constructed and commercial platforms established in the last few years, employ non-destructive methods and measurements on a large and high-throughput scale. The platforms offer to certain extent, automated measurements, using either simple single sensor analysis, or advanced integrative simultaneous analysis by multiple sensors. However, due to the complexity of the approaches used, it is not always clear what such forms of plant phenotyping can offer the potential end-user, i.e. plant biologist. This review focuses on imaging methods used in the phenotyping of plant shoots including a brief survey of the sensors used. To open up this topic to a broader audience, we provide here a simple introduction to the principles of automated non-destructive analysis, namely RGB, chlorophyll fluorescence, thermal and hyperspectral imaging. We further on present an overview on how and to which extent, the automated integrative in-house phenotyping platforms have been used recently to study the responses of plants to various changing environments.

Keywords: Plant phenotyping, RGB digital imaging, Chlorophyll fluorescence imaging, Thermal imaging, Hyperspectral imaging, Shoot growth, Biomass production

Introduction

Recently, a large number of reviews have been published on the advantages and possibilities of high-throughput plant phenotyping approaches [1-5]. Most focus on the potential of these approaches which use precise and sophisticated tools and methodologies to study plant growth and development. To review the state-of-the-art of phenotyping platforms, we present a list of recent publications in Table 1. Interestingly, in about a half of these, only one measuring tool, mostly RGB imaging, for plant phenotyping was used. In the other papers, integrative phenotyping, signifying two or more measuring tools but which are rarely automated, was used (Table 1). This illustrates that the integrative automated high-throughput phenotyping measurements/platforms are still rather rare. Greenhouse- and grow chamber-based plant phenotyping platforms are publically available and these offer their services and collaborative projects. Descriptions, methodological background and focus can be found at http://www.plant-phenotyping-network.eu/eppn/select_installation. As an example of the integrative automated high-throughput phenotyping platform, a grow chamber-based phenotyping facility installed at Palacký University in Olomouc, Czech Republic is presented in Figure 1.

High-throughput integrative phenotyping facilities provide an opportunity to combine various methods of automated, simultaneous, non-destructive analyses of plant growth, morphology and physiology, providing a complex picture of the plant growth and vigour in one run, and repeatedly during the plant's life-span. Particular methods used in integrative plant phenotyping are often not new and usually represent those which have already been used for a number of years in basic research, e.g. non-invasive methods that employ visible or fluorescence imaging (described in more detail further in the text). High-throughput then allows analysis of the plants on a

* Correspondence: lukas.spichal@upol.cz
†Equal contributors
[1]Department of Chemical Biology and Genetics, Centre of the Region Haná for Biotechnological and Agricultural Research, Faculty of Science, Palacký University, Šlechtitelů 11, Olomouc CZ-78371, Czech Republic
Full list of author information is available at the end of the article

Table 1 List of selected works describing automated high-throughput analysis to study plant stress responses

Study	Plant species	Type of stress	Type of the study	Type of automated analysis	Platform name/origin
Granier et al. 2006; [58]	*Arabidopsis*	drought-stress	methodology	RGB (top view)	PHENOPSIS
Skirycz et al. 2011; [59]	*Arabidopsis*	drought-stress	applied	RGB (top view)	WIWAM
Clauw et al. 2015; [60]	*Arabidopsis*	drought-stress	applied	RGB (top view)	WIWAM
Tisné et al. 2013; [61]	*Arabidopsis*	drought-stress	applied	RGB (top view)	PHENOSCOPE
Neumann et al. 2015; [26]	barley	drought-stress	methodology	RGB (multiple views)	LemnaTec
Pereyra-Irujo et al. 2012; [62]	soybean	drought-stress	methodology	RGB (two-views)	GlyPh (self-construction)
Honsdorf et al. 2014; [16]	barley, (wild species)	drought-stress	applied	RGB (multiple views)	LemnaTec
Coupel-Ledru et al. 2014; [63]	grapevine	drought-stress	applied	RGB (multiple views)	LemnaTec
Petrozza et al. 2014; [66]	tomato	drought-stress	applied	RGB (multiple views), hyperspectral NIR, SLCFIM	LemnaTec
Harshavardhan et al. 2014; [67]	*Arabidopsis*	drought-stress	applied	RGB (top view), hyperspectral NIR	LemnaTec
Bresson et al. 2013; [68]	*Arabidopsis*	drought-stress	applied	RGB (top view)	PHENOPSIS
Bresson et al. 2014; [69]	*Arabidopsis*	drought-stress	applied	RGB (top view), TLCFIM	PHENOPSIS
Chen et al. 2014; [64]	barley	drought-stress	methodology	RGB (multiple-views), hyperspectral NIR, SLCFIM	LemnaTec
Fehér-Juhász et al. 2014; [19]	wheat	drought-stress	applied	RGB (multiple views), thermoimaging	self-construction, semi-automated
Cseri et al. 2013; [65]	barley	drought-stress	methodology	RGB (multiple views), thermoimaging	self-construction, semi-automated
Vasseur et al. 2014 [71]	*Arabidopsis*	heat-stress, drought-stress	applied	RGB (top view)	PHENOPSIS
Rajendran et al. 2009; [73]	wheat	salt-stress	applied	RGB (multiple views)	LemnaTec
Harris et al. 2010; [74]	wheat, barley	salt-stress	applied	RGB (multiple views)	LemnaTec
Golzarian et al. 2011; [18]	barley	salt-stress	methodology	RGB (multiple views)	LemnaTec
Schilling et al. 2014; [75]	barley	salt-stress	applied	RGB (multiple views)	LemnaTec
Hairmansis et al. 2014; [76]	rice	salt-stress	applied	RGB (multiple views) SLCFIM	LemnaTec
Chaerle et al. 2006; [77]	tobacco	biotic-stress	methodology	thermoimaging, TLCFIM	self-construction
Poiré et al. 2014; [79]	*Brachypodium*	nutrient-deficiency	methodology	RGB (multiple views)	LemnaTec
Neilson et al. 2015; [80]	*Sorghum*	nutrient-deficiency	methodology	RGB (multiple views), hyperspectral NIR	LemnaTec
Chaerle et al. 2007; [81]	bean	nutrient-deficiency, biotic-stress	methodology	RGB (top view), thermoimaging, TLCFIM	self-construction
Jansen et al. 2009; [37]	*Arabidopsis*, tobacco	drought-stress, chilling-stress	methodology	RGB (top view), KCFIM	GROWSCREEN (self-construction)
Humplík et al. 2015; [20]	pea, field cultivars	cold-stress	methodology	RGB (multiple views), KCFIM	PlantScreen

large scale. This enables users to apply statistics to discover subtle but significant differences between the studied genotypes and treatment variants.

The potential users of such facilities, mostly biologists, are often not very familiar with the applied physical methods used in integrative plant phenotyping. Thus, in this mini-review, we present a simple introduction to the basis of various non-invasive sensors used in high-throughput phenotyping platforms, namely visible red-green-blue (RGB) imaging, chlorophyll fluorescence imaging (CFIM), thermoimaging, and hyperspectral imaging. Further, we describe potential applications of some of the phenotyping methods that have been used to study the responses of different plant species to various stresses.

Non-destructive analysis of growth and physiology of plant shoots

The methods for automated phenotyping and their aims have been reviewed in a number of recent reports [3,6,7]. In the following text we give a description of the basis of the automated non-invasive analysis of plant

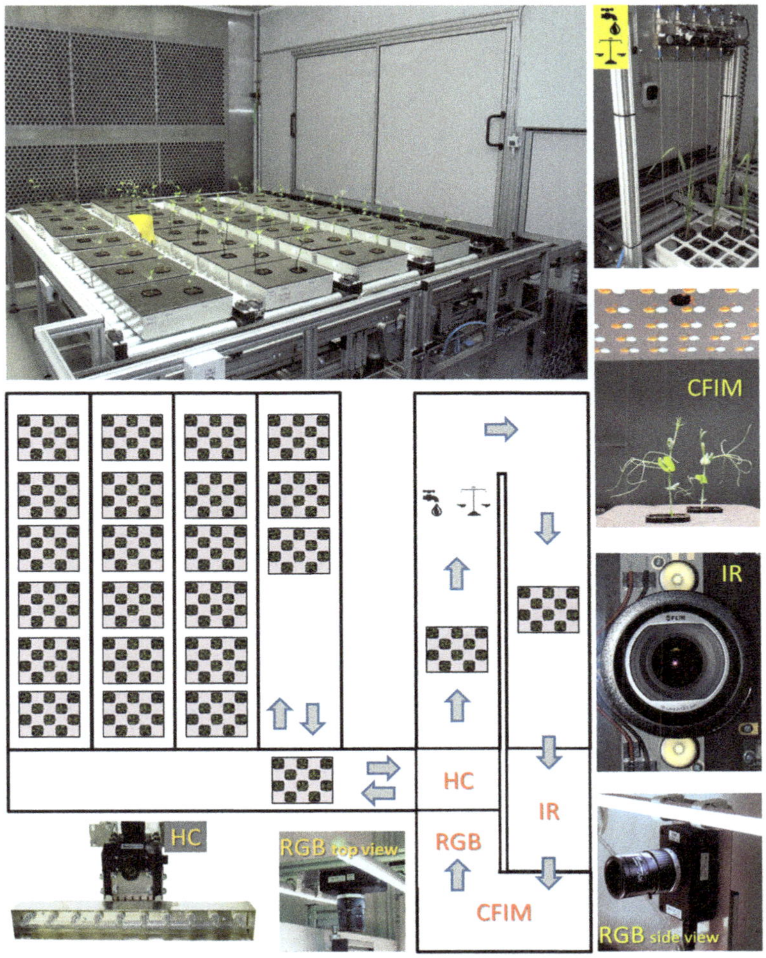

Figure 1 Scheme of the grow chamber-based automated high-throughput phenotyping platform PlantScreen™ (Photons Systems Instruments, Brno, Czech Republic), installed at Palacký University in Olomouc, Czech Republic [20]. The system is located in a growth chamber with white LED illumination (max. 1000 μmol photons m^{-2} s^{-1}) and controlled environment (10 – 40°C, 30 – 99% relative humidity). The growth area with roller conveyer has capacity of up to 640 *Arabidopsis*, cereals and other crops grown in standardized pots. The measuring cabinet contains acclimation chamber for dark adaptation of plants coupled with an automated weighting and watering area. The cabinet is equipped with KCFIM and RGB imaging (top and 2 side views), thermoimaging (IR) to measure stomata openness and SWIR hyperspectral imaging to determine water content. The platform can be controlled either from the place or *via* remote control software. The operating software enables automatic data evaluation.

shoots and appropriate sensors that have been used for studies of plant stress responses.

Visible RGB imaging of plant shoots

Apart from the importance of root-growth analysis, a key descriptive parameter in plant physiology is the growth of plant shoots. Although there are numerous secondary traits describing the morphology of shoots in particular species and their developmental stages, the primary and universal trait is biomass formation. Shoot biomass is defined as the total mass of all the above-ground plant parts at a given point in a plant's life [8]. This trait can be easily assessed by a simple weighing of the fresh (FW) and dry (DW) masses. However, this involves the destruction of the measured plant thus only allowing end-point analyses. Similarly, leaf area and consequently the plant growth rate are usually determined by manual measurements of the dimensions of plant leaves [9-11]. Such measurements are highly time consuming and thus cannot be used for large scale experiments. For this reason, plant phenotyping facilities prefer to evaluate the growth rate using imaging methods which employ digital cameras with subsequent software image analysis. This enables a faster and more precise determination of the leaf area [12-14] and other parameters called projected area (Figure 2), or hull area in the case of monocots [15,16]. In general, non-invasive techniques of shoot growth determination have proven very reliable, and high correlations between the digital area and the shoot fresh, or dry weights, respectively,

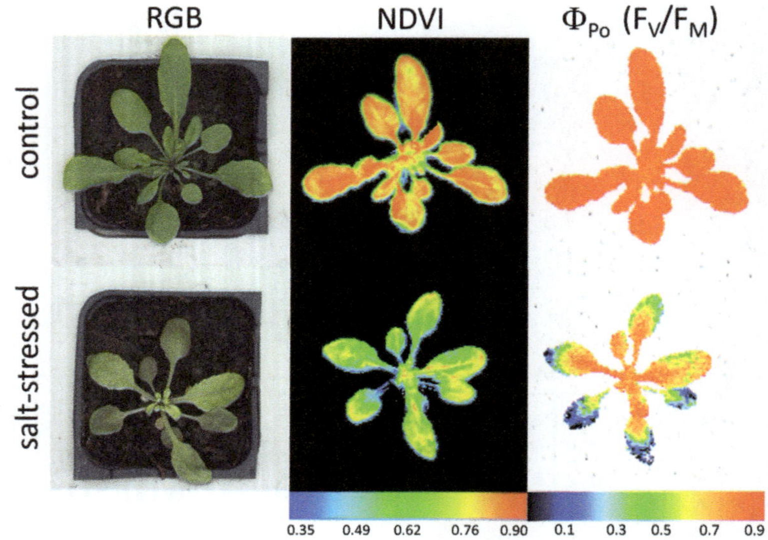

Figure 2 The illustrative figure presenting outcome of simultaneous analysis of control and salt-stressed *Arabidopsis* plants, using RGB, hyperspectral and Chl fluorescence imaging. The 18 DAG old soil-grown *Arabidospis* plants were treated with 250 mM NaCl (salt-stressed) and water (control) and after 48 hours were analysed by different sensors for comparison in: morphology (top-view RGB imaging can be used for computation of rosette area or shape parameters), spatial distribution of vegetation index reflecting changes in the chlorophyll content (NDVI) provided by VIS/NIR hyperspectral camera, and the changes in maximal quantum yield of PSII photochemistry for a dark-adapted state (Φ_{Po}, also referred as F_V/F_M) reflecting the photosynthetic activity of the plants obtained from KCFIM.

were reported in *Arabidopsis*, tobacco [17], cereals [18,19], and pea [20]. An example of a general shoot phenotyping protocol based on biomass estimation was reported by Berger et al. [21]. Similarly, other common morphometric parameters such as stem length, number of tillers and inflorescence architecture can be assessed non-destructively and manually, but again the time requirements, limit the number of plants analysed. High-throughput approaches for analyses of these rather species-specific traits would be very valuable [15], however, with the exception of *Arabidopsis* [22] the range of accessible solutions is still limited (for some emerging methods see [23-26]).

Correct determination of digital plant growth area can be distorted by overlapping leaves, leaf twisting and curling, and circadian movement, especially when the RGB image is taken only from one view (e.g. from top view). A new approach developed for *Arabidopsis* consisting of plant area estimation (which takes into account leaf overlapping), growth modelling and analysis, followed by application of a nonlinear growth model to generate growth curves, and subsequent functional data analysis, was shown to analyse the plant growth in high-throughput experiments more precisely [14]. However, due to the use of only a top-view RGB imaging, this approach cannot be applied for analyses of most of the agronomical important plants with vertical growth. A set-up that introduces more projections (e.g. side-views) into the phenotyping platforms thus can partially solve this problem. The three-views RGB imaging together with linear mathematical modelling was used for accurate estimation of plant shoot dry weight of wheat and barley from two dimensional images [18]. The accuracy of three-view approach has been recently validated in species with challenging shoot morphology such as field pea [20].

Chlorophyll fluorescence imaging (CFIM)

One of the chlorophyll (Chl) fluorescence methods is chlorophyll fluorescence induction (CFIN), i.e., the measurement of the Chl fluorescence signal during illumination of the sample following prior dark adaptation. Since the first paper on CFIN by Kautsky and Hirsch [27], CFIN has been one of the most common methods used in photosynthesis and plant physiology research: it is inexpensive, non-destructive, and above all, provides a great deal of information about the photosynthetic function of the sample (reviewed, e.g., by Lazár [28,29]). Use of pulse amplitude modulation (PAM) techniques for the measurement of CFIN together with the application of the saturation pulse (SP) method enables the separation of photochemical and non-photochemical events occurring in the sample [30]. Chl fluorescence is excited and measured with the help of weak measuring flashes, whereas photosynthesis is maintained by actinic illumination and saturation of photosynthesis is achieved by the SPs. Since Chls absorb in blue (Chl *a* at 436 nm and Chl *b* at 470 nm, respectively) and red (at about 650 nm for both Chls *a* and *b*) regions of visible spectrum, the measuring and actinic light is the light with one of the above wavelengths, usually 650-nm. The SPs are usually generated by white light. On the other hand, Chl fluorescence emission spectrum at room temperature shows two peaks centred at about 680 and 735 nm. To avoid a possible overlap of the 650-nm excitation light with Chl fluorescence emission, the Chl fluorescence signal is detected at wavelengths longer than 700 nm. To reveal spatial heterogeneity of the fluorescence signal during CFIN, imaging Chl fluorometers were developed [31,32]. In the images (for illustration see Figure 2), different colours are used to show different fluorescence intensities according to a chosen false colour scale (as mentioned above, fluorescence emission is always above 700 nm, red light). An additional advantage of the CFIM is that it provides a huge amount of data which can be thoroughly analysed and used for early detection of plant stress as shown, e.g., by Lazár et al. [33]. At present, modern CFIM instruments adopt PAM and SP methods/techniques and are thus highly suitable for high-throughput plant phenotyping (reviewed, e.g., by Gorbe and Calatayud [34], Harbinson et al. [35]). However, over the course of time, too many Chl fluorescence parameters were defined and claimed to reflect particular functions of photosynthetic apparatus. Hence, there is a problem over which parameter should be measured/evaluated and presented. Values of most of the parameters cannot be mutually compared. It is only possible to compare relative changes (caused, e.g., by a stress treatment) of a given parameter. The parameters of the so-called energy partitioning, i.e., quantum yields of processes responsible for the use of the absorbed light energy, are the best choice (reviewed by Lazár [36]) as they are all defined on the same basis and can be directly compared. Since all quantum yields sum to unity, the quantum yields express fractions of absorbed excitation light that are used for given processes (photochemical and various types of non-photochemical energy dissipations).

It is also worth mentioning here that kinetic types of CFIM (KCFIM) that measure whole CFIN and also apply the SPs which then allow computation of various Chl fluorescence parameters, and integrate signal from the whole leaf or shoot, are the most valuable for physiological studies. However, integration of KCFIM into high-throughput systems [20,37] is not very common and in the majority of recent reports, imaging systems measuring either single Chl fluorescence level (SLCFIM) or two Chl fluorescence levels (usually the minimal and maximal Chl fluorescence levels for the dark-adapted state; TLCFIM) were used (see Table 1). As intensity of Chl fluorescence depends on the amount of chlorophylls, the SLCFIM might be used, e.g. to distinguish between non-stressed and senescent leaves (when the amount of Chls is decreased) at the later stages of stress progression but it does not provide any information about early processes in photosytem II (PSII) that are not necessarily linked to the later senescence events. Further, the usual output of the TLCFIM, the F_V/F_M ratio, which estimates the maximum quantum yield of photosystem II photochemistry, provides only a limited information about photosynthetic function compared with the outputs of the KCFIMs, which also allow determination of the other quantum yields and parameters (see [36] for a review).

Thermoimaging

Plants are cooled by transpiration and when the stomata are closed, plant temperature increases. Based on this principle, thermal imaging was used for the first time to detect the changes in the temperature of sunflower leaves caused by water deficiency [38]. In addition to transpiration, stomata also drive water vapour, both parameters being typically determined by leaf gas exchange measurements. However, leaf gasometry involves contact with leaves which often interferes with their function. Further, leaf gasometry is time-consuming, limited by sample size and/or large number of samples required. In addition to heat emission, plants can lose heat by conduction and convection, which in fact represent mechanisms of a non-photochemical quenching of excited states. For this reason, it is not unexpected that an increased thermal signal correlates with an increase in non-photochemical quenching as shown by Kaňa and Vass [39]. Given the foregoing, thermoimaging is a very suitable method for plant phenotyping [19,40,41]. Like CFIM, it uses cameras to measure spatial heterogeneity of heat emissions, usually from leaves; the heat is electromagnetic radiation in the infrared region, usually between 8 – 13 μm. Generally, thermal imaging has been successfully used in a wide range of conditions and with diverse plant species. The technique can be applied to different scales, e.g., from single seedlings/leaves through whole trees or field crops to regions. However, researchers have to keep in mind that environmental variability, e.g., in light intensity, temperature, relative humidity, wind speed, etc. affects the accuracy of thermal imaging measurements and therefore the measurements and their interpretations must be done with care. Although thermal imaging sensors have been integrated into the in-house phenotyping platforms with controlled-environment (see section *The use of phenotyping methods to study plant stress responses*) the majority of studies have been performed so far in field conditions [42-44]. All aspects of thermal imaging used for the exploration of plant-environment interactions, as well as an overview of the application of thermoimaging in field phenotyping, were recently reviewed by Costa et al. [45].

Hyperspectral imaging (VIS-NIR, SWIR)

The absorption of light by endogenous plant compounds is used for calculations of many indices which reflect the composition and function of a plant. Such indices are, for example, the normalized difference vegetation index (NDVI) [46], an estimator of the Chl content, and the photochemical reflectance index (PRI) [47], an estimator of the photosynthetic efficiency. The absorption of a compound (e.g., water) at a given wavelength [48] can also be used for direct estimation of the compound contents in the plant. For practical reasons, measurement of absorbance is replaced here by measurements of reflectance. Depending on the measured wavelengths of reflected signal, various detectors are used, usually VIS-NIR (visible-near infrared region (400–750) - (750–1400 nm)) and SWIR (short wavelength infrared region; 1400–3000 nm). Measurements of the reflectance signal in VIS-NIR and SWIR regions originate from methods of remote sensing [49-51]. However, due to the high value of the information they carry, they are very suitable methods for plant phenotyping [52-54]. The reflectance signal can be detected at selected wavelengths or separated spectral bands (so-called multispectral detection). The whole spectral region can also be measured even for each pixel when cameras are applied and the hyperspectral imaging is carried out (Figure 2). Whereas the hyperspectral imaging in the VIS-NIR spectral region is used for evaluation of several indices as mentioned above, the SWIR spectral region is mainly used for the estimation of the plant's water content. Several aspects of plant reflectance were recently reviewed by Ollinger [55]. Despite the many indices that have been defined so far, based on the reflectance measurements, it is difficult to assess them accurately, similar to the situation with CFIN parameters (see above). For this reason, critical revision of all of the reflectance indices is needed to evaluate which of them provide the required information in the best way.

The use of phenotyping methods to study plant stress responses

One of the most important applications of automated plant phenotyping methods is in studies of plants' responses to various types of environmental stresses. In Table 1 we listed recent reports describing phenotyping protocols developed for indoor automated shoot phenotyping used in stress-related studies. Since the integrative approaches are a logical but rather new step in the development of phenotyping platforms, there are limited reports on the use of simultaneous analysis by multiple sensors. For this reason, we included here "single-sensor" experiments as well, that were performed in the automated platforms.

Perhaps the most widely used application of high-throughput phenotyping is in the search for drought-tolerant varieties. Objectives, traits and approaches related to automated plant selection for drought stress resistance were recently reviewed in Mir et al. [56], and Berger et al. [57]. Here, we add information from examples of the use of non-invasive plant phenotyping in this field. One of the early reports on the use of the high-throughput phenotyping platform describes the employment of the commercial-prototype system for evaluation of drought tolerance in nine *Arabidopsis* accessions [58]. The screening was based on RGB imaging, estimating rosette-leaf area and automated pot weighing and watering to assess transpiration rates. A very similar approach was later used by Skirycz et al. also in *Arabidopsis* [59]. The same platform was further used in a recent physiological study of Clauw and co-authors in which the impact of mild-drought on various *Arabidopsis thaliana* accessions was evaluated [60]. Another study on *Arabidopsis* employing top-view RGB imaging, pot weighing and automated rotation of pots was performed by Tisné et al. [61]. The phenotyping platform was designed to prevent position effect on water evaporation and authors demonstrated important improvement in the evaporation homogeneity [61].

Although these studies represent an important contribution to the development of automated phenotyping, the design of the platform for top-view experiments has limited their use to analyses of plants with leaf rosette. Further progress thus lay in development of platforms allowing RGB imaging from multiple positions. The most recent advances in the use of multiple-view RGB imaging followed by software analysis were demonstrated in a study by Neumann et al. [26]. The authors were able to automatically extract from the images of the barley plants, the plant height and width, and also leaf colours to evaluate the impact of drought on the degradation of chlorophyll. Earlier, Pereyra-Irujo et al. [62], reported a study that employed a self-constructed high-throughput platform for the RGB screening of growth and water-use efficiency (WUE) in two soybean (*Glycine max* L.) genotypes. The system with automated weighing and watering placed in the greenhouse was used to analyse the projected area of the shoots and the mass of the pots [62]. An impressive number of plants was analysed for similar traits in the study by Honsdorf et al. [16]. These authors searched for drought-tolerance QTLs in 48 wild barley introgression lines, using a commercial greenhouse based platform with multiple-view RGB imaging and automated weighing and watering [16]. A similar approach utilizing estimation of shoot biomass based on RGB imaging was used by Coupel-Ledru et al., to screen thousands of grapevine plants for drought tolerance [63]. In these studies, the plant water management was automatically analysed by simple weighing of the pots. This approach, however, begs several questions about the homogeneity of evaporation from the soil of the pots placed in different positions of the growing area. The solution to this issue usually requires an exhaustive validation process with numerous control pots and artificial plant-like objects randomly distributed throughout the growing area (Mark Tester, personal communication). A more elegant solution could be the use of the specific sensors controlling directly the plant water content [64] or transpiration [65] of each plant. Even this approach, however, requires appropriate validation.

An integrative way of analysis was employed in the study of Petrozza et al. [66]. Here, the effect of Megafol treatment on drought-stressed tomatoes was assessed using RGB imaging to distinguish shoot area, SLCFIM measurement to calculate "stress index" and NIR camera for water content estimation. Repeated measurements by NIR camera throughout the experiment allowed visualizing the drop of the high water content index that precedes the growth limitation caused by drought stress [66]. A combination of RGB and NIR imaging techniques was also used by Harshavardhan et al. for analysis of the drought-tolerance of transgenic *Arabidopsis* plants [67]. The RGB imaging was employed by Bresson et al. to study the effect of plant-bacteria interactions on plant tolerance to drought stress [68]. The integration of F_V/F_M measurement by TLCFIM provided complementary information to the growth rate and WUE analysis obtained by pot weighing [69]. A combination of RGB, SLCFIM and NIR imaging techniques was used by Chen et al. [64] to study different phenotypic traits of 18 barley genotypes. The authors used sophisticated statistics and mathematical modelling to classify genotypes based on their response to drought stress [64].

Another important trait in drought studies is the leaf surface temperature that reflects the transpiration rate of the plant (as discussed above in the section *Thermoimaging*). A combination of shoot digital imaging, thermoimaging and automated weighing and watering to study WUE was used by Fehér-Juhász et al. [19]. These authors employed a self-constructed greenhouse-based platform for the selection of drought-tolerant transgenic wheat plants. The platform allows monitoring of the mature cereal plants′ growth by multiple-view RGB imaging and assessment of the leaf surface temperature by side-view thermal camera recording the differences in temperatures of plant shoots [19]. The same platform and a similar phenotyping experimental design were used for evaluation of drought tolerance in barley. The system provides integrative analysis of plant growth and physiology, but its use for large-scale analysis is limited by a semi-automated regime requiring manual loading of the plants into the system [65].

Given that physiological responses to drought and high temperature stresses are tightly connected, similar approaches can be used to study the tolerance of plants to both drought and high temperature. The use of high-throughput phenotyping for high temperature tolerance and a description of the appropriate sensors can be found in a review by Gupta et al. [70]. More recently, the effects of the high temperature on the *Arabidopsis* plants were studied by Vasseur et al. [71]. The authors used commercial-prototype platform allowing the top-view RGB imaging and WUE analysis followed by highly-sophisticated statistical approach to reveal contrasting adaptive strategies to the high temperature and drought stresses [71].

The salinization of soil is another phenomenon often associated with drought and high temperature stress. The example of the protocol for salt stress study in various cereals combining RGB imaging with destructive leaf sampling to measure Na^+ concentration was described by Berger et al. [72]. The effect of salt stress was studied by Rajendran et al. [73] using digital RGB imaging in a greenhouse-based commercial system. This study provided deep insight into the physiological processes connected with salinity in wheat. The authors used the multiple-view RGB imaging to estimate a digital area of shoot, and to visualize changes in leaf colour for quantification of the senescent area. Using non-invasive plant phenotyping and analysis of Na^+ concentration in 4th leaf, the authors predicted a plant salinity tolerance index that showed a good correlation with the results obtained from conventional salt-tolerance measurements [73]. Simple RGB imaging in wheat and barley was carried out in the physiological study of Harris et al. [74], and described in the methodological report of Golzarian et al. [18]. Recently, Schilling et al. applied a similar approach to select a salt-tolerant line of transgenic barley [75]. The combination of digital RGB imaging (used to measure shoot growth rate) with SLCFIM (used for the assessment of senescent areas) was used for the selection of salt-tolerant cultivars of rice by Hairmansis et al. [76]. These studies of salt-stress tolerance were performed using the same commercial platform involving SLCFIM sensor. As mentioned in the section *Chlorophyll fluorescence imaging (CFIM)* this type of CFIM in fact provides only estimation of a senescent area that can be obtained using an older way of estimation based on colour detection by RGB imaging. Thus, to increase the value of the physiological evaluation, the use of KCFIM is necessary for quantification of the quantum yield of photochemistry and of the other competitive processes [36].

Combination of RGB imaging, thermoimaging and TLCFIM was used in the pioneer work of Chaerle at al. who evaluated the effects of mild mottle virus infection on tobacco and bean plants [77]. The use of high-throughput techniques in the nutrient starving stress studies have been already reported too. The principle of the method based on RGB imaging of leaf expansion was described by Moreau et al. [78]. A comprehensive study on the phenotypic effects of nitrogen and phosphorus nutrient statuses of *Brachypodium* was carried out by Poire et al. employing RGB imaging to estimate growth rate [79]. A similar approach was used in a study of Neilson et al. [80] where the responses to nitrogen deficiency and drought were evaluated by RGB imaging, NIR imaging and automated weighing, respectively. The authors also developed software that extracted from the images, additive traits such as projected plant height and the height to the ligule of the youngest fully expanded leaf, which showed very good correlations with standard manually measured agronomical parameters [80]. The multiple-sensor approach was described earlier in beans by Chaerle et al., who used RGB imaging, thermoimaging and TLCFIM to evaluate the phenotypes related to magnesium deficiency and biotic stress [81]. The impact of cold stress on plant growth and physiology is routinely studied using non-invasive methods through the analysis of Chl fluorescence, but not using fluorescence sensors integrated into complex growth-analysing platforms [82-84]. Jansen et al. studied the effects of chilling stress in *Arabidopsis* and tobacco plants using a growth chamber based system equipped with digital top-view RGB screening and KCFIM [37]. Very recently an automated screening approach based on RGB imaging and KCFIM analysis for selection of pea cultivars with different cold-sensitivity was developed by Humplík et al. [20]. The reported study was not intended only for selection of cold-sensitive/tolerant varieties of pea but also for studies of plant cold-response strategies in general. Since the CFIM analysis is not limited to plant morphology and the image analysis was sensitive enough to detect tiny tendrils of pea, the described procedure should be theoretically employed for shoot analyses of other plant species [20].

Conclusions

This mini-review focuses on recent advances towards development of integrative automated platforms for high-throughput plant phenotyping that employ multiple sensors for simultaneous analysis of plant shoots. In both basic and applied science, the recently emerging approaches have found importance as tools in unravelling complex questions of plant growth, development, responses to environment, as well as selection of appropriate genotypes in molecular breeding strategies. As far as phenotype is an interactive network of responses by the plant to its environment that affects in turn, the expression of the genotype it is worth pointing out that attention to the way the analyses are done, under precisely controlled conditions allowing for direct linking the huge amount of complex phenotyping data obtained to the particular conditions. It would also help the end user – the biologist – to narrow his/her view on the importance of various parameters and indices available from the specialized measurements (specifically CFIN and reflectance measurements) and evaluate which of them provide the required information in the best way and hence thus the most suitable for high-throughput plant phenotyping. Such information and standardized protocols applicable for the particular phenotyping methodologies should be available in the near future due to the phenotyping community efforts.

Abbreviations

Chl: Chlorophyll; CFIM: Chlorophyll fluorescence imaging; CFIN: Chlorophyll fluorescence induction; DW: Dry weight; F_M: Maximal chlorophyll fluorescence levels for dark- adapted state; FW: Fresh weight; F_V: variable chlorophyll fluorescence level for a dark-adapted state; Φ_{Po}: The maximal quantum yield of photosystem II photochemistry for a dark-adapted state; KCFIM: Kinetic chlorophyll fluorescence imaging; NDVI: Normalized difference vegetation index; PAM: Pulse amplitude modulation; PRI: Photochemical reflectance index; PSII: Photosystem II; RGB: Red-green-blue; SLCFIM: Single-level chlorophyll fluorescence imaging; SP: Saturation pulse; SWIR: Short wavelength infrared; TLCFIM: Two-level chlorophyll fluorescence imaging; VIS-NIR: Visible-near infrared; WUE: Water-use efficiency.

Competing interests

The authors declare that they have no competing interests.

Authors' contributions

JFH, DL, AH and LS drafted the manuscript. All authors read and approved the final manuscript.

Acknowledgment

This work was supported by the grant No LO1204 (Sustainable development of research in the Centre of the Region Haná) from the National Program of Sustainability I, Ministry of Education, Youth and Sports, Czech Republic.

Author details

[1]Department of Chemical Biology and Genetics, Centre of the Region Haná for Biotechnological and Agricultural Research, Faculty of Science, Palacký University, Šlechtitelů 11, Olomouc CZ-78371, Czech Republic. [2]Department of Biophysics, Centre of the Region Haná for Biotechnological and Agricultural Research, Faculty of Science, Palacký University, Šlechtitelů 11, Olomouc CZ-78371, Czech Republic.

References

1. Araus JL, Cairns JE. Field high-throughput phenotyping: the new crop breeding frontier. Trends Plant Sci. 2014;19:52–61.
2. Granier C, Vile D. Phenotyping and beyond: modelling the relationships between traits. Curr Opin Plant Biol. 2014;18:96–102.
3. Fiorani F, Schurr U. Future scenarios for plant phenotyping. Annu Rev Plant Biol. 2013;64:267–91.
4. Fiorani F, Rascher U, Jahne S, Schurr U. Imaging plants dynamics in heterogenic environments. Curr Opin Biotech. 2012;23:227–35.
5. Walter A, Studer B, Kölliker R. Advanced phenotyping offers opportunities for improved breeding of forage and turf species. Annals Bot. 2012;110:1271–9.
6. Jansen M, Pinto F, Nagel KA, van Dusschoten D, Fiorani F, Rascher U, et al. Non-invasive phenotyping methodologies enable the accurate characterization of growth and performance of shoots and roots. In: Tuberosa R, Graner A, Frison E, editors. Genomics of Plant Genetic Resources. Volume 1st ed. Netherlands: Springer; 2014. p. 173–206.

7. Pieruschka R, Poorter H. Phenotyping plants: genes, phenes and machines. Funct Plant Biol. 2012;39:813–20.
8. Roberts MJ, Long SP, Tieszen LL, Beadle CL. Measurement of plant biomass and net primary production of herbaceous vegetation. In: Hall DO, Scurlock JMO, Bolhar-Nordenkampf HR, Leegood RC, Long SP, editors. Photosynthesis and Production in a Changing Environment. Netherlands: Springer; 1993.
9. Rouphael Y, Mouneimne AH, Ismail A, Mendoza-De Gyves E, Rivera CM, Colla G. Modeling individual leaf area of rose (*Rosa hybrida* L.) based on leaf length and width measurement. Photosynthetica. 2010;48:9–15.
10. Cemek B, Unlukara A, Kurunc A. Nondestructive leaf-area estimation and validation for green pepper (*Capsicum annuum* L.) grown under different stress conditions. Photosynthetica. 2011;49:98–106.
11. Misle E, Kahlaoui B, Hachicha M, Alvarado P. Leaf area estimation in muskmelon by allometry. Photosynthetica. 2013;51:613–20.
12. Green JM, Appel H, MacNeal Rehrig E, Harnsomburana J, Chang J-F, Balint-Kurti P, et al. PhenoPhyte: a flexible affordable method to quantify 2D phenotypes from imagery. Plant Methods. 2012;8:45.
13. Zhang X, Hause RJ, Borevitz JO. Natural genetic variation for growth and development revealed by high-throughput phenotyping in *Arabidopsis thaliana*. G3-Genes Genom Genet. 2012;2:29–34.
14. Tessmer OL, Jiao Y, Cruz JA, Kramer DM, Chen J. Functional approach to high-throughput plant growth analysis. BMC Syst Biol. 2013;7 Suppl 6:S17.
15. Furbank RT, Tester M. Phenomics–technologies to relieve the phenotyping bottleneck. Trends Plant Sci. 2011;16:635–44.
16. Honsdorf N, March TJ, Berger B, Tester M, Pillen K. High-throughput phenotyping to detect drought tolerance QTL in wild barley introgression lines. PLoS One. 2014;9, e97047.
17. Walter A, Scharr H, Gilmer F, Zierer R, Nagel KA, Ernst M, et al. Dynamics of seedling growth acclimation towards altered light conditions can be quantified via GROWSCREEN: a setup and procedure designed for rapid optical phenotyping of different plant species. New Phytol. 2007;174:447–55.
18. Golzarian MR, Frick RA, Rajendran K, Berger B, Roy S, Tester M, et al. Accurate inference of shoot biomass from high-throughput images of cereal plants. Plant Methods. 2011;7:1–11.
19. Fehér-Juhász E, Majer P, Sass L, Lantos C, Csiszár J, Turóczy Z, et al. Phenotyping shows improved physiological traits and seed yield of transgenic wheat plants expressing the alfalfa aldose reductase under permanent drought stress. Acta Physiol Plant. 2014;36:663–73.
20. Humplík JF, Lazár D, Fürst T, Husičková A, Hýbl M, Spíchal L. Automated integrative high-throughput phenotyping of plant shoots: a case study of the cold-tolerance of pea (Pisum sativum L.). Plant Methods. 2015;11:1–11.
21. Berger B, de Regt B, Tester M. High-throughput phenotyping of plant shoots. In: Normanly J, editor. High-Throughput Phenotyping in Plants. New York City: Humana Press; 2012. p. 9–20.
22. Camargo A, Papadopoulou D, Spyropoulou Z, Vlachonasios K, Doonan JH, Gay AP. Objective definition of Rosette shape variation using a combined computer vision and data mining approach. PLoS One. 2014;9, e96889.
23. Crowell S, Falcão AX, Shah A, Wilson Z, Greenberg AJ, McCouch SR. High-resolution inflorescence phenotyping using a novel image-analysis pipeline, PANorama. Plant Physiol. 2014;165:479–95.
24. Paulus S, Dupuis J, Riedel S, Kuhlmann H. Automated analysis of barley organs using 3d laser scanning: an approach for high throughput phenotyping. Sensors. 2014;14:12670–86.
25. Fanourakis D, Briese C, Max JF, Kleinen S, Putz A, Fiorani F, et al. Rapid determination of leaf area and plant height by using light curtain arrays in four species with contrasting shoot architecture. Plant Methods. 2014;10:9.
26. Neumann K, Klukas C, Friedel S, Rischbeck P, Chen D, Entzian A, Stein N, Graner A, Kilian B. Dissecting spatio-temporal biomass accumulation in barley under different water regimes using high-throughput image analysis. Plant Cell Environ 2015. doi: 10.1111/pce.12516, in press.
27. Kautsky H, Hirsch A. Neue Versuche zur Kohlensureassimilation. Naturwissenschaften. 1931;19:964.
28. Lazár D. Chlorophyll *a* fluorescence induction. Biochim Biophys Acta. 1999;1412:1–28.
29. Lazár D. The polyphasic chlorophyll *a* fluorescence rise measured under high intensity of exciting light. Funct Plant Biol. 2006;33:9–30.
30. Schreiber U, Schliwa U, Bilger W. Continuous recording of photochemical and non-photochemical chlorophyll fluorescence quenching with a new type of modulation fluorometer. Photosynth Res. 1986;10:51–62.
31. Omasa K, Shimazaki K-I, Aiga I, Larcher W, Onoe M. Image analysis of chlorophyll fluorescence transients for diagnosing the photosynthetic system of attached leaves. Plant Physiol. 1987;84:748–52.
32. Daley PF, Raschke K, Ball JT, Berry JA. Topography of photosynthetic activity of leaves obtained from video images of chlorophyll fluorescence. Plant Physiol. 1989;90:1233–8.
33. Lazár D, Sušila P, Nauš J. Early detection of plant stress from changes in distributions of chlorophyll *a* fluorescence parameters measured with fluorescence imaging. J Fluoresc. 2006;16:173–6.
34. Gorbe E, Calatayud A. Applications of chlorophyll fluorescence imaging technique in horticultural research: A review. Sci Hortic. 2012;138:24–35.
35. Harbinson J, Prinzenberg AE, Kruijer W, Aarts MGM. High throughput screening with chlorophyll fluorescence imaging and its use in crop improvement. Curr Opin Biotech. 2012;23:221–6.
36. Lazár D. Parameters of photosynthetic energy partitioning. J Plant Physiol. 2015;175:131–47.
37. Jansen M, Gilmer F, Biskup B, Nagel KA, Rascher U, Fischbach A, et al. Simultaneous phenotyping of leaf growth and chlorophyll fluorescence via GROWSCREEN FLUORO allows detection of stress tolerance in Arabidopsis thaliana and other rosette plants. Funct Plant Biol. 2009;36:902–14.
38. Hashimoto Y, Ino T, Kamer PJ, Naylor AW, Strain BR. Dynamic analysis of water stress of sunflower leaves by means of a thermal image processing system. Plant Physiol. 1984;76:266–9.
39. Kaňa R, Vass I. Thermoimaging as a tool for studying light-induced heating of leaves Correlation of heat dissipation with the efficiency of photosystem II photochemistry and non-photochemical quenching. Environ Exp Bot. 2008;64:90–6.
40. Siddiqui ZS, Cho I-L, Park S-H, Kwon T-R, Ahn B-O, Lee G-S, et al. Phenotyping of rice in salt stress environment using high-throughput infrared imaging. Acta Bot Croat. 2014;73:149–58.
41. Virlet N, Lebourgeois V, Martinez S, Costes E, Labbé S, Regnard J-L. Stress indicators based on airborne thermal imagery for field phenotyping a heterogeneous tree population for response to water constraints. J Exp Bot. 2014;65:5429–42.
42. Jones HG, Serraj R, Loveys BR, Xiong L, Wheaton A, Price AH. Thermal infrared imaging of crop canopies for the remote diagnosis and quantification of plant responses to water stress in the field. Funct Plant Biol. 2009;36:978–89.
43. Costa JM, Ortuño MF, Lopes CM, Chaves MM. Grapevine varieties exhibiting differences in stomatal response to water deficit. Funct Plant Biol. 2012;39:179–89.
44. Grant OM, Davies MJ, James CM, Johnson AW, Leinonen I, Simpson DW. Thermal imaging and carbon isotope composition indicate variation amongst strawberry (Fragaria × ananassa) cultivars in stomatal conductance and water use efficiency. Environ Exp Bot. 2012;76:7–15.
45. Costa JM, Grant OM, Chaves MM. Thermography to explore plant–environment interactions. J Exp Bot. 2013;64:3937–49.
46. Rouse JWJ, Haas RH, Schell JA, Deering DW. Monitoring vegetation systems in the Great Plains with ERTS. In: Freden SC, Marcanti EP, Becker MA, editors. NASA SP-351. Proceedings of the 3rd Earth Resources Technology Satellite-1Symposium. Washington DC: NASA Scientific and Technical Information Office; 1974. p. 309–17.
47. Gamon JA, Peñuelas J, Field CB. A narrow-waveband spectral index that tracks diurnal changes in photosynthetic efficiency. Remote Sens Environ. 1992;41:35–44.
48. Carter GA. Primary and secondary effects of water content on the spectra reflectance of leaves. Am J Bot. 1991;78:916–24.
49. Huber S, Tagesson T, Fensholt R. An automated field spectrometer system for studying VIS, NIR and SWIR anisotropy for semi-arid savanna. Remote Sens Environ. 2014;152:547–56.
50. Lamb DW, Schneider DA, Stanley JN. Combination active optical and passive thermal infrared sensor for low-level airborne crop sensing. Precis Agric. 2014;15:523–31.
51. Saberioon MM, Amin MSM, Anuar AR, Gholizadeh A, Wayayok A, Khairunniza-Bejo S. Assessment of rice leaf chlorophyll content using visible bands at different growth stages at both the leaf and canopy scale. Int J Appl Earth Observ Geoinform. 2014;32:35–45.
52. Garriga M, Retamales JB, Romero-Bravo S, Caligari PDS, Lobos GA. Chlorophyll, anthocyanin, and gas exchange changes assessed by spectroradiometry in *Fragaria chiloensis* under salt stress. J Integr Plant Biol. 2014;56:505–15.

53. Mahajan GR, Sahoo RN, Pandey RN, Gupta VK, Kumar D. Using hyperspectral remote sensing techniques to monitor nitrogen, phosphorus, sulphur and potassium in wheat (*Triticum aestivum* L.). Precis Agric. 2014;15:499–522.
54. Petach AR, Toomey M, Aubrecht DM, Richardson AD. Monitoring vegetation phenology using an infrared-enabled security camera. Agri Forest Meteorol. 2014;195–196:143–51.
55. Ollinger SV. Sources of variability in canopy reflectance and the convergent properties of plants. New Phytol. 2011;189:375–94.
56. Mir RR, Zaman-Allah M, Sreenivasulu N, Trethowan R, Varshney RK. Integrated genomics, physiology and breeding approaches for improving drought tolerance in crops. Theor Appl Genet. 2012;125:625–45.
57. Berger B, Parent B, Tester M. High-throughput shoot imaging to study drought responses. J Exp Bot. 2010;61:3519–28.
58. Granier C, Aguirrezabal L, Chenu K, Cookson SJ, Dauzat M, Hamard P, et al. PHENOPSIS, an automated platform for reproducible phenotyping of plant responses to soil water de cit in. New Phytol. 2006;169:623–35.
59. Skirycz A, Vandenbroucke K, Clauw P, Maleux K, De Meyer B, Dhondt S, et al. Survival and growth of Arabidopsis plants given limited water are not equal. Nat Biotechnol. 2011;29:212–4.
60. Clauw P, Coppens F, De Beuf K, Dhondt S, Van Daele T, Maleux K, et al. Leaf Responses to Mild Drought Stress in Natural Variants of Arabidopsis thaliana. Plant Physiol. 2015;114:254–84.
61. Tisné S, Serrand Y, Bach L, Gilbault E, Ben Ameur R, Balasse H, et al. Phenoscope: An automated large-scale phenotyping platform offering high spatial homogeneity. Plant J. 2013;74:534–44.
62. Pereyra-Irujo GA, Gasco ED, Peirone LS, Aguirrezábal LA. GlyPh: a low-cost platform for phenotyping plant growth and water use. Funct Plant B iol. 2012;39:905–13.
63. Coupel-Ledru A, Lebon É, Christophe A, Doligez A, Cabrera-Bosquet L, Péchier P, et al. Genetic variation in a grapevine progeny (Vitis vinifera L. cvs Grenache × Syrah) reveals inconsistencies between maintenance of daytime leaf water potential and response of transpiration rate under drought. J Exp Bot. 2014;65:6205–218.
64. Chen D, Neumann K, Friedel S, Kilian B, Chen M, Altmann T, et al. Dissecting the Phenotypic Components of Crop Plant Growth and Drought Responses Based on High-Throughput Image Analysis. Plant Cell. 2014;26:4636–55.
65. Cseri A, Sass L, Törjék O, Pauk J, Vass I, Dudits D. Monitoring drought responses of barley genotypes with semi-robotic phenotyping platform and association analysis between recorded traits and allelic variants of some stress genes. Aust J Crop Sci. 2013;7:1560–70.
66. Petrozza A, Santaniello A, Summerer S, Di Tommaso G, Di Tommaso D, Paparelli E, et al. Physiological responses to Megafol® treatments in tomato plants under drought stress: A phenomic and molecular approach. Sci Hortic (Amsterdam). 2014;174:185–92.
67. Harshavardhan VT, Van Son L, Seiler C, Junker A, Weigelt-Fischer K, Klukas C, et al. AtRD22 and AtUSPL1, Members of the Plant-Specific BURP Domain Family Involved in Arabidopsis thaliana Drought Tolerance. PLoS One. 2014;9, e110065.
68. Bresson J, Varoquaux F, Bontpart T, Touraine B, Vile D. The PGPR strain Phyllobacterium brassicacearum STM196 induces a reproductive delay and physiological changes that result in improved drought tolerance in Arabidopsis. New Phytol. 2013;200:558–69.
69. Bresson J, Vasseur F, Dauzat M, Labadie M, Varoquaux F, Touraine B, et al. Interact to survive: *Phyllobacterium brassicacearum* improves *Arabidopsis* tolerance to severe water deficit and growth recovery. PLoS One. 2014;9, e107607.
70. Gupta PK, Balyan HS, Gahlaut V, Kulwal PL. Phenotyping, genetic dissection, and breeding for drought and heat tolerance in common wheat: status and prospects. Plant Breeding Reviews. 2012;36:85–147.
71. Vasseur F, Bontpart T, Dauzat M, Granier C, Vile D. Multivariate genetic analysis of plant responses to water deficit and high temperature revealed contrasting adaptive strategies. J Exp Bot. 2014;65:6457–69.
72. Berger B, de Regt B, Tester M. Trait dissection of salinity tolerance with plant phenomics. In: Shabala S, Cuin TA, editors. Plant Salt Tolerance. New York City: Humana Press; 2012. p. 399–413.
73. Rajendran K, Tester M, Roy SJ. Quantifying the three main components of salinity tolerance in cereals. Plant, Cell Environ. 2009;32:237–49.
74. Harris BN, Sadras VO, Tester M. A water-centred framework to assess the effects of salinity on the growth and yield of wheat and barley. Plant Soil. 2010;336:377–89.
75. Schilling RK, Marschner P, Shavrukov Y, Berger B, Tester M, Roy SJ, et al. Expression of the Arabidopsis vacuolar H^+-pyrophosphatase gene (AVP1) improves the shoot biomass of transgenic barley and increases grain yield in a saline field. Plant Biotechnol J. 2014;12:378–86.
76. Hairmansis A, Berger B, Tester M, Roy SJ. Image-based phenotyping for non-destructive screening of different salinity tolerance traits in rice. Rice. 2014;7:16.
77. Chaerle L, Pineda M, Romero-Aranda R, Van Der Straeten D, Barón M. Robotized thermal and chlorophyll fluorescence imaging of pepper mild mottle virus infection in *Nicotiana benthamiana*. Plant Cell Physiol. 2006;47:1323–36.
78. Moreau D, Schneider C, Huguet T, Salon C, Munier-Jolain N. Can differences of nitrogen nutrition level among Medicago truncatula genotypes be assessed non-destructively? Probing with a recombinant inbred lines population. Plant Signal Behav. 2009;4:30–2.
79. Poiré R, Chochois V, Sirault XRR, Vogel JP, Watt M, Furbank RT. Digital imaging approaches for phenotyping whole plant nitrogen and phosphorus response in Brachypodium distachyon. J Integr Plant Biol. 2014;56:781–96.
80. Neilson EH, Edwards AM, Blomstedt CK, Berger B, Moller BL, Gleadow RM. Utilization of a high-throughput shoot imaging system to examine the dynamic phenotypic responses of a C4 cereal crop plant to nitrogen and water deficiency over time. J Exp Bot. 2015;66:1817–32.
81. Chaerle L, Hagenbeek D, Vanrobaeys X, Van Der Straeten D. Early detection of nutrient and biotic stress in *Phaseolus vulgaris*. Int J Remot Sens. 2007;28:3479–92.
82. Devacht S, Lootens P, Baert J, Van Waes J, Van Bockstaele E, Roldán-Ruiz I. Evaluation of cold stress of young industrial chicory (Cichorium intybus L.) plants by chlorophyll a fluorescence imaging. I. Light induction curve. Photosynthetica. 2011;49:161–71.
83. Lootens P, Devacht S, Baert J, Van Waes J, Van Bockstaele E, Roldán-Ruiz I. Evaluation of cold stress of young industrial chicory (*Cichorium intybus* L.) by chlorophyll a fluorescence imaging. II. Dark relaxation kinetics. Photosynthetica. 2011;49:185–94.
84. Mishra A, Mishra KB, Höermiller II, Heyer AG, Nedbal L. Chlorophyll fluorescence emission as a reporter on cold tolerance in *Arabidopsis thaliana* accessions. Plant Signal Behav. 2011;6:301.

Quantifying spatial heterogeneity of chlorophyll fluorescence during plant growth and in response to water stress

Justine Bresson[1,2,3†], François Vasseur[4†], Myriam Dauzat[1], Garance Koch[1], Christine Granier[1] and Denis Vile[1*†]

Abstract

Background: Effects of abiotic and biotic stresses on plant photosynthetic performance lead to fitness and yield decrease. The maximum quantum efficiency of photosystem II (F_v/F_m) is a parameter of chlorophyll fluorescence (ChlF) classically used to track changes in photosynthetic performance. Despite recent technical and methodological advances in ChlF imaging, the spatio-temporal heterogeneity of F_v/F_m still awaits for standardized and accurate quantification.

Results: We developed a method to quantify the dynamics of spatial heterogeneity of photosynthetic efficiency through the distribution-based analysis of F_v/F_m values. The method was applied to *Arabidopsis thaliana* grown under well-watered and severe water deficit (survival rate of 40%). First, whole-plant F_v/F_m shifted from unimodal to bimodal distributions during plant development despite a constant mean F_v/F_m under well-watered conditions. The establishment of a bimodal distribution of F_v/F_m reflects the occurrence of two types of leaf regions with contrasted photosynthetic efficiency. The distance between the two modes (called S) quantified the whole-plant photosynthetic heterogeneity. The weighted contribution of the most efficient/healthiest leaf regions to whole-plant performance (called W_{max}) quantified the spatial efficiency of a photosynthetically heterogeneous plant. Plant survival to water deficit was associated to high S values, as well as with strong and fast recovery of W_{max} following soil rewatering. Hence, during stress surviving plants had higher, but more efficient photosynthetic heterogeneity compared to perishing plants. Importantly, S allowed the discrimination between surviving and perishing plants four days earlier than the mean F_v/F_m. A sensitivity analysis from simulated dynamics of F_v/F_m showed that parameters indicative of plant tolerance and/or stress intensity caused identifiable changes in S and W_{max}. Finally, an independent comparison of six Arabidopsis accessions grown under well-watered conditions indicated that S and W_{max} are related to the genetic variability of growth.

Conclusions: The distribution-based analysis of ChlF provides an efficient tool for quantifying photosynthetic heterogeneity and performance. S and W_{max} are good indicators to estimate plant survival under water stress. Our results suggest that the dynamics of photosynthetic heterogeneity are key components of plant growth and tolerance to stress.

Keywords: *Arabidopsis thaliana*, Chlorophyll fluorescence imaging, Heterogeneity of F_v/F_m values, Modelling, Photosynthetic performance, Pixels distribution, Plant growth, Plant survival, Sensitivity analysis

* Correspondence: denis.vile@supagro.inra.fr
†Equal contributors
[1]Laboratoire d'Ecophysiologie des Plantes sous Stress Environnementaux (LEPSE), INRA, Montpellier SupAgro, UMR759, F-34060 Montpellier, France
Full list of author information is available at the end of the article

Background

High-throughput phenotyping is increasingly used for dissecting the genetic and eco-physiological determinisms of plant performance and stress tolerance. Over the last decade, efficient automated imaging systems have been developed for the acquisition of visible, bioluminescence, fluorescence and multi-spectral images. A rising difficulty is now to extract valuable, *i.e.*, biologically meaningful, preferably quantitative, information from the large collection of images generated by these systems [1].

Chlorophyll fluorescence (ChlF) imaging has become one of the most powerful and popular tools to track changes in the photosynthetic capacities of plants in response to abiotic and biotic factors [2-4]. Pulse-amplitude modulated ChlF techniques provide non-invasive assessment of the photosystem II (PSII) efficiency to supply electrons to the photosynthetic machinery. Light energy absorbed by chlorophyll molecules can undergo one of three competing fates: (*i*) driving photosynthesis (photochemistry); (*ii*) being dissipated as heat; or (*iii*) being re-emitted as ChlF. These three processes take place in a competitive manner, and under stress conditions, the photochemistry declines whereas heat dissipation and ChlF emission increase (for recent reviews, see [5,6]). ChlF is estimated by the quantification of the light re-emitted (in the red wavebands) after the application of a saturating flash (usually for a few seconds) to the photosynthetic organs [5]. The saturating flash induces the transport of electrons through PSII centres, driving the reduction of Q_A, the primary stable electron acceptor of PSII. Once reduced, Q_A cannot accept new electrons before electrons are transferred to the next acceptor (the reaction centre is considered to be 'closed'), and the excess of energy is dissipated through heat and fluorescence.

Amongst the different ChlF parameters, the F_v/F_m ratio is a useful and rapid parameter that reflects the maximum quantum efficiency of the PSII photochemistry [7]. In dark-adapted leaves (in which all PSII reaction centres are in the 'open' state; Q_A fully oxidized), a measuring beam is applied to elicit the minimal value of ChlF, F_0 (*i.e.*, basal fluorescence). F_0 represents the energy dissipation via light-harvesting antenna pigments when excitation energy is not being transferred to the PSII reaction centres. After reaching F_0, the application of a brief saturating pulse induces a maximum value of ChlF, F_m (PSII reaction centres get 'closed' because of electron accumulation; Q_A fully reduced). The difference between F_0 and F_m is the variable fluorescence, F_v and F_v/F_m is given by $(F_m$-$F_0)/F_m$ (for more details, see [5]). Low F_v/F_m indicate substantial photoinhibition or down-regulation of PSII that occurs when plants experience stress. It has been shown that F_v/F_m is a robust indicator of plant health. Healthy photosynthetic tissues of most plant species exhibit a mean F_v/F_m at *ca.* 0.83, while lower values are indicative of an impaired physiological status [8,9]. Rapid modifications of F_v/F_m are for instance reported in response to many environmental factors, such as water stress [8,10], temperature [11-13], wounding [14], photoinhibition [11,15], biotic interactions such as pathogenic as well as beneficial bacteria [16-19].

Soil water availability is one of the most important environmental factors for plant growth and development. The impact of water deficit on the photosynthetic performance of plants depends on the severity and duration of the stress. In the short-term, decrease in water supply usually induces stomata closure to maintain a favourable leaf water status, what in turn leads to a reduction of internal CO_2 concentration [20]. Hence, stomata closure under water stress promotes an imbalance between the photochemical activity of PSII and the electron requirement for carbon fixation, leading to over-excitations and subsequent photoinhibitory damages to PSII reaction centres [21]. As a consequence, substantial decline in F_v/F_m in response to moderate water deficit is observed in various plant species (see references in [2]), and was closely related to decreased relative leaf water content [8]. With increasing stress severity or duration, carbon starvation and hydraulic failure, which strongly alter F_v/F_m at the whole-plant level, lead to partial (or total) senescence or leaf abscission [22]. Even though exacerbated leaf senescence can be lethal, sacrificing a few leaves might be a good strategy to ensure survival under severe resource limitation [23]. Growth recovery following severe water stress is then associated with the (partial) reestablishment of the photosynthetic capacities of the senescing leaves, and/or with the development of new leaves with optimal photosynthetic performance [24].

ChlF imaging has revealed that photosynthetic performance is extremely heterogeneous at the leaf surface, as well as between leaves, when plants experience environmental stresses. For examples heterogeneity in ChlF is reported in response to changing CO_2 concentration [25], light stimuli [26], ozone-induced perturbations [27], low growth temperature [28], chilling [29], pathogen attack [16], drought [10,30] or treatment with abscisic acid [31]. Spatio-temporal heterogeneity across photosynthetic areas has been assessed by visual inspection of leaves [24,26,30], by measurements at spatially different small areas on the leaf surface [10,29,30], or by visual inspection of the shape of F_v/F_m distributions across leaves [26,28,29,32]. ChlF imaging of leaves of Arabidopsis grown under water stress for instance reveals a progressive decline of F_v/F_m beginning at the leaf's tip [10]. However we still lack an automatic and standardized method for the quantification of the spatial heterogeneity of F_v/F_m values, which is crucial to compare photosynthetic performance depending on the developmental stage, the genotype, or the environmental conditions.

Here, we analysed the distribution of F_v/F_m to estimate the spatial heterogeneity of photosynthetic efficiency (S) and the weighted contribution of the most efficient/healthiest leaf regions to whole-plant photosynthetic performance (W_{max}). We first showed that the changes in S and W_{max} were related to the survival of the Arabidopsis Col-0 accession to a severe water deficit (SWD). A sensitivity analysis of S and W_{max} to simulated dynamics of F_v/F_m distributions showed to what extent S and W_{max} can vary depending on plant tolerance and/or stress intensity. Finally, we found that a significant part of the variation in biomass accumulation in six contrasted Arabidopsis accessions is explained by the variation of W_{max} in the course of plant development.

Results

Analytical framework: severe water deficit strongly affects plant growth, photosynthetic efficiency and induces plant mortality

Arabidopsis Col-0 plants were grown in the PHENOPSIS automaton [33] (Figure 1A). Plants were subjected to SWD by withholding irrigation from the four-leaves stage (L4; stage 1.04, [34]; Figure 1B) in order to progressively reach a very low soil relative water content (RWC_{soil}) of 6% g H_2O g^{-1} dry soil (corresponding to water potential *ca.* of −9.52 MPa; see Additional file 1: Figure S1). Thereafter, irrigation was resumed to progressively reach the well-watered (WW) soil condition (35% g H_2O g^{-1} dry soil; 0.07 Mpa, Additional file 1: Figure S1) maintained until the flowering of surviving plants (Figure 1B). These two soil conditions allowed the investigation of F_v/F_m heterogeneity with highly contrasted physiological status and thus, with a wide range of leaf damages and senescence. Plant growth and F_v/F_m were daily measured from early developmental stages (*i.e.*, emergence of the two first leaves, stage 1.02, [34]) to the emergence of the flowering stem (*i.e.*, bolting, stage 5.01, [34]; Figure 1C), with a high-throughput ChlF imaging system (Imaging-PAM M-Series, Maxi-version, Heinz Walz GmbH, Germany) implemented on the automaton (Figures 1A, C). We developed an ImageJ (1.47v, Rasband, Bethesda, Maryland, USA) macro "PHENOPSIS-Fluo" to semi-automatically extract the whole-rosette F_v/F_m mean, the distribution of F_v/F_m values across the rosette and the projected total leaf area from ChlF images.

Under SWD, 40% of the plants survived, resumed growth and flowered whereas the remaining plants failed to recover, perished and decomposition of tissues started (Figures 1D, E). Whole-rosette mean F_v/F_m followed the variation of RWC_{soil} and was therefore dramatically affected by the SWD (Figures 1B and 2A). Whole-rosette mean F_v/F_m of stressed plants remained stable at 0.812 ± 0.041 (n = 4–30) during the 14 days after water withholding, similar to plants grown under WW conditions (0.813 ± 0.019; n = 4–31; Figure 2A). Then, whole-rosette mean F_v/F_m of stressed plants decreased dramatically (Figure 2A). This was mainly due to the decrease of F_v/F_m in the oldest leaves of the rosette, notably with a gradient from the tip to the base of the leaves (see 3-D representations in Figure 2B and Additional file 2: Figure S2). Just before rewatering, SWD resulted in a significant 38% and 43% decrease of mean F_v/F_m in surviving and perishing plants, respectively (Figure 2A). Upon rewatering, mean F_v/F_m continued to decline steadily for three further days. Afterwards, surviving plants progressively recovered F_v/F_m values up to 88% of their initial values after 6 days following rewatering (Figure 2A). This was mainly achieved by shedding of almost all senescing leaves (Figure 2B). In contrast, mean F_v/F_m of perishing plants continued to decrease to reach undetectable threshold of photosynthetic activity (*i.e.*, plants were completely senescing or decomposing; Figure 2B and Additional file 2: Figure S2). A clear separation of mean F_v/F_m between surviving and perishing plants was visible four days after rewatering (Figure 2A).

Computing and quantifying the heterogeneity of plant photosynthetic efficiency during growth and under severe water deficit

During SWD, F_v/F_m values at the plant surface became heterogeneous, as illustrated by the changes in the mean and distribution of F_v/F_m values (Figures 2A, B). We notably observed the establishment of multimodal distributions during SWD, reflecting the spatial variability of F_v/F_m in the rosette (Figure 2B). To explore the heterogeneity of F_v/F_m values during time course, we applied the Hartigan's non-parametric significance test for unimodality [35-37]. As expected, the proportion of stressed plants showing multimodal distributions increased strongly after stress exposure (Figure 2C). Under WW conditions, the proportion of plants that displayed significant multimodal distributions also increased from < 10% to > 90% between 1 to 10 days after L4 stage. Stressed plants even displayed a slightly lower proportion of multimodal distributions compared to plants grown under WW conditions (Figure 2C).

After distinguishing the plants that exhibited significant multimodal distributions, we used the REBMIX algorithm for finite mixture models [38] to characterize each mode i of the mixture of distributions of F_v/F_m values (*i.e.*, mean μ_i, standard deviation σ_i and weight ρ_i) for each individual rosette. All distributions displaying multimodality were accurately represented by bimodal mixtures of normal distributions where the distributions are composed of two clusters of F_v/F_m values grouping in two modes. The higher mode (maximum; μ_{max}, σ_{max} and ρ_{max}; with the highest F_v/F_m values) represented the photosynthetically most efficient/healthiest parts of the rosette. The

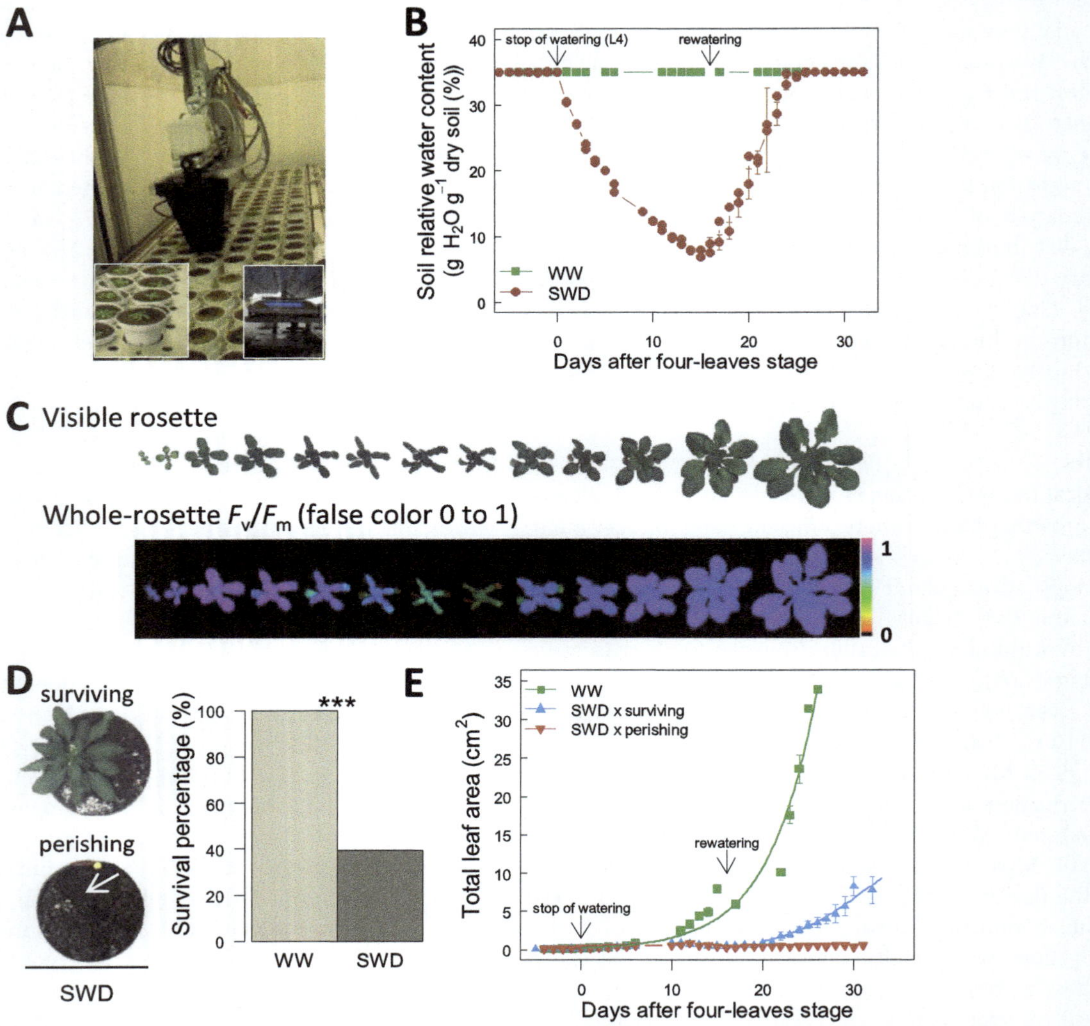

Figure 1 High-throughput analysis of Arabidopsis growth and chlorophyll fluorescence in the PHENOPSIS automaton. (A) *A. thaliana* plants are grown in controlled environmental conditions in the PHENOPSIS platform equipped with a chlorophyll fluorescence imaging system. **(B)** Dynamics of soil relative water content in two watering scenarios including constant well-watered conditions (WW) and water withdrawing from the four-leaves stage (L4; beginning of stress) followed by rewatering after 1 day at 6% g H_2O^{-1} dry soil (SWD). Data are means (± SE) of 13 and 48 plants under WW and SWD, respectively. **(C)** Plant growth (top) and whole-rosette F_v/F_m (bottom) during plant development and under SWD. F_v/F_m values are represented by false colour scale ranging from black (pixel values 0) through red, yellow, green, blue to purple (ending at 1). **(D)** Visible images of surviving and perishing plants (left) and survival percentage of plants under WW and SWD conditions (right). Asterisks indicate significant differences following Chi^2 test between plants grown in WW conditions ($n = 13$) and plants under SWD ($n = 19$ and 29 for surviving and perishing plants, respectively; ***: $P < 0.001$). **(E)** Total projected leaf area of plants under WW conditions and SWD (surviving and perishing plants) as a function of days after L4 stage until bolting. Data are means (± SE) of 13–29 plants.

second mode (minimum μ_{min}, σ_{min} and ρ_{min}; with the lowest F_v/F_m values) represented the least efficient or senescing parts of the rosette (Figure 3A). In case of unimodal distribution, the mode was considered as the single maximum mode.

For plants grown under WW conditions, each parameter was roughly constant during plant development (Figures 3B-G). In stressed plants, while μ_{max} essentially followed the same variation of whole-rosette mean F_v/F_m (Figures 2A and 3C), μ_{min} decreased to reach very low values ($\mu_{min} = 0.24 \pm 0.13$ and 0.37 ± 0.17 for perishing and surviving plants, respectively; Figure 3B). Standard deviation σ_{max} progressively increased during SWD establishment. However, while σ_{max} of surviving plants recovered values similar to WW plants after rewatering, σ_{max} continued to increase in perishing plants (Figure 3E). By contrast, standard deviation σ_{min} increased more in surviving than in perishing plants, but recovered their initial value 13 days after rewatering (Figure 3D). In addition, the weight, *i.e.* the proportion, of the minimum mode ρ_{min}

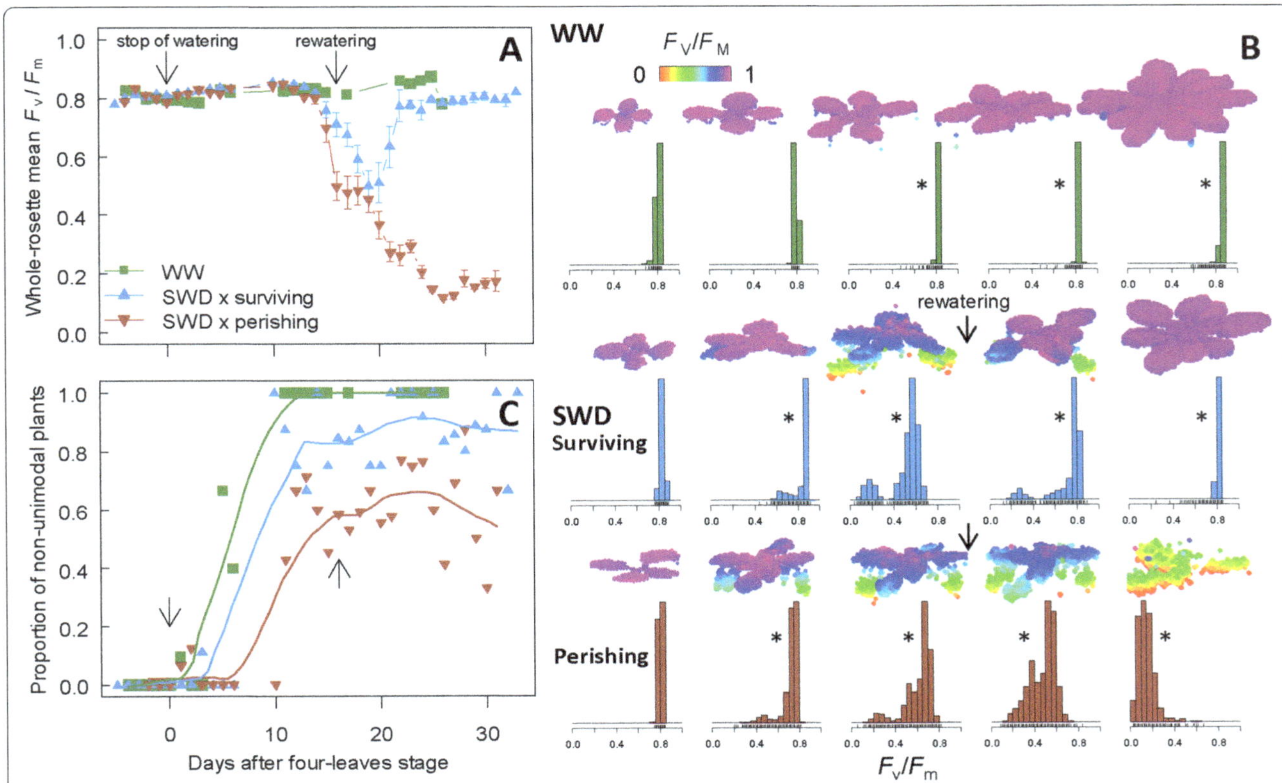

Figure 2 Effects of severe water deficit on plant photosynthetic efficiency. (A) Dynamics of whole-rosette mean F_v/F_m of well-watered (WW) plants and stressed (surviving and perishing) plants under severe water deficit (SWD) as a function of days after four-leaves stage (L4; beginning of SWD) until bolting. Data are means (± SE) of 13–29 plants. **(B)** 3-D representations of vegetative rosettes under WW and SWD conditions in F_v/F_m false colour (from black pixel values (0) through red, yellow, green, blue to purple (ending at 1)) and their corresponding F_v/F_m distributions during time courses. Asterisks indicate *p*-value < 0.01 (Hartigan's dip test) meaning significant departure from unimodality of F_v/F_m values. Arrows indicate rewatering step. **(C)** Dynamics of the proportion of non-unimodal (*i.e.*, multimodal) plants under WW and SWD after L4 stage until bolting following the Hartigan's dip test.

increased to a greater extent in perishing plants (and the weight of the maximum mode ρ_{max} decreased likewise) compared to surviving plants (Figures 3F, G).

A quantification of the disparity between the two modes of a bimodal distribution, *i.e.* the heterogeneity of the values, is given by the 'bimodal separation' $S = (\mu_{max} - \mu_{min}) / 2(\sigma_{max} + \sigma_{min})$ [39]. S is roughly the distance between the two peaks, and $S > 1$ when the two modes do not overlap. Here, the F_v/F_m heterogeneity across the plant increased regardless of soil water conditions during time course (Figure 4). However, S increased more in plants that survived the SWD than in others plants, whereas perishing plants had the same heterogeneity than those grown in WW conditions. A clear difference between S values of surviving and perishing plants was visible just before rewatering (Figure 4), *i.e.* four days earlier than mean F_v/F_m.

Quantifying the effect of photosynthetic heterogeneity on whole-plant performance: description

Under SWD, S accurately represented the photosynthetic heterogeneity and allowed deciphering surviving and perishing plants. However, it failed to quantify the effect of photosynthetic heterogeneity on plant performance and stress tolerance, as shown by the overlap of S values between WW and perishing plants (Figure 4). This is because the deviation of both modes to the photosynthetic optimum is as important as the disparity between the two modes.

It was shown from energy conversion modelling of PSII that theoretical optimum of F_v/F_m is about 0.87 in unstressed dark-adapted leaves [40,41]. However, a healthy plant displays a typical maximal mean $F_v/F_m = 0.83$ [8,9] and shows considerable variation around the mean. The theoretical optimum would be reached if a plant exhibits a unimodal distribution of mean 0.87 and variance 0. Hence, the photosynthetic deviation of each mode i to the theoretical optimum can be estimated as the bimodal separation S_i such as $S_i = (0.87 - \mu_i) / 2\,\sigma_i$ (*i.e.*, S_{max} and S_{min}; Figure 5A). High S_i represents low photosynthetic performance of the mode i. Then, the weighted deviation to the optimum, which measured the size-corrected performance of a given mode, was calculated as $S_{max} \times \rho_{max}$ and $S_{min} \times \rho_{min}$, for the maximum and the minimum

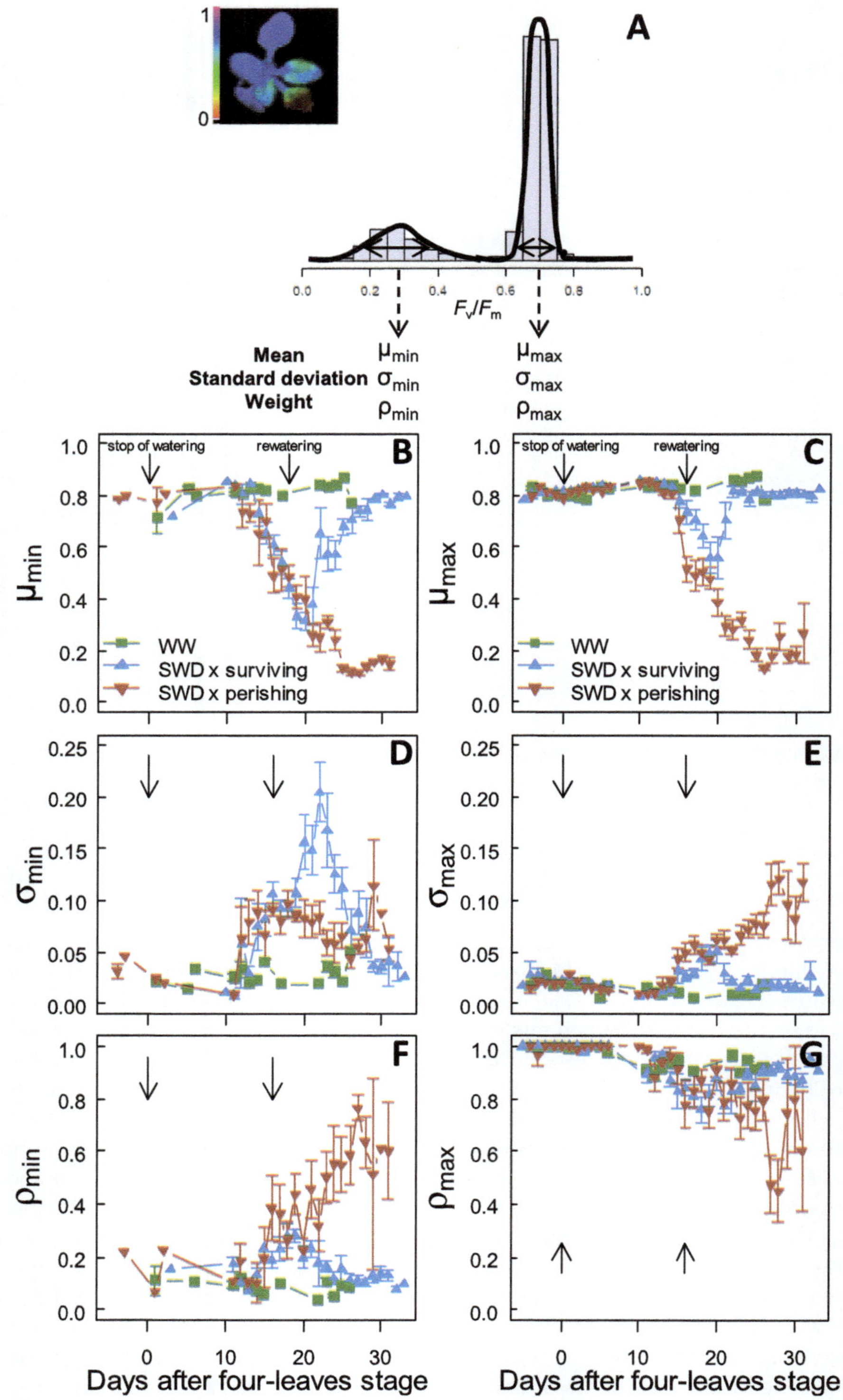

Figure 3 Dynamics of the parameters describing the bimodal distributions. (A) Example of a bimodal distribution composed of two clusters of pixels grouping in two modes. The higher mode (max; with the highest F_v/F_m values) represents the healthiest parts of the rosette whereas the second mode (min; with the lowest F_v/F_m values) represents damaged/senescing parts of the rosette. Each mode i of the mixture distribution of F_v/F_m values is characterized by mean μ_i, standard deviation σ_i and weight ρ_i. **(B-G)** Dynamics of μ_{max} and μ_{min} of F_v/F_m values, σ_{max} and σ_{min}, and, ρ_{max} and ρ_{min} in well-watered (WW) plants and under severe water stress (SWD; surviving and perishing plants) as a function of days after four-leaves stage (beginning of SWD) until bolting. Data are means (± SE) of 13–29 plants.

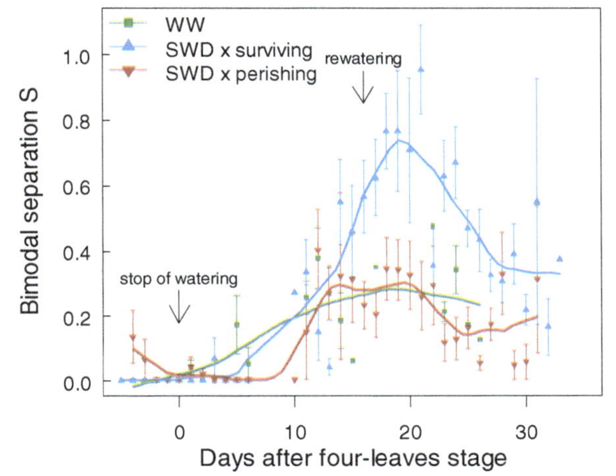

Figure 4 Dynamics of spatial heterogeneity of whole-plant photosynthetic efficiency during development and severe water deficit. Bimodal separation (S) of F_v/F_m values of well-watered (WW) plants and stressed plants (SWD; surviving and perishing) as a function of days after four-leaves stage (beginning of SWD) until bolting. S = (μ_{max} - μ_{min}) / 2 (σ_{max} + σ_{min}) measures the distance between the modes and is superior to 1 essentially if the two modes do not overlap. Data are means (± SE) of 13–29 plants.

modes, respectively (Figure 5A). To estimate the spatial efficiency of a photosynthetically heterogeneous plant to convert light energy into chemical energy (W_{max}), we calculated the proportion of $S_{max} \times \rho_{max}$ (*i.e.*, the weighted deviation to the optimum of the most efficient leaf regions) in the distribution of F_v/F_m values, as $W_{max} = (S_{max} \times \rho_{max} - S_{min} \times \rho_{min}) / S_{max} \times \rho_{max}$ (Figure 5A). By definition, for a unimodal distribution $W_{max} = 0$ because there is no spatial heterogeneity ($S_{max} \times \rho_{max} = S_{min} \times \rho_{min}$). Basically, increase or decrease in W_{max} indicates that the contribution of the most efficient/healthiest regions to the whole-plant photosynthetic performance is more or less important, respectively, than the contribution of the least efficient or senescing regions (note that W_{max} has a maximum value of 1). For a heterogeneous surface (*i.e.*, not in the first stages of plant development which display $W_{max} = 0$ because of unimodal distributions), $W_{max} = 0$ is assumed to be the compensation point, where the healthiest leaf regions compensate the negative effect of the less efficient leaf regions. Negative values of W_{max} appear when the contribution of senescing leaf regions is prevailing.

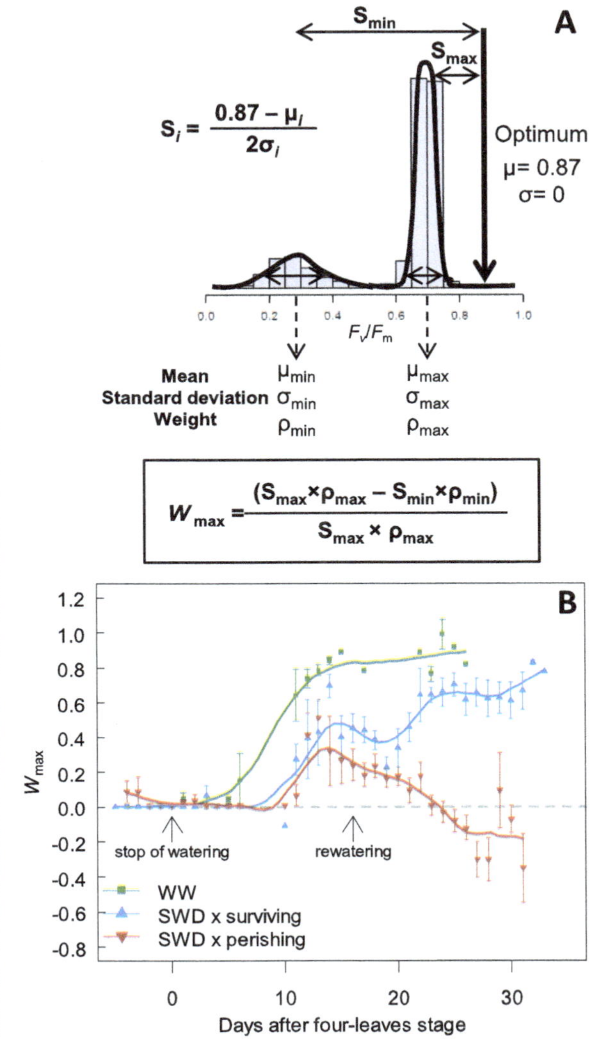

Figure 5 Dynamics of the spatial efficiency of a photosynthetically heterogeneous plant (W_{max}). (A) Illustration of the mixture parameters in the case of a bimodal distribution. W_{max} is calculated as the proportional difference in the weighted bimodal separation of each mode (S_{max} and S_{min}) to the theoretical optimum of photosynthetic performance (0.87, with standard deviation = 0), such as: $W_{max} = (S_{max} \times \rho_{max} - S_{min} \times \rho_{min}) / S_{max} \times \rho_{max}$. W_{max} estimates the relative contribution of the most efficient/healthiest leaf regions to the whole-plant photosynthetic performance. **(B)** Dynamics of W_{max} of plants under well-watered (WW) and severe water deficit (SWD; surviving and perishing) conditions as a function of days after four-leaves stage (beginning of SWD) until bolting. Data are means (± SE) of 13–29 plants.

Quantifying the effect of photosynthetic heterogeneity on whole-plant performance: applications

In plants grown in WW conditions, W_{max} increased progressively during development from 0 to *ca.* 0.85 (Figure 5B). This reflects the increase in the heterogeneity of whole-plant photosynthetic performance (*i.e.*, a switch from unimodality to bimodality) with a very low and negligible effect of the minimum mode compared to the maximum mode. In stressed plants, the increase of W_{max} was delayed and reduced (Figure 5B). In surviving plants, W_{max} started to decrease at 15 days after L4 stage, and recovered shortly (2 days) after rewatering. At bolting, surviving plants exhibited a W_{max} of *ca.* 0.65, *i.e.* 23% less than WW plants at the same developmental stage (Figure 5B). By contrast, in perishing plants, W_{max} started to decrease after 14 days following the L4 stage and became negative ten days later.

We used simple mathematical functions to model the dynamics of the parameters of the bimodal distributions in various stressing conditions, and simulate the associated variations of S and W_{max} (see Additional file 3). First, this simulation exercise was sufficient to reproduce what has been observed in this paper in plants grown under SWD that did not survive the stress. The parameters of these functions were then varied to simulate different scenarios of photosynthetic heterogeneity generated by different stress intensities. Our sensitivity analysis of W_{max} showed that it becomes as negative as (1) the rate of decrease in means and (2) the increase in proportion of damaged leaf regions, are high (*i.e.*, low stress tolerance, and/or diffuse stress effect, high stress intensity). Conversely, its decrease is delayed when the rates of decrease in means and rates of changes are low (*i.e.*, high stress tolerance, stress effects with high patchiness, and/or low stress intensity; see Additional file 3).

To explore further the possible applications of W_{max}, we performed the same analysis on two other datasets. First, we used an independent dataset (not generated with the PHENOPSIS platform) to explore the genetic variability in photosynthetic performance in six accessions of Arabidopsis from contrasted geographic locations. The plants displayed little variation during plant development in mean F_v/F_m values (Figure 6A). However, we observed an increase in photosynthetic heterogeneity S and W_{max} during plant development (see Additional file 4: Figure S3). We calculated the increase in W_{max} during development as the slope of the relationship between W_{max} and plant age. Interestingly, we found that 72% of the variability in plant dry mass at 48 days after stratification (DAS) was explained by the variation of W_{max} between 17 and 48 DAS ($P < 0.05$; $R = 0.85$; Figure 6B).

Second, we investigated the effect of soil inoculation with *Phyllobacterium brassicacearum* STM196 strain, a plant growth-promoting rhizobacteria (PGPR) that improves plant tolerance to moderate water deficit [42] and also increases plant survival under SWD [19]. Bresson *et al.*, 2014 [19] showed that STM196-inoculated surviving plants also exhibit a higher growth recovery after rewatering, leading to a higher plant biomass than non-inoculated plants [19]. Here, we showed that STM196-inoculation induced a faster and higher increase in W_{max} than non-inoculated plants from 2 days after rewatering (Additional file 5: Figure S4). This suggests that the positive effects of STM196 on growth recovery, biomass production and plant survival may be related to its effects on whole-plant photosynthetic heterogeneity.

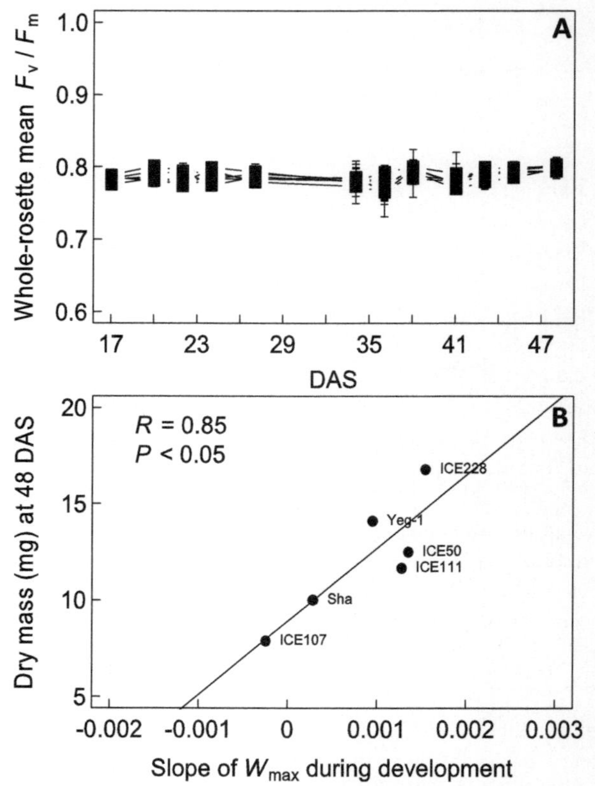

Figure 6 Variation of F_v/F_m and relationship between W_{max} and growth in six accessions of *A. thaliana*. (A) Dynamics of whole-rosette mean F_v/F_m as a function of days after stratification (DAS). Pots (n = 4) were manually watered three times per week to maintain good (non-stressing) soil moisture. **(B)** Relationship between the slope of W_{max} in the course of plant development and plant dry mass at 48 DAS. The accessions were collected from six different geographic origins (ICE107: South Italia; Sha: Kazakhstan; ICE111: South Italia; ICE50: Spain; Yeg-1: Caucasus; ICE228: South Tyrol). R: Pearson's product–moment correlation coefficient.

Discussion

Analysing the effects of environmental conditions on plant growth, survival and yield requires massive, rapid and non-invasive tools to track changes in plant performance. Non-invasive ChlF imaging has been developed to give insights into plant photosynthetic capacities and explore the ability of plants to tolerate various environmental stresses (*e.g.*, [8,16,43]). Most often the mean values of various indices of ChlF, including the widely used F_v/F_m, of an organ or a plant is used to characterize the response to a stressor (*e.g.*, [8,11]). However, a ChlF image is composed of a panel of pixels in a given range ($F_v/F_m = [0; 1]$). Hence, using mean values does not give a clear clue of the disparity of values that corresponds to contrasted physiology. Heterogeneity in the photosynthetic capacities of plants has been observed but rarely quantified in responses to a wide variety of external stimuli (*e.g.*, [10,16,25-32]). For instance, the establishment of F_v/F_m heterogeneity in response to stress has been described by sampling small areas on the leaf surface [10,29,30], visual inspection of the shape of distributions [25-29,32], or by F_v/F_m clustering [16]. However, this is prone to large variations depending on the species, experimenter and stress. ChlF heterogeneity is often admitted but its standardized, objective and reproducible quantification is still lacking. For instance, previous methods used threshold-based analysis to quantify the area and progression of senescence or damages [8,16]. Here, we proposed a method to quantify (1) the heterogeneity of F_v/F_m values at whole-plant level (S) and (2) the spatial efficiency of a photosynthetically heterogeneous plant (W_{max}). Although we applied our method to measurements of F_v/F_m in the Arabidopsis rosette under a severe water deficit scenario, we argue that the approach can be used with other ChlF parameters (*e.g.*, ΦII, NPQ) as well as in response to other stressing conditions that induce variations of the physiological status.

Photosynthetic heterogeneity is intrinsic to the development of plants

Our results showed that the distribution of F_v/F_m values shifted from unimodal to bimodal distributions both under WW and SWD conditions, and this despite a constant mean F_v/F_m in WW plants. This result indicates that heterogeneity in photosynthetic efficiency (*i.e.*, the increase in the proportion of bimodal F_v/F_m distributions) does not appear only under stress but is intrinsic to the development of plants. Importantly, S and W_{max} in WW plants also significantly increased during development. It therefore indicates that, even in the absence of visible senescence, (1) there were leaf regions exhibiting lower F_v/F_m, (2) low-efficiency leaf regions increased during development and, (3) the contribution of these latter was minor on whole-plant photosynthetic performance under WW conditions. There might be different sources of photosynthetic heterogeneity. First, at the whole-plant level, photosynthetic heterogeneity in plants might be caused by age-induced leaf senescence, *i.e.* by visible and non-visible cell death and nutrient remobilization, notably on the edges of the oldest leaves. In addition, the increase in the size of leaf veins with increasing leaf size can also induce a decrease in the mean F_v/F_m, as well as an increase in F_v/F_m heterogeneity. Second, at the sub-cellular level, some of PSII centres are inactive to linear electron transport. Functional PSII heterogeneity is for instance expected since 70-80% of PSII are located in the stacked grana region and the remaining PSII are located in the stroma-exposed region of the thylakoid membrane [44-47].

We also showed that the variation in photosynthetic heterogeneity might be a key trait related to plant growth, as suggested by the significant correlation between the increase in W_{max} during development and biomass in six contrasted Arabidopsis accessions and despite no distinct differences in the mean F_v/F_m between genotypes. The analysis of the distributions of F_v/F_m values, as proposed with S and W_{max}, allows the quantification of the whole-plant heterogeneity and may be more informative than the whole-plant mean value to investigate changes during plant development and genetic variation in plant performance.

The indicators of photosynthetic heterogeneity (S and W_{max}) are linked to plant tolerance to severe water deficit

Our analysis revealed that SWD affected the establishment of the intrinsic heterogeneity in plants during development. The heterogeneity of F_v/F_m values (quantified by S) across the rosette increased differently depending on the state of the plants. Importantly, S was a more sensitive indicator of the plant physiological status than the mean F_v/F_m. Indeed the mean F_v/F_m was stable during the first 14 days in stressed plants, while a strong photosynthetic heterogeneity was already present (Figures 2A and 4). S allows the discrimination between surviving and perishing plants earlier, *ca.* four days, than the whole-rosette mean F_v/F_m. The lag time before recovery was also shorter in S values than the mean F_v/F_m.

Surprisingly, surviving plants displayed a higher increase of S than the others plants during stress establishment, and perishing plants exhibited S dynamics similar to plants grown under WW conditions. This did not reflect the lower absolute values of F_v/F_m in perishing plants. The higher photosynthetic heterogeneity in surviving plants can be explained by the establishment of a gradient of F_v/F_m values from the tip to the base in the oldest leaves, often observed under water stress [10] and with high F_v/F_m values in the youngest leaves (as suggested in this study, see Figure 2). The analysis of the different parameters of bimodal distributions shows that SWD did not induce a global decrease of F_v/F_m, but plants rather maintained leaf regions with near-optimum F_v/F_m and sacrificed other leaf regions. Moreover, plant survival to SWD was associated to a large variability in F_v/F_m of the most damaged/senescing leaves; but to a low variability in the healthiest leaves or leaf regions (Figure 3). After rewatering, we showed that surviving plants recovered optimal mean F_v/F_m values with decreasing S, by loss of senescing leaves and/or by development of new leaves with optimal F_v/F_m. This is in accordance with the survival strategy of plants aiming at recycling and reallocating resources from the oldest or senescing leaves to active growing organs [23]. On the contrary, perishing plants displayed a decrease of F_v/F_m values in their oldest but also youngest leaves, resulting in a lower and constant value of bimodal separation S across the rosette. Perishing plants thus presented the same heterogeneity than well-watered plants.

However, the contribution of the most efficient leaf regions was more negatively impacted in the perishing plants. In surviving plants, W_{max} decreased during stress establishment and recovered rapidly after rewatering. This indicated that the contribution of the healthiest leaf regions was sufficient to compensate the negative effects of senescence and to support plant survival. By contrast, in perishing plants, W_{max} constantly decreased and reached negative values, reflecting that the contribution of low-efficiency leaf regions on whole-plant photosynthetic performance was too important and led to plant death. We also showed that increase in W_{max} is related to the improvement of plant survival and biomass production upon inoculation with a PGPR that affects photosynthesis in plants [42]. We argue that W_{max} could be a good indicator to predict plant survival under water stress, and presumably in response to other stresses.

The results of the sensitivity analysis (Additional file 3) showed that W_{max} is specifically sensitive to changes in the proportion of damaged leaf regions and in the lag time, *i.e.* the time before the appearance of the first symptoms. By contrast, S is independent of the proportion of leaf damaged but strongly impacted by the dynamics of the means. This simulation analysis suggests that the variation of W_{max} during stress (specifically, the rate of decrease and the time to reach negative values) can be a good indicator of stress tolerance in response to many environmental stresses. We argue that the parameters of the mathematical functions used to model the dynamics of mixture distributions could be used as useful tolerance/sensitivity indices. Additional work is however needed to implement physiological hypotheses under the spatio-temporal dynamics of the mixture parameters.

Together, our results showed that the dynamics of S and W_{max} allow quantifying photosynthetic heterogeneity and its relationship with plant performance, during plant development and under stress. Although the variation in mean F_v/F_m is informative about overall changes in plant performance, we showed that the modifications of F_v/F_m distributions are not accurately tractable by the modifications of mean F_v/F_m, or other threshold-based methods. For example, in comparison to Woo *et al.* [8] who used a 33% threshold for a mean $F_v/F_m = 0.80$ to discriminate surviving to perishing plants, here we showed that the bimodal separation S allowed discriminating plants four days earlier than mean F_v/F_m. Moreover, the quantitative measurement of photosynthetic heterogeneity can be followed, repeated and modelled in the course of plant development. For instance, negative values of W_{max} indicated death for individual plants; this may represent a powerful tool to screening plants to water stress. Our study highlights that the management of the spatial photosynthetic heterogeneity may be key to plant survival, and that the F_v/F_m heterogeneity is a sensitive measure of plant responses to the environment. Further studies will undoubtedly improve our capacity to predict plant tolerance, including survival, to different stressing environmental conditions using the indices of photosynthetic heterogeneity.

Conclusion

Heterogeneity in photosynthetic performance has implications for overall plant performance. In this study, we characterized the heterogeneity of chlorophyll fluorescence in leaves impacted by severe water deficit. We used a quantitative measure of (1) the heterogeneity of photosynthetic efficiency S, and (2) the spatial efficiency of a photosynthetically heterogeneous plant, W_{max}. S and W_{max} gave a more accurate indication of the dynamics of leaf senescence or damages induced by water deficit than the whole-plant mean F_v/F_m. For instance, they allowed an earlier discrimination between surviving and perishing plants. S and W_{max} were also strongly related to the genetic variability of growth between six contrasted accessions of Arabidopsis. Together our analyses suggest that S and W_{max} could be useful indicators of plant responses to their abiotic and biotic environments. Other studies are now needed to explore further the physiological causes and implications of the quantitative variations of photosynthetic heterogeneity and then improve our ability to predict plant response to the environment.

Methods

Phenotyping platform PHENOPSIS

The PHENOPSIS platform [33] includes three automatons set up in growth chambers strictly controlled for environmental conditions. In each chamber, an automated arm is equipped with a balance (XB620C; Precisa, Dietikon, Zurich, Swiss) and a complete system of irrigation (A1645-6; Electro finish, Saint-Egrève, France) to accurately weigh and irrigate up to 504 *Arabidopsis thaliana* plants in individual pots. The arm is also equipped with multiple devices for non-destructive acquisition of plant phenotypic data such as Charge Coupled Device (CCD) cameras or fluorescence sensors [1]. Light is provided by a bank of cool-white fluorescent tubes (36 W T8 Grolux, 120 cm; Sylvania) and quartz metal halide lamps (HPI-T Plus 400 W 645 E40; Philips). Light intensity is measured continuously at plant height, using a light sensor over the waveband of 400–700 nm (SKP215; Campbell Scientific, Logan, USA). Air temperature and relative humidity are measured every 20 s (HMP45C-L; Campbell Scientific). All measurements of temperature, light intensity and relative humidity are averaged and stored every 600 s in a data-logger (CR10X; Campbell Scientific) with data-logger support software (Loggernet

V4; Campbell Scientific). The climatic regulation of the growth-chambers is controlled by Loggernet software allowing the control of the desired environment by employing an air drier or a water sprayer to modify air humidity, an air-cooler or a heater to modify air temperature.

Plant material, growth conditions and irrigation treatments
The experimentations performed in the PHENOPSIS automaton used *A. thaliana* (L.) Heynh, accession Columbia-0. Five seeds were sown at the soil surface in 250 cm^3 cylindrical pots (10 cm high, 6 cm diameter) filled with a damped mixture (1:1, v:v) of loamy soil and organic compost Neuhaus N2 (see Additional file 6: Table S1 for soil chemical properties). Initial soil water content was controlled during pot filling by determining soil fresh weight (FW_{soil}) and soil dry weight (DW_{soil}, after 5 days at 80°C) every ten pots. Soil relative water content was calculated as $RWC_{soil} = (FW_{soil} - DW_{soil}) \times 100 \times DW_{soil}^{-1}$. Subsequent changes in pot weight were attributed to a change in soil water status. The pots were kept in the dark for 2 days and were damped with sprayed deionised water three times a day until germination. Then, plants were cultivated under 12 h day length (180 µmol m^{-2} s^{-1} photosynthetic photon flux density, at plant height). During germination phase (7 days), air temperature was set to 20°C day and night, and air relative humidity was adjusted in order to maintain constant water vapour pressure deficit (VPD) at 0.6 kPa. Then, plants were grown at 20/17°C day/night and 0.8 kPa of VPD. Just before the beginning of water stress, seedlings of similar sizes and developmental stages were selected and were thinned to one to four plants per pot. Each pot was daily weighed and watered with a modified one-tenth-strength Hoagland solution [48] to reach the target RWC_{soil}. RWC_{soil} was maintained at 0.35 g H_2O g^{-1} dry soil in the WW treatment (35%). SWD was started at L4 stage by stopping irrigation to decrease progressively RWC_{soil} to reach 0.06 g H_2O g^{-1} dry soil (6%). After RWC_{soil} = 6% g H_2O g^{-1} dry soil, irrigation was resumed by adding a daily constant volume of nutritive solution to reach the WW soil condition level, and was then maintained until final harvests at first flower open (stage 6.00; [34]). Soil water potential was determined by using a potentiometer (WP4-T dewpoint meter, Decagon Devices, Pullman, WA 99163, USA) during the soil drying.

An independent experiment was performed in the Max Planck Institute for Developmental Biology (Weigel lab, Tübingen, Germany) on six natural accessions from contrasted geographic origins: ICE107 (South Italia), ICE111 (South Italia), ICE228 (South Tyrol), ICE50 (Spain), Sha (Kazakhstan), Yeg-1 (Caucasus). Each accession was grown in four replicates. Five to ten seeds were sown at the soil surface of each pot and stratified during 2 days in the dark at 4°C. Plants were then grown at 16°C and under 8 h day length. At L4 stage, only one plant per pot was kept and grown until 48 DAS. Pots were manually watered 3 times a week to maintain good soil moisture. F_v/F_m was measured every 2–3 days from 17 to 48 DAS (using the same ChlF imaging system as in the PHENOPSIS automaton described below). At 48 DAS, rosettes were harvested, dried at 65°C for 4 days and weighed.

High resolution of chlorophyll fluorescence imaging
Acquisition of chlorophyll fluorescence images
ChlF measurements were performed using Imaging-PAM chlorophyll fluorometer and ImagingWin software application (ver. 2-45d, Heinz Walz GmbH) connected with PHENOPSIS automaton. ImagingWin software is driven by Optima PLC (ver. 2–14, build v323, by Optimalog SARL; Saint-Cyr-sur-Loire, France) that allows the automatic movement of Imaging-PAM implemented on the arm of robot (Figure 1A) and the ChlF measurement of each pot with identical settings. The measuring system consists of a 2/3″ Gigabit Ethernet IMAG K6-CCD camera (Manta, G-145B, ASG Allied Vision Technologies GmbH, Stadtroda, Germany), 1392 × 1040 pixel primary resolution with enhanced sensitivity by 4-pixel-binning resulting in 640 × 480 pixel images, coupled to an objective lens (F1.4/f = 12.5 mm; Cosmicar-Pentax, Hamburg, Germany) with a detector filter (RG665, 3 mm) and a short-pass interference filter ($\lambda < 770$ nm). Imaging-PAM is equipped with a powerful array of 44 high-power Luxeon LEDs for fluorescence excitation and actinic illumination with blue light (450 nm) as well as assessment of absorbed photosynthetically active radiation with the help of red light (650 nm) and near-infrared (NIR)-light (780 nm). The ChlF imaging system was equipped by a conic, black, metal shading hood of 21.5 cm height wherein the pot was lift up by the balance controlled by a hydraulic cylinder (Figure 1A). This system allows avoiding the illumination of neighbouring plants and achieving the optimal working of 18.5 cm distance from the LEDs resulting in an imaged area of approximately 9 x 12 cm with +/− 7% maximal deviation of intensity from the mean value. The image acquisition takes 30 s per plant, and different files are generated such as .PIM files (which are only usable by ImagingWin software) and .CSV files, which contain requested parameters such as F_0, F_m and F_v/F_m averaged on a region of interest. Two images of ChlF acquisition are also generated: *.JPEG files which give a representation of the image acquired in false colour (from black pixel values (0) through red, yellow, green, blue to purple (ending at 1)) and *.TIFF files with contain different stacks of fluorescence parameters (see below).

Measurement of maximum efficiency of PSII (F_v/F_m)
The ChlF measurement was initiated by exposing dark-adapted leaf to measuring light pulses (1 Hz frequency,

Intensity 2) for determination of F_0. The F_m level of fluorescence is recorded during a saturating pulse (Si 9, width 800 ms). F_v/F_m, calculated as $(F_m-F_0)/F_m$, provides the maximum quantum yield of PSII (*i.e.*, photosynthetic efficiency). In the study, F_v/F_m was daily performed, from the two first leaves to bolting, on dark-adapted plants (8–12 h of dark), under WW and SWD conditions.

Data extraction of photosynthetic efficiency and rosette expansion during time course

We developed an ImageJ macro "PHENOPSIS-Fluo" to semi-automatically extract whole-rosette F_v/F_m pixel values and total leaf area from the generated TIFF files (containing F_0 and F_m pixel values, and NIR images). The analysis of ChlF images starts with image segmentation from NIR pictures, *i.e.* the automatic separation of the region of interest (here the rosette) from the background. Then, by subtracting and dividing F_0 and F_m stacks, $[(F_m - F_0)/F_m]$, the macro generates an image of F_v/F_m pixel values, which are represented in the 255 greyscale (0 and 255 corresponding to the minimum and maximum pixel values of the selection, respectively). The macro gives the whole-rosette F_v/F_m mean and the list of pixel values across the plant. Projected area of the rosette (RA_{proj}) was also determined from plant selection. Different plants in the same pot can be independently measured. Whole-rosette mean F_v/F_m extracted with the macro "PHENOPSIS-Fluo" was highly correlated with the mean of the rosette extracted from the commercial software ImagingWin ($R^2 = 0.98$; Additional file 7: Figure S5). The "PHENOPSIS-Fluo" macro is available on the PHENOPSIS website (http://bioweb.supagro.inra.fr/phenopsis/MacroImageJ.php).

Under SWD conditions, growth dynamics of surviving plants were modelled as a sigmoid curve fitted following $RA_{proj} = a \,/\, [1 + \exp\text{-}[(d\text{-}a/2)/b]]$ where a is the maximum area, and d is the number of days after L4 stage. Under WW conditions, an exponential curve was fitted such as $RA_{proj} = \exp\,(a - b \times d)$.

Statistical analyses and modelling procedures

All analyses were performed using R 3.1[49]. Comparisons of mean trait values between treatments were performed with Kruskal-Wallis non-parametric tests. Survival percentage was calculated as the proportion of surviving plants at the end of experiment compared to initial number of plants. Plant survival was analysed by Chi^2 tests. Non-parametric significance test for unimodality, Hartigan's dip test (R package 'dip test' [35-37]) was used to identity multimodal distribution, with p-value < 0.01 as the significance threshold for departure from unimodality. Then, the REBMIX algorithm for finite mixture models ([38]; R package 'rebmix') as used to characterize each mode i of the bimodal distributions of F_v/F_m values (*i.e.*, mean μ_i, standard deviation σ_i and weight ρ_i), using the Bayesian Information Criteria (BIC).

We developed a modelling procedure of the temporal dynamics of the means, standard deviations and proportions (weights) of the mixture distributions. We then simulated the variation of these parameters and, subsequently, the variation of S and W_{max} (Additional file 3).

All meteorological and phenotypic data, ChlF files and images, R scripts are available in the PHENOPSIS web site (http://bioweb.supagro.inra.fr/phenopsis/) and database [50].

Additional files

Additional file 1: Figure S1. Soil water potential during soil drying. Soil water potential was determined using a potentiometer (WP4-T dewpoint meter, Decagon Devices, Pullman, WA 99163, USA) during soil drying (from 0.35 to 0.06 g H_2O g^{-1} dry soil).

Additional file 2: Figure S2. Representations of vegetative rosettes in F_v/F_m false colour (from black pixel values (0) through red, yellow, green, blue to purple (ending at 1)) under well-watered (WW) conditions and under severe water stress (SWD; surviving and perishing plants) during time courses.

Additional file 3: Modelling the spatial heterogeneity of photosynthetic efficiency under stress.

Additional file 4: Figure S3. Spatial heterogeneity S of whole-plant F_v/F_m and W_{max} in six accessions of *A. thaliana*. (A) Bimodal separation S and (B) the spatial efficiency of a photosynthetically heterogeneous plant W_{max} as a function of days after stratification (DAS). The six accessions were collected from different geographic origins (ICE107: South Italia; Sha: Kazakhstan; ICE111: South Italia; ICE50: Spain; Yeg-1: Caucasus; ICE228: South Tyrol).

Additional file 5: Figure S4. W_{max} under biotic and abiotic interaction: effect of inoculation by a plant growth promoting rhizobacteria (PGPR) under severe water stress. W_{max} in non-inoculated plants (NI; solid lines) and inoculated plants with the PGPR *Phyllobacterium brassicacearum* (STM196; dashed lines) under well-watered (WW) conditions and severe water stress (SWD; surviving and perishing) as a function of days after four-leaves stage (beginning of SWD) until bolting.

Additional file 6: Table S1. Soil chemical properties of the compost (Neuhaus N2), soil and two mixtures of both. Mixture 1 was sampled before experimentation and mixture 2 was sampled after experimentations. nd: not determined. Soil analysis was performed by ALFA Agricultural Service and Research Building, Soil Testing Laboratory of Auburn University.

Additional file 7: Figure S5. Correlation between extracted whole-rosette mean F_v/F_m with the macro "PHENOPSIS-Fluo" and extracted from CSV files from ImagingWin software. R^2 is the Pearson's coefficient of correlation between 164 rosettes analysed with both the PHENOPSIS-Fluo macro and the ImagingWin software.

Abbreviations

ChlF: Chlorophyll fluorescence; PSII: Photosystem II; Q_A: Quinone A, the primary stable electron acceptor of PSII centres; F_v/F_m: Maximum quantum efficiency of PSII photochemistry (photosynthetic efficiency); F_0: Minimal fluorescence emission of a dark-adapted plant; F_m: Maximum fluorescence emission after a short pulse of a saturating light; F_v: Variable fluorescence from dark-adapted plant; S: Spatial heterogeneity of F_v/F_m; W_{max}: Spatial efficiency of a photosynthetically heterogeneous plant; SWD: Severe water deficit; L4: Four-leaves stage; RWC_{soil}: Soil relative water content; WW: Well watered condition; DAS: Days after stratification; PGPR: Plant growth-promoting rhizobacteria; CCD: Charge Coupled Device; FW_{soil}: Soil fresh weight; DW_{soil}: Soil dry weight; VPD: Water vapour pressure deficit; NIR: Near-infrared; RA_{proj}: Projected area of the rosette.

Competing interests

The authors declare that they have no competing interests.

Authors' contributions

JB, DV conceived and designed the experiments in the PHENOPSIS platform. JB, MD and GK performed the experiments in the PHENOPSIS platform. FV conceived and performed the experiments on the six Arabidopsis accessions. JB, FV and DV analysed the data. DV performed the modelling. DV and CG contributed reagents/materials/analysis tools. JB, FV, DV, CG and MD contributed to the writing of the manuscript. All authors read and approved the final manuscript.

Acknowledgements

We are very grateful to G. Rolland, A. Bediee, F. Bouvery and C. Balsera for help during the experiments. We thank F. Fiorani for giving us advice on the design of the ChlF system. We thank D. Fourré, M. Desigaux, O. Meyerhoff and J. Kolbowski for helpful technical supports. We thank B. Touraine and F. Varoquaux for constructive remarks on plant-microbes interactions and for providing us the rhizobacteria (STM196). We thank B. Genty and S. Bieker for helpful comments and suggestions on this paper. We thank D. Weigel for his contribution and support at different stages of the study. We thank A. Christophe for the helpful discussions during the modelling procedure. We also thank two anonymous reviewers for helpful comments on this paper. JB was funded by French Ministry of Higher Education and Research and was supported by the Institutional Strategy of the University of Tuebingen (Deutsche Forschungsgemeinschaft, ZUK 63). FV was funded by a post-doctoral grant from the Max Planck Society. This work was made possible by a series of grants supporting the development of the plant phenotyping platform: the European Plant Phenotyping Network funded by the FP7 Research Infrastructures Program of the European Union (EPPN, grant agreement no. 284443) and the EIT Climate-KIC project AgWaterBreed.

Author details

[1]Laboratoire d'Ecophysiologie des Plantes sous Stress Environnementaux (LEPSE), INRA, Montpellier SupAgro, UMR759, F-34060 Montpellier, France. [2]Laboratoire des Symbioses Tropicales et Méditerranéennes (LSTM), UMR113, Université Montpellier 2-IRD-CIRAD-INRA-SupAgro, F-34095 Montpellier, France. [3]Center for Plant Molecular Biology (ZMBP), General Genetics, University of Tuebingen, D-72076 Tuebingen, Germany. [4]Max Planck Institute for Developmental Biology, D-72076 Tuebingen, Germany.

References

1. Granier C, Vile D. Phenotyping and beyond: modelling the relationships between traits. Curr Opin Plant Biol. 2014;18:96–102.
2. Baker NR, Rosenqvist E. Applications of chlorophyll fluorescence can improve crop production strategies: an examination of future possibilities. J Exp Bot. 2004;55:1607–21.
3. Porcar-Castell A, Tyystjärvi E, Atherton J, van der Tol C, Flexas J, Pfündel EE, et al. Linking chlorophyll *a* fluorescence to photosynthesis for remote sensing applications: mechanisms and challenges. J Exp Bot. 2014;doi:10.1093/jxb/eru191.
4. Chaerle L, Leinonen I, Jones HG, Van Der Straeten D. Monitoring and screening plant populations with combined thermal and chlorophyll fluorescence imaging. J Exp Bot. 2007;58:773–84.
5. Murchie EH, Lawson T. Chlorophyll fluorescence analysis: a guide to good practice and understanding some new applications. J Exp Bot. 2013; doi:10.1093/jxb/ert208.
6. Maxwell K, Johnson GN. Chlorophyll fluorescence - a practical guide. J Exp Bot. 2000;51:659–68.
7. Genty B, Briantais J-M, Baker NR. The relationship between the quantum yield of photosynthetic electron transport and quenching of chlorophyll fluorescence. Biochimica et Biophysica Acta (BBA)-General Subjects. 1989;990:87–92.
8. Woo NS, Badger MR, Pogson BJ. A rapid, non-invasive procedure for quantitative assessment of drought survival using chlorophyll fluorescence. Plant Methods. 2008;4:27.
9. Hunt S. Measurements of photosynthesis and respiration in plants. Physiol Plant. 2003;117:314–25.
10. Sperdouli I, Moustakas M. Spatio-temporal heterogeneity in *Arabidopsis thaliana* leaves under drought stress. Plant Biol. 2012;14:118–28.
11. Gray GR, Hope BJ, Qin XQ, Taylor BG, Whitehead CL. The characterization of photoinhibition and recovery during cold acclimation in *Arabidopsis thaliana* using chlorophyll fluorescence imaging. Physiol Plant. 2003;119:365–75.
12. Ehlert B, Hincha DK. Chlorophyll fluorescence imaging accurately quantifies freezing damage and cold acclimation responses in Arabidopsis leaves. Plant Methods. 2008;4:12.
13. Janka E, Körner O, Rosenqvist E, Ottosen C-O. High temperature stress monitoring and detection using chlorophyll *a* fluorescence and infrared thermography in chrysanthemum (*Dendranthema grandiflora*). Plant Physiol Biochem. 2013;67:87–94.
14. Quilliam RS, Swarbrick PJ, Scholes JD, Rolfe SA. Imaging photosynthesis in wounded leaves of *Arabidopsis thaliana*. J Exp Bot. 2006;57:55–69.
15. Bolhàr-Nordenkampf HR, Öquist G. Chlorophyll fluorescence as a tool in photosynthesis research. In: Hall DO, Scurlock JMO, Bolhàr-Nordenkampf HR, Leegood RC, Long SP, editors. Photosynthesis and Production in a Changing Environment. Springer Netherlands; 1993. p. 193–206.
16. Rousseau C, Belin E, Bove E, Rousseau D, Fabre F, Berruyer R, et al. High throughput quantitative phenotyping of plant resistance using chlorophyll fluorescence image analysis. Plant Methods. 2013;9:17.
17. Rincon A, Valladares F, Gimeno TE, Pueyo JJ. Water stress responses of two Mediterranean tree species influenced by native soil microorganisms and inoculation with a plant growth promoting rhizobacterium. Tree Physiol. 2008;28:1693–701.
18. Heidari M, Golpayegani A. Effects of water stress and inoculation with plant growth promoting rhizobacteria (PGPR) on antioxidant status and photosynthetic pigments in basil (*Ocimum basilicum L.*). J Saudi Soc Agric Sci. 2012;11:57–61.
19. Bresson J, Vasseur F, Dauzat M, Labadie M, Varoquaux F, Touraine B, et al. Interact to survive: *Phyllobacterium brassicacearum* improves Arabidopsis tolerance to severe water deficit and growth recovery. PLoS One. 2014;9:e107607.
20. Chaves MM. Effects of water deficits on carbon assimilation. J Exp Bot. 1991;42:1–16.
21. Krause GH. Photoinhibition of photosynthesis. An evaluation of damaging and protective mechanisms. Physiol Plant. 1988;74:566–74.
22. McDowell NG. Mechanisms linking drought, hydraulics, carbon metabolism, and vegetation mortality. Plant Physiol. 2011;155:1051–9.
23. Munné-Bosch S, Alegre L. Die and let live: leaf senescence contributes to plant survival under drought stress. Funct Plant Biol. 2004;31:203–16.
24. Hensel LL, Grbić V, Baumgarten DA, Bleecker AB. Developmental and age-related processes that influence the longevity and senescence of photosynthetic tissues in Arabidopsis. Plant Cell. 1993;5:553–64.
25. Genty B, Meyer S. Quantitative mapping of leaf photosynthesis using chlorophyll fluorescence imaging. Funct Plant Biol. 1995;22:277–84.
26. Baker NR, Oxborough K, Lawson T, Morison JI. High resolution imaging of photosynthetic activities of tissues, cells and chloroplasts in leaves. J Exp Bot. 2001;52:615–21.
27. Leipner J, Oxborough K, Baker NR. Primary sites of ozone-induced perturbations of photosynthesis in leaves: identification and characterization in *Phaseolus vulgaris* using high resolution chlorophyll fluorescence imaging. J Exp Bot. 2001;52:1689–96.
28. Oxborough K, Baker N. An instrument capable of imaging chlorophyll *a* fluorescence from intact leaves at very low irradiance and at cellular and subcellular levels of organization. Plant Cell Environ. 1997;20:1473–83.
29. Hogewoning SW, Harbinson J. Insights on the development, kinetics, and variation of photoinhibition using chlorophyll fluorescence imaging of a chilled, variegated leaf. J Exp Bot. 2007;58:453–63.
30. Calatayud A, Roca D, Martínez P. Spatial-temporal variations in rose leaves under water stress conditions studied by chlorophyll fluorescence imaging. Plant Physiol Biochem. 2006;44:564–73.
31. Daley PF, Raschke K, Ball JT, Berry JA. Topography of photosynthetic activity of leaves obtained from video images of chlorophyll fluorescence. Plant Physiol. 1989;90:1233–8.
32. Lazár D, Sušila P, Nauš J. Early detection of plant stress from changes in distributions of chlorophyll *a* fluorescence parameters measured with fluorescence imaging. J Fluoresc. 2006;16:173–6.
33. Granier C, Aguirrezabal L, Chenu K, Cookson SJ, Dauzat M, Hamard P, et al. PHENOPSIS, an automated platform for reproducible phenotyping of plant responses to soil water deficit in *Arabidopsis thaliana* permitted the identification of an accession with low sensitivity to soil water deficit. New Phytol. 2006;169:623–35.
34. Boyes DC, Zayed AM, Ascenzi R, McCaskill AJ, Hoffman NE, Davis KR, et al. Growth stage-based phenotypic analysis of Arabidopsis: a model for high throughput functional genomics in plants. Plant Cell. 2001;13:1499–510.

35. Hartigan JA, Hartigan P. The dip test of unimodality. The Annals of Statistics. 1985;13:70–84.
36. Hartigan PM. Algorithm AS 217: Computation of the Dip Statistic to Test for Unimodality. J R Stat Soc: Ser C: Appl Stat. 1985;34:320–5.
37. Maechler M, Ringach D. diptest: Hartigan's dip test statistic for unimodality–corrected code. R package version 075–74. 2012. Available online at: http://CRANR-projectorg/package=diptest [Retrieved: January 4, 2013].
38. Nagode M. rebmix. An R Package for continuous and discrete finite mixture models. 2004.
39. Zhang C, Mapes BE, Soden BJ. Bimodality in tropical water vapour. Q J Roy Meteorol Soc. 2003;129:2847–66.
40. Demmig-Adams B, Adams Iii WW, Barker DH, Logan BA, Bowling DR, Verhoeven AS. Using chlorophyll fluorescence to assess the fraction of absorbed light allocated to thermal dissipation of excess excitation. Physiol Plant. 1996;98:253–64.
41. Björkman O, Demmig B. Photon yield of O_2 evolution and chlorophyll fluorescence characteristics at 77 K among vascular plants of diverse origins. Planta. 1987;170:489–504.
42. Bresson J, Varoquaux F, Bontpart T, Touraine B, Vile D. The PGPR strain *Phyllobacterium brassicacearum* STM196 induces a reproductive delay and physiological changes that result in improved drought tolerance in Arabidopsis. New Phytol. 2013;200:558–69.
43. Barbagallo RP, Oxborough K, Pallett KE, Baker NR. Rapid, noninvasive screening for perturbations of metabolism and plant growth using chlorophyll fluorescence imaging. Plant Physiol. 2003;132:485–93.
44. Chylla R, Garab G, Whitmarsh J. Evidence for slow turnover in a fraction of photosystem II complexes in thylakoid membranes. Biochimica et Biophysica Acta (BBA)-Bioenergetics. 1987;894:562–71.
45. Chylla RA, Whitmarsh J. Light saturation response of inactive photosystem II reaction centers in spinach. Photosynth Res. 1990;25:39–48.
46. Cao J. Chlorophyll *a* fluorescence transient as an indicator of active and inactive photosystem II in thylakoid membranes. Biochimica et Biophysica Acta (BBA)-Bioenergetics. 1990;1015:180–8.
47. Lavergne J, Leci E. Properties of inactive photosystem II centers. Photosynth Res. 1993;35:323–43.
48. Hoagland DR, Arnon DI. The water-culture method for growing plants without soil. Calif Agric Exper Stn Circ. 1950;347:1–32.
49. R Development Core Team. R: a language and environment for statistical computing. Vienna, Austria: R Foundation for Statistical Computing; 2009.
50. Fabre J, Dauzat M, Negre V, Wuyts N, Tireau A, Gennari E, et al. PHENOPSIS DB: an Information System for Arabidopsis thaliana phenotypic data in an environmental context. BMC Plant Biol. 2011;11:77.

CSGM Designer: a platform for designing cross-species intron-spanning genic markers linked with genome information of legumes

Jin-Hyun Kim[1†], Chaeyoung Lee[1†], Daejin Hyung[2], Ye-Jin Jo[3], Joo-Seok Park[4], Douglas R Cook[5] and Hong-Kyu Choi[3*]

Abstract

Background: Genetic markers are tools that can facilitate molecular breeding, even in species lacking genomic resources. An important class of genetic markers is those based on orthologous genes, because they can guide hypotheses about conserved gene function, a situation that is well documented for a number of agronomic traits. For under-studied species a key bottleneck in gene-based marker development is the need to develop molecular tools (e.g., oligonucleotide primers) that reliably access genes with orthology to the genomes of well-characterized reference species.

Results: Here we report an efficient platform for the design of cross-species gene-derived markers in legumes. The automated platform, named CSGM Designer (URL: http://tgil.donga.ac.kr/CSGMdesigner), facilitates rapid and systematic design of cross-species genic markers. The underlying database is composed of genome data from five legume species whose genomes are substantially characterized. Use of CSGM is enhanced by graphical displays of query results, which we describe as "circular viewer" and "search-within-results" functions. CSGM provides a virtual PCR representation (eHT-PCR) that predicts the specificity of each primer pair simultaneously in multiple genomes. CSGM Designer output was experimentally validated for the amplification of orthologous genes using 16 genotypes representing 12 crop and model legume species, distributed among the galegoid and phaseoloid clades. Successful cross-species amplification was obtained for 85.3% of PCR primer combinations.

Conclusion: CSGM Designer spans the divide between well-characterized crop and model legume species and their less well-characterized relatives. The outcome is PCR primers that target highly conserved genes for polymorphism discovery, enabling functional inferences and ultimately facilitating trait-associated molecular breeding.

Keywords: Marker design, Genic marker, Legumes, eHT-PCR, Circular viewer

Background

Genetic markers and mapping are a principal means to study the genetic features of organisms. In the case of crop species, genetic markers can facilitate the introgression of desirable traits among accessions and species, and they provide the basis of marker-assisted trait dissection (e.g., QTL analysis). Genetic markers were first employed in 1932 based on isozyme analysis and progressed to DNA based markers as electrophoretic methods became practical in the early 1980's [1]. In the absence of DNA sequence data, anonymous genetic markers (e.g., restriction fragment length polymorphisms [RFLP] and related methods) were the dominant technologies [2]. With the advent of genome sequencing projects, it became feasible to develop genetic markers that amplified common genes among related species. Initially such approaches depended on transcript sequence (e.g., Expressed Sequence Tags), which in many cases is now supplanted by the availability of whole genome sequence (WGS) information. For sequence-characterized genomes, high throughput genotyping-by-sequencing (GBS) methods eliminate the need to create individual genetic markers. However, numerous species lack sufficient genome data for GBS methods, and in these cases the use of PCR amplification remains an important tool for marker development.

* Correspondence: hkchoi@dau.ac.kr
†Equal contributors
[3]Department of Genetic Engineering, Dong-A University, Saha-Gu Nakdong-Daero 550 beongil 37, Busan 604-714, Republic of Korea
Full list of author information is available at the end of the article

Genetic markers that target orthologous genes are particularly useful, because they permit cross-species inferences (e.g., of structural and conservation), and this approach has been used widely in legumes [3] and other species.

The enormity of genomic information currently available in public databases is both an asset and a deterrent to large-scale molecular marker development. On the one hand such information provides the basis of systematic identification of orthologous genes and corresponding tools (e.g., PCR primers), while on the other hand doing so requires organized bioinformatics activities not available for many small research groups working on under-characterized species. In the absence of such bioinformatic activities, development of genetic markers remains laborious and time-consuming. Of particular interest are cross-species gene-targeted markers, which have utility in translating marker information across different crop plants, however such markers can be technically difficult to design.

Towards this end, we focus on the Fabaceae. The Fabaceae (legume family) is the third largest family among flowering plants and the second most important group of crop plants, second only to the Poaceae (grass family). Legumes consist of approximately 20,000 species among three subfamilies, the Caesalpionoideae, Mimosoideae and Papilionoideae [4]. Within the most recently diverged Papilionoideae subfamily, the galegoid and phaseoloid/ millettioid clades contain the vast majority of agricultural species. Several model and crop legumes derive from these two clades and many of these species have substantially complete genome sequences, including *Medicago truncatula* [5], *Lotus japonicus* [6], *Glycine max* [7], *Phaseolus vulgaris* [8] and *Cajanus cajan* [9]. Because the phylogenetic space encompassed by these species includes the majority of crop legumes, their genome data provides an opportunity to identify large sets of genes that are also common to related uncharacterized legume species, and on this basis one can develop genetic markers that are operative for a range of legume crops that lack genomic resources.

Here we describe the "CSGM (cross-species genic marker) Designer" web resource, a legume-focused platform for the systematic design of cross-species genic markers. To enhance ease of use we employed graphic visualization tools. The "circular viewer" pane depicts the attributes of numerous competing PCR primer candidates to facilitate selection by the user. The eHT (electronic high throughput)-PCR pane reports the results of *in silico* PCR at the genome-wide scale, which aids the selection of primer pairs most likely to yield specific-versus-nonspecific PCR products prior to laboratory validation.

Results

Conceptual implementation of cross-species intron-spanning genic marker designing platform

CSGM Designer implements an automated platform to design gene-targeted molecular markers, with the goal of creating PCR amplicons that operate across multiple evolutionarily diverged, but related species (Figure 1). Towards this end, CSGM Designer focuses on PCR primer design within exon regions, where nucleotide sequence divergence is typically more constrained than non-protein coding regions. To accommodate the need for high rates of polymorphism within crop germplasm, which is of low genetic diversity, CSGM Designer identifies primer pairs that amplify across less constrained intron sequences, thus enhancing the success rate of marker development. The position of exon-intron junctions can be accurately detected by aligning genomic sequence with the corresponding CDS (Figure 1), permitting CSGM Designer to avoid primer design over splice junctions that would preclude PCR amplification. The best-conserved exon regions between two aligned orthologous genes are analyzed to identify DNA sequence regions with the lowest level of mismatch, and

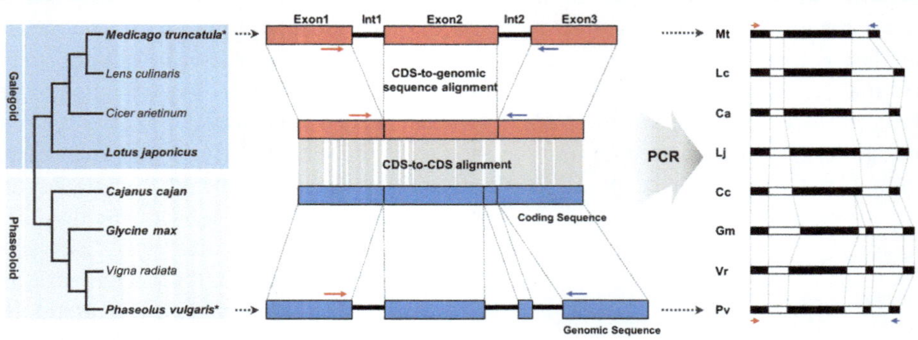

Figure 1 Conceptual design of cross-species intron-spanning genic markers. Five fully sequenced legume species used to construct the database are denoted in bold italic. Two species, *M. truncatula* and *P. vulgaris*, which were used to anchor the two clades, are highlighted with the asterisk superscript. Conserved regions in the CDS-to-CDS alignment, where are the sites for cross-amplifiable PCR primer pairs, are depicted using grey boxes.

these highly conserved intervals become the template for primer design.

Maintaining specificity in gene-targeted PCR amplification is frequently hampered by the presence of numerous close paralogs within the target genome, a particular problem for large gene families. PCR products that include multi-genomic loci are not readily distinguished by electrophoretic separation, unless primer pairs are selected to generate paralog-specific amplicon size variation. Thus CSGM Designer includes a PCR simulation module to predict PCR amplicon size with given primer pairs. In addition, the platform is delivered through an interactive web-based GUI structured to guide users to efficient selection of PCR primer pairs.

Construction of legume genomic information-linked genic marker design platform

CSGM Designer consists of three parts: a user interface, an analysis module, and the underlying database (Figure 2). The user interface is structured around two main windows, one for data input with parameter control boxes, and one for the visualization of resulting primer pairs. The database includes information for CDS and genomic sequences of five substantially sequenced legume genomes (i.e., *M. truncatula*, *G. max*, *P. vulgaris*, *C. cajan* and *L. japonicus*) and Arabidopsis. The sequence data of target genes can be retrieved from the database using locus ID as a query, or by user provided candidate gene sequences that are compared to the database content. The user interface permits the user to adjust primer design parameters, and resulting primer pair candidates are displayed in the order of scores for individual primer pairs, either in a graphical "circular viewer" or in a "tabular viewer".

CSGM Designer's analysis module consists of two parts, "Primer Maker" and the eHT-PCR module. Primer Maker is a primer design module that incorporates two alignment algorithms. Splign [10] performs CDS-to-genomic sequence alignments with default options to locate exon-intron junctions. Simultaneously, the ClustalO program [11] conducts CDS-to-CDS alignment between gene pairs of full length. Splign and ClustalO outputs are used by Structure Maker to produce a pairwise gene structure, including the location of conserved and variable regions, which typically correspond to exons and introns. The output of Structure Maker provides the template for PCR primer design, which is carried out by Primer Maker.

Virtual amplification specificity of primer pairs created within the Primer Maker module is simulated by the eHT-PCR module. The BWA algorithm [12] was adapted for this purpose and serves a central role in the eHT-PCR module. Using BWA, numerous primer pair candidates are mapped against the reference genomes and evaluted for the specificity of cross-genome PCR amplification. By employing *in silico* PCR simulation prior to experimental practice, users can estimate how selected primer pairs will perform within and among the reference genomes, and by inference have utility in target genomes not included in the database.

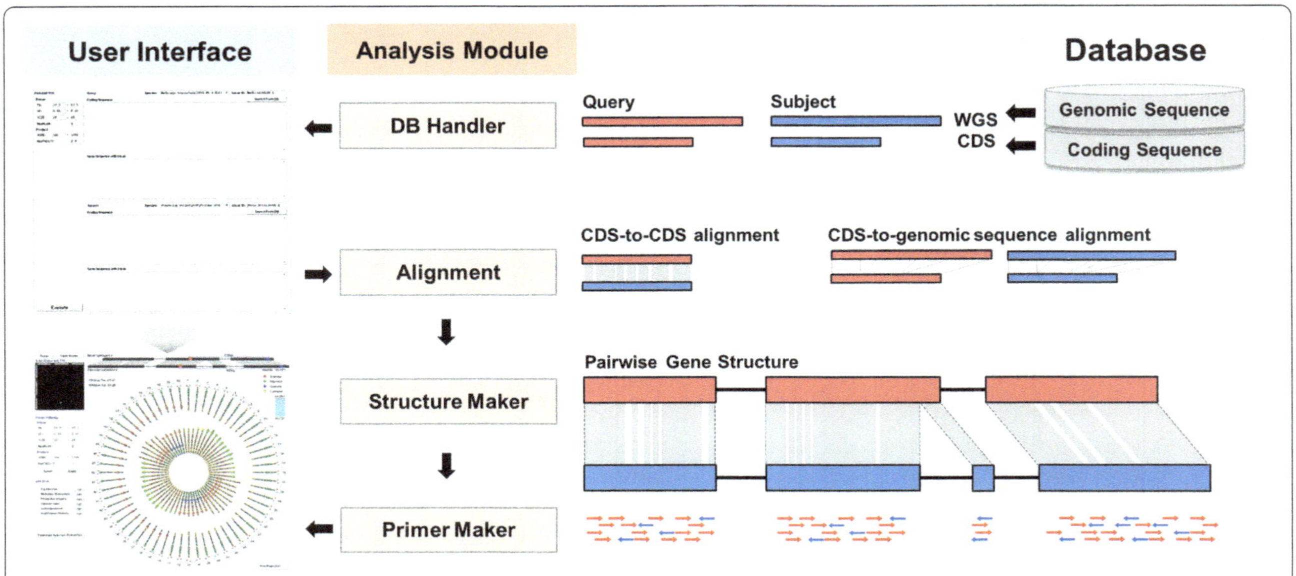

Figure 2 Organization and workflow of the CSGM Designer platform. The platform consists largely of database, analysis module and user interface. The panes on left upper and lower corners depict the windows for data input and circular viewer of results, respectively. Forward and reverse primers are denoted with red and blue arrows, respectively.

Features of CSGM Designer

CSGM Designer incorporates a dynamic GUI that creates a "circular viewer" of primer candidates (Figure 3) and an "eHT-PCR" representation of virtual PCR outcomes (Figure 4), both of which are intended to enhance the user's experience during cross-species marker development. The top-scoring 60 primer pairs are displayed in the first page of the circular viewer, providing the user with a graphical overview of otherwise complex query results. The viewer provides a range of information intended to guide the user towards the most suitable primer pairs, including primer melting point (Tm) values, oligonucleotide sequences, and mismatch distribution between primers and target sequences. Tm values permit the user to inspect differences between primer pair combinations. Colors are used to distinguish each of four different nucleotides, and mismatched positions are highlighted, providing a snapshot of the frequency and location of mismatched nucleotides (Figure 3A). Even for the best-conserved regions of candidate genes, mismatches are typical when sequences are compared among species. PCR is robust to such mismatch, assuming that mismatches have particular characteristics; here we limit the output to three mismatches, while the user can impose additional criteria. For example, mismatches in neighboring or consecutive sites and at the 3′-ends of primers, as well as AT-rich 3′ ends, can negatively affect PCR performance. The circular viewer facilitates such user-supplied criteria, enhancing success rates of PCR amplification.

CGSM Designer's "overview pane", on the upper left corner of the circular viewer, presents positions of all PCR primers within the context of target gene structure (Figure 3A). PCR product size can be visually estimated in the overview pane. The overview pane also provides a simple visual assessment of the relationship between primer positions and exon-intron distribution, revealing the number of introns included in virtual PCR products and whether their location of primer pairs is evenly distributed or clustered.

While the circular viewer provides a search-at-a-glance perspective, users can also investigate expanded details of chosen primer pairs in a "primer viewer" panel. In this viewer, users encounter further details of chosen primer pairs, including the location of primer hybridization and of mismatched bases in the context of sequence text. Genomic structure of aligned genes provides information on introns-exon structure, including nucleotide conservation/variation and PCR product sizes (Figure 3B).

PCR primer design can be further analyzed by the "search-within-result" function, to narrow primer candidates based on user provided criteria. Post-processing refinement allows the user to revisit parameters used for the first round of analysis, thereby increasing selectivity for primer candidates based on the user's needs. Users can also directly select multiple primer pairs, with the selected primer pairs automatically incorporated into CSGM's other viewer functions (Figure 3A).

Secondly, specificity of cross-species PCR amplification can be refined and guided by results from the eHT-PCR module. Reduced-specific PCR amplification can be difficult to overcome, especially in cases of DNA fragments derived from conserved gene families. The eHT-PCR analysis can be conducted using full genomic sequence information for any of the six species contained within the CSGM Designer database. Users can select all or a subset of species for analysis and the results can be seen in the primer viewer pane (Figure 3B). A specific virtual amplicon (i.e., single genomic locus) is represented with its corresponding information about genomic position of species selected for eHT-PCR, while results of non-specific PCR products (i.e., two or more genomic loci) are displayed by number of possible PCR amplifications.

eHT-PCR preprocessing pipeline is illustrated in Figure 4. The eHT-PCR simulation tool is performed in three steps: i) generation of non-redundant primer set by removing redundant primers from the initial pool of total primer candidates, ii) mapping the reduced primer set to selected genomes using the BWA program, and iii) predicting amplified genomic sites based on BWA mapping results (Figure 4). In conjunction with BWA-based genome-wide mapping of primer candidates, CSGM Designer can effectively reduce the number of possible forward/reverse primer combinations by removing redundancy in primer pools and false primer pairs (Figure 4). Subsequent combinations of forward and reverse primer pools are used to generate a list of virtual amplicons. Because eHT-PCR can be implemented for all six species in the database, users can rapidly determine which primer pairs are likely to be specific or non-specific PCR amplicons, and whether this prediction varies among species.

PCR primer pair design using CSGM platform

CSGM Designer platform is intended to be intuitive. The step-by-step primer design can be processed according to the following procedures. A graphic-aided easy-to-use tutorial is also available on the CSGM homepage (http://tgil.donga.ac.kr/CSGMdesigner/tutorial.html).

(i) To design primer pairs, the CSGM Designer requires both genomic and coding sequences from at least two species. Users can retrieve the sequences of target gene using locus IDs from the database or the user can directly input candidate gene sequences

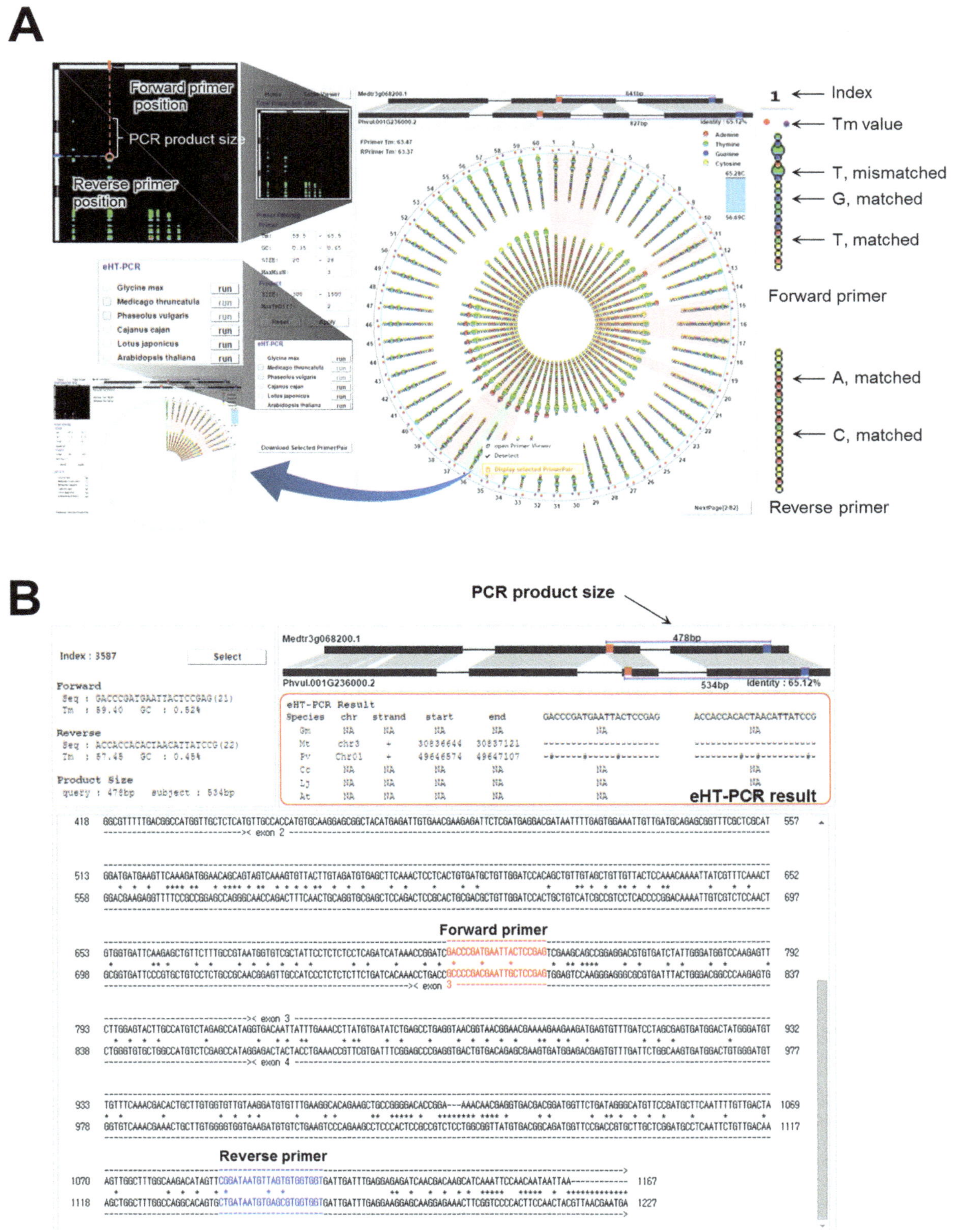

Figure 3 (See legend on next page.)

(See figure on previous page.)
Figure 3 User interface for reviewing results of designed marker candidates. **A.** Circular viewer. The circular viewer presents a maximum of 60 primer pairs per page. Outer circle represents the forward primers, while inner circle the reverse primers. Each of four different nucleotides are denoted with different colors; red-A, green-T, blue-G and yellow-C. Mismatched nucleotides are highlighted using larger circles over matched ones. The pane on left upper corner display overall positional information for primers designed under selected parameters. Tm values for forward and reverse primers are depicted with red and blue circles, respectively, within minimum-to-maximum range, and accurate Tm values can be checked by selecting one of the primer pair candidates. Individual primer pairs of user's interest can be selectively re-displayed using right button of the mouse over the circular viewer (the pane on left lower corner). **B.** Primer viewer. This viewer presents detailed information of selected primer pairs with aligned gene structure, corresponding genic sequence alignment, PCR product size and primer sequences. It also provides the results of eHT-PCR analyses.

for customized design using user-specified parameters.

(ii) By clicking "execute" button, CSGM Designer renders the pairwise gene structure, and subsequently the circular viewer allows visualization of designed primer pair candidates. Alignment results for chosen primer pairs are displayed along with detailed information about the primers and their virtual PCR products. Users can further select multiple primer candidates by clicking individual primer pairs in the circular viewer to arrive at a final decision.

(iii) To test and select primer candidates for specific PCR amplification at the genome scale, the user can invoke eHT-PCR within all or a subset of species. This *in silico* PCR is conducted in high throughput manner. Once done, primer pairs predicted to produce non-specific PCR amplicons can be removed from the candidate pool. Users can further reduce the number of primer pair candidates by applying more stringent PCR parameters.

(iv) Thereafter, users can view the full results for selected primer pairs, reconfirm the quality of individual or groups of primer pairs, and make a final decision based on amplification specificity, product size, the number of introns, Tm value and other factors. Users may download the selected primer pairs and related information by exporting the output as an Excell file.

Experimental validation of the CSGM Designer

To test reliability of CSGM Designer output, a pilot scale experiment was performed. A total of 60 genes (Additional file 1: Table S1) were selected from a gene set associated with abiotic stress responses published in previous study [13]. A corresponding set of 60 PCR primer pairs, one for each target, was obtained using CSGM Designer. For this particular anlaysis, the genomes of *M. truncatula* (galegoid) and *P. vulgaris* (phaseoloid) were selected anchor species, with the expectation that resulting primers would have functionality for species in both clades. The *M. truncatula* A17 genome was used as the template sequence. The resulting 60 primer pairs were tested in PCR amplifications using 16 legume genotypes representing 8 genera and 12 species. The selected species are well distributed between the galegoid and phaseoloid clades and include both

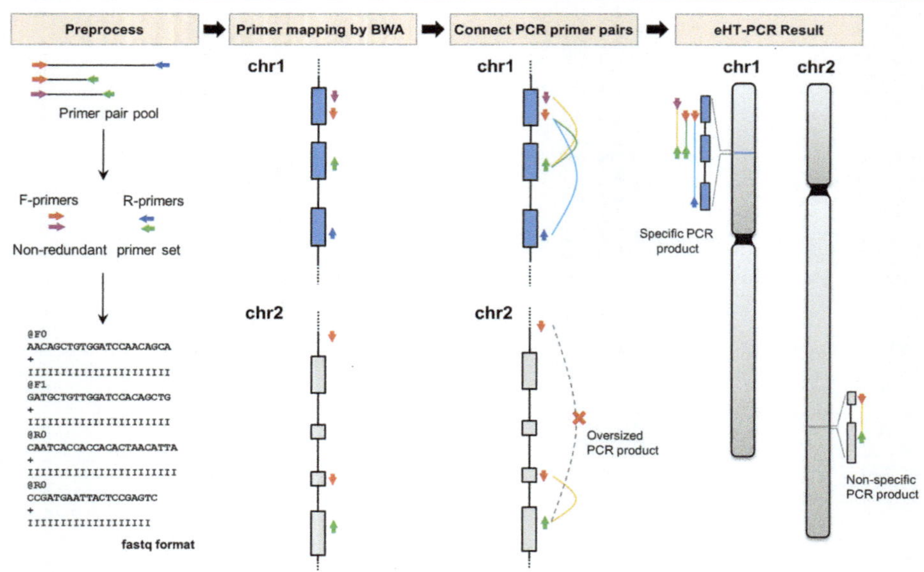

Figure 4 Concept of eHT-PCR and its workflow.

species contained within the database as well as species that lack substantive genomic resources. A total of 960 PCR reactions were performed and visualized by agarose gel electrophoresis, from which the success rate of PCR amplifications was calculated. Representative PCR amplification for 26 genes is shown in Figure 5, with success rates summarized in Figure 6. Of 960 PCR reactions, 819 PCR products were single bands, corresponding to 85.3% amplification. 25 genes (41.7%) were amplified across all 16 genotypes. PCR success rate for galegoid and phaseoloid species was 79.5% (286 amplicons/360 PCR reactions) and 88.8% (533 amplicons/600 PCR reactions), respectively (Figure 6A). These results demonstrate efficient cross-clade PCR amplification of conserved genes. Depending on species and genotypes, PCR success rates ranged from 65.0% (*C. arietinum* ICC4958) to 98.3% (*P. vulgaris* Jalo) (Figure 6B). Average success rate of *M. truncatula* and *P. vulgaris*, which were used as anchoring reference genomes, were 95.0% and 88.3%, respectively (Figure 6B).

Of 60 cross-species genic marker candidates, 9 genes (i.e., 144 PCR amplicons) were selected for the purpose of marker development. Of these, 130 sequences (i.e., 65 sets of parental lines) of high quality (GenBank accession numbers: KM047242-KM047371) were obtained and used to develop CAPS and/or SNP markers. Sequence comparisons of the 65 parental sets allowed us to detect nucleotide polymorphisms in 47 pairwise alignments (72.3%). Of these 47 SNP-bearing marker candidates, 37 parental sets (78.7%) were suitable for the

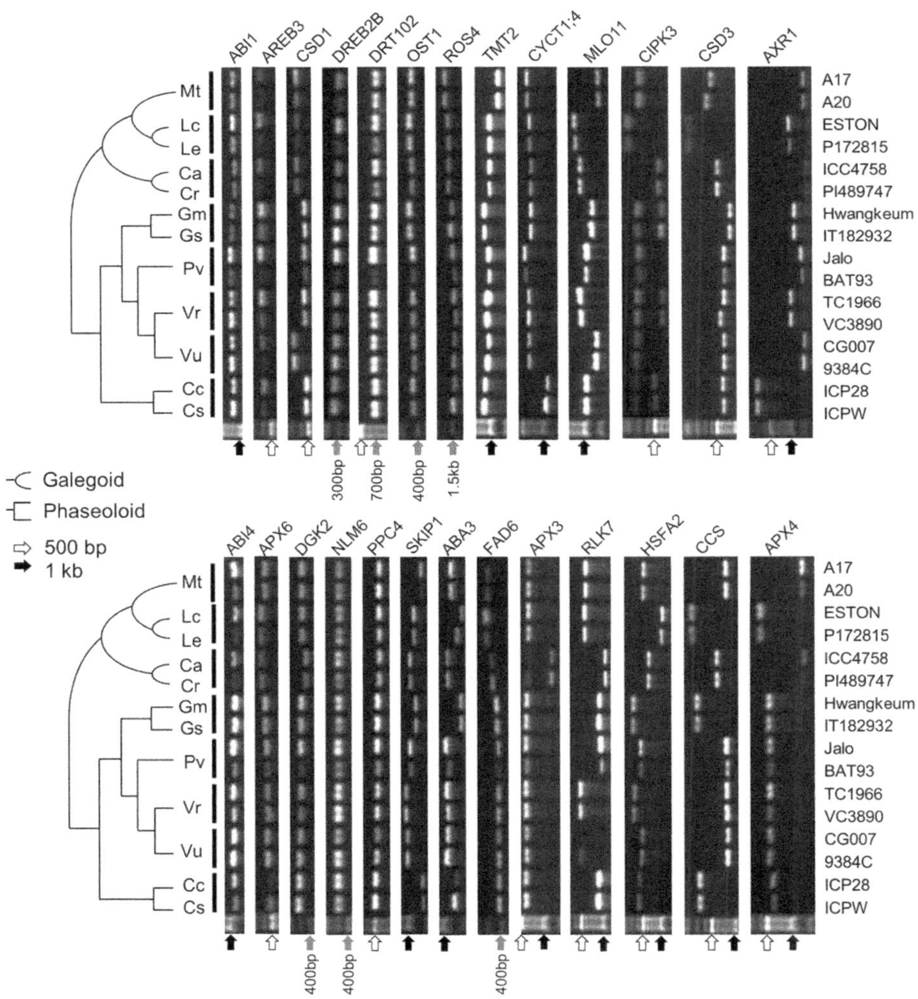

Figure 5 Cross-species PCR amplification of genic marker candidates. Twelve species employed for the experimental validation are organized according to their phylogenetic relationships on the left side. A total of sixteen genotypes were used and denoted on the right side. The abbreviations for each species denote their names as follows. Mt: *Medicago truncatula*, Lc: *Lens culinaris*, Le: *Lens Ervoides*, Ca: *Cicer arietinum*, Cr: *Cicer reticulatum*, Gm: *Glycine max*, Gs: *Glycine soja*, Pv: *Phaseolus vulgaris*, Vr: *Vigna radiata*, Vu: *Vigna unguiculata*, Cc: *Cajanus cajan*, Cs: *Cajanus scaraboides*.

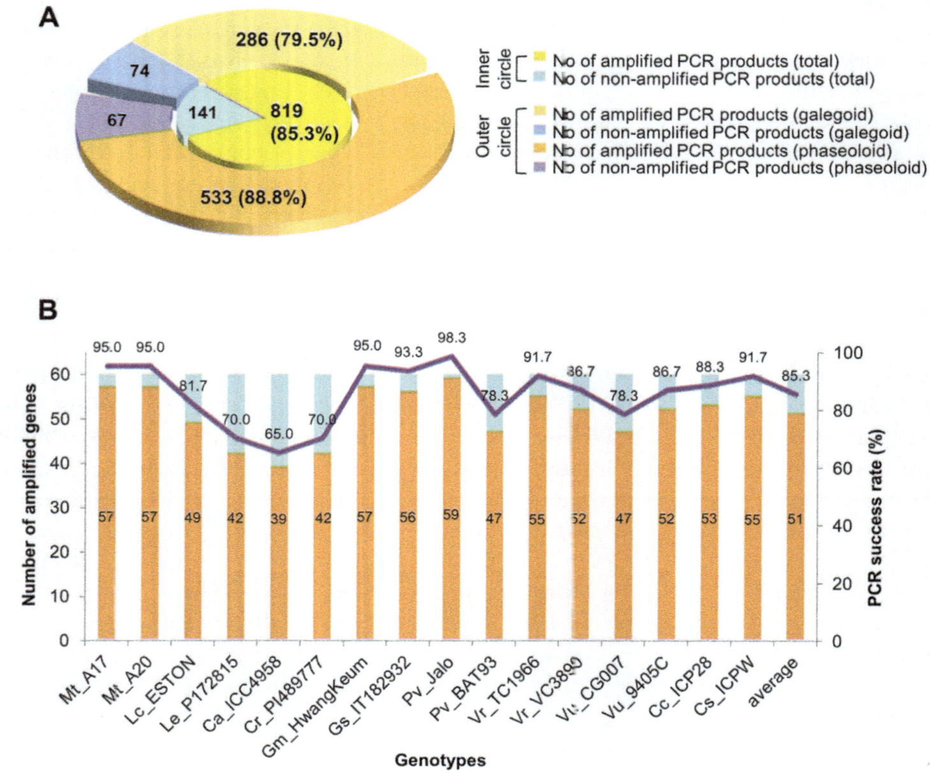

Figure 6 Summary of cross-species amplification of 60 selected abiotic stress-responsive genes. **A.** PCR success rates of overall cross-species and cross-clade amplifications. **B.** PCR success rates according to 16 genotypes. Pale brown bars and numbers indicate the number of genes amplified out of 60 selected genes. Solid line and numbers denote percent success rate of PCR amplifications. Abbreviated species names are the same as in Figure 5.

CAPS marker development. Cleavage patterns of amplified PCR fragments were predicted and reconfirmed by restriction enzyme digestions, of which three genic CAPS markers are exemplified in Additional file 2: Figure S1. Entire marker information for SNP, cleavage patterns, restriction enzymes is available in Additional file 3: Table S2.

Discussion

CSGM Designer is an easy-to-use platform that can automatically produce primer pairs, and functions efficiently as the basis for cross-species gene-based markers. Functionality and efficiency of the platform are enhanced by intuitive graphical interfaces, including a circular viewer that provides details of primer-target alignments (Figure 3) and an eHT-PCR module that represents the outcome of genome-wide *in silico* PCR amplification (Figure 4). Reference genome information from five legume species (Figure 2) is leveraged to create PCR primers that operate within the taxonomic space of the majority of legume crops. The platform's reliability was validated by testing cross-clade PCR amplification and subsequent SNP detection and genetic marker development (Figures 5 and 6, Additional file 3: Table S2) for 12 legume species. The concept of highly conserved, cross-species intron-spanning genic markers is not new, and was first widely implemented by Choi et al. [3,14]. Importantly, the utility of these markers as phylogenetic tools and their ability to routinely identify orthologous genes even for uncharacterized species has been demonstrated across the entire legume family [15]. By analogy, the markers developed using CSGM should, in a high percentage of cases, also extend across the entire legume family. Moreover, similar approaches should have relevance for other plant families of agricultural importance, where well characterized and poorly characterized species co-occur.

Similar databases have been constructed for both legumes and grasses [16], however such analyses were implemented using incomplete information of genome and/or EST sequences. The recent completion of whole genome sequencing for several important model and crop legumes provides the basis for revisiting such approaches using genome-wide data. The advent of genome-wide data underlies a range of NGS-based studies in plants, including RNAseq-based transcriptome analyses [17,18] and resequencing-based GWAS (genome-wide association study) analyses [8,19]. Here, based on a multi-species comparison, we extend these benefits to species whose genomes have not been sequenced, but are related to the sequenced reference species. Genome structures of plant species are frequently complex, in particular due to segmental duplication and polyploidization [20,21]. CSGM designer would have greatest utility of single or low copy fractions of legume genomes, though CSGM also permits the identification and *in silico* PCR testing of candidates even for large gene families.

Reference-guided genome resequencing is routinely used to discover nucleotide variation at the genome scale, and given access to appropriate populations to associate traits of interest with polymorphisms. But the power of such approaches depends in large part on access to substantial genome sequence data. Alternative methods are required for crops where genome data is lacking. Towards this end, CSGM Designer provides the infrastructure for high throughput development of gene-associated molecular markers that can link poorly characterized target species with the increasingly well-characterized genomes of the reference legumes.

Conclusions

In this study, we present the development and validation of CSGM designer, a new legume-focused bioinformatics platform for designing cross-species genic markers. Integration of reference legume genome information, interactive graphical content, and genome-wide *in silico* prediction of PCR amplification are key features of the platform. CSGM designer allow users, with minimal effort and high precision, to select marker candidates and test their efficacy *in silico* using genome-wide predictions of PCR amplification, as a prelude to actual experimentation. We expect that CSGM Designer have particular utility for poorly characterized legume species. In particular, connections to well characterized genomes through a network of highly conserved genes will enhance the potential for functional inference based on homology and synteny, and in doing so create a synergy for trait-associated molecular breeding.

Methods

Resources for the DNA sequence data

To construct WGS DB-associated marker designing platform, we used whole genomic information for five legume species (*Medicago truncatula, Glycine max, Phaseolus vulgaris, Lotus japonicas* and *Cajanus cajan*), whose genomes were recently fully sequenced and publically available. In addition to these five legumes, Arabidopsis was also included with the main intent of extracting functional annotations. Detailed genome information for six model and crop species is available in Table 1. Genome sequences for *G. max, P. vulgaris* and Arabidopsis were downloaded from Phytozome DB (http://www.phytozome.net). Genome data for *M. truncatula, Lotus japonicas* and *Cajanus cajan* was obtained from http://www.jcvi.org/medicago/, http://www.kazusa.or.jp/lotus/, and http://www.icrisat.org/gt-bt/iipg/Home.html, respectively. Genomic sequences of individual genes were extracted from the WGS data of each species provided by the general feature format (GFF) files. Coding sequences (CDS) were directly downloaded from corresponding databases and used to identify exon-intron junctions based on CDS-to-genomic sequence alignments.

Constitution of the CSGM Designer platform

The data processing pipeline of the CSGM Designer involves a combination of PHP and Javascript languages, while processed WGS and CDS data are organized within a MySQL relational database. Two programs, ClustalO (http://www.clustal.org/omega/) and Splign (http://www.ncbi.nlm.nih.gov/sutils/splign/splign.cgi), were employed to process two stepwise alignments. The ClustalO [11] was used to make CDS-to-CDS alignments and subsequent identification of the best-conserved regions within exons between two aligned coding sequences. The Splign algorithm [10] was used to align cDNA sequences with

Table 1 Summary of genomic information on six species used for CSGM designer platform

Species name	Common name	Genome size (Mbp)	Chromosome No	Release version	Number of transcript
Arabidopsis thaliana	Arabidopsis	157	n = 5	Phytozome v9.0	35,386
				Phytozome v10.0	35,386
Glycine max	Soybean	1,115	n = 20	Phytozome v9.0	73,320
				Phytozome v10.0	88,647
Medicago truncatula	Barrel medic	470	n = 8	Mt 4.0 v1	62,319
Phaseolus vulgaris	Common bean	625	n = 11	Phytozome v9.0	31,638
				Phytozome v10.0	31,638
Lotus japonicus	Birdsfoot trefoil	471	n = 6	Release 2.5	38,482
Cajanus cajan	Pigeon pea	833	n =11	Nov. 6, 2011	48,680

genomic DNA as a means to identify exon-intron junctions and exon-to-exon junctions within aligned CDSs. Structure Maker is a module that integrates the two types of alignment results and generates a corresponding pairwise gene structure. Based on the pairwise gene structure, Primer Maker was developed to extract gapless regions (i. e., continuous well conserved sequences) and to design paired primer candidates that avoid exon-intron junction sites. The Javascript was used to produce the graphic user interface (GUI) for the purpose of dynamic interactions between users and the GUI.

Development of the eHT-PCR analysis module

We adapted the Burrows-Wheeler Aligner (BWA, http://bio-bwa.sourceforge.net/) [12] to develop an electronic high throughput PCR (eHT-PCR) analysis module. BWA was developed to map short sequence reads produced by the Next Generation Sequencing (NGS). Its ability to align short DNA sequences, makes the BWA program useful for electronic PCR in that i) PCR primers are short DNA sequence, ii) single primers can be evaluated at multiple loci within the genome, iii) mismatch rates can be flexibly controlled, facilitating cross-species primer design. In conjunction with BWA, the eHT-PCR data processing pipeline was constructed as follows. First, redundant primers were removed from a large number of PCR primer pairs produced by the CSGM Designer, resulting in non-redundant primer set. Secondly, the non-redundant primers were mapped on selected genomes using BWA with a maximum of five mismatches (parameters: –N -k 2 -i 30 -d 30 -n 5). Finally, the eHT-PCR graphic displays the number and positions of electronically simulated PCR products, including amplicon size among species.

Plant materials and DNA isolation

A total of 16 legume genotypes, selected from and 12 species among 8 genera, were used for experimental validation of the CSGM Designer: *M. truncatula* cv. Jemalong A17 and A20, *Lens culinaris* ESTON, *Lens Ervoides* P172815, *Cicer arietinum* ICC4958, *Cicer reticulatum* PI489777, *Glycine max* cv Hwangkeum, *Glycine soja* IT182932, *P. vulgaris* Jalo and BAT93, *Vigna radiata* TC1966 and VC3890, *Vigna unguiculata* CG007 and 9405C, and *C. cajan* cv. ICP28 and *C. scaraboides* ICPW. Seed sterilization, scarification, germination and growth for *M. truncatula* were conducted according to publically available protocol (http://medicago.org/documents/Protocols/seedgerm.html). Other legume seeds were sterilized with 5% bleach for 3 minutes. After germination, seedlings were grown in incubators at 16 hr day period and 25°C. Young leaves were sampled for the DNA isolation. DNA was isolated using Exgene™ Plant SV kit (GeneAll, ROK) according to manufacturer's instruction and concentrations of each DNA sample were equalized to 10 ng/ul for PCR reactions.

PCR amplification & marker development

10 ul PCR reactions contained 1.0 unit of ExPrime Taq DNA polymerase (GenetBio, ROK), 1X reaction buffer, 0.25 mM of each dNTP, 2.5 pmole of each F/R-primer, 2.5 mM MgCl2 and 1 ng template DNA. PCR thermocycling was conducted with a 5 min initial denaturation step, followed by 35 cycles of 94°C for 30 sec, 55 ~ 57°C for 30 sec, and 72°C for 2 min, with a final extension for 10 min at 72°C. Amplified PCR products were purified using ExoSAP-IT PCR Product Cleanup Kit (Affymetrix, Santa Clara, CA) according to the manufacturer's protocol. BigDye terminator reagents were used for sequencing reactions as described previously [14]. The resulting sequences were aligned, edited and scored for single nucleotide polymorphisms using the MEGA5 program (http://www.megasoftware.net/). SNPs discovered between mapping parents were used to design CAPS (cleaved amplified polymorphic sequence) markers. Polymorphic sites were revealed by digesting PCR products with the appropriate restriction enzymes and subsequent agarose gel electrophoresis.

Additional files

Additional file 1: Table S1. List of genes and primer information used to validate CSGM Designer.

Additional file 2: Figure S1. Examples of cross-species genic markers. Restriction enzymes used to reveal polymorphisms are denoted in the parenthesis. Information for all other markers developed for experimental validation is available in Addional file 3: Table S2.

Additional file 3: Table S2. Information of cross-species genic markers.

Abbreviations

CSGM: Cross-species genic marker; eHT-PCR: Electronic high throughput PCR; GUI: Graphic user interface; BWA: Burrows-Wheeler aligner; CDS: Coding sequence; GFF: General feature format; WGS: Whole genome sequence; GBS: Genotyping-by-sequencing; QTL: Quantitative trait loci; GWAS: Genome-wide association study.

Competing interests

The authors declare that they have no competing interests.

Authors' contributions

HKC and JHK conceived and planned the research. CL and YJJ performed experimental validation and analyses. JHK, DH and JSP conducted database construction and program development. HKC, JHK and CL wrote the manuscript. DRC commented and edited the manuscript. All authors read and approved the final manuscript.

Acknowledgements

This study was supported by a grant from the Next-Generation BioGreen21 Program (PJ011182), Rural Development Administration and by the Basic Science Research Program through the National Research Foundation of Korea (NRF) funded by the Ministry of Education, Science and Technology (NRF-2010-0023585), Republic of Korea.

Author details

[1]Department of Medical Bioscience, Dong-A University, Saha-Gu Nakdong-Daero 550 beongil 37, Busan 604-714, Republic of Korea. [2]Department of Computer Science, Dong-A University, Saha-Gu Nakdong-Daero 550 beongil 37, Busan 604-714, Republic of Korea. [3]Department of Genetic Engineering, Dong-A University, Saha-Gu Nakdong-Daero 550 beongil 37, Busan 604-714, Republic of Korea. [4]Department of Applied Bioscience, Dong-A University, Saha-Gu Nakdong-Daero 550 beongil 37, Busan 604-714, Republic of Korea. [5]Department of Plant Pathology, University of California, One Shields Ave, Davis, CA 95616, USA.

References

1. Semagn K, Bjørnstad Å, Ndjiondjop MN. An overview of molecular marker methods for plants. Afr J Biotechnol. 2006;5:2540–68.
2. Poczai P, Varga I, Laos M, Cseh A, Bell N, Valkonen JP, et al. Advances in plant gene-targeted and functional markers: a review. Plant Methods. 2013;9:6.
3. Choi H-K, Mun J-H, Kim D-J, Zhu H, Baek J-M, Mudge J, et al. Estimating genome conservation between crop and model legume species. Proc Natl Acad Sci U S A. 2004;101:15289–94.
4. Doyle JJ, Luckow MA. The rest of the iceberg. Legume diversity and evolution in a phylogenetic context. Plant Physiol. 2003;131:900–10.
5. Young ND, Debellé F, Oldroyd GED, Geurts R, Cannon SB, Udvardi MK, et al. The Medicago genome provides insight into the evolution of rhizobial symbioses. Nature. 2011;480:520–4.
6. Sato S, Nakamura Y, Kaneko T, Asamizu E, Kato T, Nakao M, et al. Genome structure of the legume, *Lotus japonicus*. DNA Res. 2008;15:227–39.
7. Schmutz J, Cannon SB, Schlueter J, Ma J, Mitros T, Nelson W, et al. Genome sequence of the palaeopolyploid soybean. Nature. 2010;463:178–83.
8. Schmutz J, McClean PE, Mamidi S, Wu GA, Cannon SB, Grimwood J, et al. A reference genome for common bean and genome-wide analysis of dual domestications. Nat Genet. 2014;46:707–13.
9. Varshney RK, Chen W, Li Y, Bharti AK, Saxena RK, Schlueter JA, et al. Draft genome sequence of pigeonpea (*Cajanus cajan*), an orphan legume crop of resource-poor farmers. Nat Biotechnol. 2012;30:83–9.
10. Kapustin Y, Souvorov A, Tatusova T, Lipman D. Splign: algorithms for computing spliced alignments with identification of paralogs. Biol Direct. 2008;3:20.
11. Sievers F, Wilm A, Dineen D, Gibson TJ, Karplus K, Li W, et al. Fast, scalable generation of high-quality protein multiple sequence alignments using Clustal Omega. Mol Syst Biol. 2011;7:539.
12. Li H, Durbin R. Fast and accurate short read alignment with Burrows–Wheeler transform. Bioinformatics. 2009;25:1754–60.
13. Hyung D, Lee C, Kim J-H, Yoo D, Seo Y-S, Jeong S-C, et al. Cross-family translational genomics of abiotic stress-responsive genes between Arabidopsis and *Medicago truncatula*. PLoS One. 2014;9, e91721.
14. Choi H-K, Kim D, Uhm T, Limpens E, Lim H, Mun J-H, et al. A sequence-based genetic Map of Medicago truncatula and comparison of marker colinearity with M. sativa. Genetics. 2004;166:1463–502.
15. Choi H-K, Luckow MA, Doyle J, Cook DR. Development of nuclear gene-derived molecular markers linked to legume genetic maps. Mol Genet Genomics. 2006;276:56–70.
16. Fredslund J, Madsen LH, Hougaard BK, Sandal N, Stougaard J, Bertioli D, et al. GeMprospector—online design of cross-species genetic marker candidates in legumes and grasses. Nucleic Acids Res. 2006;34:670–5.
17. Hiz MC, Canher B, Niron H, Turet M. Transcriptome Analysis of salt tolerant common bean (Phaseolus vulgaris L.) under saline conditions. PLoS One. 2014;9, e92598.
18. Kudapa H, Azam S, Sharpe AG, Taran B, Li R, Deonovic B, et al. Comprehensive transcriptome assembly of Chickpea (Cicer arietinum L.) using Sanger and next generation sequencing platforms: development and applications. PLoS One. 2014;9, e86039.
19. Takagi H, Abe A, Yoshida K, Kosugi S, Natsume S, Mitsuoka C, et al. QTL-seq: rapid mapping of quantitative trait loci in rice by whole genome resequencing of DNA from two bulked populations. Plant J. 2013;74:174–83.
20. Marcussen T, Sandve SR, Heier L, Spannagl M, Pfeifer M, Jakobsen KS, et al. Ancient hybridizations among the ancestral genomes of bread wheat. Science. 2014;345. doi: 10.1126/science1250092.
21. Cannon SB, Ilut D, Farmer AD, Maki SL, May GD, Singer SR, et al. Polyploidy did not predate the evolution of nodulation in all legumes. PLoS One. 2010;5, e11630.

Image analysis of anatomical traits in stalk transections of maize and other grasses

Sven Heckwolf[1], Marlies Heckwolf[2,3], Shawn M Kaeppler[2,3], Natalia de Leon[2,3] and Edgar P Spalding[1*]

Abstract

Background: Grass stalks architecturally support leaves and reproductive structures, functionally support the transport of water and nutrients, and are harvested for multiple agricultural uses. Research on these basic and applied aspects of grass stalks would benefit from improved capabilities for measuring internal anatomical features. In particular, methods suitable for phenotyping populations of plants are needed.

Results: To meet the need for large-scale measurements of stalk anatomy features, we developed custom image processing software that utilized a variety of global thresholding, local filtering, and feature detection methods to measure rind thickness, pith area, vascular bundle counts, and individual vascular bundle size from digital images of hand-cut transections of stalks collected with a flatbed document scanner. The tool determined vascular bundle number with an average accuracy of 90% across maize genotypes that varied five-fold for this trait. The method is demonstrated on maize, sorghum, and *Miscanthus* stalks. The computer source code is staged for download.

Conclusions: Simplicity of sample preparation and semi-automated analyses enabled by this tool greatly increase measurement throughput relative to standard microscopy-based techniques while maintaining high accuracy. The tool is expected to be useful in genetic and physiological studies of the relationships between stalk anatomy and traits such as biofuel suitability, water use efficiency, or nutrient transport.

Introduction

The stalks of widely cultivated grass species such as maize and sorghum support multiple architectural and physiological functions, while contributing the most to aerial non-grain biomass. Visible in transections of such stalks are the many vascular bundles scattered throughout the parenchymatous pith. Surrounding the pith is a layer rich in collenchyma that is usually visibly distinct and commonly called the rind. The developmental mechanisms that determine the number of vascular bundles, their distribution, and the proportion of rind to total stem tissue in graminaceous crops is currently an important topic of research [1]. For example, water movement through the plant may relate to the number and size of xylem-bearing vascular bundles. Therefore, selection-based breeding for stem anatomy traits could be an effective strategy to improve water use efficiency [2,3]. Also, cell walls in the sclerenchyma surrounding the bundles or collenchyma in the rind are typically highly lignified, which limits digestibility when the stems are used for animal feedstock or industrial fermentation for ethanol production [4,5]. Plants better suited for ethanol production may be identified through surveys of natural variation for stem anatomy traits [6-8]. Another reason for studying anatomical features of grass stalks is their relationship to mechanical properties. The strength of the stalk determines the degree of wasteful crop lodging and pre-harvest breakage in the field [9-12].

These various motivations for measuring the anatomical features of grass stems create the need for a method that is efficient enough to measure hundreds if not thousands of individuals within defined populations for the purpose of mapping the genetic loci responsible for variation in the trait. Traditional methods for studying anatomy usually rely on sectioning chemically fixed tissue with a microtome followed by mounting the cut section on glass slides for examination with a microscope. These microscopic methods give superb cellular-level resolution, and have been used in large-scale studies of anatomical features [13-15], but typically their throughput is

* Correspondence: spalding@wisc.edu
[1]Department of Botany, University of Wisconsin-Madison, 430 Lincoln Drive, Madison, WI 53706, USA
Full list of author information is available at the end of the article

low. Relaxing the resolution criterion from cellular to tissue level increases the feasibility of automation. Higher throughput achieved by greater automation would improve the feasibility of acquiring the large data sets needed for some types of studies, such as statistical genetic trait mapping.

Computerized processing of digital images is an increasingly common means of quantifying plant structure. Sometimes sophisticated microscopy is called for [16-18] but simpler devices such as flatbed document scanners are perfectly adequate in many cases [19,20]. To quantify stem anatomical features as phenotypes across populations of plants suitable for statistical genetic analyses, for example, a simple imaging device and minimal sample preparation may provide the appropriate balance between resolution and throughput. The goal of the present work was to create an image analysis tool that could operate on images of hand-cut stem transections obtained on a flatbed scanner to measure anatomical features in high throughput.

Results

Image preprocessing

Example images of four stalk transections cut from two different maize genotypes are shown in Figure 1A. The images were acquired with a flatbed document scanner in three-color (red, green, blue) mode. The tool to be described here operates on such images to measure stalk diameter, rind thickness, number of vascular bundles, density of vascular bundles, and vascular bundle size from such images. The first step in the method creates a binary representation of a grayscale version of original image (Figure 1B) using Otsu's method for determining a threshold value that minimizes the intraclass variance [21,22]. A subsequent filtering step eliminates objects (dirt or debris) too small or eccentric (smaller than 15000 pixels, which is significantly smaller than the smallest stalks we have ever analyzed, or with an eccentricity greater than 0.99) to represent a stem section, producing a binary image consisting only of background

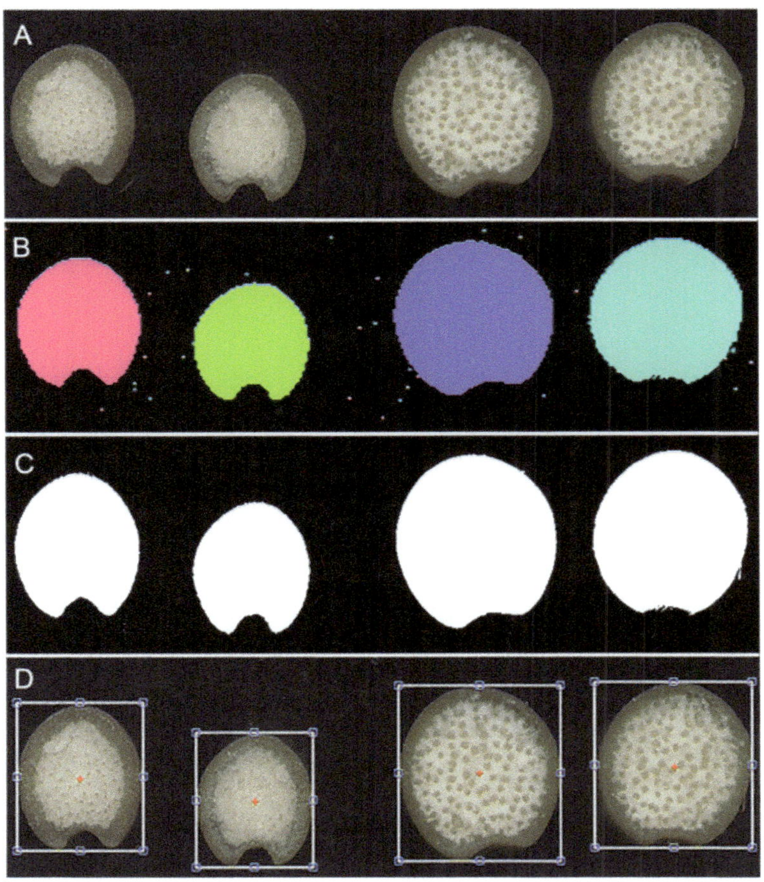

Figure 1 The initial processing steps of maize stalk transections. A) Original, unprocessed images of four stalk transections. The two on the left are the same genotype. The two on the right are the same genotype but different from those on the left. **B)** Segmenting all non-background objects by a simple thresholding technique identifies the stalk samples and smaller non-stalk objects such as scratches and debris. **C)** A filtering step based on object size and shape results in a binary image consisting of background and stalk samples. **D)** A crop box centered on each of the stalk objects is placed on the original image to allow the user to make adjustments to the scenes to be processed.

Table 1 Range of trait values observed in a sample of thirty maize genotypes

Trait	Genotype	Mean value
Stalk diameter (cm)	E2558W	3.26 ± 0.23
	LH85	1.60 ± 0.06
Rind thickness (cm)	W182BNgt	0.42 ± 0.08
	A663	0.13 ± 0.04
Vascular bundle density (cm^{-2})	OS602	57.3 ± 3.8
	W182BN	36.8 ± 2.8
Vascular bundle size (cm^2)	PHK93	$9.04 \pm 1.1 \times 10^{-4}$
	Mo39	$4.31 \pm 1.2 \times 10^{-4}$

Of the 30 genotypes sampled by collecting transections from 3 individual plants each, those which gave the largest and smallest value for each trait are listed in the middle column. The mean ± SEM of the three samples per genotype is given in the right hand column.

and stalk transection (Figure 1C). A box bounding each object in the filtered image is automatically placed on the original image for the purpose of cropping individual stem section images (Figure 1D). This box, generated automatically by the program, is the smallest possible rectangle that includes all of the identified object. The user may manually adjust the crop box if, for example, a portion of the stem section should be omitted from further analysis due to damage during sectioning or other anomaly. Standard binary object morphology operations quantify the total area, diameter (average), and perimeter of each stalk section.

The maize genotypes shown in Figure 1 were two of 30 randomly selected from the hundreds of genotypes comprising the Wisconsin Diverse Association Panel [23]. Of the 30 genotypes sampled, E2558W was found to have the widest average stalk diameter and LH85 had the narrowest (Table 1). The difference was approximately two-fold.

Rind thickness

The rind of monocotyledonous stalks is primarily composed of the epidermis plus lignified sclerenchyma and collenchyma cells beneath the epidermis. To isolate this portion of the stalk from the adjacent central pith, the green channel of the original image was filtered with a Gaussian distribution set to suppress the signal from the rind and maintain the bright signal from the pith.

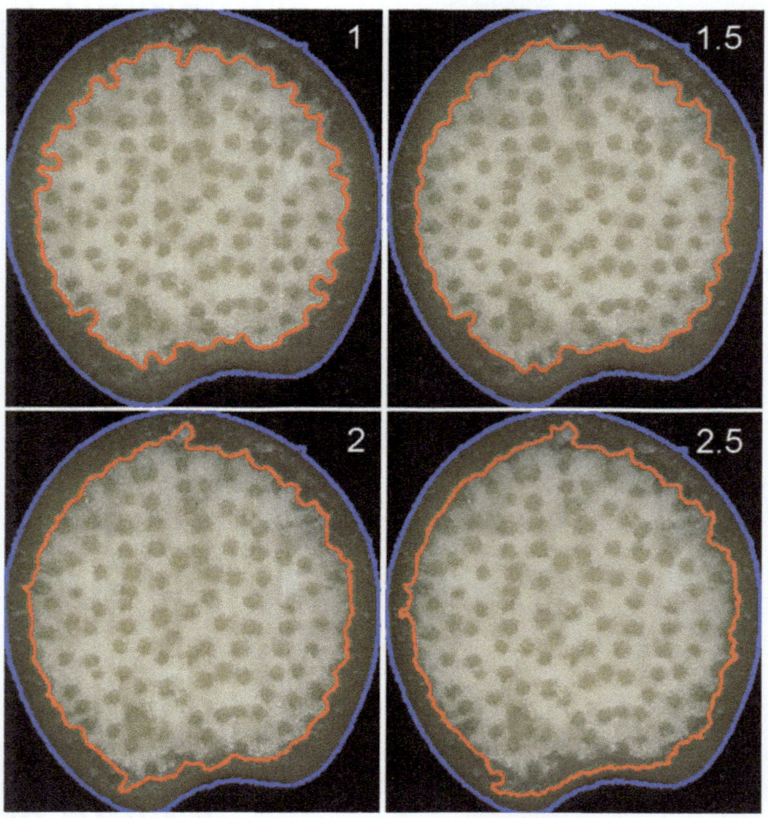

Figure 2 Discerning the pith/rind boundary. Processing of the green channel of the original image with a Gaussian filter suppresses the rind region to establish the boundary of the pith (red line). The result depends on the width of the Gaussian filter used to convolve the image. Results from filter width 1, 1.5, 2, and 2.5 are shown.

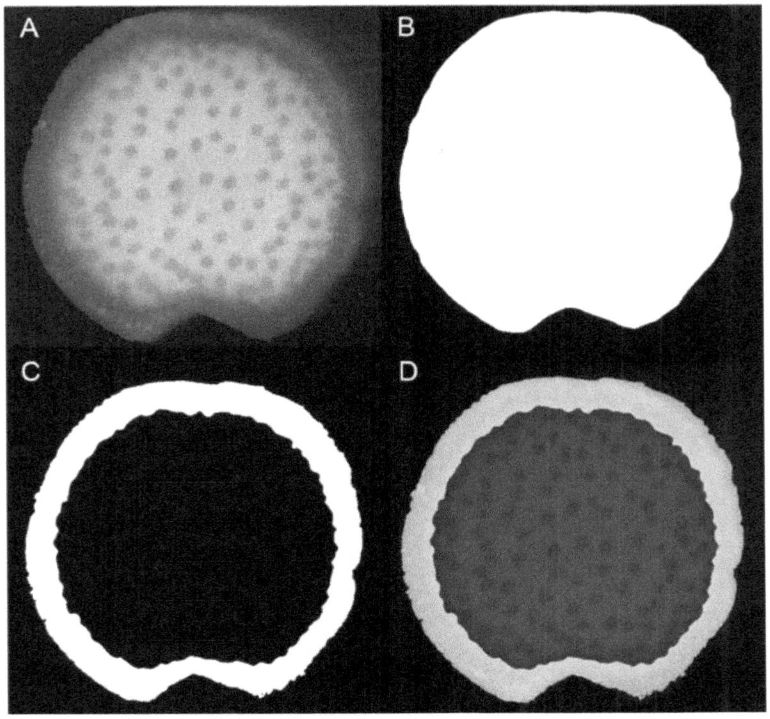

Figure 3 Determining rind width. A) An original stalk transection image. **B)** Image after binarization. **C)** Black pith mask determined from boundary found as shown in Figure 2 imposed in the binarized transection to segment the rind (white). **D)** Rind segment overlayed on the image in A. The width of the rind is the average distance between each point on the pith boundary and its nearest point on the sample perimeter.

Binarizing this filtered image by Otsu thresholding creates an image with a boundary marked by the red line in Figure 2. The rind segmentation result depends to a relatively minor degree on the filter width parameter, which the user may adjust from the default setting (Figure 2). Figure 3 shows an original transection image, the binarized image, a pith mask based on the step shown in Figure 2, and an overlay showing how the rind could be faithfully segmented from the stalk image. To measure the average width of the rind, the mean Euclidean distance between each point on the object boundary and the point nearest it on the rind/pith boundary is computed. The rind of W182BNgt plants was more than three-fold thicker than that of the A663 genotype (Table 1),

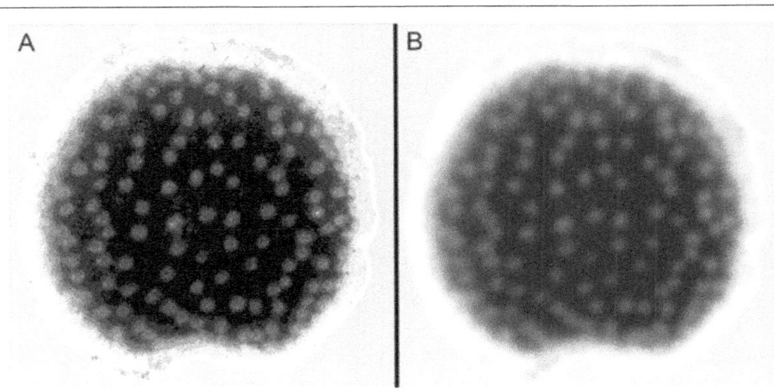

Figure 4 Processing to highlight vascular bundles. A grayscale representation of a stalk transection before **(A)** and after **(B)** anisotropic filtering. This step renders the vascular bundles as bright spots of mostly uniform intensity that will appear as single rather than split peaks in a gray value intensity plot.

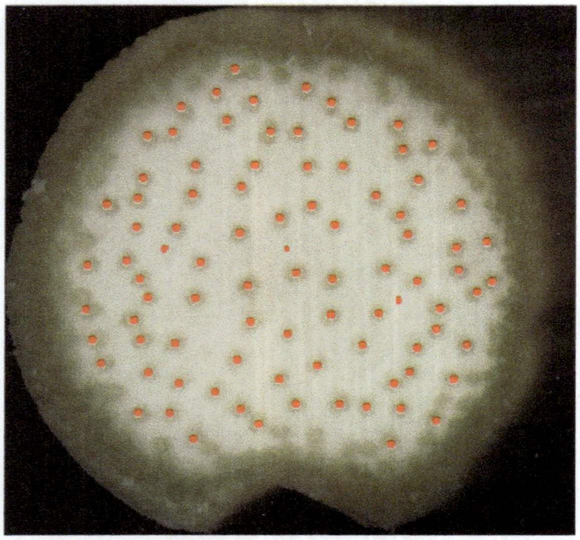

Figure 5 Vascular bundle localization. The position of each peak in grayscale value detected after the filtering step shown in Figure 4 is projected on to the original image and labeled in red. At this stage, the user may choose to select bundles not detected by the program.

indicating considerable genetic variation in the specification of this anatomical trait.

Vascular bundle number and density

To enumerate the vascular bundles specifically in the pith, a grayscale version of each image is convolved with a Gaussian kernel having a width approximately equal to the average size of a vascular bundle. A typical result is shown in Figure 4A. An anisotropic diffusion filter [22-26] is then applied to enhance the contrast and reduce noise. These steps result in localized peaks in grayscale value, fairly uniform bright areas, corresponding to the location of vascular bundles (Figure 4B). The peaks are computationally identified via non-maximum suppression [27,28]. Masking the rind (Figure 3) restricts the bundle counts to the pith area. Figure 5 shows a typical result, in which most of the vascular bundles identifiable by eye are counted by the tool. The user may manually select vascular bundles missed by the tool to add to the count. A random sample of 41 transection images was counted by a human expert and processed automatically by the tool. The expert counted 3469 bundles in the piths of these samples and the tool counted 3128, or 90% of the ground truth. Result of this accuracy test are presented in Figure 6. One reason for the consistent though minor undercounting of bundles by the tool is that the human may have included bundles that the tool determined not to be within the pith.

Dividing the number of bundles by the pith area gives vascular bundle density. This trait also varied among the 30 genotypes sampled, though relatively less than the stalk or rind thickness traits. The highest average density, observed in the OS602 genotype, was 57% greater than the lowest density observed, in the W182BN genotype (Table 1).

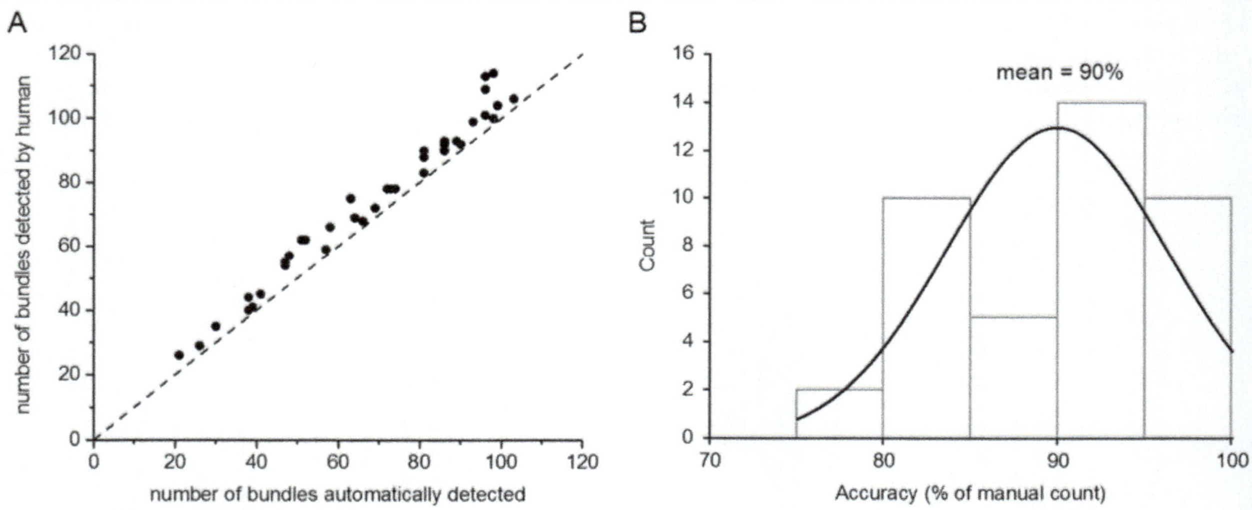

Figure 6 Accuracy of vascular bundle counting. A random set of 41 transection images was used to test the accuracy of the tool. **A)** The vascular bundles located in the pith in 41 transections were counted by a human and by the tool. Each point in the scatter plot represents the two numbers associated with each different section. The diagonal line represents perfect agreement. **B)** The number of bundles automatically counted relative to the number determined by eye was converted to a percent accuracy value for each transection and presented as a frequency histogram. Fitting a normal distribution to the histogram determined the mean accuracy value to be 90%.

Determining the size of the bundles

Despite the simplicity of the tissue preparation (hand cut transections) and imaging system (flatbed document scanner) the size of individual vascular bundles could be measured from the images. The process begins by copying a 40 × 40 pixel region surrounding the center coordinates of each previously identified bundle. A homomorphic filtering process [29-31] normalizes brightness and increases contrast of the vascular bundle by first taking the logarithm of the intensity values and then performing two-dimensional discrete Fourier transforms to identify low frequency components of the image, which are subsequently suppressed by filtering because they tend to represent reflectance artifacts more than structural features in images. A 2D Gaussian distribution fit to the grayscale values of this enhanced patch, when projected onto the image plane, produces a measure of the vascular bundle area. Figure 7 shows i) unprocessed vascular bundles of different sizes, ii) after processing, and iii) with contour lines of the fitted Gaussian distribution projected onto the image. Among the genotypes sampled, vascular bundle size varied more than two-fold between the largest in PHK93 and the smallest in Mo39 (Table 1).

Application to other species

Transections of *Sorghum bicolor*, a grass species with a maize-like stem anatomy, were similarly collected and imaged as shown in Figure 8A. The tool isolated the boundaries of the rind and enumerated the vascular bundles within the pith (Figure 8B-D). *Miscanthus giganteus*, the stalk of which is intensively studied for biofuels purposes, has considerably narrower stalks and smaller vascular bundles than sorghum or maize. Nonetheless, manually adjusting the respective parameters as discussed above enabled the tool to measure the target suite of traits automatically as demonstrated in maize (Figure 9). Thus, the tool presented here may be expected to quantify anatomical features in the stalks of many grass species having discernable rinds, pith, and vascular bundles.

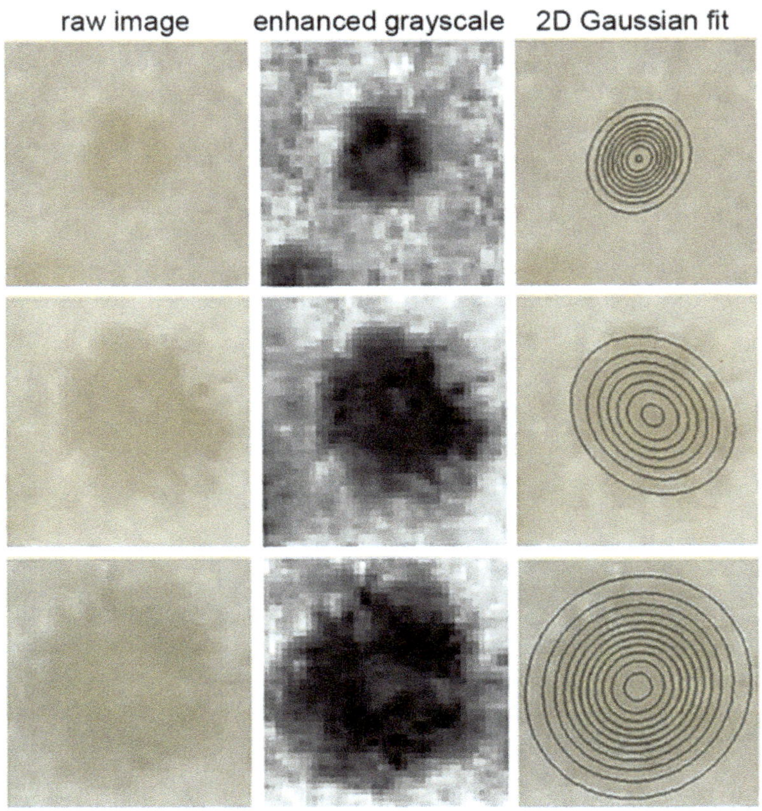

Figure 7 Measuring vascular bundle size. Each row is a different individual vascular bundle. A small, medium, and large bundle was selected for presentation. The left column shows the original unprocessed color image. The center column shows the processed grayscale image enhanced by a homomorphic filtering step. The right column shows the level contours of a 2D Gaussian distribution fit to enhanced grayscale map. The outermost ring is taken as a measure of the size of the bundle.

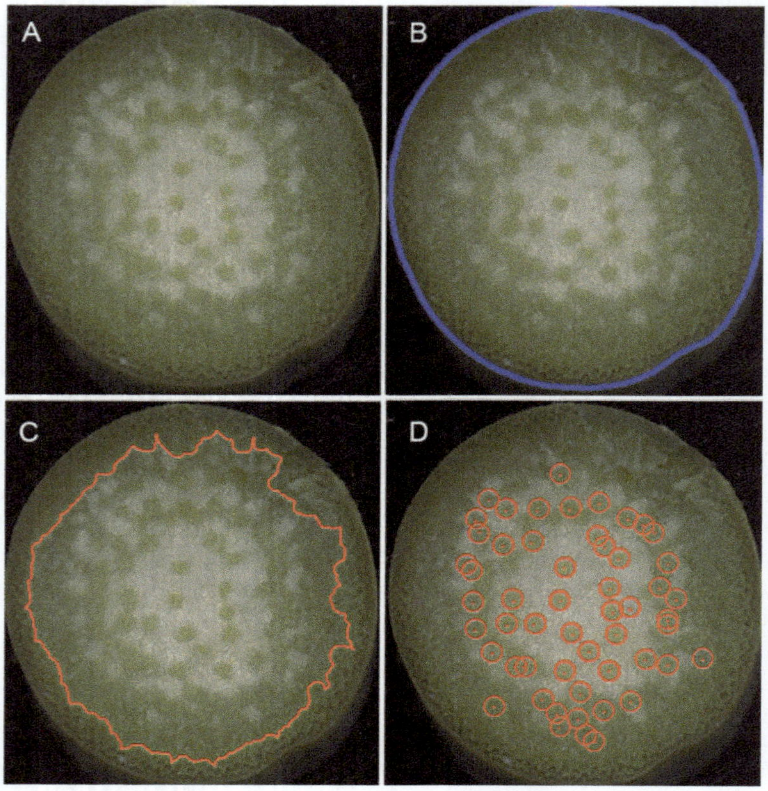

Figure 8 Applying the tool to transections of *Sorghum bicolor* stalks. A) Unprocessed image of a hand-cut transection of a sorghum stalk. **B)** Stalk perimeter (blue line) determined from a binary representation of the transection superimposed on the original image. **C)** Rind/pith boundary (red line) superimposed on the original image. **D)** Detected vascular bundles (red circles) superimposed on the original image.

Discussion

The need for a practical method for quantifying the major anatomical features in large numbers of grass crop stalks motivated this tool development project. Three competing objectives needed to be balanced to meet the need. One objective was to require only minimal sample preparation. Another was for the analysis not to limit the rate of the overall process. The third was to achieve high accuracy. Sample preparation involves hand cutting transections from stalk internodes systematically collected and tagged in the field and placing them directly on the imaging surface of a computer-controlled document scanner. Image acquisition and file saving requires only one manual input. Thus, the actions needed to prepare and image the samples are simple and efficient. Analysis of the resulting image file representing 12 transections is typically automatic, though scenarios requiring manual adjustments are encountered. For example, regions physically damaged during sectioning or by insect feeding may require cropping by manually adjusting the bounding box. The default settings of the Gaussian distribution width or the rigidity of the anisotropic diffusion parameters serve well for most maize transections but may require adjustment when strongly divergent genotypes or other species are to be studied. Even in scenarios requiring semi-automated operation, in which the user chooses some degree of supervision and parameter adjustment, the analysis step was not more time consuming than sample preparation. Achieving acceptable accuracy or fidelity in the analysis stage required development of appropriate feature extraction algorithms. Here the technical challenges to be overcome stemmed principally from the choice of simple sample preparation and imaging because without fixation, staining, or optical enhancement of contrast, the anatomical information in the images is frequently obscure. The steps described here and summarized in a flowchart (Figure 10) dealt with the inhomogenieties in lighting across individual transections, probably due to variation in sample thickness and other consequences of crude hand sectioning techniques, better than alternative image processing methodologies such as watershed [32,33] and level-set techniques [34,35], which were also tried. Lack of sample uniformity probably had a larger effect on result quality than did parameter

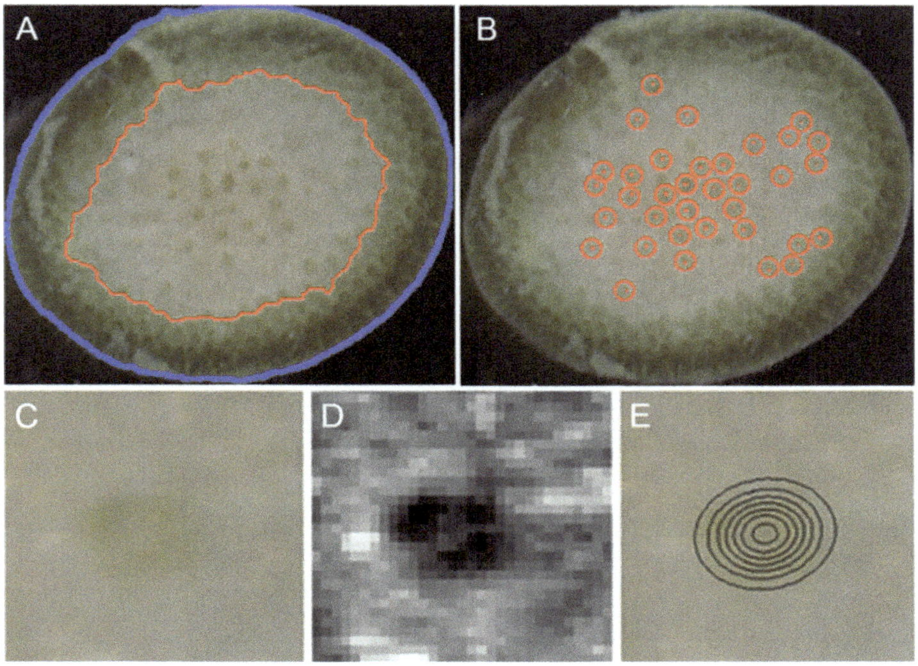

Figure 9 Applying the tool to transections of *Miscanthus gigantum* stalks. A) Stalk perimeter (blue line) and rind/pith boundary (red line) superimposed on the orginial, unprocessed color image. **B)** Detected vascular bundles (red circles). **C)** An individual vascular bundle in the unprocessed image. **D)** Grayscale representation of A after processing including homomorphic filtering. **E)** Level contours of the 2D Gaussian distribution fit to the grayscale image in D projected onto the original image. This shows the method measures the very small vascular bundles in thin stalks of *Miscanthus*.

tuning. Therefore, better methods for stalk sectioning may have a larger impact on the capability of this tool than further software efforts.

Overall, the set of design decisions and technical solutions produced an effective tool for high-throughput quantification of anatomical features in grass stalks. The tool, written in the Matlab computer language, is staged for download at http://phytomorph.wisc.edu/download/HeckwolfPlantMethods2015/ along with a composite test image representing a variety of transection phenotypes so that the performance of future tools for studying stalk anatomy can be benchmarked against that described here.

The current tool and future derivatives of it may be expected to enable systems-style projects attempting to discover links between the anatomical features and chemical or gene expression features, or large-studies of the genetic architecture affecting stalk architecture. More specifically, this tool may benefit biofuels researchers seeking a deeper understanding of the relationships between stalk anatomy, cell wall composition, and efficiency of carbohydrate conversion to ethanol because lignin content, which is correlated with features such as vascular bundle density quantified here (Table 1), limits the efficiency of biomass to ethanol conversion. Modeling of the size and number of vascular bundles in grass stalks for the purpose of achieving a quantitative understanding of stalk hydraulics is another area of research that may benefit from the tool.

This motivation for the development reported here, and therefore the resulting product, differs from a recent study that also used image analysis of maize stalk transections but to produce a 3D statistical model of vascular bundle distribution [36]. The present work emphasizes measurement throughput, to address the need for quantifying the most salient anatomical features in thousands of stalk samples, whereas Legland et al. [36] created a normalized model of vascular bundle distributions that facilitates quantitative comparisons between categories of samples.

Although the tool was developed to be effective with minimal sample preparation, it may prove useful in studies of transections that have been stained to highlight features such as high lignin content. Incorporation of such labeling or staining steps may give the method reported here more power to resolve anatomical details, for example by differentiating between sclerenchyma and collenchyma within the rind [37,38].

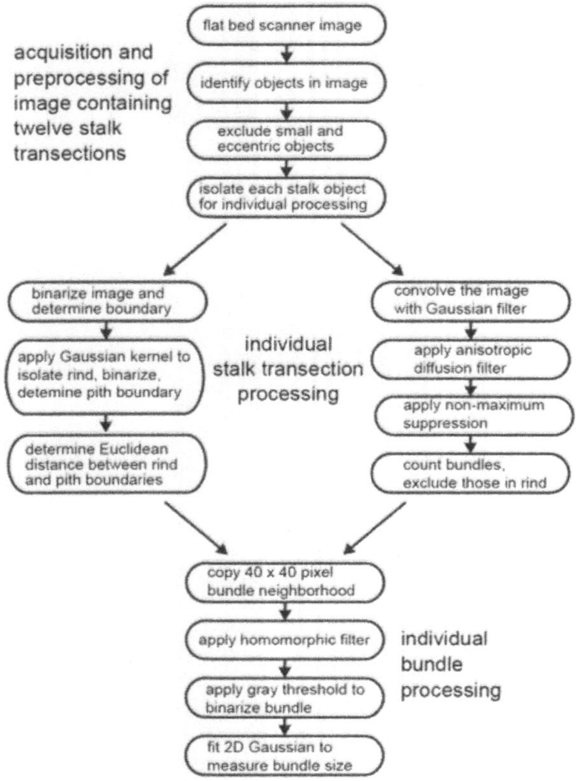

Figure 10 Flow chart showing the image processing steps from image acquisition to measurement of vascular bundle size.

Methods

Plant growth and sample collection

A set of 30 diverse maize inbred lines from the Wisconsin Diverse Association Panel [23], was grown at the Arlington Agricultural Research Station (University of Wisconsin-Madison) in 2013 in a field experiment using a randomized complete block design with two replications. Three representative plants per plot and field replication were harvested and brought to the lab 45 days after flowering for sample preparation.

Sample preparation and imaging

Transections of the third internode above the ground were cut by hand using razor blades into sections between 4 and 10 mm thick. Sections thicker than that resulted frequently in shadows or reflections on the rim of the stalk, producing a halo effect, which made the analysis of the stalks using standard parameter setting difficult or impossible. To produce an image for analysis, a total of 12 transections representing three individuals of four genotypes were placed on the horizontal imaging surface of an Epson Perfection V700 Photo Scanner and scanned at resolution of 800 dots per inch in red, green, blue color mode. Using these settings, we produced images with a height of 3800 to 3900 pixels and a width of 5000 to 5100 pixels and a resolution of 315 pixels per centimeter, with an error of at most 5 pixels. To enhance contrast between the samples and the background, the lid of the scanner was left open, resulting in a black background. The resulting files, typically larger than 30 megabytes, were saved in tagged image file format to a computer disk array.

Competing interests

The authors declare that they have no competing interests.

Authors' contributions

SH developed the software interactively with MH, who collected samples and refined the image acquisition protocol. NdL and SMK identified the need for the tool and supervised the field activities that produced the samples. EPS supervised the software development and wrote the manuscript with contributions from all others. All authors read and approved the final manuscript.

Acknowledgements

This work was funded by grant IOS-1031416 from the National Science Foundation Plant Genome Research Program to E.P.S. and in part by the DOE Great Lakes Bioenergy Research Center (DOE BER Office of Science DE-FC02-07ER64494).

Author details

[1]Department of Botany, University of Wisconsin-Madison, 430 Lincoln Drive, Madison, WI 53706, USA. [2]Department of Agronomy, University of Wisconsin-Madison, 1575 Linden Drive, Madison, WI 53706, USA. [3]DOE Great Lakes Bioenergy Research Center, 1552 University Avenue, Madison, WI 53706, USA.

References

1. Wang Y, Li J. Genes controlling plant architecture. Curr Op Biotech. 2006;17:123–9.
2. Dimond AE. Pressure and flow relations in vascular bundles of the tomato plant. Plant Physiol. 1966;41:119–31.
3. Payvandi S, Daly KR, Jones DL, Talboys P, Zygalakis KC, Roose T. A mathematical model of water and nutrient transport in xylem vessels of a wheat plant. Bull Math Biol. 2014;76:566–96.
4. Bruce DM. Mathematical modeling of the cellular mechanics of plants. Philos Trans R Soc Lond B Biol Sci. 2003;358:1437–44.
5. Zdunek A, Umeda M. Influence of cell size and cell wall volume fraction on failure properties of potato and carrot tissue. J Texture Studies. 2005;36:25–43.
6. Jung HJ, Samac DA, Sarath G. Modifying crops to increase cell wall digestibility. Plant Sci. 2012;185–186:65–77. doi:10.1016/j.plantsci.2011.10.014.
7. Saha S, Ramachandran S. Genetic improvement of plants for enhanced bio-ethanol production. Recent Pat DNA Gene Seq. 2013;7:36–44.
8. Youngs H, Somerville C. Best practices for biofuels. Science. 2014;344:1095–6.
9. Oladokun MAO, Ennos AR. Structural development and stability of rice Oryza sativa L. var. Nerica 1. J Exp Bot. 2006;57:3123–30.
10. Crook MJ, Ennos AR. Stem and root characteristics associated with lodging resistance in four winter wheat cultivars. J Agr Sci. 1994;123:167–74.
11. Schulgasser K, Witztum A. On the strength, stiffness and stability of tubular plant stems and leaves. J Theor Biol. 1992;155:497–515.
12. Schulgasser K, Witztum A. On the strength of herbaceous vascular plant stems. Ann Bot. 1997;80:35–44.
13. Zhang Y, Legay S, Barrière Y, Méchin V, Legland D. Color quantification of stained maize stem section describes lignin spatial distribution within the whole stem. J Agr Food Chem. 2013;61:3186–92.
14. Barros-Rios J, Malvar RA, Jung HJG, Bunzel M, Santiago R. Divergent selection for ester linked diferulates in maize pith stalk tissues. Effects on cell wall composition and degradability. Phytochemistry. 2012;83:43–50.

15. Olson ME, Rose JA. Vessel diameter–stem diameter scaling across woody angiosperms and the ecological causes of xylem vessel diameter variation. New Phytol. 2013;197:1204–13.
16. Méchin V, Argillier O, Rocher F, Hébert Y, Mila I, Pollet B, et al. In search of a maize ideotype for cell wall enzymatic degradability using histological and biochemical lignin characterization. J Agr Food Chem. 2005;53:5872–81.
17. Barros-Rios J, Santiago R, Malvar RA, Jung HJG. Chemical composition and cell wall polysaccharide degradability of pith and rind tissues from mature maize internodes. Animal Feed Sci Tech. 2012;172:226–36.
18. Buggenthin F, Marr C, Schwarzfischer M, Hoppe PS, Hilsenbeck O, Schroeder T, et al. An automatic method for robust and fast cell detection in bright field images from high-throughput microscopy. BMC Bioinformatics. 2013;14:297. doi:10.1186/1471-2105-14-297.
19. Fraas S, Niehoff V, Lüthen H. A high-throughput imaging auxanometer for roots and hypocotyls of Arabidopsis using a 2D skeletonizing algorithm. Physiol Plant. 2014;151:112–8. doi:10.1111/ppl.12183.
20. Smith HC, Niewohner DJ, Dewey GD, Longo AM, Guy TL, Higgins BR, et al. Using flatbed scanners to collect high-resolution time-lapsed images of the Arabidopsis root gravitropic response. J Visualized Experiments. 2014;83: e50878. doi:10.3791/50878.
21. Sezgin M, Sankur B. Survey over image thresholding techniques and quantitative performance evaluation. J Electronic Imaging. 2004;13:146–65.
22. Otsu N. A threshold selection method from gray-level histograms. IEEE Trans Sys Man Cyber. 1979;9:62–6.
23. Hansey CN, James M, Johnson JM, Sekhon RS, Kaeppler SM, de Leon N. Genetic diversity of a maize association population with restricted phenology. Crop Sci. 2011;51:704–15. doi:10.2135/cropsci2010.03.0178.
24. Weikert J. Anisotropic Diffusion in Image Processing. Stuttgart: Teubner-Verlag; 1998.
25. Kovesi PD. MATLAB and Octave Functions for Computer Vision and Image Processing. 2002. http://www.csse.uwa.edu.au/~pk/research/matlabfns/#anisodiff. Accessed 18 Jan 2015.
26. Perona P, Malik J. Scale-space and edge detection using anisotropic diffusion. Proceedings of IEEE Computer Society Workshop on Computer Vision. 1987;16–22. http://www.cs.berkeley.edu/~malik/papers/MP-aniso.pdf. Accessed 18 Jan 2015.
27. Gil J, Werman M. Computing 2-d Min, Median and Max filters. IEEE Trans Pattern Anal Mach Intell. 1993;15:504–7.
28. Neubeck A, Van Gool L. Efficient Non-Maximum Suppression. 18th International Conference on Pattern Recognition (ICPR 2006); 20–24 August 2006; Hong Kong; China; 850–855. ftp://oenone-v06.ee.ethz.ch/publications/proceedings/eth_biwi_00446.pdf. Accessed 18 Jan 2015.
29. Adelmann HG. Butterworth equations for homomorphic filtering of images. Comput Biol Med. 1998;28:169–81.
30. Eddins S. Homomorphic Filtering. 2013. http://blogs.mathworks.com/steve/2013/06/25/homomorphic-filtering-part-1/. Accessed 18 Jan 2015.
31. Ramaswamy G, Devaney N. Pre-processing, registration and selection of adaptive optics corrected retinal images. Ophthalmic Physiol Opt. 2013;33:527–39. doi:10.1111/opo.12068.
32. Beucher S, Lantuéjoul C. Use of watersheds in contour detection. International Workshop on image processing; real-time edge and motion detection/estimation; Rennes; France; 1979. http://cmm.ensmp.fr/~beucher/publi/watershed.pdf. Accessed 18 Jan 2015.
33. Nieniewski M, Serneels R. Segmentation of spinal cord images by means of watershed and region merging together with inhomogeneity correction. Machine Graphics Vision Intl. 2002;11:101–21.
34. Yi F, Moon I, Lee YH. Extraction of target specimens from bioholographic images using interactive graph cuts. J Biomed Optics. 2013;18:126015. doi:10.1117/1.JBO.18.12.126015.
35. Devaux MF, Bouchet B, Legland D, Guillon F, Lahaye M. Macro-vision and grey level granulometry for quantification of tomato pericarp structure. Postharvest Biol Tech. 2008;47:199–209. doi:10.1016/j.postharvbio.2007.06.017.
36. Legland D, Devaux MF, Guillon F. Statistical mapping of maize bundle intensity at the stem scale using spatial normalisation of replicated images. PLoS One. 2014;9:e90673. doi:10.1371/journal.pone.0090673.
37. Jung HG, Casler MD. Maize stem tissues: Cell wall concentration and composition during development. Crop Sci. 2006;46:1793–800.
38. Brauns F. The chemistry of lignin. New York: Academic; 1952.

TETX: a novel nuclear selection marker for *Chlamydomonas reinhardtii* transformation

Sergio A Garcia-Echauri[1,2] and Guy A Cardineau[1,2*]

Abstract

Background: Transformation of microalgae to obtain recombinant proteins, lipids or metabolites of economic value is of growing interest due to low costs associated with culture growth and scaling up. At present there are only three stable nuclear selection markers for the transformation of *Chlamydomonas reinhardtii*, which is the most commonly transformed microalgae, specifically: the aminoglycoside phosphotransferaseses *aph7*and *aphVIII* and the phleomycin resistance *ble* gene. As several microalgae are resistant to some of the antibiotics associated with the mentioned resistance genes, we have developed another alternative, *tet*X, a NADP-requiring Oxidoreductase that hydroxylates tetracycline substrates. We provide evidence that *tetX* can be used to obtain nuclear transformants of *Chlamydomonas reinhardtii*.

Results: We obtained nuclear transformants harbouring the *tet*X gene under the control of beta 2 tubulin or HSP70ARBCS2 promoters at an efficiency of transformation of 3.28 and 6.18 colony forming units/μg DNA respectively. This is the first report of a eukaryotic cell transformed using tetracycline as a selectable marker.

Conclusions: We developed a protocol for the nuclear transformation of *Chlamydomonas reinhardtii* using *tet*X as a selectable marker that confers stable resistance to tetracycline up to 100 μg/mL. We believe *tet*X can be used to transform *Chlamydomonas reinhardtii* chloroplasts, related microalgae and other aerobic organisms sensitive to any tetracycline antibiotic.

Keywords: *Chlamydomonas reinhardtii*, Tetracycline resistance, Glass bead transformation, *tet*X

Background

Genetic transformation of microalgae is of growing interest due to its easy growth and low cost scaling up capabilities [1]. *Chlamydomonas reinhardtii* is the most commonly used algae for genetic transformations, however it has been less frequently used to produce nucleus derived recombinant proteins due to transgene silencing [2]. At present, through the use of strong promoter/enhancer sequences like HSP70A and RBCS2 [3,4], introns of RBCS2 [5], and fusions with selectable markers [3,6-8], transgene directed protein production in the cytoplasm has increased to 0.25% of total soluble protein. Although the production levels are still on the order of 10 times lower than comparable production achieved in the chloroplast [1,3], nuclear transformation can produce recombinant proteins that are post-translationally modified and secreted to the exterior of the cell [4].

There are two main mechanisms used to select *C. reinhardtii* nuclear transformants: either generating auxotrophic mutants and then transforming them with the wild-type gene [9] or incorporating a gene that generates resistance to an antibiotic or herbicide. Generating antibiotic resistance is the most frequently used method. Although many microalgae have been transformed, few nuclear selection markers have been expressed stably in the cytoplasm. Specifically, the available selection marker genes are the *aph7* gene from *Streptomyces hygroscopicus* [10], the *Streptomyces rimosus aphVIII* [11] and the *ble* gene from *Streptoalloteichus hindustanus.* Both *aph7* and *aphVIII* confer resistance to different aminoglycoside antibiotics. The *aph7* gene only confers resistance to hygromycin B [10,12], and while *aphVIII* confers paromomycin resistance 2–10 fold higher than the minimum inhibitory concentration (MIC) to *C. reinhardtii*, which is adequate to select

* Correspondence: guy.cardineau@itesm.mx
[1]Centro de Biotecnología-FEMSA, Tecnológico de Monterrey, Monterrey, México
[2]Centro de Agrobiotecnología, Tecnológico de Monterrey, Monterrey, México

transformants, poor results have been seen with G418, kanamycin and neomycin, which require concentrations on the order of 1.2–1.3 fold higher than MIC for selection. As this is inadequate for selection, paromomycin is the agent of choice used to select transformants [11]. Other wild type microalge like *Chlorella sorokiniana*, *Picochlorum* sp. *Botrycoccus braunii*, *Tetraselmis suecica*, *Dunaliella salina* are resistant to paromomycin [13] and probably to other aminoglycoside antibiotics. The *ble* gene generates resistance to phleomycin derived antibiotics [14] such as zeocin, which is the most readily available of that group, however, all the phleomycin antibiotics are expensive. Green algae like *Pseudokirchneriella subcapitata*, the cyanobacteria like *Microcystis aeruginosa*, and *Anabaena* CPB4337 are considered sensitive to tetracycline compounds [15,16]. As there is no reported transformation protocol or selectable marker that generates tetracycline resistance in microalgae or any eukaryote, we pursued the development of such a system for the transformation of the model microalgae *C. reinhardtii*.

There are three main resistance mechanisms that confer tetracycline resistance in bacteria: efflux pumps [17], ribosomal protectors [18] and covalent modifiers [19]. Most proteins that confer resistance to tetracycline used in bacterial transformation are membrane bound effector pumps that export tetracycline to the exterior of the cell. Those could be a challenge to express properly in microalgae and other eukaryotic cells due to their association with the bacterial membrane. Ribosomal protectors are large; for example TetM and TetO are each comprised of 639 amino acid (a.a.) residues (Acc. ADV76307, YP_009080033) and because tetracycline prevents the binding of aminoacyl- tRNA with the bacterial ribosome [20], ribosomal protectors would need to be targeted to the chloroplast and/or mitochondria which contain the homologue to the bacterial ribosome in eukaryotes.

Alternatively, Tetx [21,22] is a 43.7 kDa (388 a.a.) NADP-requiring Oxidoreductase that hydroxylates a broad spectrum of tetracycline substrates resulting in unstable compounds that undergo non-enzymatic decomposition [23]. TetX degrades the following compounds: chlortetracycline, demeclocycline, doxycycline, minocycline, oxytetracycline, tetracycline and tigecycline [23,24]. This is particularly useful in the generation of transformants harboring the *tet*X gene as different organisms have distinct sensitivity to each compound, for example, the protozoan *Toxoplasma gondi* is sensitive to doxycycline and resistant to tetracycline [25], the pathogenic yeast *Candida albicans* is also resistant to tetracycline and sensitive to minocycline and tigecycline [26-28].

Tetracycline resistance used to select genetically modified organisms has been present since the dawn of recombinant DNA technology [29,30], however until this report its use as a selection marker was restricted to bacteria. We designed a synthetic *tet*X gene for expression in the eukaryotic model microalgae *C. reinhardtii* and describe a nuclear transformation protocol for *Chlamydomonas reinhardtii* using *tet*X that generates tetracycline resistance. This is the first report of a genetically modified eukaryote selected with tetracycline. We compare the transformation efficiency of *tet*X under two promoters, beta 2 tubulin and the HSP70ARBCS2, with that of two commonly used selection markers, the *aphVIII* gene controlled by the beta 2 tubulin promoter and the *ble* gene controlled by the RBCS2 promoter. We also performed tetracycline resistance stability assays of transformants grown in the absence of antibiotic.

Results

TetX genes

We synthetized a *tet*X open reading frame codon optimized for expression in *C. reinhardtii* cytoplasm, driven by the beta 2 tubulin promoter [31] and with chlamyopsin1 3′ UTR [11] We named this construct BtetX (Figure 1). A second version with *tet*X under HSP70A/RBCS2 enhancer/promoter containing one intron copy of RBCS2 intron 1 and RBCS2 3′UTR was assembled and correct fragments verified by restriction digestion (Data not shown). That plasmid was named AtetX. Plasmid characteristics can be reviewed in Table 1.

TetX transformants were obtained for both constructs

Transformations with both constructs were carried out and, 8–12 days after plating transformed cells, tetracycline resistant *C. reinhardtii* colonies appeared. Both promoters used to drive *tetX* expression yielded transformant colonies, which indicates the versatility of this system to work with low and high level expression promoters (Figure 2).

Transformation efficiency

As expected, the HSP70A/RBCS2 enhancer/promoter yielded more colonies than the β-tubulin promoter. In 3 independent transformations the average number of colonies per plate was 20 ± 2.08 and 9 ± 0.57 respectively. The calculated efficiency was 6.18 cfu/μg of DNA for AtetX and 3.28 for BtetX. The efficiency was similar to that of pKS-aphVIII at 4.51 and lower than the *ble* gene with 22.56 (Table 1). We believe the lower efficiency of AtetX compared to the *ble* gene is caused by the addition of an extra RBCS2 intron 1 within the sequence of the *ble* gene, as this extra intron increased the efficiency 3 times compared to a *ble* gene with only one intron [5].

Effect of light intensity and cell concentration on false positive transformants

As light causes tetracycline degradation, we analyzed the relationship between the appearance of resistant positive

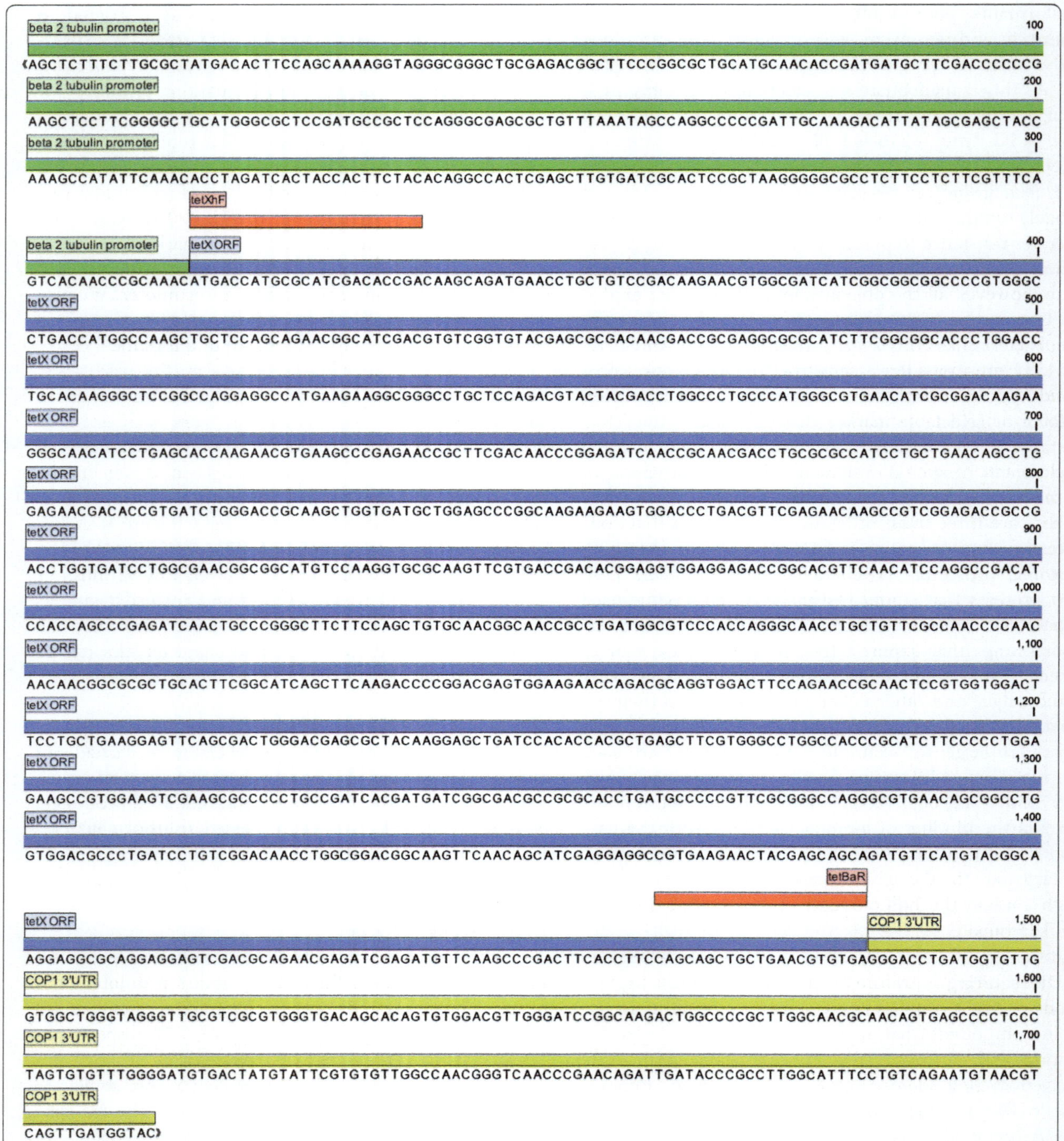

Figure 1 BtetX sequence. The BtetX gene sequence includes the beta 2 tubulin promoter, tetX Open Reading Frame (ORF) and COP 1 3'UTR. In red are primer binding sites for tetXhF and tetBaR.

or false positive colonies with respect to cell concentration per plate and light intensity (Table 2). When 5×10^6 cells/plate were grown under medium light conditions (17–25 μmoles m^{-2} s^{-1}), tetracycline at 15 μg/mL was sufficient to prevent false positives from growing on selection plates. All transformants selected at these conditions were positive. However at concentrations above 5×10^6 cells/plate, and/or light intensity above 27 μmoles m^{-2} s^{-1}, false positives appeared as either a lawn or patches of small colonies. However because positive colonies grow faster than negative transformants, they can be easily distinguished and selected from the small colony false

Table 1 Plasmids used to transform *C. reinhardtii*

Plasmid	Promoter/terminator/#RBCS2 introns	Resistance	Base pairs/MW (g/mol)	TE	Source
AtetX	HSP70A:RBCS2/RBCS2/1	Tetracycline	5238/3236640.1	6.18	This work
BtetX	β-2 tubulin/COP-1/NO	Tetracycline	4445/2746642.9	3.28	This work
pKs-aphVIII	β-2 tubulin/COP-1/NO	Paromomycin	4308/2661961.7	4.51	[38]
psP124S-ble	RBCS2/RBCS2/2	Zeocin	4133/2553770	22.56	[5]

Listing of plasmids used in this work with their corresponding promoter/terminator/number of RBCS2 introns, associated resistance, plasmid size, molecular weight (MW) and Transformation Efficiency (TE).

positives. Positive transformants can be grown on plates containing tetracycline concentrations up to 100 μg/mL. We also assayed tetracycline concentrations of 25 and 50 μg/mL to select primary transformants, however, at those concentrations the number of positive colonies decreased approximately 75% compared to those obtained with 15 μg/mL (data not shown).

Transformants grown with no antibiotic retained tetracycline resistance

Nuclear transgene silencing in *C. reinhardtii* has caused loss of resistance phenotype in half of *aadA* transformants that conferred spectinomycin or streptomycin resistance [32], which is why it is not routinely used for selection. We therefore assayed tetracycline resistance when positive transformed strains expressing the *tetX* gene were grown without antibiotic (Figure 3). Antibiotic resistance was maintained at minimum for 26 divisions, and transformants were also resistant to concentrations up to 100 μg/mL (Figure 3).

Wild type cell-walled *C. reinhardtii* strain is sensitive to tetracycline

Because we used a cell wall deficient strain for transformation and selection, we tested sensitivity to tetracycline using a wild type (WT) cell-walled strain, CC-124, in order to evaluate more general utility. We therefore grew strain CC-124 in TAP agar plates, under the same light conditions used for transformation, tetracycline concentrations of 15, 25 and 50 μg/mL, and cell concentrations from 0.5 1.0, 2.5 and 5.0 × 10^6 cells per plate. After a period of 10 days, the time required for transformants to become visible, we found no growth of wild type cells. This suggests that tetracycline uptake at the assayed concentrations is not sufficiently affected by the wild type cell-wall to alter the sensitivity of the cells to tetracycline and that tetracycline selection could work with normal cell-walled wild type cells.

Conclusions

We have developed a new stable nuclear selection marker for *C. reinhardtii* that confers resistance to tetracycline at up to 100 μg/mL. Although tetracycline is light sensitive and acts in the chloroplast, this was not an obstacle to develop an efficient method to obtain transformants. Compared with hygromycin B, paromomycin and zeocin, tetracycline is by far the least expensive antibiotic (Gold biotechnology, Inc. St. Louis, MO, USA) and provides a reasonable alternative selectable agent for transformation of microalgae. As *tet*X hydrolyzes several tetracycline analogues, [23,33] their use might favor increased light incubation to obtain transformants. We believe that the *tetX* gene produces a versatile tetracycline degrading enzyme, which suggests it could be used to transform the nucleus of other microalgae, as well as the chloroplast or mitochondria of other tetracycline sensitive cells, such as *Saccharomyces cerevisiae* or human HeLa cells that are sensitive to tetracycline concentrations above 10 μg/mL [34,35]. Codon bias

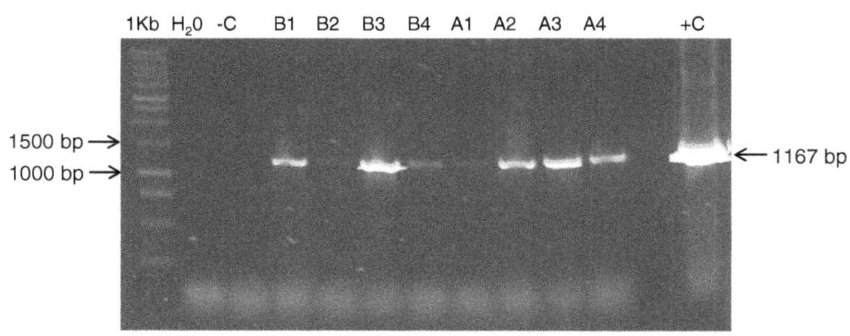

Figure 2 TetX gene presence in AtetX and BtetX transformants. 1% TAE-agarose gel of colony PCR of tetracycline resistant colonies transformed with AtetX (A1-4) or BtetX (B1-4). 1 kb: 1 kb DNA ruler, H_2O: water was used instead of DNA template, –C: DNA from an untransformed cc-849 strain was used as negative control, +C: AtetX plasmid was used as template.

Table 2 Effect of light intensity and cell concentration on false positive appearance

Cells/plate	Light intensity (μmoles $m^{-2} s^{-1}$)		
	17	24	>26
2.5×10^6	-	-	+
5.0×10^6	-	-	+
1.0×10^7	-	+	+
3.0×10^7	+	+	+

We measured the appearance of false positives on *tetX* transformed *C. reinhardtii* plates when grown at different cell concentrations and light intensities: (–) indicates no false positives, (+) indicates presence of false positives.

of the specific target host and organelle may need to be considered to optimize expression.

Methods

Algal strain and growth conditions

Cell wall deficient strain CC-849 of *Chlamydomonas reinhardtii* (Chlamydomonas Resource Center, University of Minnesota) was used in all algal transformation experiments. We chose this strain because it is readily transformed with glass beads or electroporation. Strain CC-124 (Chlamydomonas Resource Center) was used as a control to assay tetracycline sensitivity of WT cell wall strains. All algal strains were grown routinely in TAP media [36] at 25°C with a 16/8 light/dark photoperiod in a growth chamber on top of translucid glass shelves. In front of the shelves, two pairs of Sylvania GRO-LUX 40 W wide spectrum fluorescent light tubes (Osram sylvania ltd. Mississauga, ON, CA) and two OCTRON ECO 32 W fluorescent light tubes (Osram sylvania ltd) placed perpendicular to the shelve provided for light. Plates were placed at 3 to 40 cm from the light source which provided light from 78.6 to 25 μmoles $m^{-2} s^{-1}$. Light intensity was measured with a LI-250A light meter (LI-COR, Lincoln, NE, USA), readings are the sum of 15 second averages from two positions: placing the sensor on top of the glass shelf targeted at the ceiling and at the same place with the sensor targeted towards the floor. To achieve lower than 25 μmoles $m^{-2} s^{-1}$, plates were placed in front of only one pair of Sylvania GRO-LUX 40 W wide spectrum fluorescent light tubes at 30–40 cm from the light source. The lowest light setting: 1.81 μmoles $m^{-2} s^{-1}$, was achieved by placing the plates in the lowest light location in the growth chamber (17 μmoles $m^{-2} s^{-1}$) and covering the plates with two double-layers of gauze.

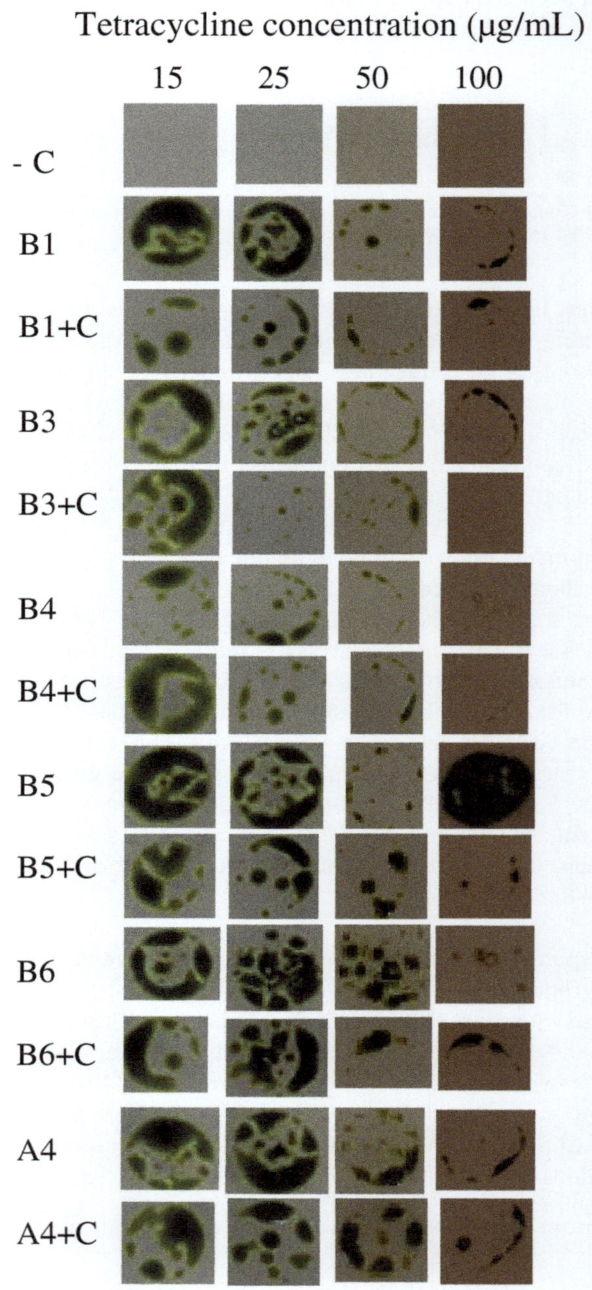

Figure 3 Tetracycline resistance phenotype. BtetX (5 strains) or AtetX (1 strain) positive transformed *C. reinhardtii*, and CC-849 as a negative control, were grown in TAP media with (+C) and without antibiotic selection for 26 cell divisions. A 10 μL droplet containing 10^4 cells of each strain was plated on TAP plates supplemented with either 15, 25, 50 or 100 μg/mL of tetracycline and incubated with a light intensity of 25 μmoles $m^{-2} s^{-1}$. The tetracycline resistance phenotype is present in all transformed strains. We expected confluent cell growth, however, there appears to be a selection for more resistant cells at higher tetracycline concentrations that grow as patches or along the periphery of the absorbed droplet. Negative controls did not grow at any assayed tetracycline concentration.

Plasmid construction

E. coli strains carrying Plasmids pHsp70A/RbcS2-cgLuc [7,37], pSP124S ble cassette [5] and pKS-aphVIII-lox [38] were obtained from the Chlamydomonas Resource Center.

TetX open reading frame [native sequence from *Enterobacteriaceae bacterium;* Genbank: JQ990987] was synthesized *de novo* at GenScript (Piscataway, NJ, USA) with codons optimized for *Chlamydomonas reinhardtii* cytoplasmic expression under the control of constitutive beta 2 tubulin promoter [31] and chlamyopsin1 3′UTR [11]; the plasmid was named Btetx (Figure 1). A second version of the construct was generated by Polymerase Chain Reaction (PCR), amplifying the open reading frame with a Veriti thermal cycler (Applied Biosystems, Foster City, CA, USA) in a 25 μL reaction volume containing 1U of the proof-reading high fidelity (1 error/100,000 bp = 0.001%) enzyme Advantage HD DNA Polymerase (Clontech, Palo alto, USA), 0.1 mM dNTP's, and 0.25 μM of each oligo tetXhF and tetBaR (Table 3) that carried *Xho*I and *Bam*HI sites in their 5′ends for cloning purposes. The reaction was carried out with an initial denaturation at 98°C 2 min. following 35 cycles of 98°C 10 sec, 70°C 10 sec, 72°C 1 min. with a final 5 min extension at 72°C. The amplicon was digested with *Bam*HI and *Xho*I (New England Biolabs, Ipswich, MA, USA), and cloned into the corresponding sites of plasmid pHsp70A/RbcS2-cgLuc, replacing the luciferase ORF with that of *tet*X. The ligation was transformed into *E. coli* stbl4 (Invitrogen, Carlsbad, CA, USA) generating plasmid AtetX.

Glass bead transformation

C. reinhardtii strain CC-849 was transformed with supercoiled plasmid DNA of Atetx, Btetx, pKS-aphVIII or pSP124S (Table 1) by the glass bead method [39]. Briefly, the algae were grown at 25°C with a 18/6 (light/dark) photoperiod in TAP media to mid-log phase (1–2 × 10^6 cells/mL), the cells were harvested by centrifugation 5 min. at 5000 × G, the growth medium was removed and fresh TAP was added to achieve a cell concentration of 2 × 10^8 cells/mL. 300 μL of the cell suspension were placed in a 1.5 mL centrifuge tube containing 0.3 g of sterile 0.4 - 0.6 mm diameter glass beads (Sigma, St. Louis, MO, USA) and 1 × 10^{-12} mols of the desired plasmid DNA. The cell/DNA/glass bead suspensions were vortexed 15 s at maximum power in a VWR mini vortex. The cells were transferred to a glass tube with 5 mL of fresh TAP media and incubated at 25°C overnight with a 8/6 (light/dark) photoperiod. After 14 hours the cells were concentrated by centrifugation at 5,000 × g for 15 minutes and resuspended in TAP media to yield either 3 × 10^7 cells/mL of pKS-aphVIII and pSP124S transformants and 2.5 × 10^6, 5 × 10^6, 1 × 10^7 or 3 × 10^7 cells/mL for *tet*X transformants. 300 μL of the cell suspensions were spread on 100 mm diameter and 15 mm depth TAP agar plates supplemented with either 150 μg/mL paromomycin, 20 μg/mL Zeocin or 15 μg/mL tetracycline. Plates were incubated at 25°C with a 18/6 (light/dark) photoperiod in low light (1.81 μmoles m^{-2} s^{-1}) for 1 day, and then *tet*X plates were transferred to medium light (25 μmoles m^{-2} s^{-1}) while paromomycin and zeocin plates were incubated in high light (78.6 μmoles m^{-2} s^{-1}) When colonies appeared, they were streaked on to selective TAP agar plates.

Table 3 Oligonucleotides used

Name	Sequence 5′-3′
aph8F	CGTGCACTGCGGGGTCGGT
aph8R	CCGCCCCATCCCACCCGC
ble1F	CCGGGTCGCGCAGGGC
ble1R	GCGCCGTTCCGGTGCTCA
tetXhF	**TCTCGAG**ATGACCATGCGCATCGACACCGA
tetBaR	**TGGATCC**TCACACGTTCAGCAGCAGCTGCTG

Listing of oligonucleotides used in this work. Bases in bold indicate recognition sites for restrictions enzymes and accessory bases used for cloning purposes.

PCR confirmation of transformants

To verify the gene presence in transformants, colony PCR [40] was performed in a 15 μL reaction volume containing 7.5 μL of GoTaq Green Master Mix (Promega, Madison Wi, USA) , 0.25 μM of each appropriate oligo pair: aph8F, aph8R; ble1F, ble1R; tetXhF, tetBaR (Table 3). An initial 5 minute denaturation at 95°C was performed, followed by 35 cycles of 95°C 20 sec, 60–70°C 15 sec, 72°C 1 min with a final 5 min extension at 72°C. The amplicons were resolved in a 1% TAE-agarose gel stained with Sybr safe (Invitrogen, Carlsbad, CA, USA). Positive colonies confirmed by PCR were counted and the efficiency reported as colony forming units (cfu) per μg of DNA.

Tetracycline sensitivity resistance assays

Five random strains of BtetX, one of AtetX and one of CC-849 untransformed control were selected from TAP-agar plates and 10^4 cells of each strain were grown in 2 mL TAP media without tetracycline. Positive controls were grown on TAP agar plates supplemented with 15 μg/mL tetracycline. After five days, cell concentration of the cultures was 1 × 10^7 cells/mL which corresponds to 26 cell divisions. At that time, cultures of strains grown with or without antibiotic were diluted with TAP media to 1 × 10^6 cells/mL, and 10 μL (10^4 cells) of each strain were grown on TAP agar plates containing tetracycline at 15, 25, 50 and 100 μg/μL. After 7 days of growth at a light intensity of 25 μmoles m^{-2} s^{-1} photographs were taken of each plate.

Wild type *C. reinhardtii* tetracycline sensitivity

C. reinhardtii strain CC-124 (wt, mt-) was grown in TAP agar supplemented with tetracycline concentrations of 15, 25 or 50 μg/mL, and cell concentrations from 0.5 1.0, 2.5 and 5.0 × 10 ^{6} cells per plate, incubated for 10 days at a light intensity of 25 μmoles m^{-2} s^{-1}.

TetX plasmid deposit

The TetXA and TetXB transformation plasmids have been deposited with the Chlamydomonas Resource Center, University of Minnesota.

Competing interests

The authors declare that they have no competing interests.

Authors' contributions

SAG-E and GAC conceived, designed the experiments, and helped to draft the manuscript. SAG-E performed all experiments. Both authors read and approved the final manuscript.

Acknowledgements

The authors thank CONACyT for scholarship 209658 to SAG-E and the Cátedra de Agrobiotecnología, Tecnologico de Monterrey, for economic support.

References

1. Specht E, Miyake-Stoner S, Mayfield SP. Micro-algae come of age as a platform for recombinant protein production. Biotechnol Lett. 2010;32:1373–83.
2. Cerutti H, Johnson AM, Gillham NW, Boynton JE. Epigenetic silencing of a foreign gene in nuclear transformants of Chlamydomonas. Plant Cell. 1997;9:925–45.
3. Rasala BA, Lee PA, Shen Z, Briggs SP, Mendez M, Mayfield SP. Robust Expression and Secretion of Xylanase1 in Chlamydomonas reinhardtii by Fusion to a Selection Gene and Processing with the FMDV 2A Peptide. PLoS One. 2012;7:e43349.
4. Eichler-Stahlberg A, Weisheit W, Ruecker O, Heitzer M. Strategies to facilitate transgene expression in Chlamydomonas reinhardtii. Planta. 2009;229:873–83.
5. Lumbreras V, Stevens D, Purton S. Efficient foreign gene expression in Chlamydomonas reinhardtii mediated by an endogenous intron. Plant J. 1998;14(February):441–7.
6. Fuhrmann M, Oertel W, Hegemann P. A synthetic gene coding for the green fluorescent protein (GFP) is a versatile reporter in Chlamydomonas reinhardtii. Plant J. 1999;19:353–61.
7. Fuhrmann M, Hausherr A, Ferbitz L, Schödl T, Heitzer M, Hegemann P. Monitoring dynamic expression of nuclear genes in Chlamydomonas reinhardtii by using a synthetic luciferase reporter gene. Plant Mol Biol. 2004;55:869–81.
8. Rasala BA, Barrera DJ, Ng J, Plucinak TM, Rosenberg JN, Weeks DP, et al. Expanding the spectral palette of fluorescent proteins for the green microalga Chlamydomonas reinhardtii. Plant J. 2013;74:545–56.
9. Debuchy R, Purton S, Rochaix JD. The argininosuccinate lyase gene of Chlamydomonas reinhardtii: an important tool for nuclear transformation and for correlating the genetic and molecular maps of the ARG7 locus. EMBO J. 1989;8:2803–9.
10. Berthold P, Schmitt R, Mages W. An Engineered Streptomyces hygroscopicus aph 7″Gene Mediates Dominant Resistance against Hygromycin B in Chlamydomonas reinhardtii. Protist. 2002;153(December):401–12.
11. Sizova I, Fuhrmann M, Hegemann P. A Streptomyces rimosus aphVIII gene coding for a new type phosphotransferase provides stable antibiotic resistance to Chlamydomonas reinhardtii. Gene. 2001;277:221–9.
12. Blochlinger K, Diggelmann H. Hygromycin B phosphotransferase as a selectable marker for DNA transfer experiments with higher eucaryotic cells. Mol Cell Biol. 1984;4:2929–31.
13. Díaz-Santos E, de la Vega M, Vila M, Vigara J, León R. Efficiency of different heterologous promoters in the unicellular microalga Chlamydomonas reinhardtii. Biotechnol Prog. 2013;29:319–28.
14. Stevens DR, Rochaix JD, Purton S. The bacterial phleomycin resistance gene ble as a dominant selectable marker in Chlamydomonas. Mol Gen Genet. 1996;251:23–30.
15. Van der Grinten E, Pikkemaat MG, van den Brandhof E-J, Stroomberg GJ, Kraak MHS. Comparing the sensitivity of algal, cyanobacterial and bacterial bioassays to different groups of antibiotics. Chemosphere. 2010;80:1–6.
16. González-Pleiter M, Gonzalo S, Rodea-Palomares I, Leganés F, Rosal R, Boltes K, et al. Toxicity of five antibiotics and their mixtures towards photosynthetic aquatic organisms: implications for environmental risk assessment. Water Res. 2013;47:2050–64.
17. Li X-Z, Nikaido H. Efflux-mediated drug resistance in bacteria. Drugs. 2004;64:159–204.
18. Dönhöfer A, Franckenberg S, Wickles S, Berninghausen O, Beckmann R, Wilson DN. Structural basis for TetM-mediated tetracycline resistance. Proc Natl Acad Sci U S A. 2012;109:16900–5.
19. Yu Z, Reichheld SE, Cuthbertson L, Nodwell JR, Davidson AR. Characterization of tetracycline modifying enzymes using a sensitive in vivo reporter system. BMC Biochem. 2010;11:34.
20. Chopra I, Roberts M. Tetracycline Antibiotics : Mode of Action, Applications, Molecular Biology, and Epidemiology of Bacterial Resistance Tetracycline Antibiotics: Mode of Action, Applications, Molecular Biology, and Epidemiology of Bacterial Resistance. Microbiol Mol Biol Rev. 2001;65:232–60.
21. Guiney DG, Hasegawa P, Davis CE. Expression in Escherichia coli of cryptic tetracycline resistance genes from bacteroides R plasmids. Plasmid. 1984;11:248–52.
22. Speer BS, Bedzyk L, Salyers AA. Evidence that a novel tetracycline resistance gene found on two Bacteroides transposons encodes an NADP-requiring oxidoreductase. J Bacteriol. 1991;173:176–83.
23. Yang W, Moore IF, Koteva KP, Bareich DC, Hughes DW, Wright GD. TetX is a flavin-dependent monooxygenase conferring resistance to tetracycline antibiotics. J Biol Chem. 2004;279:52346–52.
24. Moore IF, Hughes DW, Wright GD. Tigecycline is modified by the flavin-dependent monooxygenase TetX. Biochemistry. 2005;44:11829–35.
25. Chang HR, Comte R, Pechère JC. In vitro and in vivo effects of doxycycline on Toxoplasma gondii. Antimicrob Agents Chemother. 1990;34:775–80.
26. Lavarde V, Acar JF, Drouhet E. Effect of minocycline on Candida albicans. "In vitro" study: comparison with tetracycline. Pathol Biol (Paris). 1975;23:725–8.
27. Waterworth PM. The effect of minocycline on Candida albicans. J Clin Pathol. 1974;27:269–72.
28. Ku TSN, Palanisamy SKA, Lee SA. Susceptibility of Candida albicans biofilms to azithromycin, tigecycline and vancomycin and the interaction between tigecycline and antifungals. Int J Antimicrob Agents. 2010;36:441–6.
29. Cohen SN, Chang AC, Hsu L. Nonchromosomal antibiotic resistance in bacteria: genetic transformation of Escherichia coli by R-factor DNA. Proc Natl Acad Sci U S A. 1972;69:2110–4.
30. Cohen SN, Chang AC, Boyer HW, Helling RB. Construction of biologically functional bacterial plasmids in vitro. Proc Natl Acad Sci U S A. 1973;70:3240–4.
31. Davies JP, Weeks DP, Grossman AR. Expression of the arylsulfatase gene from the β2-tubulin promoter in Chlamydomonas reinhardtii. Nucleic Acids Res. 1992;20:2959–65.
32. Cerutti H, Johnson AM, Gillham NW, Boynton JE. A Eubacterial Gene Conferring Spectinomycin Resistance on. Genetics. 1997;145:97–110.
33. Rose WE, Rybak MJ. Tigecycline: first of a new class of antimicrobial agents. Pharmacotherapy. 2006;26:1099–110.
34. Garí E, Piedrafita L, Aldea M, Herrero E. A set of vectors with a tetracycline-regulatable promoter system for modulated gene expression in Saccharomyces cerevisiae. Yeast. 1997;13:837–48.
35. Gossen M, Bujard H. Tight control of gene expression in mammalian cells by tetracycline-responsive promoters. Proc Natl Acad Sci U S A. 1992;89:5547–51.
36. Harris EH. The Chlamydomonas Sourcebook. A Comprehensive Guide to Biology and Laboratory Use. San Diego, CA: Academic Press. 1989. xiv, 780 pp., illus.
37. Ruecker O, Zillner K, Groebner-Ferreira R, Heitzer M. Gaussia-luciferase as a sensitive reporter gene for monitoring promoter activity in the nucleus of the green alga Chlamydomonas reinhardtii. Mol Genet Genomics. 2008;280:153–62.
38. Heitzer M, Zschoernig B. Construction of modular tandem expression vectors for the green alga Chlamydomonas reinhardtii using the Cre/lox-system. Biotechniques. 2007;43:324–32.
39. Kindle KL. High-frequency nuclear transformation of Chlamydomonas reinhardtii. Proc Natl Acad Sci U S A. 1990;87:1228–32.
40. Cao M, Fu Y, Guo Y, Pan J. Chlamydomonas (Chlorophyceae) colony PCR. Protoplasma. 2009;235:107–10.

Permissions

All chapters in this book were first published in Plant Methods, by BioMed Central; hereby published with permission under the Creative Commons Attribution License or equivalent. Every chapter published in this book has been scrutinized by our experts. Their significance has been extensively debated. The topics covered herein carry significant findings which will fuel the growth of the discipline. They may even be implemented as practical applications or may be referred to as a beginning point for another development.

The contributors of this book come from diverse backgrounds, making this book a truly international effort. This book will bring forth new frontiers with its revolutionizing research information and detailed analysis of the nascent developments around the world.

We would like to thank all the contributing authors for lending their expertise to make the book truly unique. They have played a crucial role in the development of this book. Without their invaluable contributions this book wouldn't have been possible. They have made vital efforts to compile up to date information on the varied aspects of this subject to make this book a valuable addition to the collection of many professionals and students.

This book was conceptualized with the vision of imparting up-to-date information and advanced data in this field. To ensure the same, a matchless editorial board was set up. Every individual on the board went through rigorous rounds of assessment to prove their worth. After which they invested a large part of their time researching and compiling the most relevant data for our readers.

The editorial board has been involved in producing this book since its inception. They have spent rigorous hours researching and exploring the diverse topics which have resulted in the successful publishing of this book. They have passed on their knowledge of decades through this book. To expedite this challenging task, the publisher supported the team at every step. A small team of assistant editors was also appointed to further simplify the editing procedure and attain best results for the readers.

Apart from the editorial board, the designing team has also invested a significant amount of their time in understanding the subject and creating the most relevant covers. They scrutinized every image to scout for the most suitable representation of the subject and create an appropriate cover for the book.

The publishing team has been an ardent support to the editorial, designing and production team. Their endless efforts to recruit the best for this project, has resulted in the accomplishment of this book. They are a veteran in the field of academics and their pool of knowledge is as vast as their experience in printing. Their expertise and guidance has proved useful at every step. Their uncompromising quality standards have made this book an exceptional effort. Their encouragement from time to time has been an inspiration for everyone.

The publisher and the editorial board hope that this book will prove to be a valuable piece of knowledge for researchers, students, practitioners and scholars across the globe.

List of Contributors

Tobias Keplinger
Institute for Building Materials, ETH Zurich, Zurich, Switzerland
Applied Wood Research Laboratory, Empa - Swiss Federal Laboratories for Material Testing and Research, Duebendorf, Switzerland

Johannes Konnerth
Department of Materials Science and Process Engineering, BOKU-University of Natural Resources and Life Science, Konrad Lorenz Strasse 24, A-3430, Tulln, Vienna, Austria

Véronique Aguié-Béghin
INRA-UMR-614 Fractionnement des Agro-Ressources et Environnement (FARE), F-51686, Reims, France
Université de Reims Champagne-Ardenne, UMR614 Fractionnement des Agro-Ressources et Environnement (FARE), F-51686, Reims, France

Markus Rüggeberg
Institute for Building Materials, ETH Zurich, Zurich, Switzerland
Applied Wood Research Laboratory, Empa - Swiss Federal Laboratories for Material Testing and Research, Duebendorf, Switzerland

Notburga Gierlinger
Institute for Building Materials, ETH Zurich, Zurich, Switzerland
Applied Wood Research Laboratory, Empa - Swiss Federal Laboratories for Material Testing and Research, Duebendorf, Switzerland

Ingo Burgert
Institute for Building Materials, ETH Zurich, Zurich, Switzerland
Applied Wood Research Laboratory, Empa - Swiss Federal Laboratories for Material Testing and Research, Duebendorf, Switzerland

Vijaykumar Veerappan
Department of Biological Sciences, University of North Texas, 1155 Union Circle #305220, Denton, Texas 76203, USA

Khem Kadel
Department of Biological Sciences, University of North Texas, 1155 Union Circle #305220, Denton, Texas 76203, USA

Naudin Alexis
Department of Biological Sciences, University of North Texas, 1155 Union Circle #305220, Denton, Texas 76203, USA

Ashley Scott
Department of Biological Sciences, University of North Texas, 1155 Union Circle #305220, Denton, Texas 76203, USA

Igor Kryvoruchko
Plant Biology Division, The Samuel Roberts Noble Foundation, Ardmore, OK 73401, USA

Senjuti Sinharoy
Plant Biology Division, The Samuel Roberts Noble Foundation, Ardmore, OK 73401, USA

Mark Taylor
Plant Biology Division, The Samuel Roberts Noble Foundation, Ardmore, OK 73401, USA

Michael Udvardi
Plant Biology Division, The Samuel Roberts Noble Foundation, Ardmore, OK 73401, USA

Rebecca Dickstein
Department of Biological Sciences, University of North Texas, 1155 Union Circle #305220, Denton, Texas 76203, USA

Janett Göhring
Max F. Perutz Laboratories, Medical University of Vienna, Dr. Bohrgasse 9/3, Vienna, Austria

Nick Fulcher
Gregor Mendel Institute, Dr. Bohrgasse 3, Vienna, Austria

Kurt Schilcher
Upper Austria University of Applied Sciences, Campus Linz, Linz 4020, Austria

Andrea Barta
Max F. Perutz Laboratories, Medical University of Vienna, Dr. Bohrgasse 9/3, Vienna, Austria

Jaroslaw Jacak
Max F. Perutz Laboratories, Medical University of Vienna, Dr. Bohrgasse 9/3, Vienna, Austria
Upper Austria University of Applied Sciences, Campus Linz, Linz 4020, Austria

Dimitrios Fanourakis
IBG-2: Plant Sciences, Institute for Bio- and Geosciences, Forschungszentrum Jülich, 52425 Jülich, Germany

Christoph Briese
IBG-2: Plant Sciences, Institute for Bio- and Geosciences, Forschungszentrum Jülich, 52425 Jülich, Germany

Johannes FJ Max
IBG-2: Plant Sciences, Institute for Bio- and Geosciences, Forschungszentrum Jülich, 52425 Jülich, Germany
Center of Applied Biology, Department of Soil Science and Plant Nutrition, Hochschule Geisenheim University, 65366 Geisenheim, Germany

Silke Kleinen
IBG-2: Plant Sciences, Institute for Bio- and Geosciences, Forschungszentrum Jülich, 52425 Jülich, Germany

Alexander Putz
IBG-2: Plant Sciences, Institute for Bio- and Geosciences, Forschungszentrum Jülich, 52425 Jülich, Germany

Fabio Fiorani
IBG-2: Plant Sciences, Institute for Bio- and Geosciences, Forschungszentrum Jülich, 52425 Jülich, Germany

Andreas Ulbrich
IBG-2: Plant Sciences, Institute for Bio- and Geosciences, Forschungszentrum Jülich, 52425 Jülich, Germany
University of Applied Sciences Osnabrück, 49076 Osnabrück, Germany

Ulrich Schurr
IBG-2: Plant Sciences, Institute for Bio- and Geosciences, Forschungszentrum Jülich, 52425 Jülich, Germany

Wannes Voorend
Department of Plant Systems Biology, VIB, Technologiepark 927, 9052 Gent, Belgium
Department of Plant Biotechnology and Bioinformatics, Ghent University, Technologiepark 927, 9052 Gent, Belgium
Plant Sciences Unit - Growth and Development, Institute for Agricultural and Fisheries Research (ILVO), Caritasstraat 21, 9090 Melle, Belgium

Peter Lootens
Plant Sciences Unit - Growth and Development, Institute for Agricultural and Fisheries Research (ILVO), Caritasstraat 21, 9090 Melle, Belgium

Hilde Nelissen
Department of Plant Systems Biology, VIB, Technologiepark 927, 9052 Gent, Belgium
Department of Plant Biotechnology and Bioinformatics, Ghent University, Technologiepark 927, 9052 Gent, Belgium

Isabel Roldán-Ruiz
Plant Sciences Unit - Growth and Development, Institute for Agricultural and Fisheries Research (ILVO), Caritasstraat 21, 9090 Melle, Belgium

Dirk Inzé
Department of Plant Systems Biology, VIB, Technologiepark 927, 9052 Gent, Belgium
Department of Plant Biotechnology and Bioinformatics, Ghent University, Technologiepark 927, 9052 Gent, Belgium

Hilde Muylle
Plant Sciences Unit - Growth and Development, Institute for Agricultural and Fisheries Research (ILVO), Caritasstraat 21, 9090 Melle, Belgium

Hung-Yi Wu
Institute of Plant and Microbial Biology, Academia Sinica, Taipei 11529, Taiwan
Department of Plant Pathology and Microbiology, National Taiwan University, Taipei 10617, Taiwan

Kun-Hsiang Liu
Department of Molecular Biology and Center for Computational and Integrative Biology, Massachusetts General Hospital, Boston, MA 02114, USA
Department of Genetics, Harvard Medical School, Boston, MA 02114, USA

Yi-Chieh Wang
Institute of Plant and Microbial Biology, Academia Sinica, Taipei 11529, Taiwan

Jing-Fen Wu
Institute of Plant and Microbial Biology, Academia Sinica, Taipei 11529, Taiwan

Wan-Ling Chiu
Department of Molecular Biology and Center for Computational and Integrative Biology, Massachusetts General Hospital, Boston, MA 02114, USA
Department of Genetics, Harvard Medical School, Boston, MA 02114, USA
Center for the Study of Biological Complexity, Virginia Commonwealth University, Richmond, VA 23284, USA

Chao-Ying Chen
Department of Plant Pathology and Microbiology, National Taiwan University, Taipei 10617, Taiwan

Shu-Hsing Wu
Institute of Plant and Microbial Biology, Academia Sinica, Taipei 11529, Taiwan

Jen Sheen
Department of Molecular Biology and Center for Computational and Integrative Biology, Massachusetts General Hospital, Boston, MA 02114, USA
Department of Genetics, Harvard Medical School, Boston, MA 02114, USA

Erh-Min Lai
Institute of Plant and Microbial Biology, Academia Sinica, Taipei 11529, Taiwan
Department of Plant Pathology and Microbiology, National Taiwan University, Taipei 10617, Taiwan
Department of Molecular Biology and Center for Computational and Integrative Biology, Massachusetts General Hospital, Boston, MA 02114, USA
Department of Genetics, Harvard Medical School, Boston, MA 02114, USA

Huiming Zhang
Department of Horticulture and Landscape Architecture, Purdue University, West Lafayette, IN 47907, USA

Kai Tang
Department of Horticulture and Landscape Architecture, Purdue University, West Lafayette, IN 47907, USA

Bangshing Wang
Department of Horticulture and Landscape Architecture, Purdue University, West Lafayette, IN 47907, USA

Cheng-Guo Duan
Department of Horticulture and Landscape Architecture, Purdue University, West Lafayette, IN 47907, USA

Zhaobo Lang
Department of Horticulture and Landscape Architecture, Purdue University, West Lafayette, IN 47907, USA

Jian-Kang Zhu
Department of Horticulture and Landscape Architecture, Purdue University, West Lafayette, IN 47907, USA
Shanghai Center for Plant Stress Biology, Shanghai Institute of Biological Sciences, Chinese Academy of Sciences, Beijing, Shijingshan, China

Dinah Kirigia
Jomo Kenyatta University of Agriculture and Technology, Institute of Biotechnology Research, P. O. Box 62000-00200, Nairobi, Kenya

Steven Runo
Department of Biochemistry and Biotechnology, Kenyatta University, P.O. Box 43844, 00100 Nairobi, Kenya

Amos Alakonya
Jomo Kenyatta University of Agriculture and Technology, Institute of Biotechnology Research, P. O. Box 62000-00200, Nairobi, Kenya

Francisco Balao
Departamento de Biología Vegetal y Ecología, Universidad de Sevilla, Ap-1095, 41080 Sevilla, Spain
Department of Systematic and Evolutionary Botany, University of Vienna, Rennweg 14, Vienna 1030, Austria

Juan Luis García-Castaño
Departamento de Biología Vegetal y Ecología, Universidad de Sevilla, Ap-1095, 41080 Sevilla, Spain

Asmini Athman
ARC Centre of Excellence in Plant Energy Biology, University of Adelaide, Glen Osmond, SA, Australia
Waite Research Institute & School of Agriculture, Food and Wine, University of Adelaide, PMB1, Glen Osmond, SA 5064, Australia

Sandra K Tanz
ARC Centre of Excellence in Plant Energy Biology, The University of Western Australia, Perth, WA, Australia

Vanessa M Conn
ARC Centre of Excellence in Plant Energy Biology, University of Adelaide, Glen Osmond, SA, Australia
Waite Research Institute & School of Agriculture, Food and Wine, University of Adelaide, PMB1, Glen Osmond, SA 5064, Australia

Charlotte Jordans
ARC Centre of Excellence in Plant Energy Biology, University of Adelaide, Glen Osmond, SA, Australia
Waite Research Institute & School of Agriculture, Food and Wine, University of Adelaide, PMB1, Glen Osmond, SA 5064, Australia

Gwenda M Mayo
Waite Research Institute & School of Agriculture, Food and Wine, University of Adelaide, PMB1, Glen Osmond, SA 5064, Australia
Adelaide Microscopy Waite Facility, University of Adelaide, Glen Osmond, SA, Australia

Weng W Ng
ARC Centre of Excellence in Plant Energy Biology, University of Adelaide, Glen Osmond, SA, Australia
Waite Research Institute & School of Agriculture, Food and Wine, University of Adelaide, PMB1, Glen Osmond, SA 5064, Australia

Rachel A Burton
ARC Centre of Excellence in Plant Cell Walls, University of Adelaide, Glen Osmond, SA, Australia

Simon J Conn
Waite Research Institute & School of Agriculture, Food and Wine, University of Adelaide, PMB1, Glen Osmond, SA 5064, Australia
Centre for Cancer Biology, Division Immunology, Level 3, Frome Road, Adelaide, SA, Australia

Matthew Gilliham
ARC Centre of Excellence in Plant Energy Biology, University of Adelaide, Glen Osmond, SA, Australia
Waite Research Institute & School of Agriculture, Food and Wine, University of Adelaide, PMB1, Glen Osmond, SA 5064, Australia

Chantal Le Marié
Institute of Agricultural Sciences, ETH Zurich, Universitätstrasse 2, 8092 Zurich, Switzerland

Norbert Kirchgessner
Institute of Agricultural Sciences, ETH Zurich, Universitätstrasse 2, 8092 Zurich, Switzerland

Daniela Marschall
Institute of Agricultural Sciences, ETH Zurich, Universitätstrasse 2, 8092 Zurich, Switzerland

Achim Walter
Institute of Agricultural Sciences, ETH Zurich, Universitätstrasse 2, 8092 Zurich, Switzerland

Andreas Hund
Institute of Agricultural Sciences, ETH Zurich, Universitätstrasse 2, 8092 Zurich, Switzerland

Anika Oellrich
Wellcome Trust Sanger Institute, Wellcome Trust Genome Campus, Hinxton CB10 1SA, UK

Ramona L Walls
iPlant Collaborative, University of Arizona, 1657 E. Helen St., Tucson, Arizona 85721, USA

Ethalinda KS Cannon
Department of Electrical and Computer Engineering Iowa State University, 1018 Crop Informatics Lab, Ames, Iowa 50011, USA

Steven B Cannon
USDA-ARS Corn Insects and Crop Genetics Research Unit, Iowa State University, Crop Genome Informatics Lab, Iowa State University, Ames, IA 50011, USA
Department of Agronomy, Agronomy Hall, Iowa State University, Ames, IA 50010, USA

Laurel Cooper
Department of Botany and Plant Pathology, 2082 Cordley Hall, Oregon State University, Corvallis, OR 97331, USA

Jack Gardiner
Department of Genetics, Development and Cell Biology, Roy J Carver Co-Laboratory, Iowa State University, Ames, IA 50010, USA

Georgios V Gkoutos
Department of Computer Science, Aberystwyth University, Llandinam Building, Aberystwyth SY23 3DB, UK

Lisa Harper
USDA-ARS Corn Insects and Crop Genetics Research Unit, Iowa State University, Crop Genome Informatics Lab, Iowa State University, Ames, IA 50011, USA

Mingze He
Department of Genetics, Development and Cell Biology, Roy J Carver Co-Laboratory, Iowa State University, Ames, IA 50010, USA

Robert Hoehndorf
Computer, Electrical and Mathematical Sciences & Engineering Division and Computational Bioscience Research Center, King
Abdullah University of Science and Technology, 4700 King Abdullah University of Science and Technology, P.O. Box 2882, Thuwal 23955-6900, Kingdom of Saudi Arabia

Pankaj Jaiswal
Department of Botany and Plant Pathology, 2082 Cordley Hall, Oregon State University, Corvallis, OR 97331, USA

Scott R Kalberer
USDA-ARS Corn Insects and Crop Genetics Research Unit, Iowa State University, Crop Genome Informatics Lab, Iowa State University, Ames, IA 50011, USA

John P Lloyd
Department of Plant Biology, Michigan State University, 220 Trowbridge Rd, East Lansing, MI 48824, USA

David Meinke
Department of Botany, Oklahoma State University, 301 Physical Sciences, Stillwater, OK 74078, USA

Naama Menda
Boyce Thompson Institute for Plant Research, 533 Tower Road, Ithaca, NY 14853, USA

Laura Moore
Department of Botany and Plant Pathology, 2082 Cordley Hall, Oregon State University, Corvallis, OR 97331, USA

Rex T Nelson
USDA-ARS Corn Insects and Crop Genetics Research Unit, Iowa State University, Crop Genome Informatics Lab, Iowa State University, Ames, IA 50011, USA

Anuradha Pujar
Boyce Thompson Institute for Plant Research, 533 Tower Road, Ithaca, NY 14853, USA

Carolyn J Lawrence
Department of Agronomy, Agronomy Hall, Iowa State University, Ames, IA 50010, USA
Department of Genetics, Development and Cell Biology, Roy J Carver Co-Laboratory, Iowa State University, Ames, IA 50010, USA

Eva Huala
Phoenix Bioinformatics, 643 Bair Island Rd Suite 403, Redwood City, CA 94063, USA

Nejc Rački
Department of Biotechnology and Systems Biology, National Institute of Biology, Večna pot 111, SI-1000, Ljubljana, Slovenia

Tanja Dreo
Department of Biotechnology and Systems Biology, National Institute of Biology, Večna pot 111, SI-1000, Ljubljana, Slovenia

Ion Gutierrez-Aguirre
Department of Biotechnology and Systems Biology, National Institute of Biology, Večna pot 111, SI-1000, Ljubljana, Slovenia

Andrej Blejec
Department of Biotechnology and Systems Biology, National Institute of Biology, Večna pot 111, SI-1000, Ljubljana, Slovenia

Maja Ravnikar
Department of Biotechnology and Systems Biology, National Institute of Biology, Večna pot 111, SI-1000, Ljubljana, Slovenia

Graham J Etherington
The Sainsbury Laboratory, Norwich Research Park, Norwich NR4 7UH, UK
The Genome Analysis Centre, Norwich Research Park, Norwich NR4 7UH, UK

Jacqueline Monaghan
The Sainsbury Laboratory, Norwich Research Park, Norwich NR4 7UH, UK

Cyril Zipfel
The Sainsbury Laboratory, Norwich Research Park, Norwich NR4 7UH, UK

Dan MacLean
The Sainsbury Laboratory, Norwich Research Park, Norwich NR4 7UH, UK

Mark Müller-Linow
Institute of Bio and Geosciences, IBG-2: Plant Sciences, Forschungszentrum Jülich GmbH, Wilhelm-Johnen-Str, 52425 Jülich, Germany

Francisco Pinto-Espinosa
Institute of Bio and Geosciences, IBG-2: Plant Sciences, Forschungszentrum Jülich GmbH, Wilhelm-Johnen-Str, 52425 Jülich, Germany

Hanno Scharr
Institute of Bio and Geosciences, IBG-2: Plant Sciences, Forschungszentrum Jülich GmbH, Wilhelm-Johnen-Str, 52425 Jülich, Germany

Uwe Rascher
Institute of Bio and Geosciences, IBG-2: Plant Sciences, Forschungszentrum Jülich GmbH, Wilhelm-Johnen-Str, 52425 Jülich, Germany

Joana Silva-Correia
Center for Biodiversity, Functional & Integrative Genomics (BioFIG), Plant Functional Biology Center, University of Minho, Campus de Gualtar, 4710-057 Braga, Portugal
3B's Research Group – Biomaterials, Biodegradables and Biomimetics, University of Minho Headquarters of the European Institute of Excellence on Tissue Engineering and Regenerative Medicine, AvePark, 4806-909 Guimarães, Taipas, Portugal

Sara Freitas
Center for Biodiversity, Functional & Integrative Genomics (BioFIG), Plant Functional Biology Center, University of Minho, Campus de Gualtar, 4710-057 Braga, Portugal

Rui M Tavares
Center for Biodiversity, Functional & Integrative Genomics (BioFIG), Plant Functional Biology Center, University of Minho, Campus de Gualtar, 4710-057 Braga, Portugal

Teresa Lino-Neto
Center for Biodiversity, Functional & Integrative Genomics (BioFIG), Plant Functional Biology Center, University of Minho, Campus de Gualtar, 4710-057 Braga, Portugal

Herlânder Azevedo
Center for Biodiversity, Functional & Integrative Genomics (BioFIG), Plant Functional Biology Center, University of Minho, Campus de Gualtar, 4710-057 Braga, Portugal
CIBIO, Centro de Investigação em Biodiversidade e Recursos Genéticos, Universidade do Porto, Campus Agrário de Vairão, 4485-661 Vairão, Portugal

Leonardo Galindo-González
Department of Biological Sciences, University of Alberta, Edmonton, AB, Canada, T6G 2E9

David Pinzón-Latorre
Department of Biological Sciences, University of Alberta, Edmonton, AB, Canada, T6G 2E9

Erik A Bergen
Department of Biological Sciences, University of Alberta, Edmonton, AB, Canada, T6G 2E9

Dustin C Jensen
Department of Computing Sciences, Kings University College, Edmonton, AB, Canada, T6B 2H3

Michael K Deyholos
IK Barber School of Arts & Sciences, University of British Columbia, Okanagan campus, Kelowna, BC, Canada, V1V 1 V7

Céline Rousseau
PHENOTIC, SFR 4207 QUASAV, F-49045 Angers, France

Gilles Hunault
Université d'Angers, Laboratoire d'Hémodynamique, Interaction Fibrose et Invasivité tumorale hépatique, UPRES 3859, IFR 132, F-49045 Angers, France

Sylvain Gaillard
Institut de Recherche en Horticulture et Semences, UMR1345, INRA, SFR 4207 QUASAV, F-49071 Beaucouzé, France

Julie Bourbeillon
Institut de Recherche en Horticulture et Semences, UMR1345, AgroCampus-Ouest, SFR 4207 QUASAV, F-49045 Angers, France

Gregory Montiel
Université de Nantes, Laboratoire de Biologie et de Pathologie Végétales EA 1157, SFR 4207 QUASAV, F-44322 Nantes, France

Philippe Simier
Université de Nantes, Laboratoire de Biologie et de Pathologie Végétales EA 1157, SFR 4207 QUASAV, F-44322 Nantes, France

Claire Campion
Institut de Recherche en Horticulture et Semences, UMR1345, Université d'Angers, SFR 4207 QUASAV, F-49045 Angers, France

Marie-Agnès Jacques
PHENOTIC, SFR 4207 QUASAV, F-49045 Angers, France
Institut de Recherche en Horticulture et Semences, UMR1345, INRA, SFR 4207 QUASAV, F-49071 Beaucouzé, France

Etienne Belin
PHENOTIC, SFR 4207 QUASAV, F-49045 Angers, France
Laboratoire Angevin de Recherche en Ingénierie des Systèmes (LARIS), Université d'Angers, F-49000 Angers, France

Tristan Boureau
PHENOTIC, SFR 4207 QUASAV, F-49045 Angers, France
Institut de Recherche en Horticulture et Semences, UMR1345, Université d'Angers, SFR 4207 QUASAV, F-49045 Angers, France

Jan F Humplík
Department of Chemical Biology and Genetics, Centre of the Region Haná for Biotechnological and Agricultural Research, Faculty of Science, Palacký University, Šlechtitelů 11, Olomouc CZ-78371, Czech Republic

Dušan Lazár
Department of Biophysics, Centre of the Region Haná for Biotechnological and Agricultural Research, Faculty of Science, Palacký University, Šlechtitelů 11, Olomouc CZ-78371, Czech Republic

Alexandra Husičková
Department of Biophysics, Centre of the Region Haná for Biotechnological and Agricultural Research, Faculty of Science, Palacký University, Šlechtitelů 11, Olomouc CZ-78371, Czech Republic

Lukáš Spíchal
Department of Chemical Biology and Genetics, Centre of the Region Haná for Biotechnological and Agricultural Research, Faculty of Science, Palacký University, Šlechtitelů 11, Olomouc CZ-78371, Czech Republic

Justine Bresson
Laboratoire d'Ecophysiologie des Plantes sous Stress Environnementaux (LEPSE), INRA, Montpellier SupAgro, UMR759, F-34060 Montpellier, France
Laboratoire des Symbioses Tropicales et Méditerranéennes (LSTM), UMR113, Université Montpellier 2-IRD-CIRAD-INRA-SupAgro, F-34095 Montpellier, France
Center for Plant Molecular Biology (ZMBP), General Genetics, University of Tuebingen, D-72076 Tuebingen, Germany

François Vasseur
Max Planck Institute for Developmental Biology, D-72076 Tuebingen, Germany

Myriam Dauzat
Laboratoire d'Ecophysiologie des Plantes sous Stress Environnementaux (LEPSE), INRA, Montpellier SupAgro, UMR759, F-34060 Montpellier, France

Garance Koch
Laboratoire d'Ecophysiologie des Plantes sous Stress Environnementaux (LEPSE), INRA, Montpellier SupAgro, UMR759, F-34060 Montpellier, France

Christine Granier
Laboratoire d'Ecophysiologie des Plantes sous Stress Environnementaux (LEPSE), INRA, Montpellier SupAgro, UMR759, F-34060 Montpellier, France

Denis Vile
Laboratoire d'Ecophysiologie des Plantes sous Stress Environnementaux (LEPSE), INRA, Montpellier SupAgro, UMR759, F-34060 Montpellier, France

Jin-Hyun Kim
Department of Medical Bioscience, Dong-A University, Saha-Gu Nakdong-Daero 550 beongil 37, Busan 604-714, Republic of Korea

Chaeyoung Lee
Department of Medical Bioscience, Dong-A University, Saha-Gu Nakdong-Daero 550 beongil 37, Busan 604-714, Republic of Korea

Daejin Hyung
Department of Computer Science, Dong-A University, Saha-Gu Nakdong-Daero 550 beongil 37, Busan 604-714, Republic of Korea

Ye-Jin Jo
Department of Genetic Engineering, Dong-A University, Saha-Gu Nakdong-Daero 550 beongil 37, Busan 604-714, Republic of Korea

Joo-Seok Park
Department of Applied Bioscience, Dong-A University, Saha-Gu Nakdong-Daero 550 beongil 37, Busan 604-714, Republic of Korea

Douglas R Cook
Department of Plant Pathology, University of California, One Shields Ave, Davis, CA 95616, USA

Hong-Kyu Choi
Department of Genetic Engineering, Dong-A University, Saha-Gu Nakdong-Daero 550 beongil 37, Busan 604-714, Republic of Korea

Sven Heckwolf
Department of Botany, University of Wisconsin-Madison, 430 Lincoln Drive, Madison, WI 53706, USA

Marlies Heckwolf
Department of Agronomy, University of Wisconsin-Madison, 1575 Linden Drive, Madison, WI 53706, USA

DOE Great Lakes Bioenergy Research Center, 1552 University Avenue, Madison, WI 53706, USA

Shawn M Kaeppler
Department of Agronomy, University of Wisconsin-Madison, 1575 Linden Drive, Madison, WI 53706, USA
DOE Great Lakes Bioenergy Research Center, 1552 University Avenue, Madison, WI 53706, USA

Natalia de Leon
Department of Agronomy, University of Wisconsin-Madison, 1575 Linden Drive, Madison, WI 53706, USA
DOE Great Lakes Bioenergy Research Center, 1552 University Avenue, Madison, WI 53706, USA

Edgar P Spalding
Department of Botany, University of Wisconsin-Madison, 430 Lincoln Drive, Madison, WI 53706, USA

Sergio A Garcia-Echauri
Centro de Biotecnología-FEMSA, Tecnológico de Monterrey, Monterrey, México
Centro de Agrobiotecnología, Tecnológico de Monterrey, Monterrey, México

Guy A Cardineau
Centro de Biotecnología-FEMSA, Tecnológico de Monterrey, Monterrey, México
Centro de Agrobiotecnología, Tecnológico de Monterrey, Monterrey, México